REMOTE SENSING – APPLICATIONS

Edited by **Boris Escalante-Ramírez**

INTECHOPEN.COM

Remote Sensing – Applications
Edited by Boris Escalante-Ramírez

Published by InTech

Janeza Trdine 9, 51000 Rijeka, Croatia

Edition 2014

Publishing Process Manager Dragana Manestar

Technical Editor Teodora Smiljanic

Cover Designer InTech Design Team

Additional hard copies can be obtained from orders@intechopen.com

Remote Sensing – Applications, Edited by Boris Escalante-Ramírez
p. cm.
ISBN 978-953-51-0651-7

INTECH

open science | open minds

Contents

Preface

Nowadays it is hard to find areas of human activity and development that have not profited from or contributed to remote sensing. Natural, physical and social activities find in remote sensing a common ground for interaction and development. From the end-user point of view, Earth science, geography, planning, resource management, public policy design, environmental studies, and health, are some of the areas whose recent development has been triggered and motivated by remote sensing. From the technological point of view, remote sensing would not be possible without the advancement of basic as well as applied research in areas like physics, space technology, telecommunications, computer science and engineering. This dual conception of remote sensing brought us to the idea of preparing two different books. The present one is meant to display recent advances in remote sensing applications, while the accompanying book is devoted to new techniques for data processing, sensors and platforms.

Strictly speaking, remote sensing consists of collecting data from an object or phenomenon without making physical contact. In practice, most of the time we refer to satellite or aircraft-mounted sensors that use some sort of electromagnetic radiation to gather geospatial information from land, oceans and atmosphere with increasingly high spatial, spectral and temporal resolutions. Space agencies in charge of collecting remotely sensed data have shown a notorious interest in making these data available for research and social development. The confluence of remote sensing technology with other sciences has resulted in an exponential growth of knowledge, technology development and assessment of all kind of physical and natural phenomena, as well as human activities that share a common ground: geospatial information. However, the success of remote sensing influencing other areas of knowledge and human activity has not always been a paved way. The variables of great interest to scientists in different areas are not readily available from the raw remotely-sensed data. Even when the data has been processed and converted to physical-related values, or even linked to human and natural artifacts like crop fields, roads, urban areas, geomorphologic structures, vegetation indices, etc., the relationship between these and the more abstract variables that explain them such as human settlement dynamics, geophysical phenomena, climate change, etc. remain a major field of study and research.

This book intends to show the reader how remote sensing impacts other areas of science, technology, and human activity, by displaying a selected number of high quality contributions dealing with different remote sensing applications. Twenty two chapters have been carefully collected and distributed in four areas. The first part deals with land cover applications, and contains applications in vegetation indices, crop and pest monitoring, rainfall and fire relationship with vegetation, change detection, soil salinization, modeling water and heat regimes, catastrophe assessment and lava flow mapping. The second part contains contributions on climate and atmosphere, including carbon and water dynamics, ocean evaporation, and atmospheric aerosols and clouds. The third part presents oceans and cryosphere applications that include aquatic vegetation, oil spill assessment, coral reef habitat mapping, water source predictability from snow maps, cryosphere study, and maritime search and rescue. Last but not least, the last part presents contributions dealing with human activity, including population estimation, archaeology, urban growth, medicine and healthcare and military applications.

I am indebted to all authors who have contributed to this book. Without their strongest commitment this book would not have been possible. I am also thankful to InTech editorial team who has provided the opportunity to publish this book.

Boris Escalante-Ramírez
National Autonomous University of México,
Faculty of Engineering, Mexico City,
Mexico

Section 1

Land Cover

1

Narrowband Vegetation Indices for Estimating Boreal Forest Leaf Area Index

Ellen Eigemeier, Janne Heiskanen, Miina Rautiainen, Matti Mõttus, Veli-Heikki Vesanto, Titta Majasalmi and Pauline Stenberg
University of Helsinki
Finland

1. Introduction

1.1 Leaf area index

The green photosynthesizing leaf area of a canopy is an important characteristic of the status of the vegetation in terms of its health and production potential. At stand level, the amount of leaf area in a canopy is represented by a variable called the leaf area index (LAI), which is one of the key biophysical parameters in the global monitoring and mapping of vegetation by satellite remote sensing (Morisette et al., 2006). In this paper we adopt the, by now widely accepted, definition of LAI as the hemi-surface or half of the total surface area of all leaves or needles in the vegetation canopy divided by the horizontal ground area below the canopy. The definition is in line with the original definition of LAI, formulated for flat and (assumedly) infinitely thin leaves (Watson, 1947), as the one-sided leaf area per unit ground area. For coniferous canopies, the question arose on how to define the "one-sided" area of non-flat needles. While projected needle area formerly often has been used erroneously as a synonym to one-sided flat leaf area, it is now commonly accepted that the hemi-surface needle area represents the logical counterpart to the one-sided area of flat leaves (e.g. Chen & Black, 1992; Stenberg, 2006).

LAI controls many biological and physical processes, driving the exchange of matter and energy flow. Because LAI responds rapidly to different stress factors and changes in climatic conditions, monitoring of LAI yields a dynamic indicator of forest status and health. The link between forest productivity and LAI, in turn, lies in that LAI is the main determinant of the fraction of incoming photosynthetically active radiation absorbed by the canopy (fAPAR). The absorbed photosynthetically active radiation (APAR) quantifies the energy available for net primary production (NPP) and is thus a critical variable in NPP and carbon flux models. NPP is related to APAR by the light-use-efficiency originally introduced by Monteith (1977) for agricultural crops.

Traditionally, ground-based measurements of LAI have typically involved destructive sampling and determination of allometric relationships, e.g. between leaf area and the basal area of stem and/or branches carrying the leaves (the pipe model theory) (Shinozaki et al., 1964; Waring et al., 1982). However, such "direct methods" are quite laborious and indirect measurements of LAI using optical instruments are today the preferred choice (Welles &

Cohen, 1996; Jonckheere et al., 2004). They provide inverse estimates of LAI based on the fraction of gaps through the canopy in different directions, which can be measured using devices such as the LAI-2000 Plant Canopy Analyzer (LI-COR, 1992) or hemispherical photography. A vast body of classical literature exists on the dependency between LAI and canopy gap fraction underlying these techniques (e.g. Wilson, 1965; Miller, 1967; Nilson, 1971; Lang, 1986). In short, the inversion methods rely upon the assumption that leaves are randomly distributed in the canopy, in which case Beer's law can be applied to plant canopies (Monsi & Saeki, 1953). However, as the organization of leaves (needles) in forest canopies is typically more aggregated ("clumped") than predicted by a purely random distribution, the technique causes underestimation of LAI, especially in coniferous stands (e.g. Smith et al., 1993; Stenberg et al., 1994). Instead of the true LAI, the inversion of gap fraction data without correction for clumping yields the quantity commonly referred to as the "effective leaf area index" (Black et al., 1991).

Monitoring LAI in a spatially continuous mode and on a regular basis is possible only using remote sensing. Estimation of LAI from optical satellite images is considered feasible because LAI is closely linked to the spectral reflectance of plant canopies in the shortwave solar radiation range (Myneni et al., 1997). The physical relationships between canopy spectral reflectances and LAI form the basis of retrieval algorithms used in current Earth observation programs (e.g. MODIS, CYCLOPES, GLOBCARBON products) for mapping LAI at global scales. They produce bi-weekly and monthly vegetation maps that are widely used by biologists, natural resources managers, and climate modelers, e.g. to track seasonal fluctuations in vegetation or changes in land use. The arrival of narrowband reflectance data (also known as hyperspectral or imaging spectroscopy data) opens up new possibilities for satellite-derived estimation/monitoring of variables connected to the status and structure of vegetation, including LAI.

1.2 Spectral properties of boreal forests

The boreal forest zone, which spreads through Fennoscandia, Russia, Canada and Alaska, is the largest unbroken forest zone in the world and accounts for approximately one fourth of the world's forests. The boreal zone is a major store of carbon and thus plays an important role in determining global albedo and climate.

The reflectance spectra of coniferous forests (even if they have the same leaf area) are very distinct from similar broadleaved forests. The reasons for the special spectral behaviour of coniferous forests are versatile, yet primarily related to their structural, not optical, properties. Firstly, a high level of within-shoot scattering of conifers was originally noted nearly four decades ago (Norman & Jarvis, 1975). More recently, Landsat ETM+ data and a forest reflectance model were used to show that the low near infrared (NIR) reflectances observed in coniferous areas can largely be explained simply by within-shoot scattering (Rautiainen & Stenberg, 2005). Secondly, absorption by coniferous needles is higher than that by broadleaved species (Roberts et al., 2004; Williams, 1991), a phenomenon which can partly contribute to the lower reflectances of conifer-dominated areas. Other explanations include, for example, that the tree crown surface of coniferous stands is more heterogeneous than in broadleaved stands (Häme, 1991; Schull et al., 2011). In other words, when surface roughness (i.e. crown-level clumping) increases, the shaded area within the canopy increases, thus leading to lower reflectances. Overall, these results highlight the importance of various geometric properties as the main reason for the reflectance differences between broadleaved and coniferous stands.

Remote sensing of the biophysical properties, such as LAI, of a boreal coniferous forest canopy layer is further complicated by the often dominating role of the understory in the spectral signal (Rautiainen et al., 2011; Rautiainen et al., 2007; Eriksson et al., 2006; Eklundh et al., 2001; Chen & Cihlar, 1996; Spanner et al,. 1990). Coniferous forests that are regularly treated according to forest management practices tend to have relatively clumped and open canopies. Thus, the role of the understory vegetation in forming boreal forest reflectance cannot be neglected (Pisek et al., 2011).

1.3 Vegetation indices in LAI estimation

Canopy biophysical variables, such as LAI, can be estimated from remotely sensed data by two types of algorithms: empirical models and methods that use physically-based radiative transfer (RT) models. In empirical algorithms, the estimation is based on statistical relationships modelled between concurrent ground reference measurements and surface reflectance data. These relationships are typically expressed in the form of vegetation indices (VI). VIs include various combinations of spectral bands designed to maximize the sensitivity to vegetation characteristics while minimizing it to atmospheric conditions, background, view and solar angles (Baret & Guyot, 1991; Myneni et al., 1995). Operational LAI algorithms at global-scale typically make use of RT models, but the empirical models usually outperform them in more localized applications.

The design of a VI that is optimally correlated with a particular vegetation property requires good physical understanding of the factors affecting the spectral signal reflected from vegetation. The sensitivity of a VI to a vegetation characteristic is typically maximized by including bands with high sensitivity (e.g. high absorption) to the monitored entity and bands mostly unaffected by the same entity. The simplest forms of VIs are simple differences (R_{B1}-R_{B2}), ratios (R_{B1}/R_{B2}) and normalized differences [(R_{B1}-R_{B2})/(R_{B1}+R_{B2})] of the reflectances of two spectral bands (R_{B1}, R_{B2}). (In Table 2 we give examples of common VIs used in this study.) The most apparent characteristic of the green vegetation spectrum is the pronounced difference between the red and NIR reflectances, the so called red-edge around 700 nm. For example, the normalized difference vegetation index (NDVI) utilizes this difference and has been shown to correlate with many interrelated vegetation attributes, such as chlorophyll content, LAI, fractional cover, fAPAR and productivity.

The most commonly used VIs were designed for broadband sensors (one spectral band spans about 50 nm or more) having red and NIR bands, such as NOAA AVHRR and Landsat MSS (e.g. Tucker, 1979). However, the basic VIs in red and NIR spectral range suffer from three well-known problems in LAI estimation: (1) they are not sensitive to LAI over its natural range but tend to saturate already at moderate levels of LAI, (2) they are sensitive to canopy background variability, and (3) the VI-LAI relationships are dependent on the vegetation type. These VIs are also sensitive to atmospheric noise and correction.

The saturation of NDVI occurs typically at LAI levels of 2 to 6 depending on the vegetation type and environmental conditions (e.g. Sellers, 1985; Myneni et al., 1997). In general, NDVI saturates as the fractional cover of vegetation approaches one, although LAI still increases (e.g. Carlson & Ripley, 1998). Over conifer-dominated boreal forests, NDVI varies typically in a narrow range and shows poor relationships with canopy LAI (Chen & Cihlar, 1996; Stenberg et al., 2004). The reason for this is the green understory, which results in a non-contrasting background in the visible part of the spectrum (Nilson & Peterson, 1994; Myneni et al., 1997).

Many modifications of basic VIs have been suggested to give better sensitivity to LAI. Typical modifications use other visible bands than red (e.g. the green vegetation index, GNDVI, Gitelson et al., 1996), try to reduce soil effects based on the soil line concept (e.g. the soil adjusted vegetation index, SAVI, Huete, 1988), or include short wave infrared (SWIR) bands. Many modifications also attempt to reduce atmospheric effects (e.g. the enhanced vegetation index, EVI, Huete et al., 2002). The soil line is based on the observation that soil reflectances fall in a line in the red-NIR spectral space (e.g. Huete, 1988). Many VIs utilize the parameterized soil line in their calculation, but these VIs have not been successful in boreal forests as bare soil is rarely visible (e.g. Chen, 1996).

The sensitivity of shortwave infrared (SWIR) reflectance to forest biophysical variables has been recognized for a long time (e.g. Butera, 1986; Horler & Ahern, 1986) and several VIs utilizing the SWIR band have been designed. Rock et al. (1986) showed that the moisture stress index (MSI), i.e. the ratio of SWIR reflectance to NIR reflectance, was an indicator of forest damage. Later, the ratio has commonly been referred to as the infrared simple ratio (ISR, Chen et al., 2002; Fernandes et al., 2003). The SWIR reflectance has also been used for adjusting NDVI (Nemani et al., 1993) and SR (Brown et al., 2000). The reduced simple ratio (RSR) has been used specifically for estimating LAI (Brown et al., 2000; Stenberg et al., 2004) and has been employed also in regional and global-scale operational algorithms (Chen et al., 2002; Deng et al., 2006). RSR seems to reduce the sensitivity to the type and amount of understory vegetation, because background reflectance varies less in SWIR than in visible and NIR (Brown et al., 2000; Chen et al., 2002). RSR has also some capability to unify coniferous and broadleaved forest types, which reduces the need for land cover type specific LAI algorithms. However, in comparison to ISR, the use of red band makes RSR sensitive to atmospheric effects (Fernandes et al., 2003). However, although inclusion of SWIR reflectance increases the sensitivity of VIs to LAI, these indices also have a tendency to saturate at high levels of LAI (e.g. Brown et al., 2000; Heiskanen et al., 2011).

Imaging spectroscopy provides much narrower spectral bands than typical multispectral sensors. Due to the more detailed sampling of the vegetation spectra, such data can detect specific absorption features of vegetation and therefore improve the estimation of vegetation biochemical properties. For example, the SPOT 5 HRG sensors capture a spectral range from 500 nm to 1750 nm with four broad bands, in comparison to Hyperion's 242 (10 nm wide) bands between 400 nm and 2500 nm. At the canopy scale, the contents of biochemical components and LAI are highly inter-related (e.g. Asner, 1998; Roberts et al., 2004). Therefore, imaging spectroscopy could potentially improve LAI estimates. Furthermore, there is potentially complementary information outside the typical spectral bands of broadband sensors.

One way to utilize imaging spectroscopy data is to calculate narrow-band VIs in a similar fashion as for broadband data but using narrower bands. The aim is to improve the sensitivity of the VI to a specific vegetation biochemical property. For example, Ustin et al. (2009) give a comprehensive review on VIs used as indicators of plant pigments (chlorophyll, carotenoids and anthocyanin). The methods of estimating the non-pigment biochemical composition of vegetation (water, nitrogen, cellulose and lignin), on the other hand, are reviewed by Kokaly et al. (2009). Many of the developed indices have been designed to work at leaf level and do not necessarily upscale to canopy level, because of the high sensitivity to canopy structure, background, solar and view geometry. Another approach is to find iteratively the simple combinations of bands that give the best correlation with empirical data (e.g. Mutanga & Skidmore, 2004; Schlerf et al., 2005).

Most chlorophyll indices exploit the information in the red edge around 700 nm (Ustin et al., 2009). Imaging spectroscopy data also enables the estimation of the red edge position (REP), which is particularly sensitive to changes in chlorophyll content (e.g. Dawson & Curran, 1998). Water indices, on the other hand, utilize the water absorbing regions in the SWIR region of the spectrum (e.g. Gao, 1996; Zarco-Tejada et al., 2003). Those indices seem particularly interesting for LAI estimation considering the importance of the SWIR spectral region in estimating LAI using broadband indices.

There is growing evidence that imaging spectroscopy data can improve LAI estimates in comparison to broadband data by reducing the saturation effects. Depending on the vegetation type and range of LAI, different types of VIs have been found useful. However, the red edge indices have been most effective in estimating LAI of crops (Wu et al., 2010), grasslands (Mutanga & Skidmore, 2004) and thicket shrubs (Brantley et al., 2011). On the other hand, indices based on NIR and SWIR bands have been successful in broadleaved (le Maire et al., 2008) and coniferous forests (Gong et al., 2003; Schlerf et al., 2005; Pu et al., 2008). The importance of the SWIR spectral region in estimating boreal forest LAI has also been emphasized by multivariate regression analysis (e.g. Lee et al., 2004). However, broadband sensors can also have advantages over narrowband sensors in LAI estimation, for example, by being less sensitive to noise due to the sensor, atmosphere and background (e.g. Broge & Leblanc, 2000). Although there are case studies from different biomes, the performance of narrowband VIs has been poorly assessed over European boreal forests.

2. Case study

2.1 Aims

The aim of the study is to establish the extent to which vegetation indices can be used to measure variation in LAI based on a test site in southern boreal forest in Finland. We explore different VIs in LAI estimation during full leaf development. We compare the performance of narrowband VIs to traditional broadband VIs. The objective is to identify VIs, which are least sensitive to species composition and, on the other hand, perform well in coniferous stands.

2.2 Materials and methods

2.2.1 Study area

The study area, Hyytiälä, is located in the southern boreal zone in central Finland (61° 50'N, 24°17'E) and has an annual mean temperature of 3°C and precipitation of 700 mm. Dominant tree species in the Hyytiälä forest area are Norway spruce (*Picea abies* (L.) Karst), Scots pine (*Pinus sylvestris* L.) and Silver birch (*Betula pendula* Roth). Understory vegetation, on the other hand, is composed of two layers: an upper understory layer (low dwarf shrubs or seedlings, graminoids, herbaceous species) and a ground layer (mosses, lichens). The growing season typically begins in early May and senescence in late August. We measured twenty stands from the Hyytiälä forest area in July 2010 (see Section 2.2.2, Table 1). The stands represented different species compositions that are typical to the southern boreal forest zone in Finland.

Site	Vegetation	Site type	Tree height, m	Basal area, m2/ha	LAI
A4	Pine	mesic	15.8	20.4	1.77
A5	pine, understory broadleaf	mesic	18.6	24.3	2.67
B2	spruce, understory birch	mesic	7.5	10	2.64
D3	pine, understory spruce & birch	sub-xeric	17.8	20.5	2.37
D4	spruce, 25% birch	mesic	16.5	27.5	3.72
E1	birch, spruce understory	mesic	19.1	10.7	2.58
E5	50% spruce, 50% birch	mesic	23.1	27.2	4.12
E6	50% spruce, 40% birch, 10% pine	mesic	10.2	22.2	3.34
E7	Spruce	mesic	13.3	31.7	3.91
F1	birch, spruce understory	mesic	13.8	20.9	3.37
G4	spruce, 15% birch, 10% pine	herb-rich	15.5	29.1	4.57
H3	Birch	herb-rich	14.9	10.7	2.63
H5	Birch	herb-rich	14.1	20.6	2.77
I4	birch, understory pine, spruce seedlings	mesic	2.4	4	2.61
T	Spruce	mesic	24.6	56	3.43
U16	Birch	mesic	14	21	2.69
U17	birch, 10% spruce	herb-rich	11.7	27	3.35
U18	65% pine, 25% spruce, 10% birch	sub-xeric	16.5	26	3.45
U26	20% pine, 70% spruce, 10% birch	mesic	16.8	24.9	2.43
U27	5% pine, 90% spruce, 5% birch	mesic	15.2	20.9	2.63

[pine = Scots pine, spruce = Norway spruce, birch = Silver birch]

Table 1. Study stands.

2.2.2 Ground reference measurements

The LAI-2000 Plant Canopy Analyzer (PCA) is one of the most commonly used optical devices to measure LAI. The PCA's optical sensor includes five concentric rings of different zenith angles (θ) (together covering almost a full hemisphere), which measure diffuse sky radiation between 320-490 nm (LI-COR, 1992). Measurements by the PCA performed below and above the canopy yield canopy transmittances, T(θ), for each ring. Finally, LAI is calculated by numerical approximation of the integral (Miller, 1967):

$$LAI = -2 \int_{0}^{\pi/2} \ln[T(\theta)] \cos\theta \sin\theta d\theta \qquad (1)$$

There are four fundamental assumptions behind the LAI calculation method: 1. leaves (needles) are optically black in the measured wavelengths (implying that canopy transmittance closely corresponds to canopy gap fraction), 2. leaves (needles) are randomly distributed inside the canopy volume, 3. leaves (needles) are small compared to the area of view of the PCA's rings, and 4. leaves (needles) are azimuthally randomly oriented. The LAI estimate produced by Eq. 1 is commonly called effective LAI as the foliage elements are not randomly organized but typically clumped (or grouped) together, which causes the estimate produced by the PCA to be smaller than the "true" LAI (Chen et al., 1991; Deblonde et al., 1994).

The LAI measurements can be done either with one or two PCA instruments. One PCA is used for small plants such as crops, but for taller plants (e.g. trees), two units are necessary. When only one instrument is used, the measurement is at first taken below and then above the canopy. If two instruments are used, one instrument remains above the canopy and the other one below the canopy. The use of two instruments is preferable since data are logged nearly simultaneously with both sensors. The LAI estimate is calculated by combining below and above canopy data. The measurements should be conducted under diffuse light conditions; for example, when the sky has a full cloud cover or the sun angle is low (less than 16 degrees). The radius of the sample plot should be at least three times the dominant tree height as the PCA instrument has a relatively large opening angle.

In this study, the ground reference LAI (Table 1) was acquired by operating two LAI-2000 PCA instruments simultaneously. The instruments were intercalibrated before measurements were performed. The reference sensor was located above the forest canopy and set at a 15-second logging interval, while the other sensor was used inside the forest. The sampling scheme was a 'VALERI-cross' (Validation of Land European Remote Sensing Instruments, VALERI) which consists of two perpendicular 6-point transects. The distance between two measurement points was four meters, so that the sampling scheme corresponded roughly to a 20 m x 20 m plot. Measurement height was kept constant at 0.7 meters.

2.2.3 Satellite data

In this study, we used narrowband spectral data obtained from a Hyperion satellite image. Hyperion is a narrowband imaging spectrometer aboard the National Aeronautics and Space Administration (NASA) Earth Observer-1 (EO-1) satellite launched in 2000. Hyperion captures data in the 'pushbroom' manner in 7.7 km wide strips using 242 spectral bands. The spectral range of Hyperion is 356-2577 nm with each band covering a nominal spectral range of 10 nm. Each pixel in a Hyperion image corresponds to an area of 30 m x 30 m on the ground. During an acquisition, a scene with a length of either 42 km or 185 km is recorded. Hyperion is in a repetitive, circular, sun-synchronous, near-polar orbit at an altitude of 705.3 km measured at the equator. Thus, it can image almost any point on Earth and it flies over all locations at approximately the same local time. The nominal revisit time is 16 days, but due to the possibility of tilting the sensor, the potential revisit frequency is higher. The scene used in this study was captured on 03 July 2010, and was provided courtesy of the U.S. Geological Survey (USGS) Earth Explorer service.

Out of the potential 242 spectral bands, several lack illumination (due to the absorption in the atmosphere or a decrease of incident solar spectral irradiance in the longer infrared wavelengths) or have a very low spectral response. This leaves the user with 198 usable spectral bands: bands 8-57 in the visible and NIR (wavelengths 436-926 nm) and bands 77-224 in SWIR (wavelengths 933-2406 nm) (Pearlman et al., 2003). Hyperion images have several known deficiencies which can be corrected using algorithms given in scientific literature. Firstly, Hyperion suffers from systematic striping in along-track direction of the image. The stripes are characteristic to all pushbroom sensors. Instruments belonging to this broad class have a different receiving element for each image line. Hyperion has thus 256 radiation-sensitive elements for each spectral band, each seeing a separate 30 m strip of the ground, thus producing the 7.7 km wide image. The striping can be broadly divided into two classes, completely missing lines (due to non-functioning receiving elements) and actual stripes (arising from slightly different sensitivities of the 256 receivers). We removed the actual striping using Spectral Moment Matching (SpecMM), outlined by Sun et al. (2008), which uses the average and standard deviation statistics between highly correlated bands to remove stripes. Next, the missing lines containing no information were identified and corrected using the values from spatially adjacent pixels using local destriping (Goodenough et al., 2003). The results of the destriping can be seen in Figure 1.

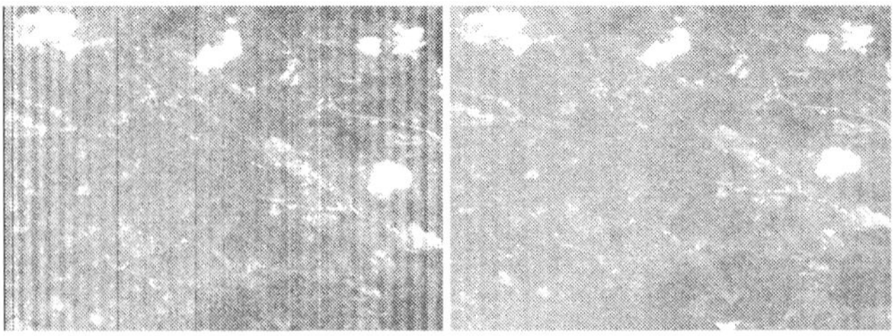

Fig. 1. Hyperion band 8 (436nm) uncorrected image (left), and corrected using Spectral Moment Matching and local destriping (right).

The second known defect in Hyperion imagery is a shift in the wavelength of each column in the across track direction from the band central wavelength. This shift, known as spectral smile, is also characteristic to pushbroom sensors and is a result of different optical paths leading to the different receiving elements. The shift is a function of wavelength and the position of the receiving element in the receiving array. As is the case for most instruments, the "smile" manifests itself in Hyperion imagery as a "frown", with the wavelengths of the columns near the edges of each band shifting negatively from the bands average wavelength (Figure 2). The smile was corrected using the pre-launch laboratory measured spectral shift (Barry, 2001). We used interpolation to bring each individual pixel to a common central wavelength based on the pre-launch calibration measurements.

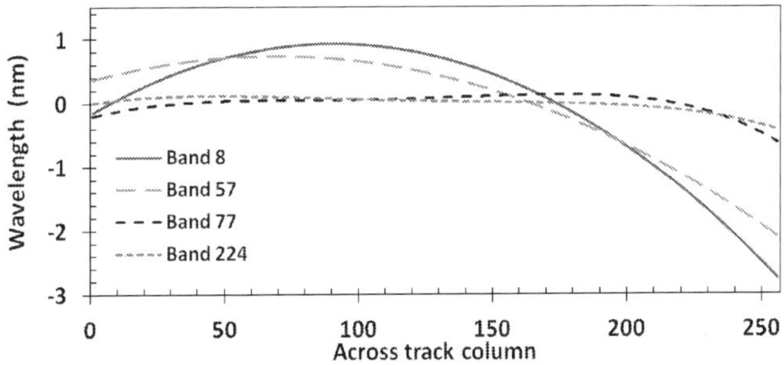

Fig. 2. Laboratory measured spectral shift of Hyperion (Barry, 2001).

The signal received by the Hyperion instrument consists of the photons scattered by the atmosphere as well as the ground surface. To study surface reflectance, the influence of the atmosphere needs to be eliminated in a process commonly known as atmospheric correction. We performed this correction using an algorithm known as Fast Line-of-sight Atmospheric Analysis of Spectral Hypercubes (FLAASH, Matthew et al, 2000). FLAASH is an absolute atmospheric correction that incorporates the MODTRAN4 radiation transfer code to model the scattering and transmission properties of the atmosphere at the time of image capture (San & Suzen, 2010). The FLAASH algorithm is incorporated into the ITT Visual Information Solutions (ITT VIS) ENVI software. For processing, FLAASH requires an input value for visibility to estimate atmospheric aerosol levels, in addition to basic geographic and temporal details about the scene. The visibility can be recalculated by FLAASH, using a ratio between dark pixels at 600 nm and 2100 nm. However, a more accurate estimate of visibility was achieved using ground based optical measurements from a weather station in the area.

The final processing stage is to resample the image pixels into a geographic coordinate system, known as geocorrection. This was done using a polynomial transformation to a vector base map from the National Land Survey of Finland. The Hyytiälä area contains numerous roads, providing a large number of easily identifiable potential ground control points (GCPs) at intersections. Around 20 GCPs were selected, with a root mean square error of 0.4 pixels being achieved. Bilinear interpolation was chosen for resampling the image pixels due to the better geometric accuracy over nearest neighbour.

The final product is a geocorrected image of the surface hemispherical-directional reflectance factors (HDRF) of the Hyytiälä area. To validate the atmospheric correction, we compared the HDRF to a field measured reflectance factor. A soccer field of about 130 m by 60 m in the area was sampled during the summer of 2010 every two to three weeks using an ASD handheld portable spectroradiometer covering a spectral range from 325-1075 nm. The sampling was done using a transect approach with 42 measurements at around 1 meter intervals. The final hemispherical-conical reflectance factor (HCRF) used for the comparison is an average of the transect representing the average for the whole field. While no ground measurements fell on the exact date of the Hyperion image, the ground measured spectra was interpolated to dates between two measurements. After interpolation the ground measured HCRF was binned into corresponding Hyperion bands using the spectral response of each band.

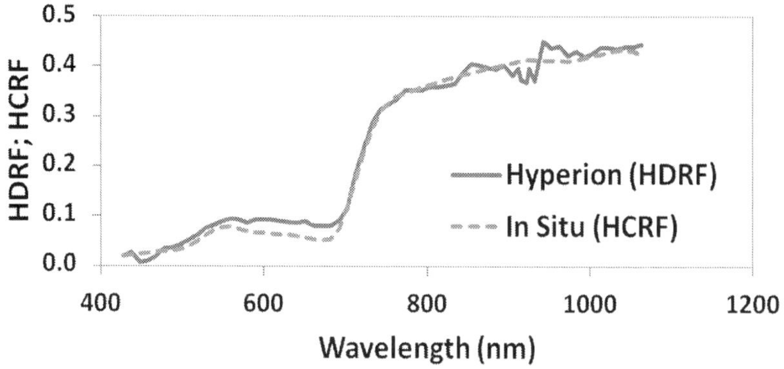

Fig. 3. Comparison of a soccer field's spectral reflectance factors from in situ radiometric measurements and corrected Hyperion data.

Overall, there is a very good correlation between the field measured reflectance and the fully processed Hyperion reflectance (Fig. 3). An overall RMSE of 1.8% is achieved, which gives us confidence in the validity of the pre-processing and atmospheric correction. However, as the *in situ* spectrum is considerably smoother than the one measured from the satellite, a considerable amount of noise is also present in the satellite-derived HDRF.

2.2.4 Vegetation indices and statistical analysis

First, we studied how HDRFs in single bands are correlated with LAI. Next, in order to evaluate narrow-band VIs for estimating LAI, we did regression analyses between various VIs and LAI. We used two approaches to select narrowband indices: 1) We made a literature survey for narrow-band VIs that have been designed to estimate foliage biochemical components. (A collection of VIs showing the highest R^2 with LAI are shown in Table 2.) 2) We calculated all the possible Ratio Indices (RI) and Normalized Difference Indices (NDI) of Hyperion bands and correlated them with LAI. In other words, the first approach also contains VIs combining several bands and the second approach aims to identify the simple two-band VIs that best correlate with LAI.

To facilitate the comparison of narrowband VIs with broadband indices, we calculated synthetic HDRFs based on Landsat 7 ETM+ bands. The HDRFs were calculated according to Jupp et al. (2002) using the ETM+ spectral sensitivity functions, and Hyperion's central wavelengths and bandwidths. Four broadband indices were calculated for comparison, SR, NDVI, ISR and RSR (Table 2). All these indices have been used for LAI estimation in various biomes. SR and NDVI were included for reference, and ISR and RSR because they have shown best performance over conifer-dominated boreal forests (see 1.3).

We analyzed the data both by grouping all the sample plots together and separately for coniferous plots (> 75% of the trees were Scots pines or Norway spruces). In the birch-dominated stands, the variation in LAI was too small for reliable regression analysis.

We studied only linear relationships. The strength of the relationship was assessed by the coefficient of determination (R^2) and the root mean square error (RMSE).

Abbr.	Index	Formula	Reference	Bands applied
Indices concentrating on the red-edge				
SR	Simple Ratio	$SR = R_{ETM+4}/R_{ETM+3}$	Rouse et al. (1974), Birth & McVey (1968)	ETM+3, ETM+4
NDVI	Normalized Difference Vegetation Index	$NDVI = (R_{ETM+4}-R_{ETM+3})/(R_{ETM+4}+R_{ETM+3})$	Rouse et al. (1974)	ETM+3, ETM+4
REP	Red Edge Position	$REP = 700+ (((R_{773} +1,5 *R_{662}) - R_{692}) / (R_{733}-R_{692})) *(740-700)$	Danson & Plummer (1995)	773, 662, 692, 733
Indices concentrating on pigment content				
PSSRa	Pigment-Specific Simple Ratio - chla	$PSSRa = R_{803}/R_{681}$	Blackburn (1998)	681, 803
Water sensitive indices				
MSI = ISR	Moisture Stress Index = Infrared Simple Ratio	$ISR = R_{ETM+5}/R_{ETM+4}$	Rock et al. (1986), Fernandes et al. (2002)	ETM+4, ETM+5
RSR	Reduced Simple Ratio	$RSR = (R_{ETM+4}/R_{ETM+3}) * ((R_{ETM+5\_min} - R_{ETM+5}) / (R_{ETM+5\_max} - R_{ETM+5\_min}))$	Brown et al. (2000)	ETM+3, ETM+4, ETM+5

Table 2. Vegetation indices investigated in this study. The symbol R refers to the HDRF. Subscripts refer to the applied ETM+ band or the central wavelength (in nm) of the Hyperion band

2.3 Results

2.3.1 General characteristics of forest spectra

Two examples of forest reflectance factors (HDRFs) are presented in Figure 4. To allow relating the vegetation spectra to satellite signals, the sensitivity functions of the corresponding ETM+ bands are shown. Note the correspondence of ETM+2 with the green peak, ETM+3 with the red local minimum and ETM+4 with the plateau in the NIR. The red-edge slope (between ETM+ bands 3 and 4) is not covered by ETM+ bands. ETM+5 and ETM+7 catch the signal in the shortwave infrared region (SWIR-1 (here: 1470-1800 nm) and SWIR-2 (here: 2030-2360 nm) respectively), avoiding the two strong water absorption bands in-between.

The average reflectance of coniferous stands is slightly lower in the green region and decidedly lower in the NIR than the reflectance of birch stands. In SWIR-1 (covered by ETM+5) the reflectances become more comparable, and in SWIR-2 (covered by ETM+7) the signals almost meet.

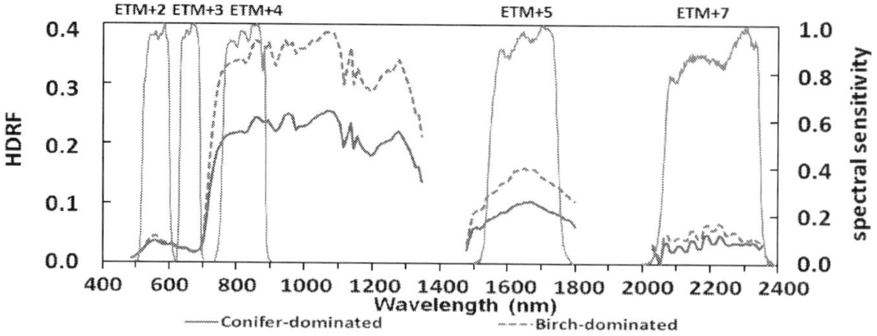

Fig. 4. Average conifer and birch-dominated stand spectra. The grey lines show the spectral sensitivity of the ETM+ bands.

2.3.2 Regression analysis for single bands

The different average HDRF for the two forest types (Fig. 4) results in different correlations of the satellite bands to LAI (Fig. 5).

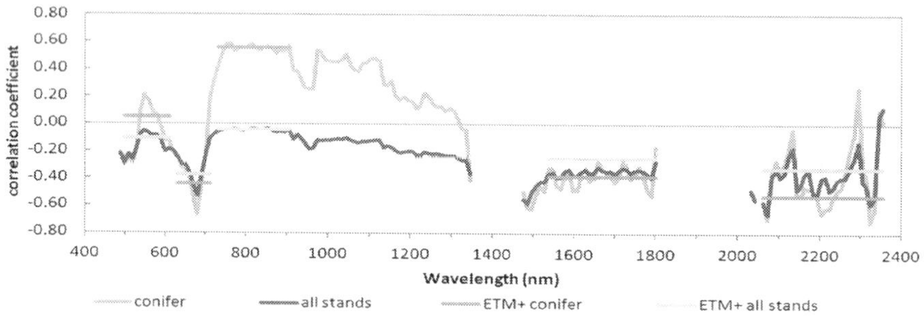

Fig. 5. Correlation coefficient of LAI with ETM+ and Hyperion spectral bands for all sample stands, and separately for conifer sample stands.

The correlation coefficients for all stands varied between -0.6 and -0.038. All correlations were negative, except for the two Hyperion bands centred at 2345 nm and 2355 nm. Two important regions (green and NIR) had almost no correlation with LAI. Only the absorption peak of chlorophyll produced a strong negative correlation at 681nm. The SWIR correlations were also mostly negative.

For conifer stands, correlation coefficients varied between -0.7 and 0.6. The first peak was at 549 nm, in the middle of the green band, followed by a strong negative correlation in the red with a peak at 681 nm. In the NIR a strong positive correlation was observed again. A slight shoulder began at 712 nm, with a plateau at 752 nm. In the SWIR, correlation coefficients were very close to those of all stands.

Fig. 5 also shows the correlation of the ETM+ bands to LAI. The lower spectral resolution averages wider wavelength ranges and therefore shows less variation in correlation coefficients.

2.3.3 Correlation of vegetation indices to LAI for all sample plots

The best broadband index analysed here was the Infrared Simple Ratio (ISR, R^2 = 0.56), followed by the Reduced Simple Ratio (RSR, R^2 = 0.40) (Table 3). The best narrowband combinations (either RI or NDI) showed more potential with R^2s exceeding 0.65 (Table 3, Fig. 6). If there were several indices based on neighbouring bands (within 10 nm) we chose the best one to Table 3.

VI	Bands applied	R^2	RMSE	RMSE Conifer	RMSE Broadleaf
broadband indices using simulated ETM+					
ISR	ETM+4, ETM+5	0.56	0.44	0.42	0.25
RSR	ETM+3, ETM+4, ETM+5	0.40	0.52	0.59	0.31
NDVI	ETM+3, ETM+4	0.09	0.64	0.68	0.51
SR	ETM+3, ETM+4	0.04	0.66	0.73	0.46
narrowband indices using Hyperion					
RI	1134, 1790	0.71	0.36	0.34	0.38
NDI	1134, 1790	0.68	0.38	0.36	0.39
RI	732, 1790	0.67	0.38	0.42	0.31
RI	1074, 1790	0.67	0.38	0.40	0.34
RI	885, 1790	0.67	0.39	0.37	0.35
RI	854, 1790	0.66	0.39	0.37	0.34
RI	1003, 1639	0.66	0.39	0.39	0.26
RI	1044, 1790	0.66	0.39	0.39	0.37
NDI	732 1790	0.66	0.39	0.42	0.33
NDI	1084, 1286	0.66	0.39	0.43	0.22

Table 3. Indices most correlated with LAI for all sample plots. RMSE was also calculated separately for each forest class. Bands for Hyperion refer to the central wavelength (in nm).

The best band combinations for RI and NDI indices were very similar (Fig. 6). A strong correlation with LAI existed for bands combining the region between 730 to 900 nm and 1130 to 1350 nm. Another interesting region was within SWIR-1; especially strong was the correlation around 1780 and 1790 nm. These bands also showed up in the best performing indices for all forest classes combined (Table 3).

The two best narrowband indices for all forest plots were the RI (R^2 = 0.71, RMSE = 0.36) and NDI (R^2 = 0.68, RMSE = 0.38) based on bands centred at 1134 and 1790 nm (Table 3). This is consistent with the best broadband index (ISR) which also combines NIR and SWIR. The same spectral regions are used by all the other best indices except two cases including a band in the red-edge (732 nm). Examples of the strongest relationships are shown in Fig. 7. However, when looking at the RMSE for conifer and broadleaf stands (Table 3) it became apparent that for some indices (e.g. NDI based on 1084 nm and 1286 nm: RMSE = 0.43 for conifers and RMSE = 0.22 for broadleaf) their LAI was correlated differently to the same VI.

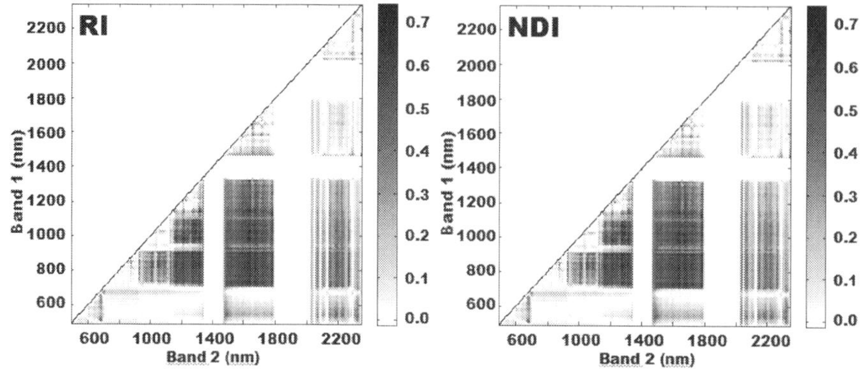

Fig. 6. Matrixes showing the R^2 between LAI and simple narrowband indices calculated for all possible combinations of Hyperion bands. The indices are defined as follows: RI=Band1/Band2, and NDI=(Band1-Band2)/(Band1+Band2).

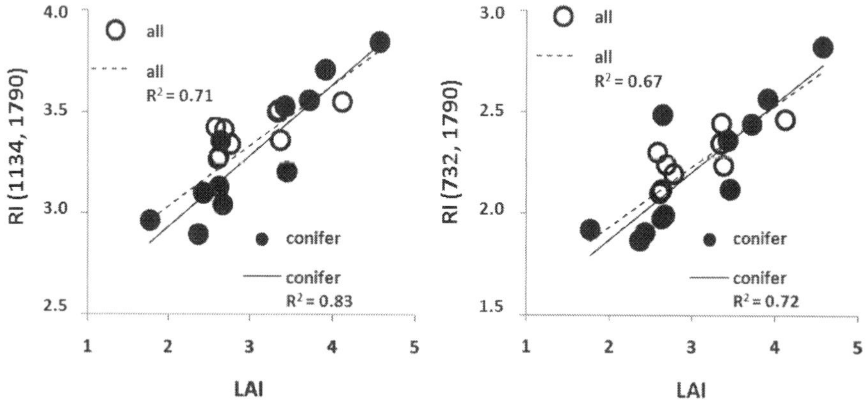

Fig. 7. The relationship of LAI and two best ratio indices (RI).

2.3.4 Correlations for coniferous dominated forest plots

The performance of the broadband indices for conifer-dominated stands was much better than over all sample stands. R^2 now ranged from 0.60 to 0.79, and NDVI showed the best correlation with LAI, followed by SR.

The best performing narrowband index over coniferous forest was neither RI nor NDI but REP (R^2 = 0.89) calculated according to the method of Danson & Plummer (1995) (Table 2). This index combined four bands in the visible and NIR; an area also represented in several of the other indices which best correlated with LAI in coniferous stands.

The matrixes for all band combinations of Hyperion bands over conifer-dominated stands (Fig. 8) showed wider spectral regions of high correlation than for all stands (Fig. 6).

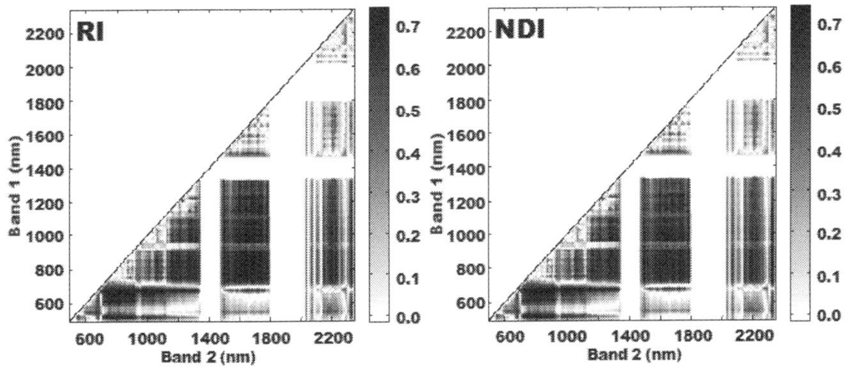

Fig. 8. Matrixes showing the R^2 between LAI and two narrowband indices calculated for all possible combinations of Hyperion bands for conifer-dominated stands.

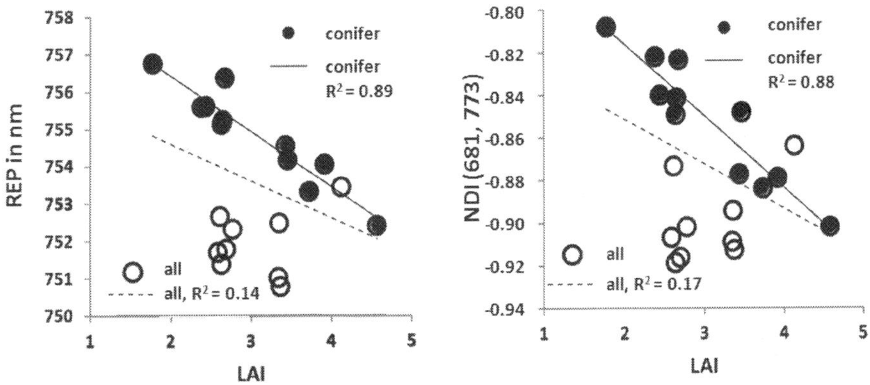

Fig. 9. The relationship of LAI and the two best performing narrowband indices for conifer-dominated stands.

Most of the indices with the highest correlations to LAI in coniferous stands used bands around the red-edge. Almost all of them (e.g. the Pigment-Specific Simple Ratio Index for chlorophyll a, PSSRa) applied the Hyperion band centred at 681nm, the peak of chlorophyll a absorption. Exceptions were the RI and NRI using the bands centred at 1185 and 1790 nm (i.e. combining NIR and SWIR), and RI and NDI using bands centred at 518 and 773 nm (i.e. combining carotene absorption and NIR).

Scatterplots for the two best indices for coniferous stands are shown in Fig. 9. In both cases, coniferous plots differed considerably from the other plots. This was indicated also by the high RMSE for all stands (up to 1.42, Table 4). However, for indices using NIR and SWIR (e.g. RI and NDI based on 1185 and 1790 nm) the differences were less pronounced. The VI showing the lowest RMSE for all stands (0.49) was the RI (1185 and 1790 nm) with an R^2 for conifer stands of 0.86 and RMSE 0.29.

VI	Bands applied	R^2	RMSE	RMSE All stands
broadband indices using simulated ETM+				
NDVI	ETM+3, ETM+4	0.79	0.36	1.20
SR	ETM+3, ETM+4	0.78	0.36	1.56
ISR	ETM+4, ETM+5	0.71	0.42	0.44
RSR	ETM+3, ETM+4, ETM+5	0.60	0.50	0.90
narrowband indices using Hyperion				
REP	671, 702, 742, 783	0.89	0.26	1.29
NDI	681, 773	0.88	0.27	1.02
RI	681, 773	0.88	0.28	1.01
RI	1185, 1790	0.86	0.29	0.49
NDI	1185, 1790	0.86	0.30	0.50
NDI	681, 742	0.85	0.30	1.01
NDI	681, 824	0.85	0.30	0.98
RI	681, 742	0.85	0.31	0.99
NDI	518, 773	0.85	0.31	1.42
PSSRa	803, 681	0.85	0.31	1.30
RI	518, 773	0.85	0.31	1.39

Table 4. Indices most correlated with LAI in conifer-dominated plots. R^2 and RMSE for conifer-dominated stands, and RMSE separately for all stands. Bands for Hyperion refer to the central wavelength (in nm).

2.4 Discussion

In our case study, the narrowband VIs provided more accurate LAI estimates than the broadband VIs synthesized from the same data in a boreal forest study site. The best narrowband combinations showed relatively strong linear relationships with LAI (R^2 > 0.65), although the Hyperion image was acquired in the middle of the growing season when LAI is the highest. The relationships were even stronger if the analysis was restricted to the conifer stands (R^2 > 0.85). The results are promising as common broadband VIs tend to saturate at the highest LAI values. The improvement of estimation accuracy is in agreement with the previous studies, which have emphasized the potential of narrowband VIs for estimating forest canopy LAI (e.g. Lee et al., 2004; Schlerf et al., 2005; Brantley et al., 2011; Wu et al., 2010).

Most of the narrowband VIs showing the strongest relationships with LAI were based on reflectances in the far red and at the red edge (680–740 nm), NIR (e.g. 885 and 1134 nm) and SWIR (e.g. 1639 nm and 1790 nm) wavelength regions (Figure 10). Many of the most important spectral regions are not covered by the ETM+ spectral bands, and the spectral regions are very narrow in comparison to the ETM+ bands.

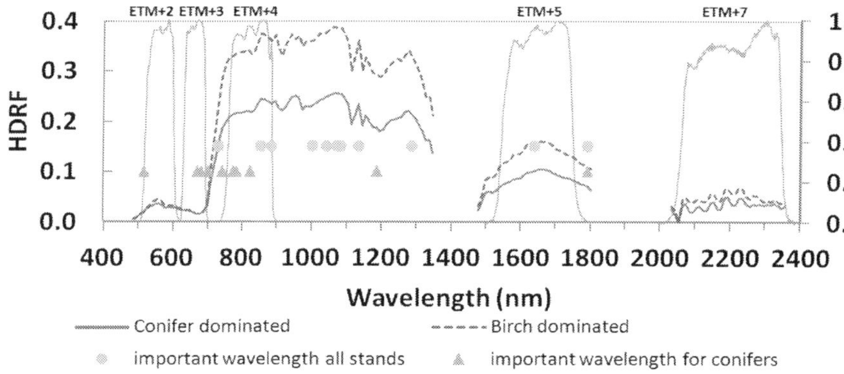

Fig. 10. Spectral regions used by the indices showing the strongest relationships with LAI over all sample stands and conifer stands.

The NIR and SWIR spectral bands were particularly important when all sample plots were analyzed together. This is in agreement with the best broadband indices, ISR and RSR. The importance of NIR and SWIR bands has been emphasized also in previous studies testing narrowband VIs for estimating forest LAI (e.g. Lee et al., 2004; Schlerf et al., 2005). The leaf (needle) reflectance at those wavelengths is mainly controlled by water absorption, although leaf biochemical components such as proteins, cellulose and lignin also contribute to absorption in the infrared (e.g. Curran, 1989). The amount of water at the canopy level is directly related to LAI, which explains strong correlations. The bands centered at 1134 nm and 1790 nm are among the Hyperion bands, which are closest to the water absorption regions centered at approximately 1200 nm and 1940 nm. The spectral bands close to the water absorption regions at 970 nm and 1400 nm are also employed in some of the best indices. The spectral bands of the broadband sensors are usually placed in the middle of the atmospheric windows to avoid atmospheric absorption. However, it seems that narrow spectral bands close to the water absorption regions are particularly interesting for estimating LAI. In these wavelength regions, the reflectance seems to be relatively insensitive to tree species or composition of the understory vegetation, as suggested earlier by the studies using broadband indices (e.g. Brown et al., 2000).

When pure coniferous stands were studied separately, the relationships became stronger and the far red and red edge spectral bands were included in several of the best VIs. However, the improvement in accuracy relative to the best VI based on NIR and SWIR reflectance (RI based on bands centered at 1185 nm and 1790 nm) was rather modest. The best broadband indices were NDVI and SR, which are based on ETM+ red and NIR bands. Usually, NDVI has shown relatively weak relationships with LAI in conifer dominated boreal forest (e.g. Stenberg et al., 2004).

The strongest relationship with LAI was provided by the red edge position (REP) calculated by the method proposed by Danson and Plummer (1995). In general, the REP is considered to be sensitive to leaf and canopy chlorophyll content, so that increasing the amount of chlorophyll, or LAI, is related to the longer REP wavelength because of the widening of the chlorophyll absorption region at approximately 680 nm (Danson & Plummer, 1995; Dawson and Curran, 1998; Sims & Gamon, 2002; Pu et al., 2003). In comparison to SWIR spectral bands, the far red and red edge spectral region is sensitive to species composition, shown as poor relationships over mixed vegetation. However, sometimes poor relationships between the REP and LAI have been reported even for pure coniferous stands (Blackburn, 2002). However, although the REP calculated in this study showed strong correlation with coniferous LAI, the estimated wavelengths do not correspond to the Red Edge Inflection Point (REIP), i.e. the steepest slope of the red-edge. The wavelengths are considerably longer. Therefore, the unusual inverse relationship between REP and LAI in this study is explained by the calculation method (Danson and Plummer, 1995). Alternative calculation methods for REP are summarized, for example, by Pu et al. (2003).

Although many studies testing narrowband VIs for LAI estimation have stressed the potential of the red edge and SWIR spectral regions, the specific spectral bands providing the strongest relationships with LAI vary between the studies. Also in our case study, the optimal band combinations provided stronger relationships with LAI than VIs collected from the literature. This is somewhat expected, as the number of spectral bands and their possible combinations is so large that empirically determined optimal band combinations are likely to depend heavily on the local environmental conditions and type of satellite image data. For example, approximately 150 useful spectral bands of Hyperion make more than 20,000 two-band combinations. Because of this, the optimal indices cannot necessarily be generalized very well. Furthermore, a large number of spectral bands combined with a small number of sample plots increase the risk that the regression models are overfitted. However, this should be mostly a problem with multivariate approaches (e.g. Lee et al., 2004). Moreover, when comparing broadband and narrowband indices, it should be noted that we used only synthesized ETM+ data and the results could differ to some extent if true ETM+ data would have been used instead (Lee et al., 2004). This is because the synthetic broadband data is affected by the lower signal-to-noise ratio of the narrow spectral bands, even if data are averaged.

3. Future perspectives

Wider use of imaging spectroscopy data is hampered by the availability of the data. Today, mostly airborne instruments are used to produce remote sensing data with high spectral resolution. Airborne measurements are associated with relatively small spatial coverage and high operating costs falling directly to data users. The Hyperion sensor used in this case study is a rare exception: it is the only true imaging spectrometer in orbit today, providing wide spectral coverage with uniform spectral resolution and contiguous bands. The scene, however, is about to change. At the end of the decade (i.e., around 2020), NASA is planning to launch the HyspIRI mission, providing narrowband data with routine global coverage (Samiappan et al., 2010). Before HyspIRI, several national space programs are striving to launch satellites with capability to produce narrowband data (e.g. the EnMAP instrument, Segl et al., 2010). Therefore, the need for developing algorithms that would make use of the advanced properties of narrowband data, compared to the more traditional multispectral data, is evident.

In this case study, we used narrowband VIs to relate forest LAI to remotely sensed reflectance signals. Historically, vegetation indices have been among the very first tools in interpreting multispectral remote sensing data from vegetated areas. Later, physically-based reflectance modelling has taken over the role of the preferred method in large-scale retrievals of vegetation biophysical variables. Similar developments may take place in the interpretation of narrowband imaging spectroscopy data. However, let us first take a closer look at narrowband indices as they are used in the current study.

As discussed above (section 1.3), VIs are usually treated as empirical (or, at least semi-empirical) tools in remote sensing. However, it has been known for a long time that the reflectance indices convey also some information on the physical processes related to the interaction of light with plant elements. Indeed, Myneni et al. (1995) showed that the common indices are actually derivatives of canopy reflectance and are physically related to abundances of absorbing pigments. For this reason, indices commonly make use of two spectral regions: one inside the spectral region where the absorption of a pigment is strong, and one outside the absorption band. The use of red and near-infrared wavelengths thus corresponds to measuring the abundance of one of the most vital plant pigments, chlorophyll.

Can such an interpretation be extended to narrowband indices? From the point-of-view of the physics of radiative transfer, there is no fundamental difference between broad- and narrowband indices. However, for calculating a spectral derivative, there is little use of well-tuned and potentially much noisier narrow spectral bands. For detecting pigments whose absorption spectra span tens, if not hundreds of nanometers, broadband indices seem a much more robust tool. Further, vegetation indices, especially early ones like the NDVI, have been shown both empirically and on the basis of theoretical studies, to be sensitive to factors others than those of interest, such as soil brightness changes and atmospheric effects. Most narrowband indices can be viewed as finely tuned versions of their older broadband counterparts. Site-specific selection of wavelengths leads to a better explanatory power of narrowband VIs as we also demonstrated in this case study. Unfortunately, the fine tuning for eliminating environmental effects makes narrowband indices potentially even more site-specific than broadband ones.

The comparison of narrowband and broadband VIs presented above did not concern indices capturing truly narrowband effects, e.g. the photochemical reflectance index PRI (Gamon et al., 1992) or various red edge parameters. Intrinsically narrowband VIs are based on effects that cannot be detected from broadband data. These indices are not more site-specific than broadband indices and do indeed, due to a finer spectral resolution, provide additional information on vegetation cover on all scales. Similarly, the red edge parameters calculated above make use of the high spectral resolution of narrowband data in a manner which is not site-specific. Therefore, it is not surprising that they provide a good fit for estimating forest stand variables regardless of dominating species.

An alternative to using narrowband indices would be to invert a full canopy reflectance model: the goals of both methods are to retrieve information on some biophysical variable of interest (Rautiainen et al., 2010). As discussed in this chapter, the theoretical foundations of the two approaches are somewhat similar. However, obvious limitations of index-based inversions lie in that it is not possible to define a spectral index sensitive to only one process, nor is it possible to design a universal spectral index which would be optimal for all applications everywhere and all the time (Verstraete & Pinty, 1996). Further, since vegetation indices carry only part of the information available in the original channel reflectances, they assume that the information of interest is contained exclusively in the observed spectral variations. VIs also often neglect the effects of surface anisotropy associated with the specific geometry of illumination and observation at the time of the measurements (Govaerts et al., 1999). Last, but not least, a fundamental shortcoming of the index-based approach lies in its potentially wide application area. A user not directly working in the field of remote sensing science may be distracted by a statistically strong dependence between a variable of interest (e.g. an ecological parameter describing diversity) and a vegetation index. However, canopy reflectance signals can carry information only on what are known as state variables of radiative transfer (abundances of optically active substances, canopy amount and structure, etc.). Other variables may be correlated with one or more of the state variables, but before drawing conclusions based on such correlations, the nature and application range of the correlation should be clarified.

Naturally, physical canopy reflectance models are immune to the problems listed above. When working in the forward mode, a modern reflectance model can reliably predict the spectral reflectance signal of a vegetation canopy given the required inputs (e.g. Widlowski et al., 2007). When run in inverse mode, the models should be able to produce an estimate of the state variables of radiative transfer based on measured spectral reflectance values. Unfortunately, due to the large number of the state variables and the mathematical nature of the inverse problem, a robust result is difficult to achieve (Baret & Buis, 2008). Despite the present-day problems with inverting canopy reflectance models, it is clear that physical models hold a clear advantage over index-based biophysical parameter estimation, especially when using imaging spectroscopy data. Physical models account for changes in environmental conditions and estimate all state variables simultaneously. They also have the advantage of failing if unphysical data is fed to them (e.g. due to sensor failure or preprocessing error) instead of producing unrealistic results. The problem with the large number of state variables can be solved by the larger information content of imaging spectroscopy data (compared with that produced by multispectral sensors) and development of novel physically based parameterizations allowing a more efficient description of canopy structure. However, until the full potential of imaging spectroscopy has been utilized by the developers of physical models, narrowband vegetation indices remain valuable tools in exploring the richness of high spectral resolution data.

4. Acknowledgment

We thank Anu Akujärvi for assisting in the field measurements. This study was funded by Emil Aaltonen Foundation, University of Helsinki Research and Postdoctoral Funds, and the Academy of Finland. Hyperion EO-1 data was available courtesy of the U.S. Geological Survey.

5. References

Asner, G.P. (1998). Biophysical and biochemical sources of variability in canopy reflectance. *Remote Sensing of Environment*, Vol. 64, No. 3, (June 1998), pp. 234-253, ISSN 0034-4257

Baret, F., Buis & S. (2008), Estimating canopy characteristics from remote sensing observations: review of methods and associated problems, In: *Advances in Land Remote Sensing: System, Modeling, Inversion and Application*, S. Liang (ed.), pp. 173-201, Springer-Verlag, ISBN 978-1-4020-6449-4, Heidelberg

Baret, F., Guyot, G. (1991). Potentials and limits of vegetation indices for LAI and APAR assessment. *Remote Sensing of Environment*, Vol. 35, No. 2-3, (February-March 1991), pp. 161-173, ISSN 0034-4257.

Barry, P. (2001). *EO-1/ Hyperion Science Data User's Guide, Level 1_B*, TRW Space, Defense & Information Systems, Redondo Beach, California, USA

Birth, G. S. & McVey, G. R. (1968) Measuring the color of growing turf with a reflectance spectrometer. *Agronomy Journal, Vol. 60 No. 6, (March 1968), pp. 640-643*, ISSN: 0002-1962

Black, T.A., Chen, J.M., Lee, X. & Sagar, R.M.. (1991). Characteristics of shortwave and longwave irradiances under a Douglas fir forest stand. *Canadian Journal of Forest Research* Vol. 21, No. 7, (July 1991), pp. 1020–1028, ISSN 0045-5067

Blackburn, G. A. (1998). Spectral indices for estimating photosynthetic pigment concentrations: a test using senescent tree leaves. *International Journal of Remote Sensing*, Vol. 19, No. 4, (March 1998), pp. 657-675, ISSN 0143-1161

Blackburn, G. A. (2002). Remote sensing of forest pigments using airborne imaging spectrometer and LIDAR imagery. *Remote Sensing of Environment*, Vol. 82, No. 2-3, (October 2002), pp. 311-321, ISSN 0034-4257.

Brantley, S.T., Zinnert, J.C. & Young, D.R. (2011). Application of hyperspectral vegetation indices to detect variations in high leaf area index temperate shrub thicket canopies. *Remote Sensing of Environment* Volume 115, Issue 2, (February 2011), pp. 514-523, ISSN 0034-4257

Broge, N.H. & Leblanc, E. (2000) Comparing prediction power and stability of broadband and hyperspectral vegetation indices for estimation of green leaf area index and canopy chlorophyll density. *Remote Sensing of Environment*, Vol. 76, No. 2, (May 2000), pp. 156-172, ISSN 0034-4257

Brown, L., Chen, J.M., Leblanc, S.G. & Cihlar, J. (2000). A shortwave infrared modification to the simple ratio for LAI retrieval in boreal forests: An image and model analysis. *Remote Sensing Environment*, Vol. 71, No. 1, (January 2000), pp. 16-25, ISSN 0034-4257.

Butera, M.K. (1986). A Correlation and Regression Analysis of Percent Canopy Closure Versus TMS Spectral Response for Selected Forest Sites in the San Juan National Forest, Colorado. *IEEE Transactions on Geoscience and Remote Sensing*, Vol. GE-24, No. 1, (January 1986), pp. 122-129, ISSN 0196-2892.

Carlson, T.N. & Ripley, D.A. (1998). On the Relation between NDVI, Fractional Vegetation Cover, and Leaf Area Index. *Remote Sensing of Environment*, Vol. 62, No. 3, (December 1997), pp. 241-252, ISSN 0034-4257.

Chen, J.M. (1996). Evaluation of vegetation indices and modified simple ratio for Boreal applications. *Canadian Journal of Remote Sensing*, Vol. 22, No. 3, (June 1996), pp. 229-242, ISSN 0703-8992.

Chen, J.M. & Black T.A. (1991). Measuring leaf-area index of plant canopies with branch architecture. *Agricultural and Forest Meteorology* Vol. 57, No. 1-3, (December 1991), pp. 1-12, ISSN 0168-1923

Chen, J.M. & Black, T.A. (1992). Defining leaf area index for non-flat leaves. *Plant Cell & Environment,* Vol. 15, No. 4, (May 1992), pp. 421-429, Online ISSN 1365-3040

Chen, J. & Cihlar, J. (1996). Retrieving leaf area index of boreal conifer forests using Landsat TM images. *Remote Sensing of Environment* Vol. 55, No. 2, (February 1996), pp. 153-162, ISSN 0034-4257

Chen, J., Pavlic, G., Brown, L., Cihlar, J., Leblanc, S., White, P., Hall, R., Peddle, D., King, D., Trofymow, J., Swift, E., Van der Sanden, J., & Pellikka, P. (2002). Derivation and validation of Canada-wide coarse-resolution leaf area index maps using high-resolution satellite imagery and ground measurements. *Remote Sensing of Environment,* Vol. 80, No. 1, (April 2002), pp. 165-184, ISSN 0034-4257.

Curran, P.J. (1989). Remote Sensing of Foliar Chemistry. *Remote Sensing of Environment,* Vol. 30, No. 3, (December 1989), pp. 271-278, ISSN 0034-4257.

Danson, F. M., & Plummer, S. E., 1995, Red edge response to forest leaf area index. *International Journal of Remote Sensing,* 1995, Vol. 16, No. 1, (January 1995), pp. 183-188, ISSN 0143-1161

Dawson, T.P. & Curran, P.J. (1998). A new technique for interpolating the reflectance red edge position. *International Journal of Remote Sensing,* Vol. 19, No. 11, (July 1998), pp. 2133-2139, ISSN 0143-1161

Deblonde, G., Penner, M. & Royer, A. (1994). Measuring Leaf Area Index with LI-COR LAI-2000 in Pine Stands. *Ecology.* Vol. 75, No. 5, (July 1994), pp. 1507-1511, ISSN 00129658

Deng, F., Chen, J., Plummer, S., Chen, M. & Pisek, J. (2006). Algorithm for global leaf area index retrieval using satellite imagery. *IEEE Transactions on Geoscience and Remote Sensing,* Vol. 44, No. 8, (August 2006), pp. 2219-2229, ISSN 0196-2892

Eklundh, L., Harrie, L. & Kuusk, A. (2001). Investigating relationships between Landsat ETM+ sensor data and leaf area index in a boreal conifer forest. *Remote Sensing of Environment,* Vol. 78, No. 3, (December 2001), pp. 239-251, ISSN 0034-4257

Eriksson, H., Eklundh, L., Kuusk, A., & Nilson, T. (2006). Impact of understory vegetation on forest canopy reflectance and remotely sensed LAI estimates. *Remote Sensing of Environment,* Vol. 103, No. 4, (August 2006), pp. 408–418, ISSN 0034-4257

Fernandes, R., Butson, C., Leblanc, S. & Latifovic, R. (2003). Landsat-5 TM and Landsat-7 ETM+ based accuracy assessment of leaf area index products for Canada derived from SPOT-4 VEGETATION data. *Canadian Journal of Remote Sensing,* Vol. 29, No. 2, (April 2003), pp. 241-258, ISSN 0703-8992.

Fernandes, R., Leblanc, S., Butson, C., Latifovic, R. & Pavlic, G. (2002). Derivation and evaluation of coarse resolution LAI estimates over Canada. In: *Proceedings of Geosciences and Remote Sensing Symposium, 2002.* IGARSS 02.2002, IEEE International Publications, Vol. 4, pp. 2097-2099, ISBN 0-7803-7536-X

Gamon, J., Penuelas, J., & Field, C. (1992). A narrow-waveband spectral index that tracks diurnal changes in photosynthetic efficiency. *Remote Sensing of Environment,* Vol. 41, No. 1, (July 1992), pp. 35-44, ISSN 0034-4257

Gao, B-C. (1996). NDWI–A normalized difference water index for remote sensing of vegetation liquid water from space. *Remote Sensing of Environment,* Vol. 58, No. 3, (December 1996), pp. 257-266, ISSN 0034-4257.

Gitelson, A. A., Kaufmann, Y. J. & Merzlyak, M. N. (1996) Use of a green channel in remote sensing of global vegetation from EOS-MODIS. *Remote Sensing of Environment,* Vol. 58, No. 3 (December 1996), pp. 289-298, ISSN 0034-4257

Gong, P., Pu, R., Biging, G.S. & Larrieu, M.R. (2003). Estimation of forest leaf area index using vegetation indices derived from Hyperion hyperspectral data. *IEEE Transactions on Geoscience and Remote Sensing* Vol. 41, No. 6, (June 2003), pp. 1355–1362, ISSN 0196-2892

Goodenough, D.G., Dyk, A., Niemann, K.O., Pearlman, J.S., Chen, H., Han, T. Murdoch, M. & West, C. (2003). Processing Hyperion and ALI for Forest Classification. *IEEE Transactions on Geoscience and Remote Sensing,* Vol. 41, No. 6, (June 2003), pp. 1321-1331, ISSN 0196-2892

Govaerts, Y. M., Verstraete, M. M., Pinty, B. & Gobron, N. (1999). Designing optimal spectral indices: a feasibility and proof of concept study. *International Journal of Remote Sensing,* Vol. 20, No. 9, (June 1999), pp. 1853-1873, ISSN 0143-1161

Häme, T. (1991). Spectral interpretation of changes in forest using satellite scanner images. *Acta Forestalia Fennica,* Vol. 222, The Society of Forestry in Finland & The Finnish Forest Research Institute, ISBN 951-651-092-2, Helsinki

Heiskanen, J., Rautiainen, M., Korhonen, L., Mõttus, M., Stenberg, P. (2011). Retrieval of boreal forest LAI using a forest reflectance model and empirical regressions. *International Journal of Applied Earth Observation and Geoinformation,* Vol. 13, No. 4, (August 2011), pp. 595-606, ISSN 0303-2434.

Horler, D.N.H. & Ahern, F.J. (1986). Forestry information content of Thematic Mapper data. *International Journal of Remote Sensing,* Vol. 7, No. 3, (January 1986), pp. 405-428, ISSN 0143-1161.

Huete, A.R. (1988) A soil-adjusted vegetation index (SAVI). *Remote Sensing of Environment,* Volume 25, Issue 3, (August 1988) , Pages 295-309, ISSN 0034-4257

Huete, A., Didan, K., Miura, T., Rodrigues, E.P., Gao, X. & Ferreira, L.G. (2002). Overview of the radiometric and biophysical performance of the MODIS vegetation indices. *Remote Sensing of Environment,* Vol. 83, No. 1-2, (November 2002), pp. 195-213, ISSN 0034-4257.

Jupp, D.L.B., Datt, B., Lovell, J., Campbell, S., King, E., et al. (2002). *Discussions around Hyperion data: background notes for the Hyperion data users workshop.* CSIRO Earth Observation Centre, Canberra. (accessed: 10.10.2011), Available from ftp://ftp.eoc.csiro.au/pub/djupp/Hyperion/Workshop/Minimal_Set/Documents/Hyp_Wsn.pdf

Jonckheere, I., Fleck, S., Nackaerts, K., Muys, B., Coppin, P., Weiss, M. & Baret, F. (2004). Reviews of methods for in situ leaf area index determination. Part I. Theories, sensors, and hemispherical photography. *Agricultural and Forest Meteorology,* Vol. 121, No. 1-2, (January 2004), pp. 19-35, ISSN 0168-1923

Kokaly, R. F., Asner, G.P., Ollinger, S.V., Martin, M.E. & Wessman, C.A. (2009). Characterizing canopy biochemistry from imaging spectroscopy and its application to ecosystem studies. *Remote Sensing of Environment*, Vol. 113, Supplement 1, (September 2009), pp. S78-S91, ISSN 0034-4257

Lang, A.R.G. (1986). Leaf area and average leaf angle from transmission of direct sunlight. *Australian Journal of Botany*, Vol. 34, No. 3, (May 1986), pp. 349 - 355. ISSN 0067-1924

Lee, K.-S., Cohen, W. B., Kennedy, R. E., Maiersperger, T. K. & Gower, S. T. (2004) Hyperspectral versus multispectral data for estimating leaf area index in four different biomes. *Remote Sensing of Environment*, Vol. 91, No. 3-4, (June 2004), pp. 508-520, ISSN 0034-4257.

LI-COR (1992). LAI-2000 *Plant Canopy Analyzer: Instruction Manual*. Lincoln, Nebraska, LI-COR, Inc.

le Maire, G., François, C., Soudani, K., Berveiller, D., Pontailler, D., Bréda, N., Genet, H.,Davi, H. & Dufrêne, E. (2008). Calibration and validation of hyperspectral indices for the estimation of broadleaved forest leaf chlorophyll content, leaf mass per area, leaf area index and leaf canopy biomass. *Remote Sensing of Environment*, Vol. 112, No. 10, (October 2008), pp. 3846-3864, ISSN 0034-4257

Matthew, M. W., Adler-Golden, S. M., Berk, A., Richtsmeier, S. C., Levine, R. Y., Bernstein, L. S., Acharya, P. K., Anderson, G. P., Felde, G. W., Hoke M. P., Ratkowski, A., Burke, H.-H., Kaiser, R. D., & Miller, D. P., (April 2000). Status of Atmospheric Correction Using a MODTRAN4-based Algorithm. In: *SPIE Proceedings, Algorithms for Multispectral, Hyperspectral, and Ultraspectral Imagery VI*. Vol. 4049, Accessed (10.10.2011), Available from: http://www.dtic.mil/cgi-bin/GetTRDoc?AD=ADA447767&Location=U2&doc=GetTRDoc.pdf

Miller, J.B. (1967). A formula for average foliage density. *Australian Journal of Botany*, Vol. 15, No. 1, (February 1967), pp. 141 - 144, ISSN 0067-1924

Monsi, M. & Saeki, T. (1953). Über den Lichtfactor in den Pflanzengesellschaften und seine bedeutung für die Stoff-production. *Japanese Journal of Botany*, Vol. 14, (March 1953) pp. 22-52, ISSN 0075-3424

Monteith, J.L. & Moss, C.J. (1977). Climate and the efficiency of crop production in Britain. *Philosophical Tranactions of the Royal Soiety B*. Vol. 281, No. 980, (November 1977), pp. 277-294. Online ISSN 1471-2970

Morisette, J.T., Baret, F., Privette, J. L., Myneni, R.B., Nickeson, J., Garrigues, S., Shabanov, N., Weiss, M., Fernandes, R., Leblanc, S., Kalacska, M., Sánchez-Azofeifa, G. A., Chubey, M., Rivard, B., Stenberg, P., Rautiainen, M., Voipio, P., Manninen, T., Pilant, A., Lewis, T., Iiames, J., Colombo, R., Meroni, M., Busetto, L., Cohen, W., Turner, D., Warner, E.D., Petersen, G.W., Seufert, G. & Cook, R. (2006). Validation of global moderate resolution LAI Products: a framework proposed within the CEOS Land Product Validation subgroup. *IEEE Transaction on Geosciences and Remote Sensing*, Vol. 44, No. 7, (July 2006), pp. 1804-1817, ISSN 0196-2892

Mutanga, O. & Skidmore, A.K. (2004). Narrow band vegetation indices overcome the saturation problem in biomass estimation. *International Journal of Remote Sensing*, Vol. 25, No. 19, (October 2004), pp. 3999-4014, ISSN 0143-1161

Myneni, R. B., Hall, F. G., Sellers, P. J. & Marshak, A. L. (1995) Interpretation of spectral vegetation indexes. *IEEE Transactions on Geoscience and Remote Sensing*. Vol. 33, No. 2, (March 1995), pp. 481-486, ISSN 0196-2892

Myneni, R.B., Nemani, R.R. & Running, S.W. (1997). Estimation of global leaf area index and absorbed PAR using radiative transfer models. *IEEE Transactions on Geoscience and Remote*, Vol. 35, No. 6, (November 1997), pp. 1380-1393, ISSN 0196-2892.

Nemani, R., Pierce, L., Running, S. & Band, L. (1993). Forest ecosystem processes at the watershed scale: sensitivity to remotely-sensed Leaf Area Index estimates. *International Journal of Remote Sensing*, Vol. 14, No. 13, pp. 2519-2534, ISSN 0143-1161

Nilson, T. (1971). A theoretical analysis of the frequency of gaps in plant stands. *Agricultural Meteorology*, Vol. 8, pp. 25-38, ISSN 0002-1571

Nilson, T. & Peterson, U. (1994). Age dependence of forest reflectance: analysis of main driving factors. *Remote Sensing of Environment*, Vol. 48, No. 3, (June 1994), pp. 319-331, ISSN 0034-4257.

Norman, J. & Jarvis, P. (1975). Photosynthesis in Sitka spruce (Picea sitchensis (Bong.) Carr.). V. Radiation penetration theory and a test case. *Journal of Applied Ecology*, Vol. 12, No. 3. (December 1975), pp. 839-877, ISSN 00218901

Pearlman, J. S., Barry, P.S., Segal, C. C., Shepanski, J., Beiso, D. & Carman, S. L. (2003) Hyperion, a space-based imaging spectrometer. *IEEE Transactions on Geoscience and Remote Sensing*, Vol. 41, No. 6, (June 2003), pp. 1160-1173, ISSN 0196-2892

Pisek, J., Rautiainen, M., Heiskanen,J. & Mõttus, M. (2012). Retrieval of seasonal dynamics of forest understory reflectance in a Northern European boreal forest from MODIS BRDF data. *Remote Sensing of Environment*, Vol. 17, pp. 464-468, ISSN 0034-4257

Pu, R., Gong, P., Biging, G.S. & Larrieu, M.R. (2003). Extraction of Red Edge Optical Parameters From Hyperion Data for Estimation of Forest Leaf Area Index. *IEEE Transaction on Geosciences and Remote Sensing* Vol. 41, No. 4, (April 2003), pp. 916-921, ISSN 0196-2892

Pu, R., Gong, P. & Yu, Q. (2008) Comparative analysis of EO-1 ALI and Hyperion, and Landsat ETM+ data for mapping forest crown closure and leaf area index. *Sensors* 2008, Vol. 8, No. 6, (June 2008), pp. 3744-3766, ISSN 1424-8220

Rautiainen, M., Heiskanen, J., Eklundh, L., Mõttus, M., Lukeš, P. & Stenberg, P. (2010), Ecological applications of physically based remote sensing methods. *Scandinavian Journal of Forest Research* Vol. 25, No. 4, (July 2010), pp. 325–339, ISSN 0282-7581

Rautiainen, M. & Stenberg, P. (2005). Application of photon recollision probability in simulating coniferous canopy reflectance. *Remote Sensing of Environment*, Vol. 96, No. 1, (May 2005), pp. 98-107, ISSN 0034-4257

Rautiainen, M.; Mõttus, M.; Heiskanen, J.; Akujärvi, A.; Majasalmi, T. & Stenberg, P. (2011). Seasonal reflectance dynamics of common understory types in a northern European boreal forest. *Remote Sensing of Environment*, in press, (available online since 27 July 2011), ISSN 0034-4257

Rautiainen, M., Suomalainen, J., Mõttus, M., Stenberg, P., Voipio, P., Peltoniemi, J. & Manninen, T. (2007). Coupling forest canopy and understory reflectance in the Arctic latitudes of Finland. *Remote Sensing of Environment*, Vol. 110, No. 3, (October 2007), pp. 332-343, ISSN 0034-4257

Roberts, D., Ustin, S., Ogunjemiyo, S., Greenberg, J., Dobrowski, S., Chen, J. & Hinckley, T. (2004). Spectral and structural measures of northwest forest vegetation at leaf to landscape scales. *Ecosystems*, Vol. 7, No. 5, (May 2004), pp. 545-562. ISSN 1432-9840

Rock, B. N., Vogelmann, J. E., Williams, D. L., Vogelmann, A. F. & Hoshizaki, T. (1986). Remote Detection of Forest Damage. *Bio Science*, Vol. 36, No. 7, (July-August 1986), pp. 439-445, ISSN 0006-3568

Rouse, J. W., Haas, R. H., Schell, J. A. & Deering, D. W. (1974). Monitoring Vegetation Systems in the Great Plains with ERTS. In: *Third ERTS Symposium*, pp. 309-317, NASA SP-351 I, United States

Samiappan, S., Prasad, S., Bruce, L.M. & Robles, W. (2010) NASA's upcoming HyspIRI Mission - Precision vegetation mapping with limited ground truth. *Geoscience and Remote Sensing Symposium (IGARSS), 2010 IEEE International*, (July 2010), pp. 3744-3747. ISSN 2153-6996

San, B. T., & M. L. Suzen, (2010), Evaluation of Different Atmospheric Correction Algorithms for EO-1 Hyperion Imagery, *International Archives of the Photogrammetry, Remote Sensing and Spatial Information Science*, Volume XXXVIII, Part 8, (August 2010), pp. 392-397, ISSN 1682-1777

Schlerf, M. et al. (2005). Remote sensing of forest biophysical variables using HyMap imaging spectrometer data. *Remote Sensing of Environment*, Vol. 95, No. 2, (March 2005), pp. 177¬–194, ISSN 0034-4257

Schull, M., Knyazikhin, Y., Xu, L., Samanta, A., Carmona, P., Lepine, L., Jenkins, J., Ganguly, S. & Myneni, R. (2011). Canopy spectral invariants, Part 2: Application to classification of forest types from hyperspectral data. *Journal of Quantitative Spectroscopy and Radiative Transfer*, Vol.112, No.4, (March 2011), pp. 736-750, ISSN 0022-4073

Segl, K., Guanter, L., Kaufmann, H., Schubert, J., Kaiser, S., Sang, B. & Hofer, S. (2010) Simulation of spatial sensor characteristics in the context of the EnMAP hyperspectral mission. *IEEE Transactions on Geoscience and Remote Sensing*, Vol.48, No. 7, (July 2010), pp. 3046-3054, ISSN 0196-2892

Sellers, P.J. (1985). Canopy reflectance, photosynthesis and transpiration. *International Journal of Remote Sensing*, Vol. 6, No. 8, pp. 1335–1372, ISSN 0143-1161

Shinozaki, K., Yoda, K, Hozumi, K. & Kira, T. (1964). A quantitative analysis of plant form - the pipe model theory. I. Basic analyses. *Japanese Journal of Ecology*, Vol. 14, No. 3, (June 1964), pp. 97-105, ISSN 0021-5007

Sims, D.A. & Gamon, J.A. (2002). Relationships between leaf pigment content and spectral reflectance across a wide range of species, leaf structures and developmental stages. *Remote Sensing of Environment, Vol.* 81, No. 2–3, (August 2002), pp. 337–354, ISSN 0034-4257

Smith, N.J., Chen, J.M. & Black, T.A. (1993). Effects of clumping on estimates of stand leaf area index using the LI-COR LAI-2000. *Canadian Journal of Forest Research, Vol.* 23, No. 9, pp. 1940-1943, ISSN 0045-5067

Spanner, M.A., Pierce, L.L., Peterson, D.L. & Running, S.W. (1990). Remote sensing of temperate coniferous forest leaf area index: the influence of canopy closure, understory vegetation and background reflectance. *International Journal of Remote Sensing*. Vol. 11, No.1, pp. 95-111, ISSN 0143-1161

Stenberg, P. (2006). A note on the G-function for needle leaf canopies. *Agricultural and Forest Meteorology*, Vol. 136, No. 1-2, (January 2006), pp 76-79, ISSN 0168-1923

Stenberg, P., Linder, S., Smolander, H. & Flower-Ellis, J. (1994). Performance of the LAI-2000 plant canopy analyzer in estimating leaf area index of some Scots pine stands. *Tree Physiology,* Vol. 14, No. 7-8-9, (July 1994), pp. 981-995, Online ISSN 1758-4469

Stenberg, P., Rautiainen, M., Manninen, T., Voipio, P. & Smolander, H. (2004). Reduced Simple Ratio better than NDVI for estimating LAI in Finnish pine and spruce stands. *Silva Fennica,* Vol. 38, No. 1, (March 2004), pp. 3-14, ISSN 0037-5330.

Sun, L., Neville, R., Staenz, K., & White, H.P. (2008). Automatic destriping of Hyperion imagery based on spectral moment matching. *Canadian Journal for Remote Sensing.* Vol. 34, Supplement 1, (May 2008), pp. 68-81, ISSN 0703-8992

Tucker, C. J. (1979) ETM+3 and photographic infraETM+3 linear combinations for monitoring vegetation. *Remote Sensing of Environment,* 8, 127-150, ISSN 0034-4257

Ustin, S. L., Gitelson, A.A., Jacquemoud, S., Schaepman, M., Asner, G.P., Gamon, J.A., & Zarco-Tejada, P. (2009). Retrieval of foliar information about plant pigment systems from high resolution spectroscopy. *Remote Sensing of Environment,* Vol. 113, Supplement 1, (September 2009), pp. S67-S77, ISSN 0034-4257.

Validation of Land European Remote Sensing Instruments (VALERI) Network. Cited: 16.8.2011. Available from: http://www.avignon.inra.fr/valeri/

Verstraete, M. M. & Pinty, B. (1996) Designing optimal spectral indexes for remote sensing applications. *IEEE Transactions on Geoscience and Remote Sensing,* Vol. 34, No. 5, (September 1996), pp. 1254-1265, ISSN 0196-2892

Waring, R.H., Schroeder, P.E. & Oren, R. (1982). Application of the pipe model theory to predict canopy leaf area. *Canadian Journal of Forest Research. Vol.* 12, No. 3. (September 1982), pp. 556-560. ISSN 0045-5067

Wilson, J.W. (1965). Stand structure and light penetration. I. Analysis by point quadrats. *Journal of Applied Ecology,* Vol. 2, No. 2, (November 1965), pp. 383-390, ISSN: 00218901

Watson, D. (1947). Comparative physiological studies in the growth of field crops. I. Variation in net assimilation rate and leaf area between species and varieties, and within and between years. *Annales of Botany,* Vol. 11, No. 1, (January 1947), pp. 41-76, ISSN 0305-7364

Welles, J.M. & Cohen, S. (1996). Canopy structure measurement by gap fraction analysis using commercial instrumentation. *Journal of Experimental Botany,* Vol. 47, No. 9, (September 1996), pp. 1335-1342, ISSN 0022-0957

Widlowski, J. L., Taberner, M., Pinty, B., Bruniquel-Pinel, V., Disney, M., Fernandes, R., Gastellu-Etchegorry, J. P., Gobron, N., Kuusk, A., Lavergne, T., Leblanc, S., Lewis, P. E., Martin, E., Mottus, M., North, P. R. J., Qin, W., Robustelli, M., Rochdi, N., Ruiloba, R., Soler, C., Thompson, R., Verhoef, W., Verstraete, M. M. & Xie, D. (2007). Third Radiation Transfer Model Intercomparison (RAMI) exercise: Documenting progress in canopy reflectance models. *Journal of Geophysical Research--Atmospheres* 112, D09111, (May 2007), 28 pages, ISSN 0148-0227

Williams, D. (1991). A comparison of spectral reflectance properties at the needle, branch, and canopy level for selected conifer species. *Remote Sensing of Environment* Vol. 35, No. 2-3, (February-March 1991), pp. 79-93, ISSN 0034-4257

Wu, C., Han, X., Niu, Z. & Dong, J. (2010). An evaluation of EO-1 hyperspectral Hyperion data for chlorophyll content and leaf area index estimation. *International Journal of Remote Sensing*, Vol. 31, No. 4, (February 2010), pp. 1079-1086, ISSN 0143-1161

Zarco-Tejada, P.J., Rueda, C.A. & Ustin, S.L. (2003). Water content estimation in vegetation with MODIS reflectance data and model inversion methods. *Remote Sensing of Environment*, Vol. 85, No. 1, (April 2003), pp. 109-124, ISSN 0034-4257.

2

Crop Disease and Pest Monitoring by Remote Sensing

Wenjiang Huang, Juhua Luo, Jingcheng Zhang, Jinling Zhao, Chunjiang Zhao, Jihua Wang, Guijun Yang, Muyi Huang, Linsheng Huang and Shizhou Du
Beijing Research Center for Information Technology in Agriculture, Beijing China

1. Introduction

Plant diseases and pests can affect a wide range of commercial crops, and result in a significant yield loss. It is reported that at least 10% of global food production is lost due to plant diseases (Christou and Twyman, 2004; Strange and Scott, 2005). Excessive pesticides are used for protecting crops from diseases and pests. This not only increases the cost of production, but also raises the danger of toxic residue in agricultural products. Disease and pest control could be more efficient if disease and pest patches within fields can be identified timely and treated locally. This requires obtaining the information of disease infected boundaries in the field as early and accurately as possible. The most common and conventional method is manual field survey. The traditional ground-based survey method requires high labor cost and produces low efficiency. Thus, it is unfeasible for large area. Fortunately, remote sensing technology can provide spatial distribution information of diseases and pests over a large area with relatively low cost. The presence of diseases or insect feedings on plants or canopy surface causes changes in pigment, chemical concentrations, cell structure, nutrient, water uptake, and gas exchange. These changes result in differences in color and temperature of the canopy, and affect canopy reflectance characteristics, which can be detectable by remote sensing (Raikes and Burpee 1998). Therefore, remote sensing provides a harmless, rapid, and cost-effective means of identifying and quantifying crop stress from differences in the spectral characteristics of canopy surfaces affected by biotic and abiotic stress agents.

This chapter introduces some successful studies about detecting and discriminating yellow rust and aphid (economically important disease and pest in winter wheat in China) using field, airborne and satellite remote sensing.

2. Detecting yellow rust of winter wheat by remote sensing

Yellow rust *(Biotroph Puccinia striiformis),* also known as stripe rust, is a fungal disease of winter wheat *(Triticum aestivum L.).* It produces leaf lesions (pustules), which are yellow in color and tend to be grouped in patches. Yellow rust often occurs in narrow stripes, 2-3 mm wide that run parallel to the leaf veins. Yellow rust is responsible for approximately 73-85%

of recorded yield losses, and grain quality is also significantly reduced (Li et al. 1989). Consequently, effective monitoring of the incidence and severity of yellow rust in susceptible regions is of great importance to guide the spray of pesticides and to provide data for the local agricultural insurance services. Fortunately, remote sensing technology provides a possible way to detect the incidence and severity of the disease rapidly.

The interaction of electromagnetic radiation with plants varies with the wavelength of the radiation. The same plant leaves may exhibit significant different reflectance depending on the level of health and or vigor (Wooley 1971, West et al. 2003, Luo et al., 2010). Healthy and vigorously growing plant leaves will generally have

1. Low reflectance at visible wavelengths owing to strong absorption by photoactive pigments (chlorophylls, anthocyanins, carotenoids).
2. High reflectance in the near infrared because of multiple scattering at the air-cell interfaces in the leaf's internal tissue.
3. Low reflectance in wide wavebands in the short-wave infrared because of absorption by water, proteins, and other carbon constituents.

The incidence and severity of yellow rust can be monitored according to the differences of spectral characteristics between healthy and disease plants. In this chapter, we will report several successful studies on the detection and identification of yellow rust in winter wheat by remote sensing.

2.1 Detecting and discriminating yellow rust at canopy level

Hyperspectral remote sensing is one of the advanced and effective techniques in disease monitoring and mapping. However, the difficulty in discriminating a disease from common nutrient stresses largely hampers the practical use of this technique. This is because some common nutrient stresses such as the shortage or overuse of nitrogen or water could have similar variations of biochemical properties and plant morphology, and therefore result in similar spectral responses. However, for the remedial procedures for stressed crops, there is a significant difference between disease and nutrient stresses. For example, applying fungicide to water-stressed crops would lead to a disastrous outcome. Therefore, to discriminate yellow rust from common nutrient stresses is of practical importance to crop growers or landowners.

The specific objectives of this study are to: (1) systematically test the sensitivity and consistency of several commonly used spectral features to yellow rust disease during major growth stages; (2) for those spectral features that are consistently sensitive to yellow rust disease, we will further examine their sensitivity to nutrient stresses to determine whether there are specifically sensitive to yellow rust disease, but insensitive to water and nitrogen stresses.

2.1.1 Materials and methods

2.1.1.1 Experimental design and field conditions

The experiments were conducted at Beijing Xiaotangshan Precision Agriculture Experimental Base, in Changping district, Beijing (40°10.6′N, 116°26.3′E) for the growing seasons of 2001-2002 and 2002-2003. Table 1 summarizes the soil properties including organic matter, total nitrogen, alkali-hydrolysis nitrogen, available phosphorus and available potassium for both growing seasons. Three cultivars of winter wheat used in 2001-2002 experiment (2002 Exp) were Jingdong8, Jing9428 and Zhongyou9507, while the cultivars used in 2002-2003 (2003 Exp) were Xuezao, 98-100 and Jing411. All the cultivars applied in both growing seasons included erective, middle and loose with respect to the canopy morphology.

Items		Disease inoculation experiment	Nutrient stress experiment
Growth period		Sep 2002-Jun 2003	Sep 2001-Jun 2002
Top soil nutrient status (0-0.3m depth)	Organic matter	1.42%-1.48%	1.21%-1.32%
	Total nitrogen	0.08%-0.10%	0.092%-0.124%
	Alkali-hydrolysis nitrogen	58.6-68.0 mg kg^{-1}	68.8-74.0 mg kg^{-1}
	Available phosphorus	20.1-55.4 mg kg^{-1}	25.2-48.3 mg kg^{-1}
	Rapidly available potassium	117.6-129.1 mg kg^{-1}	96.6-128.8 mg kg^{-1}
Cultivars		Xuezao, 98-100, Jing411	Jingdong8, Jing9428, Zhongyou9507
Treatments		Normal; YR1: 3mg 100^{-1} ml spores solution; YR2: 9mg 100^{-1} ml spores solution; YR3: 12mg 100^{-1} ml spores solution (all treatments applied 200 kg ha^{-1} nitrogen and 450 m^3 ha^{-1} water)	Normal: 200 kg ha^{-1} nitrogen, 450 m^3 ha^{-1} water; W-SD: 200 kg ha^{-1} nitrogen, 225 m^3 ha^{-1} water; W-SED: 200 kg ha^{-1} nitrogen, 0 m^3 ha^{-1} water; N-E: 350 kg ha^{-1} nitrogen, 450 m^3 ha^{-1} water; N-D: 0 kg ha^{-1} nitrogen, 450 m^3 ha^{-1} water; W-SED+N-E: 350 kg ha^{-1} nitrogen, 0 m^3 ha^{-1} water; W-SED+N-D: 0 kg ha^{-1} nitrogen, 0 m^3 ha^{-1} water;
Spectral reflectance measurements (on day after sowing)		207, 216, 225, 230, 233	196, 214, 225, 232, 239

Table 1. Basic information of disease inoculation experiment and nutrient stress experiment

For 2002 Exp, six stress treatments of water and nitrogen were applied, and the treatments were based on local conditions, which usually suffered from yellow rust in the northern part of China. Each treatment was applied on 0.3 ha area, and the treatments were 200 kg ha^{-1} nitrogen and 225 m^3 ha^{-1} water (slightly deficient water, W-SD),200 kg ha^{-1} nitrogen and no irrigation (seriously deficient water, W-SED), 350 kg ha^{-1} nitrogen and 450 m^3 ha^{-1} water (excessive nitrogen, N-E), no fertilization and 450 m3 ha-1 water (deficient nitrogen, N-D), 350 kg ha^{-1} nitrogen and no irrigation (seriously deficient water and excessive nitrogen, W-SED+N-E), and no fertilization and no irrigation (seriously deficient water and deficient nitrogen, W-SED+N-D). A 0.3 ha reference area (Normal) was applied with the recommended rate which received 200 kg ha^{-1} nitrogen and 450 m^3 ha^{-1} water. Three cultivars were evenly distributed in each treatment plot.

For 2003 Exp, according to the National Plant Protection Standard (Li et al. 1989), three levels of concentration of summer spores of yellow rust were applied, and they were 3 mg 100^{-1} ml^{-1} (Yellow rust 1, YR1), 9 mg 100^{-1} ml^{-1} (Yellow rust 2, YR2) and 12 mg 100^{-1} ml^{-1} (Yellow rust 3, YR3), with a dosage of 5 ml spores solution per square meter. The reference area (Normal) that was not inoculated yet was applied with the recommended amount of fungicide to prevent the occasional infection. Each treatment involved 1.2 ha area, with even constitution of three cultivars. All plots in 2003 Exp received the recommended rates of nitrogen (200 kg ha^{-1}) and water (450 m^3 ha^{-1}).

2.1.1.2 Canopy spectral measurements

A high spectral resolution spectrometer, ASD FieldSpec Pro spectrometer (Analytical Spectral Devices, Boulder, CO, USA) fitted with a 25 field of view fore-optic, was used for in-situ measurement of canopy spectral reflectance for both 2002 Exp and 2003 Exp. All canopy spectral measurements were taken from a height of 1.3m above ground (the height of the wheat is 90±3 cm at maturity). Spectra were acquired in the 350-2,500 nm spectral range at a spectral resolution of 3 nm between 350 nm and 1,050 nm, and 10 nm between 1,050 nm and 2,500 nm. A 40 cm × 40 cm BaSO4 calibration panel was used for calculation of reflectance. All irradiance measurements were recorded as an average of 20 scans at an optimized integration time. Prior to subsequent preprocessing, all spectral curves were resampled with 1 nm interval. All measurements were made under clear blue sky conditions between 10:00 and 14:00 (Beijing Local Time).

The spectral measurements were taken 5 times from 196 days after sowing (DAS) to 239 DAS for 2002 Exp, which covered the growth stages of stem elongation, booting, anthesis and milk development. For 2003 Exp, the spectral measurements were taken 5 times from 207 DAS to 233 DAS, which covered the growth stages of booting, anthesis and milk development. The detailed measurement dates for both experiments were given in Table 1. The stem elongation and anthesis stages are essential for the control of yellow rust development, whereas the milk development stage is important for yield loss assessment.

2.1.1.3 Selection of spectral features

The spectral features that we adopted were related to several commonly used vegetation indices (VIs), which were proved to be sensitive to variations of pigments and stresses. Furthermore, in order to conduct a thorough investigation of various types of spectral features, we also included a number of spectral features that were based on derivative transformation and continuum removal transformation (Gong et al. 2002; Pu et al. 2003;2004). Therefore, the total 38 spectral features are shown in Table 2.

Variable	Definition	Description	Literatures
Derivative transformed spectral variables			
D_b	Maximum value of 1st derivative within blue edge	Blue edge covers 490-530nm. D_b is a maximum value of 1st order derivatives within the blue edge of 35 bands	Gong et al., 2002
λ_b	Wavelength at D_b	λ_b is wavelength position at D_b	Gong et al., 2002
SD_b	Sum of 1st derivative values within blue edge	Defined by sum of 1st order derivative values of 35 bands within the blue edge	Gong et al., 2002
D_y	Maximum value of 1st derivative within yellow edge	Yellow edge covers 550-582nm. D_y is a maximum value of 1st order derivatives within the yellow edge of 28 bands	Gong et al., 2002
λ_y	Wavelength at D_y	λ_y is wavelength position at D_y	Gong et al., 2002
SD_y	Sum of 1st derivative values within yellow edge	Defined by sum of 1st order derivative values of 28 bands within the yellow edge	Gong et al., 2002
D_r	Maximum value of 1st derivative within red edge	Red edge covers 670-737nm. D_r is a maximum value of 1st order derivatives within the red edge of 61 bands	Gong et al., 2002
λ_r	Wavelength at D_r	λr is wavelength position at D_r	Gong et al., 2002
SD_r	Sum of 1st derivative values within red edge	Defined by sum of 1st order derivative values of 61 bands within the red edge	Gong et al., 2002
Continuous removal transformed spectral features			
DEP550-750	The depth of the feature minimum relative to the hull	In the range of 550nm-750nm	Pu et al., 2003;2004
DEP920-1120		In the range of 920nm-1120nm	
DEP1070-1320		In the range of 1070nm-1320nm	
WID550-750	The full wavelength width at half DEP (nm)	In the range of 550nm-750nm	Pu et al., 2003;2004
WID920-1120		In the range of 920nm-1120nm	
WID1070-1320		In the range of 1070nm-1320nm	
AREA550-750	The area of the absorption feature that is the product of DEP and WID	In the range of 550nm-750nm	Pu et al., 2003;2004
AREA920-1120		In the range of 920nm-1120nm	
AREA1070-1320		In the range of 1070nm-1320nm	

Variable	Definition	Description	Literatures
VI-based variables			
GI	Greenness Index	R_{554}/R_{677}	Zarco-Tejada et al., 2005
MSR	Modified Simple Ratio	$(R_{800}/R_{670}-1)/(R_{800}/R_{670}+1)^{1/2}$	Chen, 1996; Haboudane et al., 2004
NDVI	Normalized Difference Vegetation Index	$(R_{NIR}-R_R)/(R_{NIR}+R_R)$, where R_{NIR} indicates 775-825nm, R_R indicates 650nm-700nm, that include most key pigments	Rouse et al., 1973
NBNDVI	Narrow-band normalised difference vegetation index	$(R_{850}-R_{680})/(R_{850}+R_{680})$	Thenkabail et al., 2000
NRI	Nitrogen reflectance index	$(R_{570}-R_{670})/(R_{570}+R_{670})$	Filella et al., 1995
PRI	Photochemical Physiological Reflectance Index	$(R_{531}-R_{570})/(R_{531}+R_{570})$	Gamon et al., 1992
TCARI	The transformed chlorophyll Absorption and Reflectance Index	$3*[(R_{700}-R_{670})-0.2*(R_{700}-R_{550})*(R_{700}/R_{670})]$	Haboudane et al., 2002
SIPI	Structural Independent Pigment Index	$(R_{800}-R_{445})/(R_{800}-R_{680})$	Peñuelas et al., 1995
PSRI	Plant Senescence Reflectance Index	$(R_{680}-R_{500})/R_{750}$	Merzlyak et al., 1999
PhRI	The Physiological reflectance index	$(R_{550}-R_{531})/(R_{550}+R_{531})$	Gamon et al., 1992
NPCI	Normalized Pigment Chlorophyll ratio Index	$(R_{680}-R_{430})/(R_{680}+R_{430})$	Peñuelas et al., 1994
ARI	Anthocyanin Reflectance Index	$ARI=(R_{550})^{-1}-(R_{700})^{-1}$	Gitelson et al., 2001
TVI	Triangular Vegetation Index	$0.5[120(R_{750}-R_{550})-200(R_{670}-R_{550})]$	Broge and Leblanc, 2000; Haboudane et al., 2004
CARI	Chlorophyll Absorption Ratio Index	$(\|(a670+R_{670}+b)\|/(a^2+1)^{1/2})x(R_{700}/R_{670})$ $a = (R_{700}-R_{550})/150$, $b = R_{550}-(a \times 550)$	Kim et al., 1994

Variable	Definition	Description	Literatures
DSWI	Disease Water Stress Index	$(R_{802}+R_{547})/(R_{1657}+R_{682})$	Galvão et al., 2005
MSI	Moisture Stress Index	R_{1600}/R_{819}	Hunt and rock, 1989; Ceccato et al., 2001
SIWSI	Shortwave Infrared Water Stress Index	$(R_{860}-R_{1640})/(R_{860}+R_{1640})$	Fensholt and Sandholt, 2003
RVSI	Red-Edge Vegetation Stress Index	$[(R_{712}+R_{752})/2]-R_{732}$	Merton and Huntington, 1999
MCARI	Modified Chlorophyll Absorption in Reflectance Index	$(R_{701}-R_{671})-0.2(R_{701}-R_{549})]/(R_{701}/R_{671})$	Daughry et al., 2000
WI	Water Index	R_{900}/R_{970}	Peñuelas et al., 1997

Table 2. Definitions of spectral features used in this study

2.1.1.4 Preprocessing and normalization of spectral reflectance data

Aggregating spectral reflectance data

As the first step, all spectra were processed with the following transformation to suppress possible difference in illumination. The spectral regions with wavelength of 1330-1450 nm, 1770-2000 nm and 2400-2500 nm were removed due to strong absorption by water vapor. We then normalized the spectral curves by dividing the mean band reflectance of the curve (Yu et al., 1999). The normalized reflectance for the $band_i$ is given as:

$$Ref_i' = \frac{Ref_i}{\frac{1}{n}(\sum_{i=1}^{n} Ref_i)}$$

where Ref_i' is the normalized reflectance for $band_i$; Ref_i is the original reflectance of the band; n is the total number of bands. Fig. 1(a) shows a plot of unnormalized Ref_i versus band wavelength for six observations (three YR3 curves and three Normal curves) on 233 DAS. Fig. 1(b) shows the corresponding curves in Fig.1(a) after normalization. The normalization clearly separated the diseased spectra from the normal spectra especially over the near infrared region (approximately from 770 nm to 1300 nm). The benefit of eliminating spectral difference caused by the change of illumination conditions was also mentioned by Yu et al. (1999).

Normalization of the difference in measuring dates

As shown in Table 1, although both experiments conducted in five growth stages in 2002 and 2003, most measurement dates were not consistent, except for 255 DAS. Hence, to improve the comparability of two datasets, we adapted the 2002 Exp data to match the dates

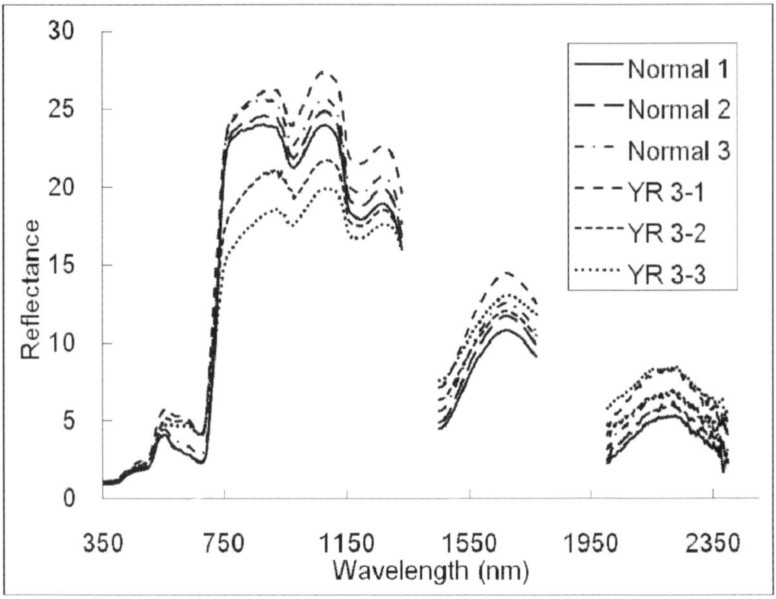

(a) Original spectra on 233 days after sowing

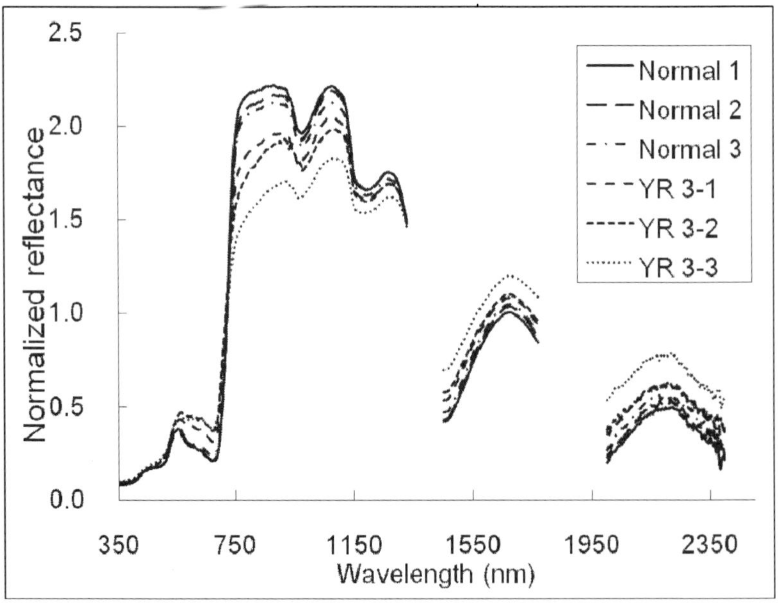

(b) Normalized spectra on 233 days after sowing

Fig. 1. Comparison between original spectra and normalized ones of 2003 Exp, by using a linear interpolation method. The reflectance curve of a certain date could be obtained based on the spectra from the adjacent data before and after the measurement date (using days after sowing as a time scale). Each band of the spectra should be processed as:

$$Ref_{current} = Ref_{before} - \frac{DAS_{current} - DAS_{before}}{DAS_{after} - DAS_{before}}(Ref_{before} - Ref_{after})$$

where $Ref_{current}$ represents the reflectance transformed from the date corresponding to an ideal date in 2003 Exp; Ref_{before} and Ref_{after} represent reflectances, respectively, from DAS_{before} and DAS_{after}; $DAS_{current}$ indicates an ideal date in 2003 Exp while DAS_{before} and DAS_{after} are the adjacent dates in 2002 Exp before and after the ideal date in 2003 Exp.

Fig. 2 provides an example of the progress of the normalization of measurement dates. The averaged reflectance at central wavelengths of green band (560 nm) and near-infrared band (860 nm) of Landsat-5 TM for normal samples were plotted against the measured dates in both 2002 Exp and 2003 Exp. The date normalized reflectance values were marked as triangle symbol in the graph. Through this step, the datasets collected in these two years could be considered as acquired in the same dates, which thereby facilitated the subsequent comparisons and analysis.

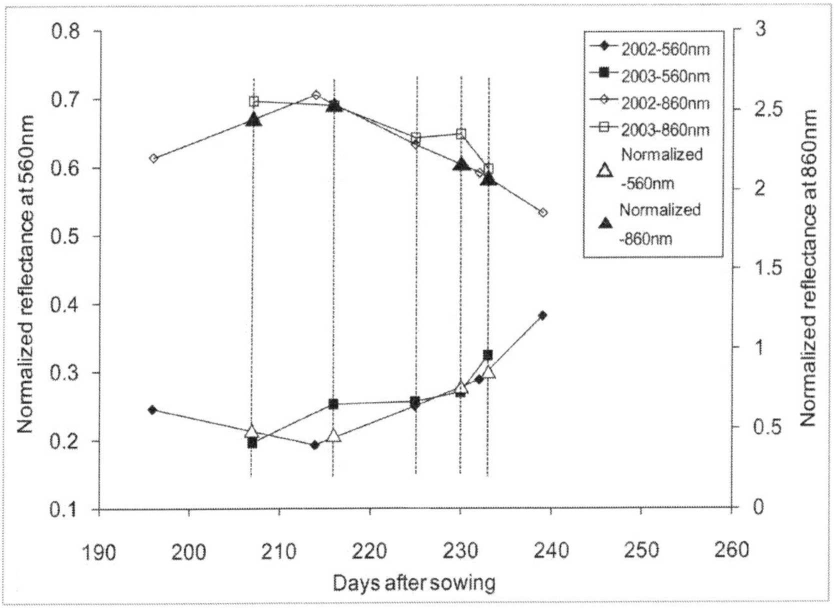

Adaptation of average reflectance of normal samples at 560 nm (central wavelengths of green band of Landsat-5 TM) and 860 nm (central wavelengths of near-infrared band of Landsat-5 TM) to match the dates of 2003 Exp, by using a linear interpolation method

Fig. 2. An example for normalization of measuring dates

Normalization of the difference from cultivars and soil backgrounds

The canopy spectra of winter wheat were not only supposed to respond to stresses, but are also determined and influenced by several other aspects such as cultivars and soil properties. Although the both 2002 Exp and 2003 Exp were conducted in the same fields that had approximately identical climate and environmental conditions, the difference in cultivars and soil properties between 2002 Exp and 2003 Exp should not be ignored (Table 1). To minimize this discrepancy, we calculated a ratio spectral curve for each of measured dates (after the normalization of the measuring dates) by the averaged spectral curve from normal samples in 2002 Exp divided by the averaged spectral curve from normal samples in 2003 Exp, resulting in a total of five ratio curves corresponding to each growth stage (Fig. 3). After that, all the spectral data measured at different growth stages were multiplied by the corresponding ratio curves to yield a set of normalized spectra. It should be pointed out that the present normalization processing to raw spectral measurements will only enhance the comparability between the 2002 Exp and 2003 Exp with little change in internal relations among different treatments because all the spectral data at one growth stage were processed with the same ratio curve. The ultimate goal of all these preprocessing and normalization steps above is to mitigate effects of the variation of illumination conditions, measurement dates, cultivars and soil properties between the 2002 Exp and 2003 Exp on target spectra.

2.1.1.5 Spectral features calculation and statistical analysis

With the spectra normalized using the methods above, we calculated 38 spectral features. An analysis of variance (ANOVA) was employed to investigate the spectral differences between the normal samples and all forms of stressed samples. Firstly, on different measured dates, both the yellow rust disease data and nutrient stressed data were compared with the normal data by ANOVA. For those spectral features that were consistently sensitive to yellow rust disease, we not only tested their differences between the normal treatment and different forms of stresses, but also tested the differences between various kinds of nutrient stresses and varying levels of disease stresses with ANOVA. Statistical analyses were conducted using SPSS 13.0 procedure.

2.1.2 Results

2.1.2.1 Spectra after normalizations

The spectral ratio curves in Fig 3 reflect the deviations between 2002 Exp and 2003 Exp's reflectance datasets at different wavelength positions. The ratio value close to 1.0 indicates no difference in reflectance exists between the two years. Generally, the ratio values ranged from 0.7 to 1.3, with an uneven distribution along the wavelength axis (Fig 3). The ratio tended to deviate from 1.0 in the regions of 350 - 730 nm, 1450 - 1570 and 2000 - 2400 nm, but stayed around 1.0 in the regions of 730 - 1330 nm and 1570 - 1770 nm. To assess the improvement in comparability, we examined the difference of normalized datasets of normal samples between 2002 Exp and 2003 Exp through an ANOVA with all 38 spectral features. The result showed that the differences of all spectral features were insignificant at all growth stages (p-value>0.05), with an average p-value (for all measuring dates) of 0.94, indicating a relatively high level of similarity between two datasets. Therefore, we confirmed that such normalization processes minimized the spectral difference originated

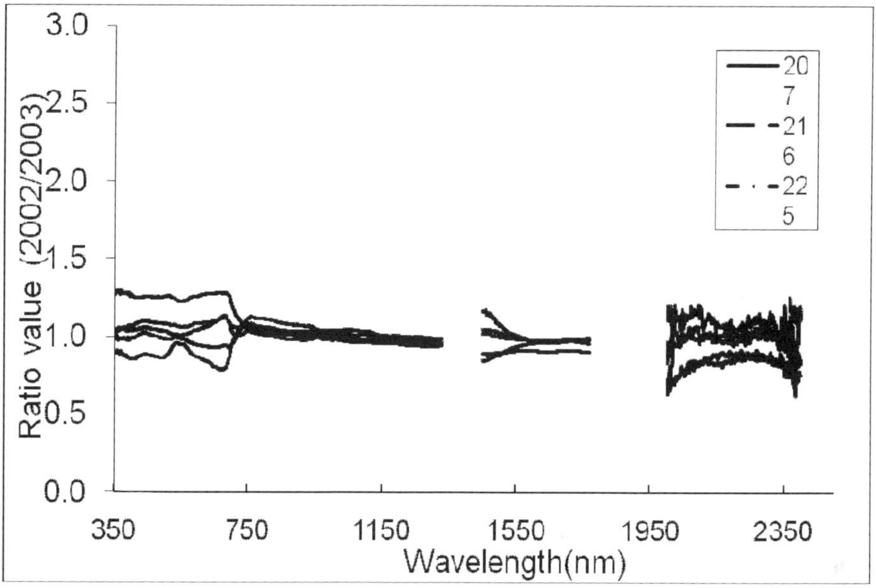

Fig. 3. Ratios of spectra for normalization with different years and varieties

from variation of illumination and different measurement dates, etc., and enabled more rational comparisons among different treatments.

2.1.2.2 Spectral responses to different forms of stresses

The result of ANOVA between normal samples and different forms of stress samples indicated that all spectral features had a response (defined as p-value<0.05) to at least one type of stresses at one growth stage, except for the WID1070-1320, which had no response to any form of stresses at all growth stages. Total 37 spectral features responded to water associated stresses (W-SD, W-SED, W-SED+N-E, W-SED+N-D) at least at one growth stage, followed by 35 spectral features to yellow rust disease, whereas only15 spectral features had a response to solely nitrogen stress (N-E, N-D). As summarized in Table 3, most spectral features were sensitive to yellow rust infection at least at one growth stage, except for λ_b, λ_r and WID1070-1320. In addition, most spectral features tended to be more sensitive at later growth stages than at the early stages. For example, several features such as DEP920-1120, AREA920-1120, D_y, GI, NDVI and Triangular Vegetation Index (TVI) only had a response to yellow rust at the last growth stage in our study (233 DAS). However, for the sake of diagnosis, the spectral features with a consistent response to yellow rust during the important growing period would be much more valuable. Therefore, those spectral features that were sensitive to the yellow rust at 4 out of 5 growth stages were selected as candidates for disease diagnosis. This yielded four vegetation indices (VIs): PRI, PhRI, NPCI and ARI.

2.1.2.3 One way ANOVA of four disease sensitive spectral features

Particularly for the four identified VIs that closely associated with yellow rust disease, a throughout one way ANOVA was conducted to compare their differences between the

Spectral features	Days after sowing				
	207	216	225	230	233
DEP550-770	√			√	√
AREA550-770	√			√	√
WID550-770			√	√	√
DEP920-1120					√
AREA920-1120					√
WID920-1120					√
DEP1070-1320					√
AREA1070-1320					√
Db			√	√	
SDb			√	√	√
Dy					√
λy					√
SDy					√
Dr				√	
SDr				√	√
GI					√
MSR				√	√
NDVI					√
NBNDVI				√	√
NRI					√
PRI		√	√	√	√
TCARI			√	√	
SIPI					√
PSRI	√			√	√
PhRI		√	√	√	√
NPCI	√		√	√	√
ARI	√		√	√	√
TVI					√
CARI			√	√	√
DSWI					√
MSI					√
SIWSI					√
RVSI			√	√	
MCARI			√	√	√
WI					√

Table 3. Responses of spectral features to yellow rust

normal sample and various kinds of stressed samples. Moreover, their differences among each pairs of stress forms were also compared. We conducted this ANOVA based on the data on 207 DAS, 225 DAS and 233 DAS respectively, which were essential growth stages for carrying out fungicide spraying and yield loss assessing procedures. In addition to the *p*-value of ANOVA, we also provided the change direction of spectral features. Positive sign indicates the average spectral feature value of diseased or nutrient stressed samples is greater than that of normal samples, and negative sign indicates the opposite cases to the positive sign. As shown in Table 4, it was observed that for the treatments of N-E and N-D, all four VIs failed to show any response at all growth stages. For the results of other treatments, the responses of four VIs behaved in a varied pattern at three growth stages.

For the results on 207 DAS (Table 4a), compared to the normal samples, the NPCI and ARI had responses to all three levels of yellow rust treatments (YR 1, YR 2, YR 3), and appeared to be more sensitive than PRI and PhRI. For nutrient stresses, the PRI, NPCI and ARI were sensitive to W-SED and W-SED+N-E treatments. Among them, NPCI and ARI showed stronger responses (p-value<0.01) to W-SD, W-SED, W-SED+N-E and W-SED+N-D treatments than the other two VIs. For the comparisons between diseased samples and nutrient stressed samples, significant differences between W-SED and W-SED+N-E treatments and YR2 and YR3 treatments were identified for PRI, NPCI and ARI. Moreover, the change directions of the three VIs for diseased and nutrient stressed samples were identical. At this 207 DAS growth stage, PhRI did not show a significant response to any of three levels of disease treatments, but responded to W-SD, W-SED and W-SED+N-E treatments. It is interesting that the change direction of diseased samples of PhRI was contrary to that of the nutrient stressed samples, suggesting a discriminating potential of the index.

For the results on 225 DAS (Table 4b), compared to the normal samples, all four VIs revealed a clear response to level 2 and level 3 of yellow rust treatments (YR2, YR3). For nutrient stresses, PRI, NPCI and ARI also appeared to be sensitive to W-SD, W-SED, W-SED+N-E and W-SED+N-D treatments. However, PhRI was insensitive to all nutrient stresses. In addition, when we looked at the difference of those VIs between diseased samples and nutrient stressed samples, only PhRI showed clear differences between YR2 and YR3 treatments and W-SD, W-SED, W-SED+N-E, and W-SED+N-D treatments. Although a significant difference between YR3 treatment and W-SED treatment also existed for ARI and NPCI, the change directions of both treatments were identical. However, for PhRI, the change directions of all levels of disease treatments were different from those of the nutrient stress treatments.

For the results on 233 DAS (Table 4c), with further development of disease symptoms, compared to the normal samples, all four indices showed responses to all three levels of disease treatments. Comparing to YR1 treatment, the four VIs had shown a stronger significant level (p-value<0.01) for YR2, YR3 treatments. For nutrient stresses, PRI, NPCI and ARI exhibited clear responses to W-SED, W-SED+N-E and W-SED+N-D treatments as well. For comparisons between diseased and nutrient stressed samples, PRI and NPCI appeared to be significantly different between YR2 and YR3 treatments and W-SD treatment. However, the change directions of both treatments were identical. Unlike the other three VIs, PhRI remained insensitive to the nutrient stresses, but was significantly different among all levels of disease treatments (YR1, YR2, and YR3) and all forms of nutrient stresses. More

Treatments	YR 1				YR 2				YR 3				Normal			
	PRI	PhRI	NPCI	ARI	PRI	PhRI	NPCI	ARI	PRI	PhRI	NPCI	ARI	PRI	PhRI	NPCI	ARI
Normal	(-)	(-)	(+) *	(-) *	(-) *	(+)	(+) *	(-) *	(-)	(+)	(+) **	(-) ***				
W-SD	(-)	(-)	(+)	(+)	(-)	(-) *	(+)	(+)	(-)	(-) *	(+)	(-)	(-)	(-) *	(+) **	(-) **
W-SED	(+)	(-)	(+) *	(+) *	(+)	(-) *	(+) *	(+) *	(+) *	(-) *	(+) *	(+)	(-) ***	(-) *	(+) ***	(-) ***
N-E	(-)	(+)	(-)	(-)	(-) *	(-)	(-)	(-)	(-)	(-)	(-) *	(-) **	(+)	(-)	(-)	(+)
N-D	(-)	(+)	(-)	(-)	(-)	(+)	(-)	(-)	(-)	(-)	(-) *	(-) **	(-)	(+)	(-)	(+)
W-SED+N-E	(+)	(-)	(+) *	(+) *	(+)	(-) *	(+) *	(+) *	(+) *	(-) *	(+)	(+)	(-) ***	(-) *	(+) ***	(-) ***
W-SED+N-D	(+)	(-)	(+)	(+)	(+)	(-)	(+)	(+)	(+)	(-) *	(+)	(+)	(-) *	(-)	(+) **	(-) ***

(a) 207 DAS

Treatments	YR 1				YR 2				YR 3				Normal			
	PRI	PhRI	NPCI	ARI	PRI	PhRI	NPCI	ARI	PRI	PhRI	NPCI	ARI	PRI	PhRI	NPCI	ARI
Normal	(+)	(+)	(+)	(+)	(+) **	(+) ***	(+) *	(+) *	(+) ***	(+) ***	(+) **	(+) **				
W-SD	(+) **	(-) *	(+) **	(+) **	(-)	(-) **	(+)	(+)	(-)	(-) **	(+)	(+)	(+) ***	(-)	(+) ***	(+) **
W-SED	(+) **	(-)	(+) ***	(+) **	(-)	(-) **	(+)	(+)	(+)	(-) *	(+) **	(+) **	(+) ***	(-)	(+) ***	(+) ***
N-E	(-) *	(-)	(-)	(-)	(-) *	(-) *	(-)	(-)	(-) *	(-) *	(-) *	(-) **	(-)	(-)	(-)	(-)
N-D	(-)	(+)	(-)	(-)	(-)	(-) *	(-)	(-)	(-) *	(-) *	(-) *	(-) **	(+)	(+)	(+)	(-)
W-SED+N-E	(+) *	(-)	(+) *	(+) *	(-)	(-) **	(+)	(+)	(-)	(-) *	(+)	(+)	(+) ***	(-)	(+) ***	(+) **
W-SED+N-D	(+) *	(-)	(+) *	(+) *	(-)	(-) **	(+)	(+)	(-)	(-) *	(+)	(+)	(+) ***	(-)	(+) **	(+) **

(b) 225 DAS

Treatments	YR 1				YR 2				YR 3				Normal			
	PRI	PhRI	NPCI	ARI	PRI	PhRI	NPCI	ARI	PRI	PhRI	NPCI	ARI	PRI	PhRI	NPCI	ARI
Normal	(+) **	(+) *	(+) *	(+) *	(+) ***	(+) **	(+) ***	(+) ***	(+) ***	(+) **	(+) ***	(+) ***				
W-SD	(-)	(-) **	(-)	(+)	(-) ***	(-) **	(-) *	(-)	(-) **	(-) **	(-) **	(-) *	(+)	(-)	(+) *	(+) *
W-SED	(+)	(-) **	(+)	(+)	(-) *	(-) **	(+)	(-)	(-) *	(-) **	(-)	(-)	(+) ***	(-)	(+) ***	(+) ***
N-E	(-) *	(-) **	(-)	(-)	(-) ***	(-) **	(-) **	(-) *	(-) ***	(-) **	(-) ***	(-) **	( )	(-)	(-)	(+)
N-D	(-) *	(-) *	(-)	(-)	(-) ***	(-) **	(-) **	(-) *	( ) ***	(-) **	(-) **	(-) **	(-)	(-)	(+)	(+)
W-SED+N-E	(+)	(-) **	(+)	(+)	(-) **	(-) **	(+)	(+)	(-) *	(-) **	(-)	(+)	(+) ***	(-)	(+) ***	(+) ***
W-SED+N-D	(+)	(-) **	(+)	(+)	(-) ***	(-) **	(+)	(-)	(-) *	(-) **	(-)	(-)	(+) ***	(-)	(+) ***	(+) ***

(c) 233 DAS

*mean difference is significant at 0.950 confidence level; **mean difference is significant at 0.990 confidence level;*** mean difference is significant at 0.999 confidence level. (+) means the average spectral feature value of diseased or nutrient stressed samples greater than that of normal samples; or means the average spectral feature value of nutrient stressed samples greater than that of diseased samples; (-) means the opposite cases to the case of (+). The definitions of treatments are as follows: "Normal" represents normal samples; "W-SD" represents samples treated with slightly deficient water; "W-SED" represents samples treated with seriously deficient water; "N-E" represents samples treated with excessive nitrogen; "N-D" represents samples treated with deficient nitrogen; "W-SED+N-E" represents samples treated with seriously deficient water and excessive nitrogen; "W-SED+N-D" represents samples treated with seriously deficient water and deficient nitrogen

Table 4. ANOVA for four VIs separately on 207 DAS, 225 DAS and 233 DAS

importantly for the PhRI, the change directions of diseased samples were opposite to those of nutrient stressed samples throughout the entire analysis.

In summary, all four VIs showed a significant sensitivity to yellow rust disease on 207 DAS, 225 DAS and 233 DAS. However, most of them also appeared to be sensitive to water associated stresses to a varing extent, except for PhRI, which was only sensitive to disease yet insensitive to any forms of nutrient stresses on 225 DAS and 233 DAS. More importantly, the change directions of PhRI to disease treatments were always opposite to those to the nutrient stress treatments at all relevant growth stages. This further confirmed the discriminating characteristic of PhRI.

2.1.3 Conclusion

Combining with a dataset of yellow rust disease inoculation and a dataset of various forms of nutrient stress treatments, we examined the responses of 38 commonly used spectral features at five important growth stages from booting stage to milk development stage using a one-way analysis of variance (ANOVA). There were 37 spectral features sensitive to water associated stresses, 35 spectral features sensitive to yellow rust disease and only 15 spectral features sensitive to sole nitrogen stresses in at least one growth stage. It was observed that more spectral features appeared to have a response to yellow rust disease at later growth stages. A throughout ANOVA was conducted particularly on PRI, PhRI, NPCI and ARI, which showed a consistent response to yellow rust disease at 4 out of 5 growth stages. However, PRI, NPCI and ARI were also responsible for water associated stresses, suggesting a risk of confusion in detecting yellow rust disease. Only PhRI was sensitive to yellow rust disease, but insensitive to different forms of nutrient stresses. The discriminative response of PhRI could provide a means of identifying and detecting yellow rust disease under complicated farmland circumstances. This finding can serve the basis of remote sensing system for detecting yellow rust disease.

2.2 Detecting yellow rust using field and airborne hyperspectral data

The aim of this study was to evaluate the accuracy of the spectro-optical, photochemical reflectance index (PRI) for quantifying the disease index (DI) of yellow rust in wheat using in-situ spectral reflectance measurements, and its applicability in the detection of the disease using hyperspectral imagery.

2.2.1 Materials and methods

2.2.1.1 Experimental design and field conditions

Experimental design and field conditions was same as 1.1.1. Experimental data from 2002 Exp were used to establish the statistical models, and the data for 2003 Exp were used to validate the models developed.

2.2.1.2 Inspection of disease severity

To quantify the severity of the disease of yellow rust, the leaves of plants were grouped into one of 9 classifications of disease incidence (x): 0,1, 10, 20, 30, 45, 60, 80 and 100% covered by rust. 0% represented no incidence of yellow rust, and 100% was the greatest incidence. The disease index (DI) was then calculated using (Li et al. 1989):

$$DI(\%) = \frac{\sum(x \times f)}{n \times \sum f} \times 100$$

where f is the total number of leaves of each degree of disease severity and n is the degree of disease severity observed (in this work, n ranged from 0 to 8). In each plot, 20 individuals were randomly selected for check.

2.2.1.3 Canopy spectral measurements

The method of canopy spectral measurements and data was same as the part 1.1.1.2 above.

2.2.1.4 Airborne hyperspectral imaging

Airborne hyperspectral images of the trial field were acquired in 2003 using the Pushbroom Hyperspectral Imager (PHI) designed by the Chinese Academy of Science (CAS) and flown onboard a Yun-5 aircraft (Shijiazhuang Aircraft Manufacturing Company, China). The PHI comprises a solid state, area array, and silicon CCD device of 780 × 244 elements. It has a field of view of 21°, and is capable of acquiring images of 1 m × 1 m spatial resolution at an altitude of 1000 m above ground. The wavelength range is 400–850 nm with a spectral resolution of 5 nm. Images of the target field were acquired in 2003 at the phenological growth stages of stem elongation (April 18, 2003, Zadoks stage 3), anthesis (May 17, 2003, Zadoks stage 5) and milky maturity (May 31, 2003, Zadoks stage 8). The inoculated wheat was adequately infected by rust on April 18, obviously infected by May 17, and seriously infected by May 31. Measurements of DI were made and in situ canopy reflectance spectra were also acquired on the same dates. All images were geometrically and radiometrically corrected using an array of georeferenced light and dark targets (5 m x 5 m) located at the extremes of the field site. The aforementioned field spectrometer was used to calibrate these targets relative to BaSO4. The location of each target, as well as field measurements of DI were recorded using a differential global positioning system (Trimble Sunnyvale California, USA).

2.2.1.5 Photochemical reflectance index (PRI)

Because yellow rust epiphyte reduced foliar physiological activity by destroying foliar pigments, the photochemical reflectance index (PRI) was selected as the spectrophotometric method of estimating the disease index. PRI was calculated by the formula in Table 2.

2.2.2 Results

2.2.2.1 PRI versus DI

Fig. 4 shows a plot of the measured DI as a function of PRI for all varieties. The data points associated with the variety Xuezao dominate in the top-left region of the scatter plot (relatively high range of DI), while those associated with the variety 98-100 are located in the mid region (mid-range DI) and those associated with Jing 411 dominate the lower right region. This distribution trend is consistent with the relative susceptibility of these varieties to rust; Xuezao is the least resistant and Jing 411 has the greatest resistance. The regression equation of DI using PRI in 2002 Exp was obtained as following (n = 64):

$$\mathrm{DI}(\%) = -721.22(PRI) + 2.40 \qquad \left(-0.14 \le PRI \le 0.02; r^2 = 0.91\right)$$

An important feature in, the associated regression equation (Fig. 4) was that the spectrally-derived PRI explained 91% of the variance observed in the disease index. This explanation also encompassed the three varieties of wheat as well as the four stages of crop development for each variety. In the subsequent validation of the PRI-DI regression equation with the 2003 Exp data (Fig. 5), the coefficient of determination (R^2) between the estimated and measured values was 0.97 (n = 80).

In Fig. 5, the locations of data points associated with individual varieties wew consistent with the levels of resistance to rust. Xuezao dominated the top right-hand region of the scatter plot (relatively high range of DI), the variety 98-100 had points scattered all along the

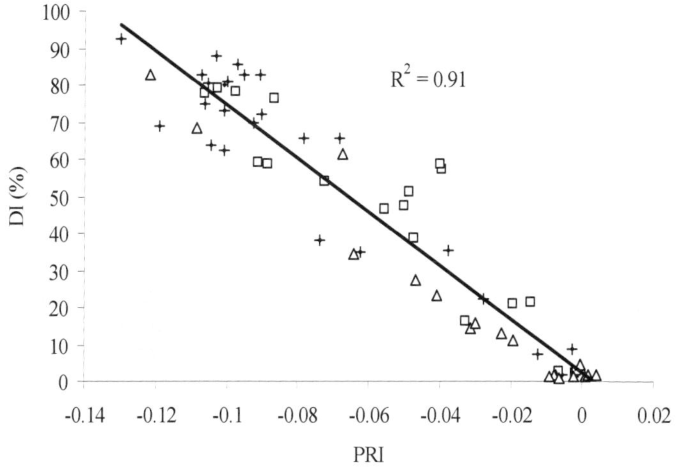

Fig. 4. Plot of measured disease index (DI) as a function of measured photochemical reflectance index(PRI) for all varieties combined in 2002 Exp. Δ: Jing 411; +: Xuezao; □: 98-100

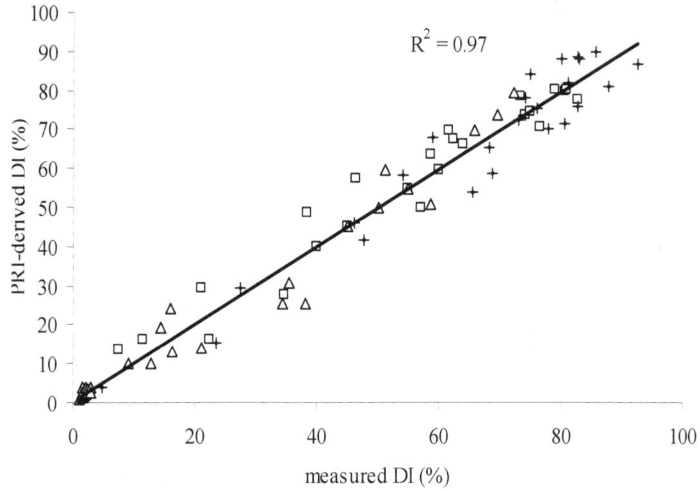

Fig. 5. Comparison of measured DI and PRI-estimated DI for 2003 Exp; 'Δ' = Jing 411; '+' = Xuezao; '□' = 98-100

regression line (predominantly mid-range DI), and Jing 411 was concentrated in the central lower-left region (lower range DI).

2.2.2.2 Application of multi-temporal PHI images for DI estimation

The DI was estimated on a pixel-by-pixel basis in each of the acquired PHI images using the regression equation. To map the degree of yellow rust infection in the trial field, the DI was binned into the following classes; very Serious (DI > 80%), serious (45% < DI ≤ 80%), moderate (10% < DI ≤ 45%), slight (1% < DI ≤ 10%) and none (0 < DI ≤ 1%) (Fig. 6).

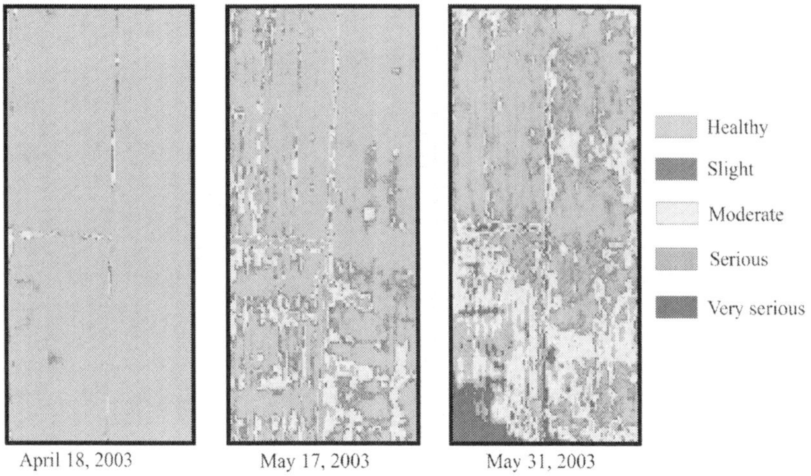

Fig. 6. Classified DI images derived from PHI airborne images of the trial site in 2003 Exp

Fig. 7 shows the relationship between the DI calculated from the multi-temporal PHI images and the actual measured DI from the 120 sample sites located within the field (R^2=0.91).

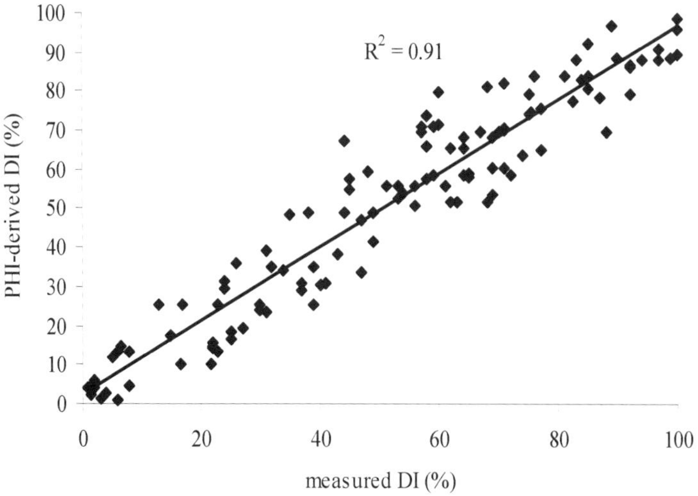

Fig. 7. Comparison of PHI-derived estimates of DI and actual DI values for 2002 Exp. Data were extracted from all three imaging times, although the DI values were< 20% for the April 18 image

2.2.3 Conclusion

The results of this work confirm PRI is a potential candidate for monitoring of yellow rust, and could form the basis of an on-the-go sensor and variable-rate spray applicator or remote detection and mapping process.

2.3 Detecting yellow rust in winter wheat by spectral knowledge base

In most cases, statistical models for monitoring the disease severity of yellow rust are based on hyperspectral information. The high cost and limited cover of airborne hyperspectral data make it impossible to apply such data for large scale monitoring. Furthermore, the established models of disease detection cannot be used for most satellite images because of the wide range of wavelengths in multispectral images (Zhang et al., 2011).

To resolve this dilemma, the study presents a novel approach by constructing a spectral knowledge base (SKB) of winter wheat diseases, which takes the airborne images as a medium and links the disease severity with band reflectance from moderate resolution remotely sensed data, such as environment and disaster reduction small satellite images (HJ-CCD) accordingly. To achieve this goal, several algorithms and techniques for data conversion and matching are adopted in the proposed system, including minimum noise fraction (MNF) transformation and pixel purity index (PPI) function. The performance of SKB is evaluated with both simulated data and field measured data.

2.3.1 materials and methods

Experimental design and field conditions was same as the part of 1.1.1.1

2.3.1.1 Inspection of disease severity

Please refer to the part of 1.2.1.2 above.

2.3.1.2 Airborne hyperspectral imaging

Please refer to the part of 1.2.1.4 above about airborne hyperspectral imaging and image processing.

2.3.1.3 Acquisition of moderate resolution satellite images

In this study, the SKB is designed to fit the charge coupled device (CCD) sensor, which is on the environment and disaster reduction small satellites (HJ-1A/B). The basic parameters of the CCD sensor (using 'HJ-CCD' in the following) are given in Table.5. The four bands of

Properties of HJ-CCD				
Band	Wavelength range (nm)	Spatial resolution (m)	Swath (km)	Revisit time (day)
Blue	0.430-0.520	30	360	2
Green	0.520-0.600			
Red	0.630-0.690			
Near-infrared	0.760-0.900			

Table 5. Properties of the environment and disaster reduction small satellites (HJ-CCD)

HJ-CCD covered the visible and near infrared spectral regions. The HJ-CCD sensor has spectral and spatial characteristics that are similar to those of Landsat-5 TM, but the HJ-1A/B satellites have more frequent revisit capability (2 days) than the Landsat-5 satellite (16 days), which is of great importance for agricultural monitoring.

2.3.1.4 Construction of the spectral knowledge base

The SKB in this study can be interpreted as a pool of relationships between spectral characteristics and prior knowledge. Here, prior knowledge stands for the degree of severity of yellow rust, and the spectral characteristics are the reflectance of the initial four bands of the HJ-CCD image. Hence, there are two major steps involved in constructing the SKB. First, the relationship between hyperspectral information and severity is obtained with a stable empirical reversion model. Then, through the RSR function of the HJ-CCD sensor, the hyperspectal data can be transferred to the wide-band reflectance. In this way, a one-to-one correspondence between the disease severity of yellow rust and reflectances from the HJ-CCD sensor is established at the pixel level. The SKB can represent disease severity in two ways: the DI (%) value and the class of disease severity. The following sections describe each step for establishing the SKB, including data selection, the reversion model, simulation of the wide-band reflectance and estimating the degree of severity. A technical flow diagram of SKB construction is summarized in Fig. 8.

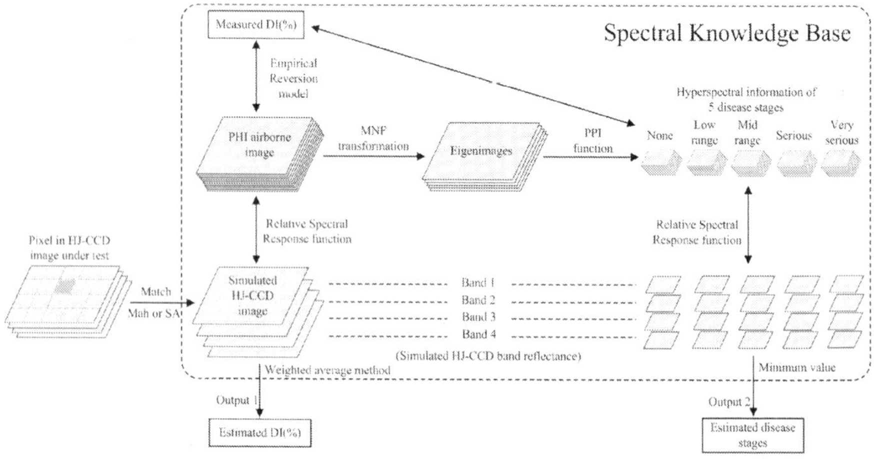

Fig. 8. The flow chart for monitoring of DI(%) of winter wheat stripe rust, b1-b4 represented the reflectance of the four bands of HJ-CCD images

As noted above, the SKB in this study comprised PHI pixels. The predicted accuracy obtained by the SKB was determined primarily by the amount of prior knowledge, which indicated the heterogeneity of disease severity. The design of the yellow rust fungus inoculation ensured a considerable variation in disease severity within the experimental field, from healthy plants to very diseased plants. In addition, to avoid using pixels on or near the ridge in the field that are considered as mixed signals, we chose three rectangular shaped areas that were within the field and comprised 7918 'crop-only' pixels for constructing the SKB.

2.3.1.5 Reversion model

The reversion model construction was the first step of establishing the SKB. Based on the conclusion of the part above, PRI was a suitable vegetation index for monitoring the severity of yellow rust disease in winter wheat. Therefore, in this study, PRI was used to establish the linkage between the disease severity and the hyperspectral data. Specifically, the yellow rust infection would be apparent at anthesis stage, and this should be closely related with the subsequent yield loss. Therefore, we chose the PHI image at this stage to form the SKB. To obtain a better fitting model, we reanalyzed the PHI-PRI and corresponding DI (%) data at the anthesis stage specifically, and obtained a linear regression model. It should be noted that the data range of DI must be between 0 and 100%. Any predicted DI results that were>100% or <0% were redefined as DI = 100% and DI = 0% to represent very infected plants and healthy plants, respectively.

2.3.1.6 Simulation of the wide band reflectance

The second step of constructing the SKB is to transform the hyperspectral reflectance of PHI-pixels to wide band reflectance of HJ-pixels. To achieve this goal, the best approach is the inherent relative spectral response (RSR) function of the HJ-CCD sensor. By integrating the hyperspectral reflectance of PHI-pixels on the RSR function, the band reflectance of HJ-CCD sensor was thus obtained. Besides, although the wavelength range of the fourth band of HJ-CCD sensor (760 nm-900 nm) was slightly exceeded the maximum wavelength of PHI sensor (850 nm), for most ground measured spectra, the reflectance basically kept on steady from 760 nm to 900 nm. Hence, the simulating results generated using the incomplete range of wavelength (760nm-850nm) should approach to the true value. The integration can be shown as follows:

$$R_{TM} = \int_{b_{start}}^{b_{end}} f(x)dx$$

where R_{TM} is the simulated reflectance of a certain band; b_{start} and b_{end} indicate the beginning and the end wavelength of this band respectively; *f(x)* indicates the RSR function, which is obtained from CRESDA.

2.3.1.7 Spectral characteristics of different degrees of disease severity

Another way to define the disease severity of an undefined pixel, apart from the DI (%) value, is to quantify disease severity by severity classes. The criterion of severity class provided by Huang et al. (2007) was adopted, which corresponded to the major physiological alteration of diseased plants. The DI (%) thresholds for each severity class were: DI<1% indicated not infected (NI), 1%<DI<10% indicated a low degree of infection (LI), 10%<DI<45% indicated mid-range infection (MI), 45%<DI<80% indicated seriously infected (SI) and DI (%)>80% indicated very seriously infected (VI). The MNF transformation and PPI function, which are used for noise reduction and end-member identification, were applied here to select the most representative pixels from the PHI image, and to form the typical spectrum for each severity class.

2.3.1.8 Spectral matching algorithms

The basic idea of spectral matching is to identify a set of pixels in the SKB that are the closest to the undefined pixel in terms of spectral characteristics. Before matching, each pixel should be standardized to eliminate systematic variation caused by aerosol conditions or other factors as follows:

$$R_{nor} = \frac{R - R_{\min}}{R_{\max} - R_{\min}}$$

where R_{nor} is the standardized reflectance of a certain band, R is the original reflectance, and R_{min} and R_{max} are the minimum and maximum band reflectance values, respectively, of the corresponding pixel.

Mahalanobis distances (Mah) and Spectral angle (SA) were selected as the distance measurement criterion. Both types of distance measurements had been proved to be with high efficiency in reflecting the spectral discrepancy (South et al., 2004; Goovaerts et al., 2005; Becker et al., 2007). The Mah distance can be written as:

$$D_M(x) = \sqrt{(x - x_R)\sum{}^{-1}(x - x_R)^T} \quad x=(x_1,x_2,x_3,x_4),\ x_R=(x_{R1}, x_{R2}, x_{R3}, x_{R4})$$

where $x_{1\text{-}4}$ are the reflectance of the pixel under test in band1 to band4, respectively; $x_{R1\text{-}4}$ are the simulated reflectance of a specific pixel in SKB. $\sum$ is the covariance matrix between x and x_R.SA can be calculated by the following formula:

$$\theta = \arccos \frac{\sum_{i=1}^{4} x_i x_{Ri}}{\sqrt{\sum_{i=1}^{4} x_i^2}\sqrt{\sum_{i=1}^{4} x_{Ri}^2}} \qquad \theta \in \left[0, \frac{\pi}{2}\right]$$

To determine the DI (%) or class of disease severity of an undefined pixel, we have to calculate the Mah and spectral angle from this pixel to each pixel or class in the SKB. A longer distance or larger angle indicates that the pixel deviated from the undefined pixel, whereas a shorter distance or smaller angle indicates that it is similar to the undefined pixel. By selecting the most similar pixel, the severity class of an undefined pixel can be determined. To determine the DI (%) of a certain pixel, the weighted average method was used. According to the distance criteria above, the five most similar pixels were selected from the SKB. For each band of these pixels (here we used the hyperspectral bands extracted from the PHI image), the reflectance was processed according to the following equation:

$$R_E = \frac{\sum_{i=1}^{k} R_i \times \frac{1}{d_i}}{\sum_{i=1}^{k} \frac{1}{d_i}}$$

where R_E is the estimated reflectance of a certain pixel through k-NN estimation; R_i is the reflectance of the i_{th} nearest pixel according to the ranking order of the distance; d_i is the distance between the pixel under test to the i_{th} nearest pixel.

2.3.1.9 Verification

To verify the performance of SKB in identifying and monitoring the severity of yellow rust diseases, two datasets were used: the simulated data and the field-measured data with corresponding satellite images.

1. Verification of SKB using simulated data
   The simulated data comprised 50 randomly selected pixels in the same experimental field, but outside the three regions selected for constructing the SKB. The hyperspectral information of each pixel was used to create the reference DI (%) and severity class with the empirical model and the corresponding threshold for each severity class. To test the performance of SKB in terms of DI (%) value, we estimated the DI value with both distance criteria described above. The samples were split into two: the pixels with a reference DI between 1 and 100%, i.e. the 'diseased' pixels, and those with a reference DI<1%, i.e. 'healthy' pixels. For the diseased pixels, the estimated DIs were compared with the reference DI by Pearson correlation analysis and the normalized root mean square error (NRMSE). For the healthy pixels, we used 'yes or no' to determine whether the estimated value indicated infection or not, which also provided an accuracy ratio. The estimation of severity class was verified by overall accuracy and the kappa coefficient.
2. Verification of SKB using field surveyed data
   The field surveyed data sets included the ground investigation of disease severity and the corresponding HJ-CCD images. Between June 1-3, 2009, when the winter wheat was at the anthesis stage, we conducted a survey in the southeast of GanSu Province. The climate of the area surveyed is characterized by high humidity and rainfall, and yellow rust disease occurs almost every year. This area has similar environmental conditions and cultivation customs to those where we constructing the SKB in Beijing, and this makes it an appropriate place for model verification. With the aid of the local Department of Plant Protection, 26 plots were randomly selected and surveyed in the area (Fig. 9). To relate the surveyed value to the pixel value of the HJ-CCD image, we defined the plot as a uniformly planted winter wheat region with an area no less than 30 m in radius. The geographical coordinates of each plot were measured by GPS at the centre of the plot. Disease severity was measured as described above. We repeated the measurement in five evenly-distributed sections in each plot, and 20 individual plants were included in each measurement. The HJ-CCD images (ID: 122516, 122518) acquired on June 2, 2009 completely covered the surveyed area. The raw data from the HJ-CCD imagery was calibrated based on the corresponding coefficients provided by CRESDA. The calibrated data were atmospherically corrected with the algorithm provided by Liang et al. (2001), which estimated the spatial distribution of atmospheric aerosols and retrieved surface reflectance under general atmospheric and surface conditions. The images were also geometrically corrected against historical reference images with the same geographical coordinates. The images were rectified with a root mean square error of less than 0.5 pixels. The spectrum of the each plot was extracted from the image according to the GPS records. The estimated accuracy in this step followed the same process as the simulated data.

2.3.2 Results

There were 7918 pixels included in the process of constructing the SKB. The linear regression model between DI (%) and PRI at anthesis stage could be illustrated as follows:

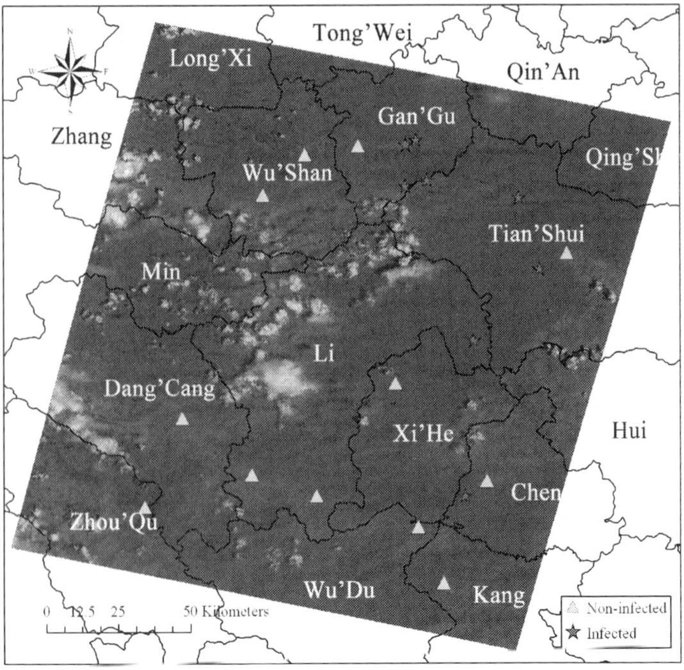

Fig. 9. The field surveyed area in Gansu Province. The base image is the HJ-CCD image acquired on June 2, 2009

$$DI(\%) = -538.98 \times PRI + 2.0983 \quad (R^2=0.88)$$

The pairs of DI (%) and PRI were plotted in Fig.4, which showed a significant correlation (R^2 = 0.88). Based on the model, there were 85 pixels with a DI of 100% and 3991 pixels with a DI between 1%and 100%, indicating 51.5%pixels infected to a varied degree of severity, whereas the other 48.5% pixels (DI = 0%) were healthy plants. In the experimental field, the variation in the degree of severity of yellow rust from totally healthy plants to very infected plants provided the essential diversity or heterogeneity of infection, which then enabled establishment of the SKB. The MNF transformation resulted in 9 leading eigenvectors with eigenvalues greater than 4.0 (Fig. 10), and these were used for further analysis.

2.3.2.1 Performance of SKB for simulated data

In the simulated dataset, there were six healthy pixels and 44 diseases affected ones. When estimating DI (%), one pixel with no infection was estimated as infected by the Mah distance criterion, whereas with the SA criterion two were mislabeled. Fig.11 shows the scatter of the disease affected pixels plotted in relation to reference DI and estimated DI; the average reference DI is 36%. The reference DIs and estimated DIs were strongly and linearly correlated for both the Mah distance (R^2 = 0.90) and SA (R^2 = 0.84) criteria. Further, the NRMSE of Mah distance and SA were 0.20 and 0.24, respectively, indicating that the SKB can estimate DIs accurately from the simulated multi-band reflectance.

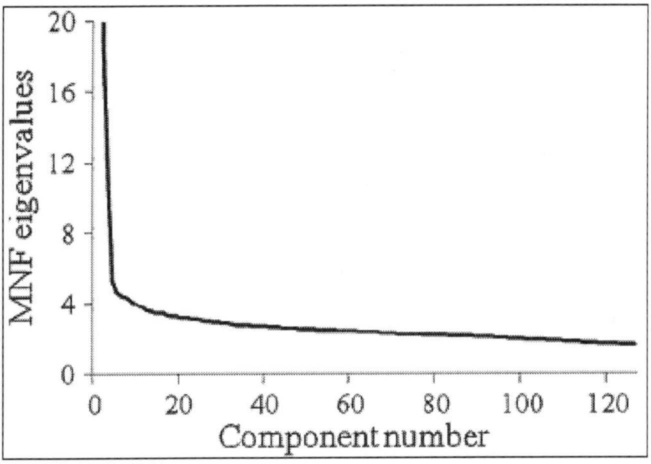

Fig. 10. MNF eigenvalues variation trend

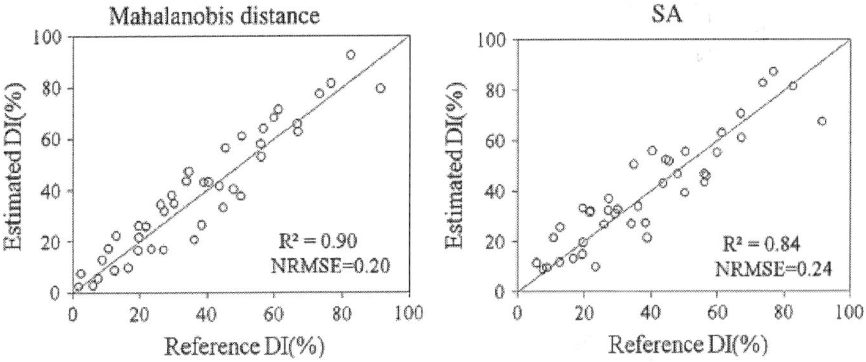

Fig. 11. Estimated DI(%) using simulated data

Table 6 gives the reference class of disease severity and the estimated class in the form of an error matrix. The overall accuracy with Mah distance and the SA criterion were 0.80 and 0.76, respectively, whereas the kappa coefficients were 0.71 and 0.65, respectively. However, we noticed that all the misclassified pixels were assigned to no more than one class adjacent to the reference class. Therefore, for simulated data, the classification accuracy was satisfactory in determining the severity class of yellow rust by SKB.

2.3.2.2 Performance of SKB for field surveyed data

Apart from the verification against simulated data, more importantly, the field surveyed data can be also used to assess the performance of the SKB. The field investigation showed that eight out of 26 plots were infected with DI ranged from 4 to 90%, whereas the other 18 plots were not affected by yellow rust. The estimation by DI (%) successfully identified the eight infected plots when the Mah distance criterion was used, whereas the SA criterion

		Reference					
		None	Low range	Mid range	Serious	Very serious	Total
Estimation (Mah)	None	6	0	0	0	0	6
	Low range	0	5	2	0	0	7
	Mid range	0	1	20	2	0	23
	Serious	0	0	1	10	1	12
	Very serious	0	0	0	1	1	2
	Total	6	6	23	13	2	50
Estimation (SA)	None	5	1	0	0	0	6
	Low range	1	4	1	0	0	6
	Mid range	0	1	20	2	0	23
	Serious	0	0	2	9	1	12
	Very serious	0	0	0	2	1	3
	Total	6	6	23	13	2	50

Table 6. Error matrix for simulated data

resulted in one misestimated plot. Figure 7 shows the scatter of the eight data plotted in relation to reference DI and estimated DI for both distance criteria. There was a significant linear trend in graphs based on both the Mah distance and SA criteria. The R^2 of Mah distance and SA were 0.80 and 0.67, respectively, whereas the NRMSE were as high as 0.46 and 0.55. In real circumstances, approximately 50% error in the estimated disease index is unsatisfactory. On the other hand, however, most of the uninfected plots were correctly identified according to DI (%) estimates (i.e. a DI<1%). For both the Mah distance and SA criteria, 15 out of 18 non-infected plots had been identified correctly, resulting in an accuracy of 77.8%. The results for estimating disease severity by severity class were even more encouraging. The overall accuracy for the Mah distance and SA criteria were 0.77 and 0.73, respectively, whereas the kappa coefficients are 0.58 and 0.49, respectively. Table 3 gives the error matrix for both criteria. The misclassified pixels were also assigned exclusively to the adjacent class.

In general, the above results demonstrate that the proposed SKB scheme has great potential for detecting the incidence and severity of yellow rust through multispectral images. As shown from several previous studies, the image processing method of MNF transformation was efficient in extracting the principle information from the images related to wheat disease infection (Zhang et al. 2003; Franke and Menz 2007). For the present study, we found that coupling MNF transformation with the PPI function was an appropriate way of extracting the principle information on yellow rust disease. To estimate disease severity by DI (%), the proposed SKB has achieved a satisfactory accuracy for simulated data. However, the estimated accuracy for field surveyed data was unsatisfactory, implying that the method tends to underestimate or overestimate the disease severity in practice. Nevertheless, to estimate disease severity through disease severity class has achieved a satisfactory accuracy for both simulated data and field surveyed data. Therefore, the disease severity class seems to be more robust in determining the disease severity. This might be because it is more rough estimation than DI (%). It is understandable that for the same sample, the less precise the criterion, the greater accuracy it would achieve. Moreover, the 5-class disease severity quantification is enough to guide field applications. We suggest that DI (%) should be used for detecting the disease severity of yellow rust by SKB. For the distance criteria used in the process of matching with SKB, the Mah distance criterion might be more appropriate because it performed better than SA in all the analyses conducted in this study (Figs. 11, 12, Tables 6, 7). Some previous studies have already emphasized the potential of hyperspectral imagery (Bravo et al. 2003; Moshou et al. 2004; Huang et al. 2007) and the high-resolution of multispectral imagery (Franke and Menz 2007) for detecting yellow rust disease. The development of SKB in the present study can be viewed as a scaling up method, which has extended the capability of detecting yellow rust disease from hyper- spectral imagery to the moderate resolution of multispectral imagery. However, it should be noted that the task of monitoring the occurrence and degrees of infection of crop diseases is far more complex than the cases described in this study. The spectral characteristics of yellow rust infection might appear similar to other sources of stress. In addition, the impact of phenology, cultivation methods, fragmentation of farmlands and other environmental conditions would also increase the difficulty and uncertainty of the estimation process. Therefore, the SKB developed in this study should correspond to the situation at the anthesis stage exclusively, and is only suitable for those regions with similar environmental characteristics and cultivation methods. For other regions with significantly different environmental characteristics, this purposed SKB may not work well. The possible solution to these problems may include incorporating suitable priors, which would require integration strategies and understanding of the mechanisms underlying some fundamental processes. Further research is required to address the problems mentioned above.

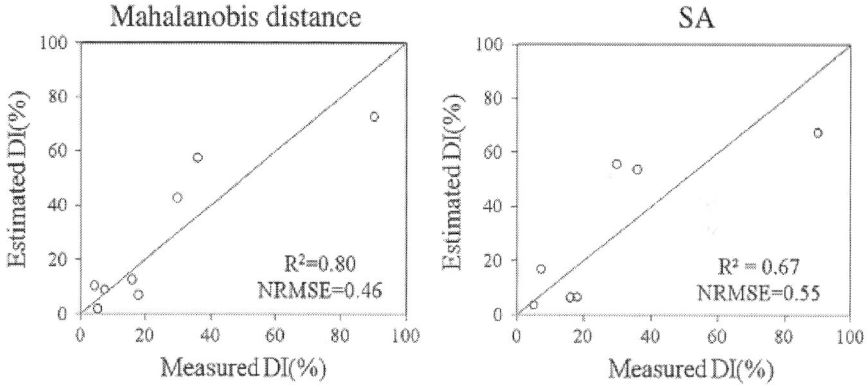

Fig. 12. Estimated DI(%) using field measurements

2.3.3 Conclusion

The low spatial resolution and few spectral bands have limited the application of moderate resolution satellite images for monitoring yellow rust disease. The spectral knowledge base developed enabled disease incidence and severity to be detected by moderate resolution satellite images. The SKB supported two ways of estimating disease severity: the disease

		Reference					
		None	Low range	Mid range	Serious	Very serious	Total
Estimation (Mah)	None	16	0	0	0	0	16
	Low range	2	2	1	0	0	5
	Mid range	0	1	3	0	0	4
	Serious	0	0	0	0	1	1
	Very serious	0	0	0	0	0	0
	Total	18	3	4	0	1	26
Estimation (SA)	None	15	0	0	0	0	15
	Low range	3	2	1	0	0	6
	Mid range	0	1	3	0	0	4
	Serious	0	0	0	0	1	1
	Very serious	0	0	0	0	0	0
	Total	18	3	4	0	1	26

Table 7. Error matrix for ground measured data

index and disease severity class. Both methods of estimation achieved a satisfactory level of accuracy for simulated data. For field surveyed data, estimation by DI (%) resulted in an unsatisfactory level of accuracy, whereas it was satisfactory for severity class. The Mah criterion performed better than spectral angle in all analyses. Therefore, the former should be considered as the more appropriate distance criterion.

Generally, the purposed SKB has a great potential in extending the capability of detecting yellow rust to multispectral remote sensing data, especially when the region of interest has similar environmental conditions to where the SKB was developed. The uncertainties caused by environmental differences should be further investigated in future studies.

2.4 Detecting yellow rust of winter wheat using land surface temperature (LST)

The air temperature and humidity are the most direct and important indicators of occurrence of yellow rust fungal. Generally, weather stations can provide the dynamic pattern of meteorological data for site sampled, yet not able to include the information of spatial heterogeneity. Fortunately, remote sensing technology has great potential for providing spatially continuous observations of some variables over large areas (Luo et al., 2010). The aim of the study was to study preliminarily on the relationship between the occurrence of wheat yellow rust and land surface temperature (LST) derived from moderate-resolution imaging spectroradiometer (MODIS) in order to predict and monitor incidence of the yellow rust on large scale.

2.4.1 Materials and methods

2.4.1.1 Survey area and field investigations acquisition

Field experiments of winter wheat were conducted during the growing seasons (form April to June) of winter wheat in 2008 and 2009. The investigation locations included Longnan district, Tianshui district, Dingxi district and Pingliang district in GanSu province and Qingyang district in ShanXi province as well as Linxia district in Ningxia Hui Autonomous Region (Fig.1), where the climates are semiarid and subhumid. Survey areas are located between latitude 32º40′N to 35º39′N and longitude 103º10′E to 107º40′E, and the mean altitude is over 2000 meter. The climate condition of surveyed area is characterized by high humidity and rainfall, and yellow rust disease almost occurs every year. It is reported that Longnan district is an important overwintering and oversummering area of yellow rust fungal (Zeng, 2003).

With the aid of the local Department of Plant Protection, 151 plots, including 68 plots from April to June in 2008, and 83 plots from April to June in 2009, were randomly selected and surveyed in the areas. The geographical coordinates of each plot were measured by GPS navigator at the middlemost of the plot. In addition, the disease severity was inspected.

2.4.1.2 MODIS land surface temperature (LST) products (MOD11)

Product description

MODIS Land Surface Temperature and Emissivity (LST/E) products (named starting with MOD11) provide per-pixel temperature and emissivity values. Temperatures are extracted in Kelvin with a view-angle dependent algorithm applied to direct observations. This method yields the error less than 1 K for materials with known emissivity. The view angle information is included in each LST/E product.

MOD11 acquisition and processing

24 MOD11A2 images (MODIS/Terra land surface temperature/emissivity 8-day L3 global 1km SIN grid v005) were acquired for free from Web (http://edc.usgs.gov/#/Find_Data) from April to July in 2008 and 2009, which covered completely the survey area, and 4 scenes images were acquired in every month. The raw data of MOD11A2 imagery were processed and transformed by MRT tool, and LST products were extracted from MODII A2 images. Then the survey area was cut by ENVI from LST images. Followed by that step, 4 scenes 8-day LST images of every month were all averaged, and 6 average LST images, including April, May, June in 2008 and 2009, were obtained. Finally, LST of 151 investigation points were respectively extracted from 6 average LST images.

2.4.2 Result

2.4.2.1 Determining LST threshold of infected points

The spatial resolution of MODIS temperature products is 1 km, while the DI of every investigation point only stands for the incidence of 30 m in semi diameter plots. Therefore, the scale of MODIS temperature products seemed not satisfied the investigation points for proper relationship between them. However, spatial variability of LST is slim, and the law still exists. A series of results could be found by establishing a two-dimensional spatial coordinate based on DI and LST, in which all investigation points were displayed (Fig 13). Firstly, the DI ranged from 0% to 100%, and most of infected points ranged from 0% to 60%. The LST values were between 292K and 310K with most of infected points distributed in the range from 298K to 306K. In addition, the points in the region of less than 298K were not infected by yellow rust basically; DI were less than 1% expect for one point (296.29K, 16%), which was thought as abnormal point. In addition, the LST values of all investigation points were less than 306K expect for one point (310.09K, 24%), which was abnormal because its LST was far away from LST values of others.

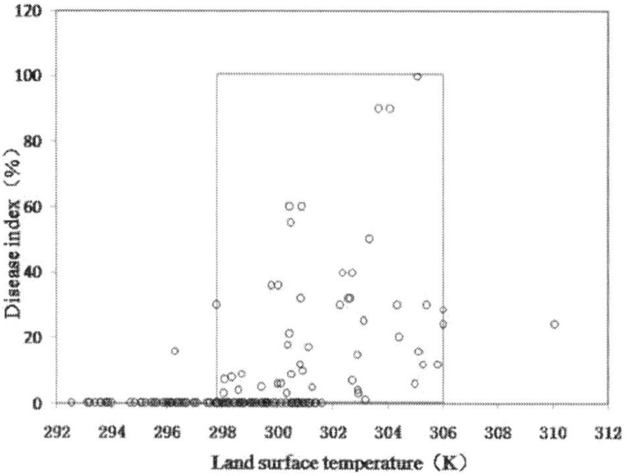

Fig. 13. The distribution of the investigation points

Therefore, without considering other factors, It is concluded that yellow rust can occur when LST is in the region from 298K to 306K.

2.4.2.2 Yellow rust incidence analysis based on LST

According to the results illustrated above, the advanced analysis was performed for incidence and possible area of yellow rust. The points in different LST range were done statistical analysis with all points' numbers and the infected points' number, and finally, the incidences were obtained by the number of the infected points dividing the number of all points in the different LST range (Table.8). The result showed that all investigation points in the region of less than 298K were not infected by yellow rust, except for the abnormal point (296.29K, 16%). On the other hand, in the LST region of more than 306K, there was only one point, which was viewed as abnormal point (310.085K, 24%). Thereby, it is quite possible that yellow rust fungus can not survive in the region of more than 306K. The conclusion was consistent with the above result (Fig. 13).

LST (K)	LST≥296	LST≥297	LST≥298	LST≥299	LST≥300	LST≥301	LST≥302	LST≥303	LST≥304	LST≥305	LST≥306
Total number	126	112	99	79	61	34	25	16	12	8	1
Number of infected points	49	48	47	42	39	27	25	16	12	8	1
Incidence (%)	38.89	42.86	47.47	53.16	63.93	79.41	100	100	100	100	100

Table 8. Statistic analysis in different LST range

Furthermore, there was an increasing trend of incidences with the rising of LST in the region from 296K to 302K. The incidence of yellow rust reached up to 100% when the LST was graeter than 302K (Fig. 14).

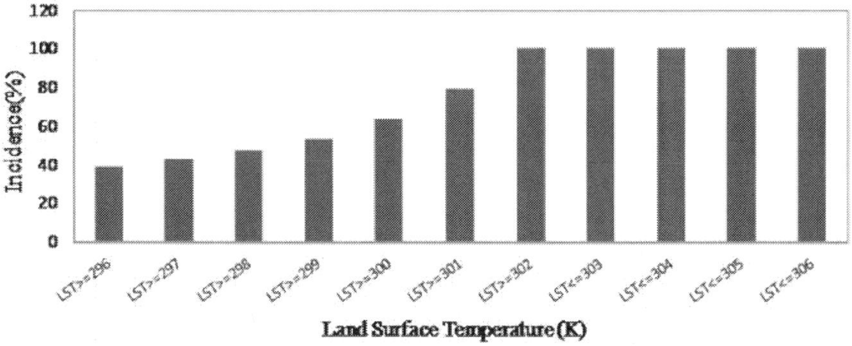

Fig. 14. The incidence of yellow rust in different LST range

2.4.2.3 Dividing yellow rust suitable occurrence region based on LST

According to Table 8 and Fig. 14, the survey areas could be divided into yellow rust unsuitable area (NSA), of which LST ranged from 298K to 306K, and yellow rust suitable area (SA), of which the LST was less than 298K and more than 306K. Moreover, the SA was divided into 3 levels according to the infected of yellow rust incidence and LST, and the LST thresholds for each level were: 298K ≤ LST ≤ 299K the low suitable area (LSA), on which the yellow rust occurs with very low possibility (incidence < 60%), 299K ≤ LST ≤ 301K the medium suitable area (MSA), which had moderate possibility for the occurrence of yellow rust (60% <incidence < 100%), and 302K ≤ LST ≤ 306K high suitable area (HSA), of which the environment was highly favorable to yellow rust (incidence=100%).

2.4.2.4 Verification

Total 26 points (from May 2008) were applied for the verification the method of estimating the incidence of yellow rust. It should be noted that those points were not used for the defining of the LST thresholds. (Fig. 15). These 26 points were constituted by 18 infected points and 8 non-infected points. Results showed the infected points were all in different suitable areas of wheat yellow rust, while the non-infected points were all in the unsuitable area. Thus the infected situation of yellow rust of these 26 points was consistent with forecast results. Geographically, it seemed that the yellow rust was prone to be prevalent in the northeast of Pingliang, southwest of Qingyang, northeast of Dingxi, the center part of Tianshui, and the west of Longnan, because they all were located in MSA and HAS. This result was consistent with the previous study (Xiao, et al, 2007). To prevent yellow rust from prevalence, more efforts should be placed on the farmlands located in the MSA, HAS and LSA.

2.4.3 Conclusions

Plant disease is governed by a number of factors, and the habitat factors play a major role in the development and propagation of fungal pathogens (Sutton et al., 1984; Hélène et al.,

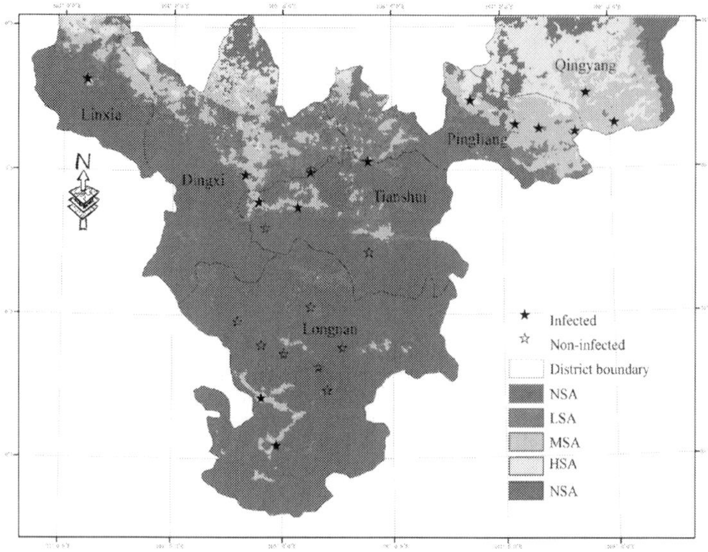

Fig. 15. Forecast map of yellow rust and distribution of measured points in May, 2008 based on LST

2002; Cooke et al., 2006). The yellow rust is no exception. The weather station can only offer points data, and remote sensing, however, can be a promising means for acquiring spatially continuous observations over large area. It has not been reported, if any, that the LST derived from remote sensing data is used to forecast the development of yellow rust.

The study tried to present a method that could forecast the suitable areas of wheat yellow rust by MODIS temperature products in a large scale. And it was proved that LST derived from remote sensing data had potential for predicting the occurrence and development of wheat yellow rust in a large area. From our results, it is clear that preventive measures of yellow rust can been made over large scale area accordingly with different real-time prediction methods based on LST derived from remote sensing data.

3. Detecting and discriminating winter wheat aphid by remote sensing

Wheat aphid, Sitobion avenae *(Fabricius)*, is one of the most destructive pests in agricultural systems, especially in temperate climates of the northern and southern hemispheres. Wheat aphid appears annually in the wheat planting area of China, causing great economic damage to plant crops as a result of their direct feeding activities. In high enough densities, wheat aphids can remove plant nutrients, and potentially reduce the number of heads, the number of grains per head, and overall seed weight. The damage is especially high when wheat aphid occurs in the flowering and filling stage of wheat. It is reported that average densities over 20 aphids per plant can cause substantial losses of yield and quality of wheat (Basky & Fónagy, 2003). There are also indirect damages including excretion of honeydew from aphids and as a vector of viruses, most notably two strains of the Luteovirus Barley Yellow Dwarf Virus (BYDV-MAV and BYDV-PAV) (Susan et al, 1992). To prevent the occurrence and prevalence of aphid, large amounts of insecticides are used, causing environment pollution. Therefore, large-scale, real-time prediction and monitoring of wheat aphid incidence and damage degree using remote sensing technology are extremely important.

3.1 Detecting winter wheat aphid using hyperspectral data

The study aimed to identify spectral characteristics of wheat leaf and canopy infected by aphid and find the sensitive bands to aphid at canopy level in filling stage of wheat, and to establish an aphid damage hyperspectral index (ADHI) based on those sensitive bands for detecting aphid damage levels in wheat canopy level in filling stage of wheat.

3.1.1 Materials and methods

3.1.1.1 Field experiments and field inventory

The field experiment plot was located at Xiaotangshan Precision Agriculture Experiment Base, Changping distract, Beijing (40°10.6′N, 116°26.3′E). The experimental field was about 250 m in length and 80 m in width. The winter wheat was planted in the study area from Oct 3, 2009, and harvested from June 25, 2010. Field inventory was conducted on June 7, 2010 when wheat was in the filling stage. Twenty five ground investigations including different aphid damage levels were selected. Aphid damage level was surveyed according to the investigation rule.

3.1.1.2 Canopy spectral measurements

Please refer to 1.1.1.2 part above.

3.1.2 Results

3.1.2.1 Leaf spectral characteristics of wheat infested by aphid

Representative reflectance measured from wheat aphid-infested and uninfested wheat leaves are shown in Fig. 16. It was evident that the spectral response of the wheat leaf was significantly affected by wheat aphid feeding (Fig. 16). The reflectance of wheat leaf infested by aphid was higher in the visible spectrum and short-wave infrared region and lower in near-infrared region than that of uninfested leaf. A significant increase in the reflectance from the wheat aphid-infested leaf in the visible region (400-700 nm) was observed, evidently due to reduction of photosynthetic pigment concentrations in particular chlorophylls caused by wheat aphid feeding (Richardson et al., 2004).

3.1.2.2 Canopy spectral characteristics of wheat infested by aphid

Compared with the canopy spectra of the healthy wheat, the canopy reflectance of aphid-infested wheat was gradually decreased in the range from 350 nm to 1750 nm, especially in the near infrared region (Fig. 17). Previous researches indicated that wheat had higher reflectance at visible wavelengths than the healthy vigorously growing wheat because the photoactive pigments (chlorophylls, anthocyanins, carotenoids) were destroyed. In this study, aphid occurred in the filling stage of wheat and the honeydew excreted by aphid absorbed dust or others from surrounding environment and contaminated (darkened) the leaf surface. As a result, the absorption at light slight wavelengths became stronger instead of weaker.

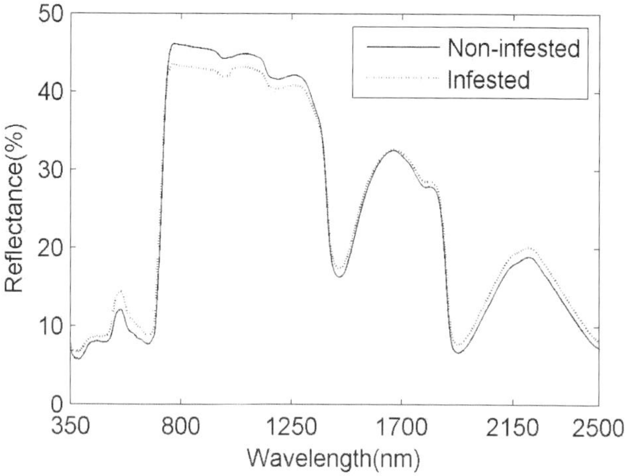

Fig. 16. The spectral reflectance of winter wheat leaf uninfested and infested by aphid

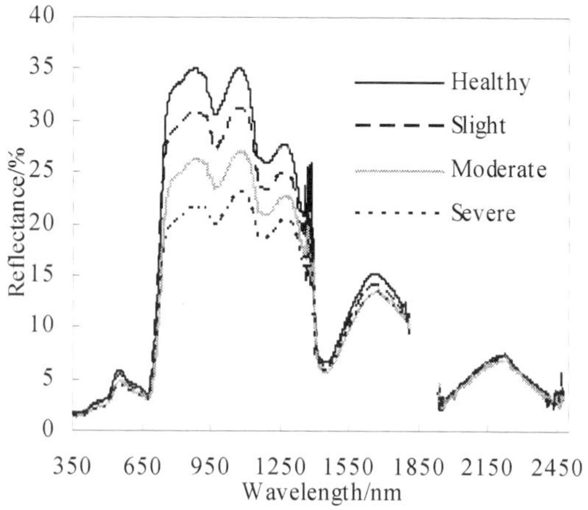

Fig. 17. The spectral reflectance of healthy wheat and wheat infested by various aphid damage levels. (Healthy: the average spectra of healthy wheat samples; Slight: the average spectra of aphid damage level 1and 2; Moderate: he average spectra of aphid damage level 3and 4; Severe: the average spectra of aphid damage level 5 and 6).

3.1.2.3 Aphid damage hyperspectral index for detecting aphid damage degree

Sensitive band selection of aphid infestation based on canopy reflectance

The sensitive bands were selected out by relevance analysis between reflectance and aphid damage levels. The reflectance ranges were from 400 nm to 690 nm, from 700 to 1300 nm and from 1500 to 1800 nm. The most sensitive bands to aphid were 551 nm (R^2=0.741) in visible light, 823 nm (R^2=0.865) in near infrared (NIR) and 1654 nm in short-wave infrared (SWIR) (R^2=0.668), respectively (Fig. 18).

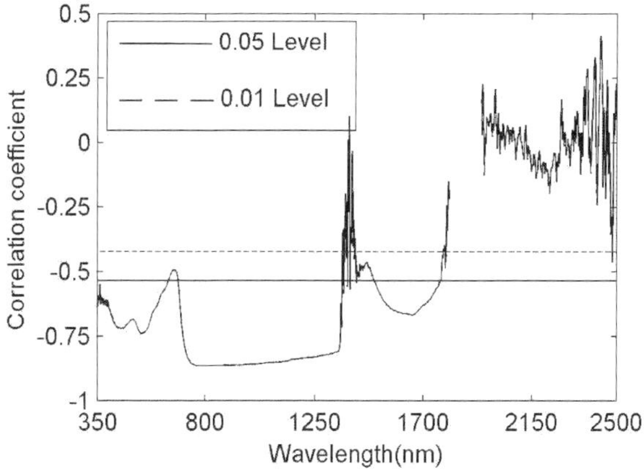

Fig. 18. Correlation coefficient between reflectance and aphid damage levels

Aphid damage hyperspectral index (ADHI) was established based on the most sensitive bands from hyperspectral data in the visible light region, NIR and SWIR and weight coefficient calculated according to rate of change of reflectance between healthy wheat and aphid-infected wheat, respectively.

$$ADHI = 0.32 \times \frac{R551_{normal} - R551_{infested}}{R551_{normal}} + 0.51 \times \frac{R823_{normal} - R823_{infested}}{R823_{normal}} + 0.17 \times \frac{R1654_{normal} - R1654_{infested}}{R1654_{normal}}$$

where $R551_{normal}$, $R823_{normal}$ and $R1654_{normal}$ are reflectance in 551 nm, 823 nm and 1654 nm of healthy wheat; $R551_{infested}$, $R823_{infested}$, $R1654_{infested}$ are reflectance in 551 nm, 823 nm and 1654 nm of aphid-infected wheat; 0.32, 0.51 and 0.17 are weight coefficients calculated by the contribution to change rates.

Further more, the correlation analysis was conducted between ADHI and aphid damage level from 25 investigation points (Fig. 19). It was concluded that ADHI exhibited high relationship with aphid damage levels (R^2=0.839). Therefore, ADHI was an important index to estimate aphid damage level in winter wheat.

3.1.3 Conclusions

Hyperspectral remote sensing has gone through rapid development over the past two decades and there is a trend toward the use of hyperspectral image in the application of remote sensing for precision farming. The study analyzed the spectral characteristics of wheat infested by aphid and selected the sensitive bands to aphid damage level. Then, an ADHI was developed using the most sensitive bands in visible light region, NIR and SWIR.

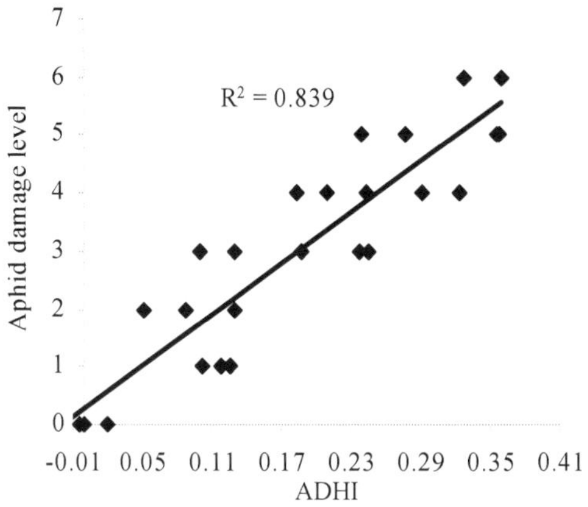

Fig. 19. The correlation between ADHI and aphid damage level

It was concluded that ADHI was a sensitive index to aphid damage levels, and could be used to retrieve aphid damage levels in the filling stage of wheat.

Crop growth is very dynamic processes and monitoring the condition of agricultural corps is a complex issue. It is possible that wheat damage symptoms caused by aphids and its response of canopy reflectance are different in different wheat growth stages. This study revealed that the reflectance of wheat infested by aphid was lower than healthy wheat in filling stage probably because of honeydew excreted by aphid. This was not consistent with previously published results in early detection of aphid infestation. Therefore, whether the ADHI can effectively retrieve aphid damage levels in other wheat growth stages remains as a task of future studies.

3.2 Detecting winter wheat aphid incidence using Landsat 5 TM

Wheat aphid occurrence and damage degrees are related to many factors including temperature, humidity, precipitation, field management, enemies, etc.. Most of the present studies on aphid prediction have been conducted based on meteorological data acquired from weather stations, and aphid density was monitored using the spectral characteristics of wheat infested by aphid. However, it is rare to investigate the relationship between environmental parameters, vegetable information derived from satellite images and aphid damage degrees. The aim of the present study is to investigate the relationships of aphid occurrence and damage degree to LST, NDWI, and MNDWI, which are related to vegetation water content derived from multi-temporal Landsat 5 TM. Another goal of the current research is to distinguish the degrees of aphid damage using 2-dimension feature spaces established by LST-NDWI and LST-MNDWI.

3.2.1 Materials and methods

3.2.1.1 Study areas

The study areas are selected in Shunyi district (116°28'—116°58' E，40°00'—40°18' N) and Tongzhou district (116° 32'—116°56' E, 39°36' —40°02' N,) of Beijing, China (Fig.20-a). The

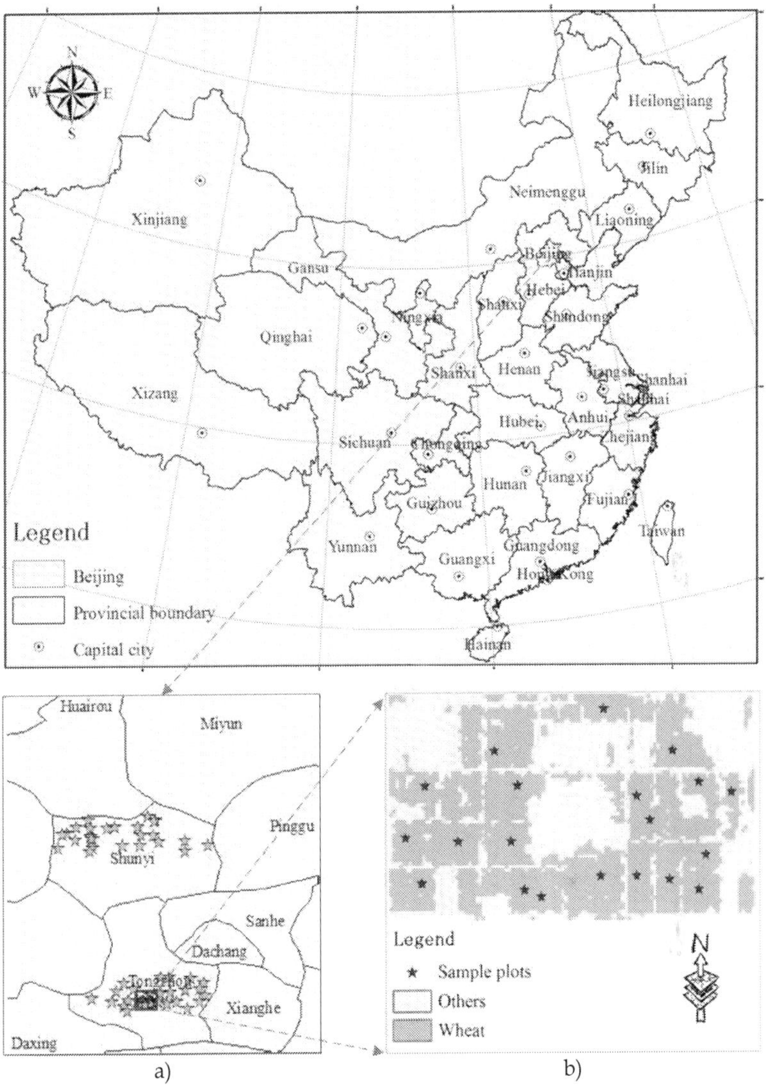

Fig. 20. The study area and the spatial distribution of sample plots

study areas have flat topography, with elevation ranging from 20 m to 40 m. The study areas have semi-humid warm temperate climate with yearly precipitation of 625 mm and mean temperature of 11.5°C in the Shunyi district and yearly precipitation 620 mm and mean temperature of 11.3°C in the Tongzhou district. Both districts are considered main winter wheat planting areas in Beijing, and aphid infestations occur in both areas almost every year.

3.2.1.2 Field inventory and data pre-processing

Field inventory was conducted during the growing seasons of winter wheat in 2010. The winter wheat in the study areas were planted between September 25 and October 7, 2009, and harvested between June 19 and June 25, 2010. Based on the combination of representative sampling and random sampling scheme, 70 sample plots with size of 0.09 ha (30 m × 30 m) each were collected as in Fig 1-a. These sample plots had different site conditions, plant densities, and management conditions. Aphid density surveys were carried out respectively on May 4 and May 6 for jointing stage, May 20 and May 21 for the heading stage, and June 3 and June 4 for the filling stage. The geographical coordinates of each plot were measured by global positioning system (GPS) (GeoExplorer 3000 GPS, with the error within 1m) at the middlemost of the plot.

Each sample covered with an area of 1 m^2. Then, 10 tillers in each sample plot were randomly selected, and the number of aphids was counted. The aphid densities were then estimated as follows: total aphids / 10 tillers.

The survey results were divided into three aphid damage degrees according to the aphid density investigated for facilitating the study. They were S0: non-infested by aphid and no damage to wheat, S1: aphid abundance/per tiller was about 3-10 and damage degree to wheat was slight, and S3: aphid abundance/per tiller was more than 20 and damage degree to wheat was severe.

3.2.1.3 Satellite image acquisition and pre-processing

Three Landsat-5 Thematic Mapper (TM) images (path 123/row 32) and three MOD 02 1 KM-Level 1B Calibrated Radiances Production (MOD 02) were acquired on May 4, May 20 and June 5, 2010, respectively. And all images were more than 90% cloud-free.

The Landsat-5 TM images were spectrally corrected to reflectance using the Landsat TM calibration tool and FLAASH (Fast line-of-sight Atmospherics Analysis of Spectral Hypercubes) was used to correct the image for atmospheric effects in ENVI 4.5. The Landsat-5 TM images were geometrically corrected versus a reference IKONOS image (equivalent scale map 1:10000) of the same area, available from a previous study. The resulting root mean square error (RMSE) did not exceed 0.3 pixels, which was adequate for the purposes of the present study.

3.2.1.4 Derivation of LST, NDWI and MNDWI from Landsat 5 TM

NDWI and MNDWI are both sensitive to changes in liquid water content of vegetation canopies (Hunt and Rock, 1989). In the current research, both NDWI and MNDWI were used to determine the threshold of aphid occurrence and the aphid damage degree. The indices are of the general form, as shown in the following:

$$NDWI = \frac{R_{NIR} - R_{SWIR}}{R_{NIR} + R_{SWIR}} \quad MNDWI = \frac{R_{GREEN} - R_{SWIR}}{R_{GREEN} + R_{SWIR}}$$

where R_{GREEN} ,R_{NIR} and R_{SWIR} are the reflectance in the green band, near-infrared band and short wave infrared band, respectively. For Landsat TM/ETM+, R_{GREEN} ,R_{NIR} and R_{SWIR} correspond to band2, band4 and band5, respectively.

LST is the radioactive skin temperature of the land surface, which plays an important role in farm and ecological environment. The present paper aims to discuss the relationship between LST and aphid occurrence and spread. LST was derived from the thermal infrared band (10.4-12.5μm) data of Landsat-5 TM using generalized single-channel algorithm developed by Jiménez-Muñoz and Sobrino (Jiménez-Muñoz and Sobrino, 2004). Surface emissivity (ε) and atmospheric water vapor content (w) were important parameters in the generalized single-channel algorithm. In the study, w was derived from the reflectance of band2 and band19 of MOD02, (Kaufman and Gao, 1992), and ε was calculated by vegetation coverage (Carlson and Ripley, 1997).

The NDWI, MNDWI and LST of all sample points were calculated and extracted from the Landsat images.

3.2.1.5 Subset image selection and wheat extraction

We resized the subset areas with size of 7.2 km^2 (3 km × 2.4 km) from the study area image located in Tongzhou district and covered with 20 evenly distributed sample points, and the aphid densities of the sample points were surveyed on May 6, May 20 and June 4, 2010, respectively. The survey results showed that the aphid damage degree of all sample points were S0 on May 6, 18 points for S1 and 2 points for S0 on May 20, and 16 points for S2 and 4 points for S0 on June 4, respectively. The subset areas were small enough and 20 sample points evenly distributed, According to the survey result, the aphid damage degree of the sample plots was basically same. Thus, the change of the aphid damage degree of wheat pixels in the wheat plots was slim or even basically the same as the sample plots. The wheat area of subset image selection area was extracted using classification of decision tree in ENVI 4.5 (Fig 20-b). The LST, NDWI and MNDWI of 2000 wheat pixels were extracted.

3.2.1.6 Methods of accuracy assessment

One basic accuracy assessment currently being used is overall accuracy, which is calculated by dividing the correctly classified pixels by the total number of the pixels checked. The Kappa coefficient is a measure of the overall agreement of a matrix introduced to the remote sensing community in early 1983. It has since become a widely used measure for classification accuracy. In contrast to overall accuracy, the Kappa coefficient takes non-diagonal elements into account (Rosenfield and Fitzpatrick-Lins, 1986), and it is calculated by the formula:

$$K = \frac{N\sum_{i=1}^{r} X_{ii} - \sum_{i=1}^{r} X_i + X_{+i}}{N^2 - \sum_{i=1}^{r} X_i + X_{+i}}$$

where r is the number of rows and columns in the error matrix; N is the total number of observations; Xii is the observation in row i and column i; Xi+ is the marginal total of row I; X+i is the marginal total of column i.

3.2.2 Results

3.2.2.1 2-dimensional feature space based on LST-VI

The minimum value, maximum value, mean values and standard deviations of LST, NDWI and MNDWI with aphid damage degrees of wheat pixels in subset image selection were listed in Table 9 and Table 10. And 2-dimensional feature space coordinates were established with LST as the abscissa and NDWI and MNDWI as the vertical axis, respectively (Figs. 2, 3). LST ranged from 287.5879 to 313.3448, NDWI ranged from 0.0226 to 0.5591 and MNDWI ranged from -0.3402 to -0.1077, respectively.

It is clear that LST was increasing from S0 to S1 to S2. LST was an important driving factor for aphid occurrence and could distinguish wheat non-infected from infested by aphids (Fig. 21 and Table 9). The general trend of NDWI increased firstly and reduced afterward, whereas MNDWI reduced firstly and increased afterward from S0 to S1 to S2.

Aphid Damage Degree	LST		NDWI		MNDWI	
	Minimum value	Maximum value	Minimum value	Maximum value	Minimum value	Maximum value
S0	287.5879	296.2498	0.0226	0.4405	-0.3402	-0.1077
S1	297.8084	306.0133	0.2083	0.5591	-0.6506	-0.3326
S2	300.5391	313.3448	0.0473	0.4542	-0.4117	-0.1159

Table 9. Minimum and maximum values of LST, NDWI and MNDWI in S0, S1 and S2

Aphid Damage Degree	LST		NDWI		MNDWI	
	Mean value	Standard deviation	Mean value	Standard deviation	Mean value	Standard deviation
S0	290.8578	1.4740	0.3029	0.0574	-0.2293	0.0296
S1	299.9236	1.0834	0.3998	0.0587	-0.4940	0.0362
S2	303.9424	1.7121	0.2979	0.0458	-0.2672	0.0402

Table 10. Mean value and standard derivation of LST, NDWI and MNDWI in S0, S1 and S2

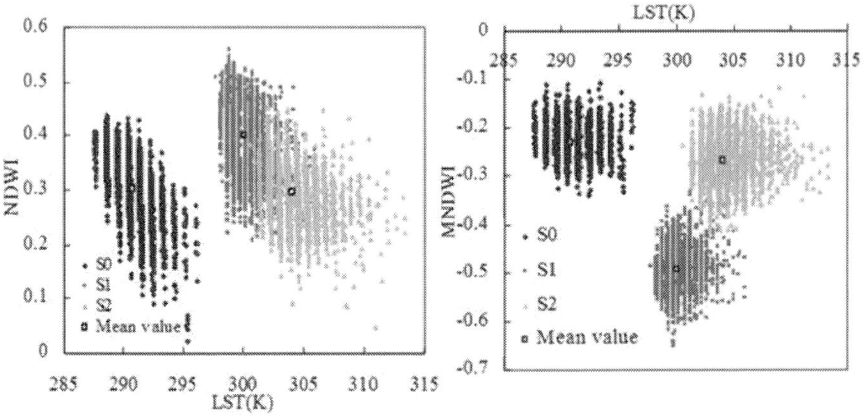

Fig. 21. The distribution of S0, S1 and S2 in the LST-NDWI (left) and LST-MNDWI (right) feature space

3.2.2.2 Discriminating aphid damage degrees using LST and MNDWI

In the 2-dimensional feature space coordinate system that was composed by LST and MNDWI, the S0 samples mainly scattered on the left part of the coordinate system, whereas S1 and S2 samples were distributed on the right part. As shown in Fig. 22, when LST was lower than the certain value, aphid did not occur, suggesting that LST served as a key factor of aphid occurrence and the MNDWI was sensitive to aphid damage degree.

Furthermore, LST_0 and $MNDWI_0$, which were the cutoff value of threshold values of LST and MNDWI of S0, S1 and S2, were determined by mean values and standard deviations. LST0 and MNDWI0 were calculated by formula as follows:

$$LST_0 = LST\_M1 - 2 \times LST\_SD1$$

$$MNDWI_0 = (M\_M1 + 3 \times M\_SD1) + [(M\_M1 + 3 \times M\_SD1) - (M\_M2 - 3 \times M\_SD2)]/2$$

where LST_M1 and LST_SD1 are the mean value and standard deviation of LST for S1; M_M1and M_SD1 are the mean value and standard deviation of MNDWI for S1; and M_M2 and M_SD2 are the mean value and standard deviation of MNDWI for S2.

According to Table 3, $LST_0 = 297.7568$ and $MNDWI_0 = -0.3866$. Wheat was not infested by aphid when LST< 297.7568, and aphid damage degree was S1 when LST≥297.7568K and -0.6506≤MDNWI ≤-0.3866 and S2 when LST≥297.7568K and -0.3866 ≤MDNWI ≤-0.1077 (Fig. 22).

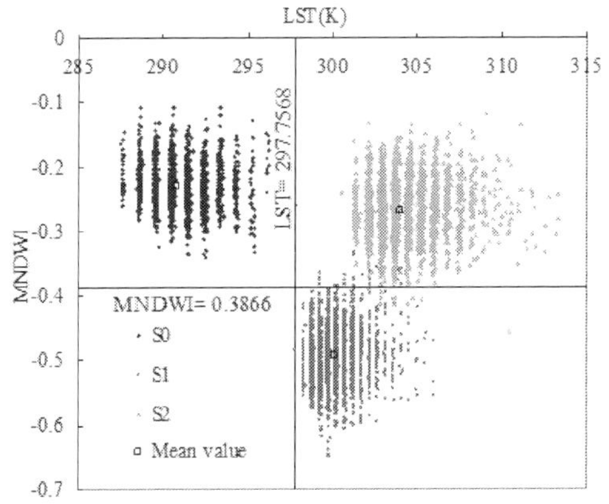

Fig. 22. Discriminating aphid damage degrees using LST and MNDWI

3.2.2.3 Verification

All survey samples, except 20 samples in the subset selection image were used to test the aphid prediction accuracy of 2-dimensional feature space based on LST and MNDWI (Fig. 23).

The discrimination accuracy was assessed using overall accuracy and kappa coefficient (Table 11). The results showed that the overall accuracy was 84%, and the Kappa accuracy was 75.67%.

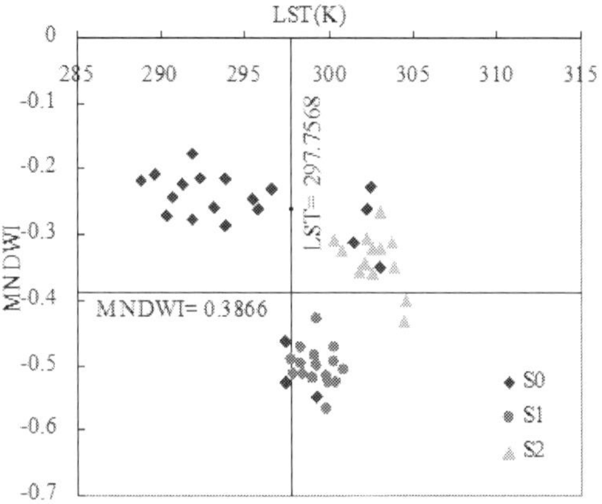

Fig. 23. Distribution of test sample plots in LST-MNDWI feature space

	S0	S1	S2	Total
S0	17	0	0	17
S1	2	14	0	16
S2	4	2	11	19
Total	23	16	11	50

Kappa coefficient = 0.7567

Table 11. Error matrices of the verification samples

3.2.3 Conclusions

This study successfully investigated the relationship between aphid damage degrees and several spectral features, such as NDWI, MNDWI and LST, through 2-dimensional feature space method. The results indicated that LST was the key factor in predicting the occurrence of aphid, and MNDWI was more sensitive to aphid damage degree than NDWI. In the 2-dimension feather space composed by LST and MNDWI, the result showed that S0, S1 and S2 were divided into three regions; S0 was distributed on the left of the space, and S1 and S2 on the right. Further, LST0 and MNDWI0 were calculated according the mean and derivation of S1, S2 as the cutoff value of threshold value to discriminate S0, S1 and S0. Through the verification of discrimination threshold value, it confirmed that the overall accuracy of discrimination was 84% and Kappa coefficient was 0.7567, suggesting that LST and MNDWI were of great potential in discriminating and monitoring the aphid damage degree over a large area, only using thermal infrared band and multi-spectral satellite images.

4. References

Basky Z. & Fónagy A. (2003).Glutenin and gliadin contents of flour derived from wheat infested with different aphid species. *Pest Management Science*, 59, 426-430.

Becker, B. L., David, P. L., & Qi, J. G. (2007). A classification-based assessment of the optimal spectral and spatial resolutions for Great Lakes coastal wetland imagery. *Remote Sensing of Environment*, 108, 111–120

Bravo, C., Moshou, D., West, J., McCartney, A., & Ramon, H. (2003). Early disease detection in wheat fields using spectral reflectance. *Biosystems Engineering*, 84, 137–145.

Broge, N. H., & E. Leblanc. 2000. Comparing prediction power and stability of broadband and hyperspectral vegetation indices for estimation of green leaf area index and canopy chlorophyll density. *Remote Sensing of Environment*, 76, 156–72.

Carlson T. N. & Ripley D. A.. On the relation between NDVI, fractional vegetation cover, and leaf area index. *Remote Sensing of Environment*, 1997, 62(3): 241-252.

Ceccato, P., N. Gobron, S. Flasse, B. Pinty, & S. Tarantola. 2002. Designing a spectral index to estimate vegetation water content from remote sensing data: Part 1. Theoretical approach. *Remote Sensing of Environment*, 82, 188–97.

Christou, P., Twyman, R.M., (2004). The potential of genetically enhanced plants to address food insecurity. *Nutrition Research Reviews* 17, 23–42.

Daughtry, C. S., C. L. Walthall, M. S. Kim, E. Brown de Colstoun, & J. E. McMurtrey.(2000). Estimating corn leaf chlorophyll concentration from leaf and canopy reflectance. *Remote Sensing of Environment*, 74, 229-239.

Fensholt, R., & I. Sandholt.(2003). Derivation of a shortwave infrared water stress index from MODIS near-and shortwave infrared data in a semiarid environment. *Remote Sensing of Environment*, 87, 111–21.

Filella, I., Serrano, L., Serra, J., & Penuelas, J. (1995). Evaluating wheat nitrogen status with canopyreflectance indices and discriminant analysis. *Crop Science*, 35, 1400–1405.

Franke, J., & Menz, G. (2007). Multi-temporal wheat disease detection by multi-spectral remote sensing. *Precision Agriculture*, 8, 161–172.

Galvão, L. S., A. Formaggio, R. & Tisot, D. A. (2005). Discrimination of sugarcane varieties in Southeastern Brazil with EO-1 Hyperion data. *Remote Sensing of Environment*, 94, 523–34.

Gamon, J. A., Penuelas, J., & Field, C. B. (1992). A narrow-waveband spectral index that tracks diurnal changes in photosynthetic efficiency. *Remote Sensing of Environment*, 41(1), 35–44.

Gitelson, A. A., Merzlyak, M. N., & Chivkunova, O. B. (2001). Optical properties and nondestructive estimation of anthocyanin content in plant leaves. *Photochemistry and Photobiology*, 74, 38–45.

Goovaerts, P., Jacquez, G. M., & Marcus, A. (2005). Geostatistical and local cluster analysis of high resolution hyperspectral imagery for detection of anomalies. *Remote Sensing of Environment*, 95, 351-367.

Gong, P., Pu R., Heald R.C.(2002). Analysis of in situ hyperspectral data for nutrient estimation of giant sequoia. *International Journal of Remote Sensing*, 23, 1827-1850.

Haboudane, D., Miller, J. R., Tremblay, N., Zarco-Tejada, P. J., & Dextraze, L. (2002). Integrated narrowband vegetation indices for prediction of crop chlorophyll content for application to precision agriculture. *Remote Sensing of Environment*, 81, 416–426.

Haboudane, D., J. R. Miller, E. Pattery, P. J. Zarco-Tejad, & I. B. Strachan. (2004). Hyperspectral vegetation indices and novel algorithms for predicting green LAI of crop canopies: Modeling and validation in the context of precision agriculture. *Remote Sensing of Environment*, 90, 337–52.

Hélène L., Frédéric F., Pierre D., Gaétan B., Isztar Z. (2002),Estimation of the spatial pattern of surface relative humidity using ground based radar measurements and its application to disease risk assessment. *Agricultural and Forest Meteorology*, 111, 223-231.

Huang, W. J., David, W. L., Niu, Z., Zhang, Y. J., Liu, L. Y., & Wang, J. H. (2007). Identification of yellow rust in wheat using in situ spectral reflectance measurements and airborne hyperspectral imaging. *Precision Agriculture*, 8, 187–197.

Hunt E.R., Rock B.N. (1989). Detection of changes in leaf water content using Near- and Middle-Infrared reflectances. *Remote Sensing of Environment*, 30, 43-54.

Jiménez-Muñoz J. C., Sobrino J. A. & Leonardo P. (2004). Land Surface Temperature Retrieval from Landsat TM 5. *Remote Sensing of Environment*, 90, 434-440.

Kaufman Y. J. & Gao B. C. (1992). Remote Sensing of Water Vapor in the Near IR from EOS/MODIS. *IEEE Transactions on Geoscience and Remote Sensing*, 30, 871-884.

Kim, M. S., C. S. T. Daughtry, E. W. Chappelle, & J. E. McMurtrey. 1994. The use of high spectral resolution bands for estimating absorbed photosynthetically active radiation (APAR). *In Proceedings of the 6th International Symposium on Physical Measurements and Signatures in Remote Sensing*, 299–306.

Li, G. B., Zeng, S. M., & Li, Z. Q. (1989). Integrated Management of Wheat Pests (pp. 185–186). Beijing: Press of Agriculture Science and Technology of China (in Chinese).

Luo, J. H., Zhang, J. C., Huang, W. J., Xu, X. G., Jin, N. (2010). Preliminary study on the relationship between land surface temperature and occurrence of yellow rust in winter wheat. *Disaster Advances*, 3, 288-292.

Luo, J. H., Huang, W. J., Zhang, J. C., Xu X. G. & Wang D. C. (2011). The preliminary study on spectral response of wheat under different stresses between field and satellite remote sensing. *Sensor Letters*, 9, 1225-1228.

Merton, R., & J. Huntington.(1999). Early simulation of the ARIES-1 satellite sensor for multi-temporal vegetation research derived from AVIRIS. In *Summaries of the Eight JPL Airborne Earth Science Workshop*, 9–11 February, 299–307. Pasadena, CA: JPL Publication 99-17.

Moshou, D., Bravo, C., West, J., Wahlen, S., McCartney, A., & Ramon, H. (2004). Automatic detection of 'yellow rust' in wheat using reflectance measurements and neural networks. *Computers and Electronics in Agriculture*, 44, 173–188.

Peñuelas, J., Baret, F., & Filella, I.(1995). Semi-empirical indices to assess carotenoids/chlorophyll a ratio from leaf spectral reflectance. *Photosynthetica*, 31, 221–230.

Merzlyak, M. N., Gitelson, A. A., Chivkunova, O. B., & Rakitin, V. Y. (1999). Non-destructive optical detection of pigment changes during leaf senescence and fruit ripening. *Physiologia Plantarum*, 106, 135–141.

Peñuelas, J., Gamon, J. A., Fredeen, A. L., Merino, J., & Field, C. B. (1994). Reflectance indices associated with physiological changes in nitrogen- and water-limited sunflower leaves. *Remote Sensing of Environment*, 48, 135–146.

Peñuelas, J., J. Piñol, R. Ogaya, & I. Filella. (1997). Estimation of plant water concentration by the reflectance water index WI (R900/R970). *International Journal of Remote Sensing,* 18, 2869–75.

Pu, R., Ge S., Kelly N.M., Gong P. (2003). Spectral absorption features as indicators of water status in coast live oak (Quercus agrifolia) leaves. *International Journal of Remote Sensing,* 24, 1799-1810.

Pu, R., Foschi L., Gong P. (2004). Spectral feature analysis for assessment of water status and health level in coast live oak (Quercus agrifolia) leaves. *International Journal of Remote Sensing,* 25, 4267-4286.

Rouse, J. W., R. H. Haas, J. A. Schell, & D. W. Deering.(1973). Monitoring vegetation systems in the Great Plains with ERTS. *Proc 3rd ERTS Symp* 1, 48–62.

Rosenfield G. & Fitzpatrick-Lins K. (1986). A coefficient of agreement as a measure of thematic classification accuracy. *Photogrammetric Engineering and Remote Sensing,* 52, 223-227.

Rules for Resistance Evaluation of Wheat to Diseases and Insect Pests Part 7: Rule for Resistance Evaluation of Wheat to Aphids. NY/T 1443.7-2007

Raikes, C., & L. L. Burpee. (1998). Use of multispectral radiometry for assessment of Rhizoctonia blight in creeping bentgrass. *Phytopathology,* 88, 446-449.

South, S., Qi, J. G., & Lusch, D. P. (2004). Optimal classification methods for mapping agricultural tillage practices. *Remote Sensing of Environment,* 91, 90–97

Susan E. Halbert, June Connelly B., Bishop G. W., et al. (1992).Transmission of barley yellow dwarf virus by field collected aphids (Homoptera: Aphididae) and their relative importance in barley yellow dwarf epidemiology in southwestern Idaho. *Annals of Applied Biology,* 121, 105-121.

Strange, R.N., Scott, P.R., (2005). Plant Disease: A Threat to Global Food Security. *Annual review of Phytopathology,* 40, 83–116.

Sutton, J.C., Gillespie, T.J., Hildebrand, P.D. (1984). Monitoring weather factors in relation to plant disease. *Plant Disease,* 68, 78–84.

Thenkabail, P. S., Smith, R. B., & De Pauw, E. (2000). Hyperspectral vegetation indices and their relationships with agricultural crop characteristics. *Remote Sensing of Environment,* 71, 158–182.

West, J. S., Bravo, C., Oberti, R., Lemaire, D., Moshou, D., & McCartney, H. A. (2003). The potential of optical canopy measurement for targeted control of field crop disease. *Annual Reviews of Phytopathology,* 41, 593–614.

Xiao, Z. Q., Li, Z. M., Fan M., Zhang, Y. Ma, S. j. (2007), Prediction model on stripe rust influence extent of winter wheat in Longnan Mountain area. *Chinese Journal of Agrometeorology,* 28, 350-353.

Yu, B., Ostland, I.M., Gong, P., Pu, R. L. (1999) Penalized discriminant analysis of in situ hyperspectral data for conifer species recognition. *IEEE Transactions on geosciences and remote sensing,* 37, 2569-2577.

Zarco-Tejada, P. J., A. Berjón, R. López-Lozano, J. R. Miller, P. Martín, V. Cachorro, M. R. González, & A. Frutos. (2005). Assessing vineyard condition with hyperspectral indices: Leaf and canopy reflectance simulation in a row-structured discontinuous canopy. *Remote Sensing Environment,* 99:271–87.

Zeng, S. M. (2003). Simulation study on oversummering process of wheat stripe rust caused Puccinia striiformis west. *In China, Acta Phytopathologica Sinica,* 33, 267-278.

Zhang, M. H., Qin, Z. H., Liu, X., & Ustin, S. L. (2003). Detection of stress in tomatoes induced by late blight disease in California, USA, using hyperspectral remote sensing. *International Journal of Applied Earth observation and Geo information,* 4, 295-310.

Zhang, J. C., Huang, W. J., Li, J. Y., Yang, G. J., Luo, J. H., Gu, X. H., & Wang, J. H. (2011). Development, evaluation and application of a spectral knowledge base to detect yellow rust in winter wheat. *Precision Agriculture,* 12, 716-731.

3

Seasonal Variability of Vegetation and Its Relationship to Rainfall and Fire in the Brazilian Tropical Savanna

Jorge Alberto Bustamante, Regina Alvalá and Celso von Randow
Brazilian National Institute for Space Researches
Brazil

1. Introduction

The Brazilian savanna, named locally Cerrado, is the second largest Brazilian biome, covering approximately two million km^2, especially in the Central Highlands (Ratter *et al.*, 1997). This biome is composed predominantly of tropical savanna vegetation and is considered as one of the world's biodiversity hotspots, a priority area for biodiversity conservation in the world (Myers et al., 2000). The Cerrado region is considered the last agricultural frontier in the world (Borlaug, 2002), which has been converted in the last 50 years especially for agriculture and pasture purposes, where natural and mainly anthropogenic annual burning is a common practice. Currently, around 50% of natural vegetation in the Cerrado region has been converted to pastures and crops (PROBIO-MMA, 2007).This conversion has impacted the biological diversity, the hydrological cycle, the energy balance, the climate and the carbon dynamics at local and regional scales due to habitat fragmentation, invasive alien species, soil erosion, pollution of aquifers, degradation of ecosystems and changes in fire regimes (Klink & Machado, 2005; Aquino & Miranda, 2008). The knowledge of spatial distribution, temporal dynamics and biophysical characteristics of the vegetation types, are important elements to improve the understanding of what is the interaction like between vegetation, precipitation and fire.

The objective of this study is to determine the relationship of environmental variables, such as precipitation and fire, with spatial and temporal distribution patterns of main vegetation type of the Brazilian tropical savanna. Thus, we seek to answer the question: how environmental variables, like rain and fire, influence the main vegetation types, like herbaceous, shrubs, deciduous trees and evergreen trees, in the Cerrado biome taking in account the seasonal patterns of the variables involved?

In this study, the potential of multi-temporal satellite data, like TRMM data for precipitation, MODIS vegetation indices products for land cover mapping, and others sensors like GOES and MODIS for fire detection is explored by the use of remote sensing and geographic information systems (GIS) techniques.

1.1 Seasonality of Cerrado vegetation

Phenological parameters of vegetation, such as start and end of the growing season, are strongly influenced by atmospheric conditions (like precipitation, temperature and humidity) at different time scales (intrannual, inter-annual, interdecadal, and so on). Atmospheric conditions at intrannual scale influence the main phenological events that the plant experiences during the annual cycle of growth (Reed et al. 1994). At greater time scales, climate influence on the spatial and temporal distribution of vegetation (Schwartz, 1994). On the other hand, the vegetation influence atmosphere while maintaining or modifying the flows of matter and energy, albedo, roughness, CO_2, which in turn affect the regional and/or global climate.

Savanna ecosystems that cover approximately 20% of the global land surface have mechanisms that control the flow of matter and energy in tropical savannas. These ecosystems are not well understood, which has hindered the inclusion of this biome in studies of regional and global modeling (Law et al., 2006).

1.2 Climate and precipitation regime

Climate patterns from intra-seasonal to decadal and century scales directly influence the timing, magnitude (productivity), and spatial patterns of vegetation growth cycles, or phenology (Reed et al., 1994; Schwartz, 1994).

The Savanna biome has a wet/dry climate. Its Köppen climate group is **Aw**. The ***A*** stands for a tropical climate, and the *w* for a dry season in the winter and the rainy season in the summer. During the dry season of a savanna, most of the plants shrivel up and die. Some rivers and streams dry up (Parker, 2000; Ritter, 2006). In the wet season all of the plants are lush and the rivers flow freely. The temperature of the savanna climate ranges from 20° to 30° C. In the winter, it is usually about 20° to 25° C. In summer the temperature ranges from 25° to 30° C. The savanna temperature does not change a lot, although when it does, it is very gradual and not drastic.

Because of its latitudinal position, the Brazilian savanna region is characterized by the transition between the warm climates of low latitudes and mesothermal climates of middle latitudes (Nimer, 1989). This region is considered almost homogeneous on the length and location of the dry and rainy periods (Rao & Hada, 1990). However, Castro et al. (1994) show that this region has a certain degree of heterogeneity due to the variation of length in the dry and rainy periods. This heterogeneity is determined by the interaction of atmospheric circulation systems in the lower and upper troposphere over the region. Some of these systems are: The South Atlantic anticyclone also known as South Atlantic Convergence Zone (SACZ), Polar anticyclone and Chaco low. SACZ is one of the main phenomena that determine the rainfall across the region (Satyamurty et al., 1998). In general, rainfall in the region ranges from 1000 to 1500 mm.

The climate of the Cerrado is tropical warm and semi-humid, with just two seasons, a dry one from May to September and a rainy one from October to April. Monthly rainfall in dry season (that include fall and winter) reduces considerably, reaching zero, resulting in a dry period that varies from three to five months duration (Coutinho, 2000). The rainy season (spring and summer) sometimes has short dry periods named locally "*veranicos*". The mean annual temperatures vary between 22 and 27°C and the mean annual precipitations between 600 and 2.200 mm.

1.3 Fire regime and detection

Fire is one of the most important drivers that influence vegetation function and structure. Fire incidence, in a given area or ecosystem, is part of a fire regime which has specific patterns of fire occurrences, frequency, size, severity, and sometimes vegetation and fire effects as well. For example, savanna fires are often of low intensity and high frequency (often annual), while forest fires are often of low frequency (once every few centuries) and very high intensity (Bowman & Murphy, 2010). Most of the wildland fires occur by the combination of edaphic, climatic and human activities (Roy, 2004). Natural fires are generally started by lightning, with a very small percentage started by spontaneous combustion of dry fuel such as sawdust and leaves. This kind of fire is insignificants in comparison to number of fires started by humans (Roy, 2004). Most tropical fires are set intentionally by humans (Bartlett 1955, 1957, 1961) and are related to several main causative agents (Goldammer, 1988): deforestation activities (conversion of natural vegetation to other land uses, e.g. agricultural lands pastures, exploitation of other natural resources); traditional, but expanding slash-and-burn agriculture; grazing land management (fires set by graziers, mainly in savannas and open forests with distinct grass strata); use of non-wood forest products (use of fire to facilitate harvest or improve yield of plants, fruits, and other forest products, predominantly in deciduous and semi-deciduous forests); wildland/residential interface fires (fires from settlements, e.g. from cooking, torches, camp fires etc.); other traditional fire uses (in the wake of religious, ethnic and folk traditions; tribal warfare) and socio-economic and political conflicts over questions of land property and land use rights.

Satellite-borne sensors can detect fires in the visible, thermal and mid-infrared bands. These sensors have been used most extensively for detecting and monitoring fire activity from landscape to global scales (Justice et al., 2003; Diaz-Delgado et al., 2004; Allan et al., 2003; Brandis & Jacobson, 2003; Miller et al. 2003; Rollins et al., 2004; Bowman et al., 2003). Justice et al. (2003) analyzed global remote sensing data and showed that occurrence of landscape fire is not random across the world, which is strongly influenced by climatic variables, like moisture deficit, wind speed, relative humidity and air temperature.

2. Methodology

2.1 Study area

The study area represents almost all (more than 90%) of the Brazilian savanna (Cerrado) biome, excluding only the southern region, which is characterized by few small isolated patches of savannas with intense anthropic activities like agriculture and ranching. The Cerrado vegetation exhibits a wide range of physiognomies. Following the "forest-ecotone-grassland" concept (Coutinho, 1978), the Cerrado ranges from *campo limpo*, a grassland, to *cerradão*, a tall woodland. The intermediate physiognomies (*campo sujo* - a shrub savanna, *campo Cerrado* - a savanna woodland, and *Cerrado sensu stricto* - a woodland) are considered ecotones of the two extremes.

The soil surface dries out during the dry season, leading the herbaceous and sub shrub plants suffering water stress. Thus, leaves dry out and die, while the underground plant structures are kept alive. The presence of dead leaves by water stress and also by frost greatly increases the litterfall and, consequently, the risk of fire (Nimer, 1977; Coutinho, 2000).

2.2 Methodology

The methodology involves the use of two spatial approaches, regional and local, to analyze the spatio-temporal relationships between environmental variables (precipitation and fire) and vegetation (NDVI).

The analysis unit at the local approach is the point, a specific pixel, which is obtained from the grid of points that were selected using a stratified random sampling. This grid contains separately the following types of vegetation: herbaceous, shrubs, deciduous trees, and evergreen trees of the Brazilian savanna in our study area.

At the regional approach, the entire region is considered another analysis unit, which means the Cerrado vegetation was not classified into four vegetation types. In this case, we calculated a NDVI mean value, keeping together all vegetation types (from grassland to forest) to each 16-days composite of the NDVI time series data.

The procedure applied to the vegetation data is also applied to the precipitation and fire data. The results are seasonal profiles to each variable along the annual cycle which were related using correlation and regression techniques. These seasonal profiles allow calculating a gradient of vegetation seasonality, which is defined by the difference of highest and lowest values of NDVI, precipitation, or fire. In the case of vegetation, the degree of seasonality is directly related to the degree of deciduousness, that is, the degree of leaf biomass loss during the dry season, when most plants suffer some degree of water stress.

The spatial and temporal resolutions of the data are: 250m and 16-day, 1km and 1-day, ~20km and 3 hours, for MODIS NDVI, fire hotspot and precipitation, respectively. These data are arranged to standardize them in the same 16-day temporal scale. Data from 2002, 2005 and 2008 were collected since they are considered as years under normal climatic condition, without the influence of El Niño-Southern Oscillation events.

2.2.1 Vegetation seasonality

The Normalised Difference Vegetation Index (NDVI), normalised ratio between near infrared reflectance (NIR) and red reflectance (red), has been widely used in satellite-based vegetation monitoring and modelling. NDVI is computed as:

$$NDVI = (NIR - red)/(NIR + red) \qquad (1)$$

Index values can range from -1.0 to 1.0, but vegetation values typically range between 0.1 and 0.7. Higher index values are associated with higher levels of healthy vegetation cover, whereas clouds and snow will cause index values near zero, making it appear that the vegetation is less green.

Six Moderate Resolution Imaging Spectroradiometer (MODIS) Normalized Difference Vegetation Index (NDVI) tiles (h12v09, h12v10, h12v11, h13v09, h13v10 e h13v11) were joined to create a mosaic for the entire study area. Three annual time-series were prepared for the following years: 2002, 2005 and 2008. Each annual dataset consists of 23 MODIS NDVI data, at 16-day composite intervals, and 250 m spatial resolution. These data were used to classify and analyze seasonal and phenology profiles of the Brazilian Cerrado vegetation.

At this point, the methodology consists first in the classification of vegetation, from the annual NDVI time-series data and using the tree decision technique, into four types: grasses and herbs, shrubs, deciduous trees and evergreen trees. This classification uses the phenological parameter named end of vegetation growing season, which corresponds to the following ranges of NDVI for each vegetation type: grasses and herbs (E1) from 100 to174, shrubs (E2) from 175 to 199, deciduous trees (E3) from 200 to 219 and evergreen trees (E4) from 220 to 255. Ground truth data was used to validate this classification.

The second part consists of selecting representative spatial points of vegetation types (Figure 1), which are obtained from the vegetation classification image. Each point corresponds to a pixel on the image and is defined as our unit of analysis. A stratified random sampling technique was used for the selection of points in the classification image. The number of points to each vegetation types was proportional to its spatial coverage in the study area. So, herbaceous (E1) represents 52% of the points, shrubs (E2) 24%, deciduous trees (E3) 15% and evergreen trees (4) 9%.

The total number of points identified in the study area was N = 639, which are distributed as follows: 251 points of herbaceous, 318 of shrubs, 59 of deciduous trees and 11 of evergreen trees (Figure 1).

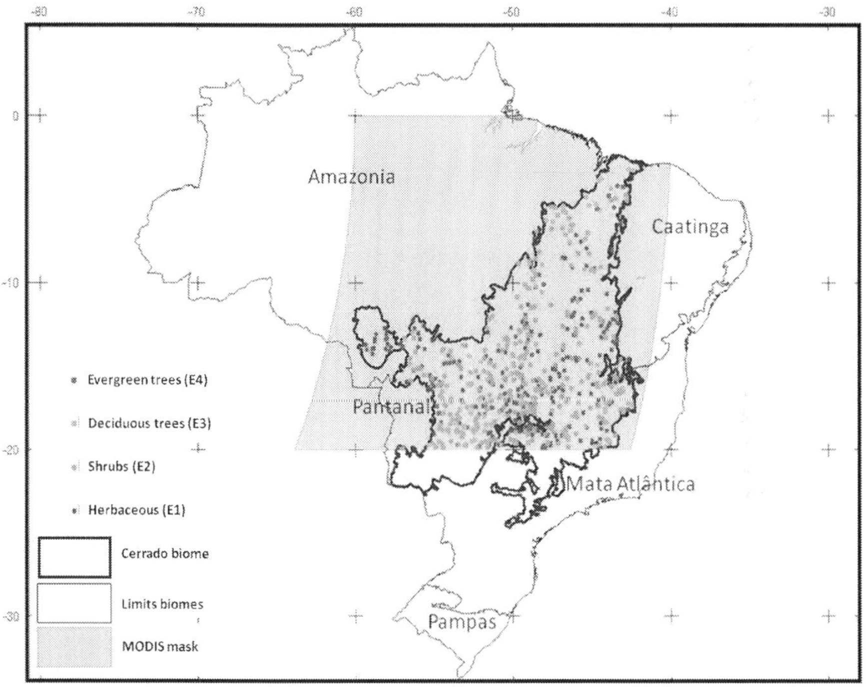

Fig. 1. Location of Brazilian biomes highlighting the savanna (Cerrado) biome. The shaded area is a mosaic of 4 MODIS-13Q1 tiles. Dots of different colors correspond to stratified random sampling of the following vegetation types: herbaceous (E1), shrubs (E2), deciduous trees (E3), and evergreen trees (E4).

2.2.2 Fire seasonality

First, daily data of fire hot spot obtained as latitude/longitude coordinates, in ASCII format, are converted in XYZ vector format data using a geographical information system (GIS) tool. These daily vectors were used to create a new vector data set of 16-days composites accumulating these daily data. Each composite were used to create a raster data set of fire hotspot density by the use of the Kernel density estimator, according to the following equation (Silverman, 1986):

$$\hat{f}(x;H) = n^{-1}\sum_{i=1}^{n} K_H(x - X_i) \tag{2}$$

Where:

- $X_1, X_2, \ldots, X_n$ is sample of n data points (fire hot spot)
- H is bandwith matrix
- $K_H x - X_i$ is normal probability density function (pdf) with mean X_i and variance H

Kernel Density calculates the density of point features around each output raster cell. The kernel function is based on the quadratic kernel function as described in Silverman (1986). Conceptually, a smoothly curved surface is fitted over each point. The surface value is highest at the location of the point and diminishes with increasing distance from the point, reaching zero at the Search radius distance from the point. Only a circular neighborhood is possible. The volume under the surface equals the Population field value for the point, or 1 if NONE is specified. The density at each output raster cell is calculated by adding the values of all the kernel surfaces where they overlay the raster cell center.

2.2.3 Precipitation

We used two kinds of data for precipitation in the study area for the years 2002, 2005 and 2008. First, Tropical Rainfall Measuring Mission (TRMM) multisatellite rainfall data (3B42 product), which has 0.25 degree spatial resolutions and 3-hours temporal resolution. Second, meteorological station rainfall data scattered throughout the study area, which has 1-hour temporal resolution.

These two datasets (TRMM and observed data) are combined following the approach of Vila et al. (2009), which use the Barnes objective analysis (Barnes, 1973; Koch et al., 1983) for data interpolation. This analysis allows the incorporation of observed data in a grid of estimated data and also improves its spatial resolution. As result, the new precipitation data has 0.2-degree spatial resolution and 1-day temporal resolution.

3. Results and discussion

3.1 Regional analysis

Figure 2 shows seasonal profiles of vegetation and precipitation in the Cerrado region for the three years analyzed (2002, 2005 and 2008). These results show that the Cerrado vegetation seasonality is well defined, which in turn has a direct relationship to the seasonality of precipitation. However, there is a time lag ranging from 1 (16 days) to 3 (48

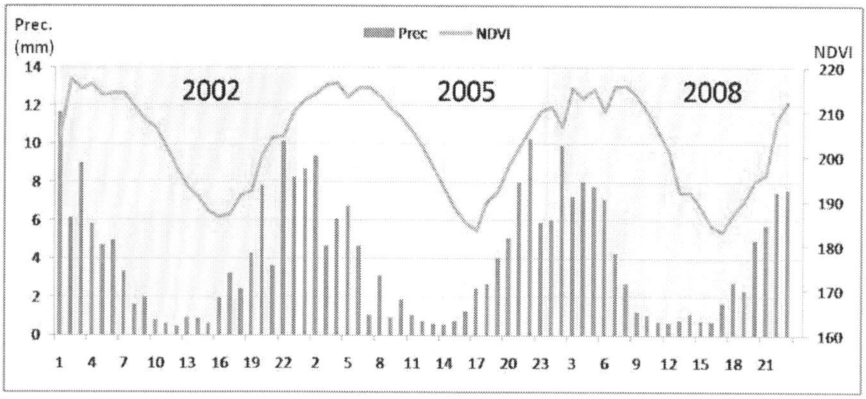

Fig. 2. Annual seasonality of vegetation and precipitation in the years 2002, 2005 and 2008 for the Cerrado biome. Each year consists of 23 16-days composite periods. Precipitation is the daily mean rainfall values for a 16-days composite period in mm (first y axis) and vegetation the mean NDVI values for the same period (second y axis).

days) periods between the beginning of the rainy season and the beginning of the vegetation growing season.

Figure 3 shows seasonal profiles of vegetation and fire in the Cerrado region for the three years analyzed (2002, 2005 and 2008). These results show, as in Figure 2, that the fire occurrence in the Cerrado has well-defined seasonality, which in turn has a direct negative relationship to the seasonality of vegetation. That means, the highest fire occurrence during the growing cycle of fire is related to the greatest loss of plant cover during the dry season, with a time lag ranging from 0 to 3 periods (0 to 48 days).

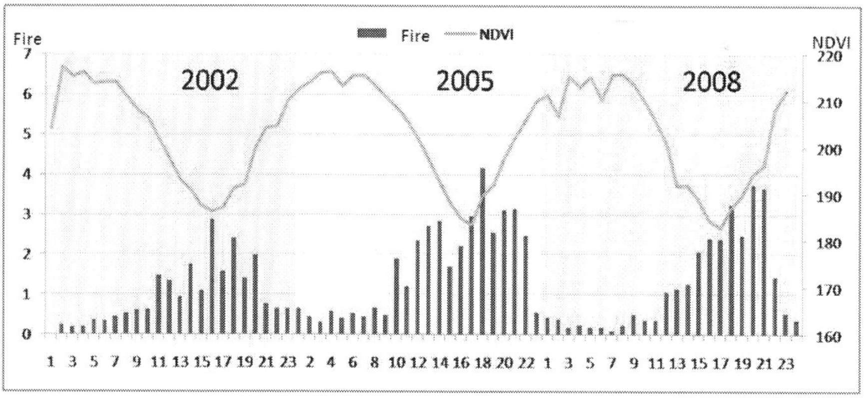

Fig. 3. Annual seasonality of vegetation and fire in the years 2002, 2005 and 2008 for the Cerrado biome. Each year consists of 23 16-days composite periods. Fire is the daily mean value of the density of hotspot within a 10km radius for a 16-days composite period and vegetation the mean NDVI values for the same period (second y axis).

3.2 Local analysis

The results presented show the seasonality of vegetation, rainfall and fire in places (points) defined by the grid points representing the four vegetation types analyzed in the study.

3.2.1 Seasonality of vegetation

Figure 4 shows the seasonal profile of four vegetation types over the three years analyzed. These results show a clear difference, regarding the degree of vegetation seasonality, among the four types of vegetation analyzed, according to the following gradient: herbaceous (E1), with strong seasonality, shrubs (E2), deciduous trees (E3), and evergreen trees (E4), with weak seasonality.

The vegetation phenology metrics are shown in Figure 5. Figure 5a shows annual maximum and minimum NDVI values, indicating the highest and lowest vegetation productivity respectively, for each type of vegetation in the three years analyzed. Figure 5b shows the difference between the maximum and minimum NDVI as a percentage, indicating the degree of seasonality. Also Figure 5a shows a slight difference between the maximum NDVI values, high plant productivity in the four vegetation types, while the difference between the minimum NDVI values, lower productivity, in the four vegetation types is significant. In general, the degree of seasonality of the vegetation (Figure 5b) was consistently detected in the four vegetation types. That is, small plants with low canopy (shrubs and herbaceous) have higher degree of seasonality than tall one with high canopy, which in turn have lower degree of seasonality.

3.2.2 Seasonality of precipitation

Figure 6 shows the seasonal profile of rainfall recorded in the same sampling points of the four vegetation types over the three years analyzed. These results show, in the beginning of the year during the rainy season, lower rainfall at sites where herbaceous and shrubs were registered than at sites where deciduous and evergreen trees are predominant. This result is a first indicator that shows a relationship between rainfall gradient and vegetation cover gradient. These gradients range from sites with higher precipitation, associated with high canopy plants (evergreen trees), to those with less precipitation, associated with a lower canopy plants (herbaceous).

3.2.3 Seasonality of fire

Figure 7 shows the pattern of the fire season recorded in the same sampling points of the four vegetation types over the three years analyzed. The results of fire occurrence throughout the Cerrado region show that there is a pronounced seasonality in all vegetation types analyzed with a peak in the months of greatest drought in the dry season.

The results show a well-defined gradient of fires in the four types of vegetation. This gradient varies from lower fire density in evergreen trees (E4), with shorter periods of time (12 to 22) throughout the annual cycle, to higher fire density in herbaceous plants (E1), with more periods of time (1 to 23, except 2), as seen in Figure 7. Most of the fire occurrences in the four types of vegetation were recorded in 2005 and 2008 indicating the occurrence of an inter-annual variability of fire. The higher fires were recorded between the periods from 15 to 21 taking into account the four vegetation types and the three years analyzed.

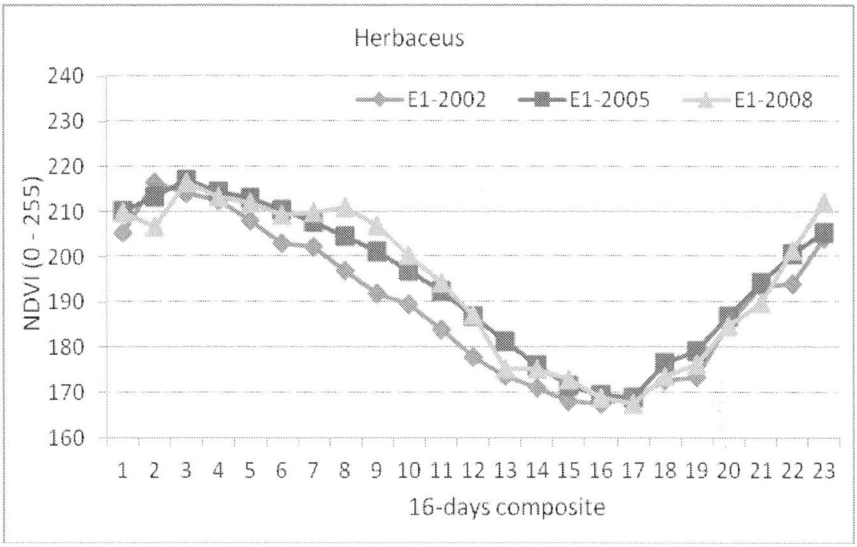
Herbaceus
E1-2002
E1-2005
E1-2008
NDVI (0 - 255)
240
230
220
210
200
190
180
170
160
1 2 3 4 5 6 7 8 9 10 11 12 13 14 15 16 17 18 19 20 21 22 23
16-days composite

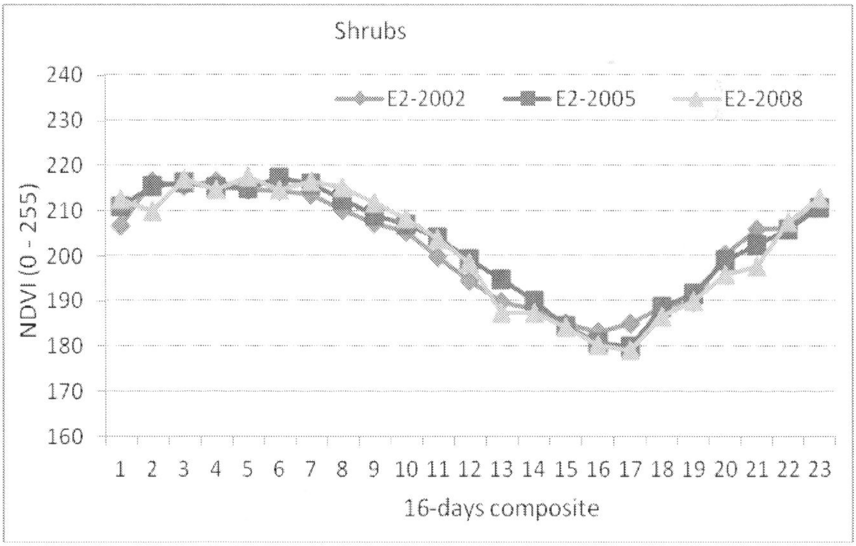
Shrubs
E2-2002
E2-2005
E2-2008
NDVI (0 - 255)
240
230
220
210
200
190
180
170
160
1 2 3 4 5 6 7 8 9 10 11 12 13 14 15 16 17 18 19 20 21 22 23
16-days composite

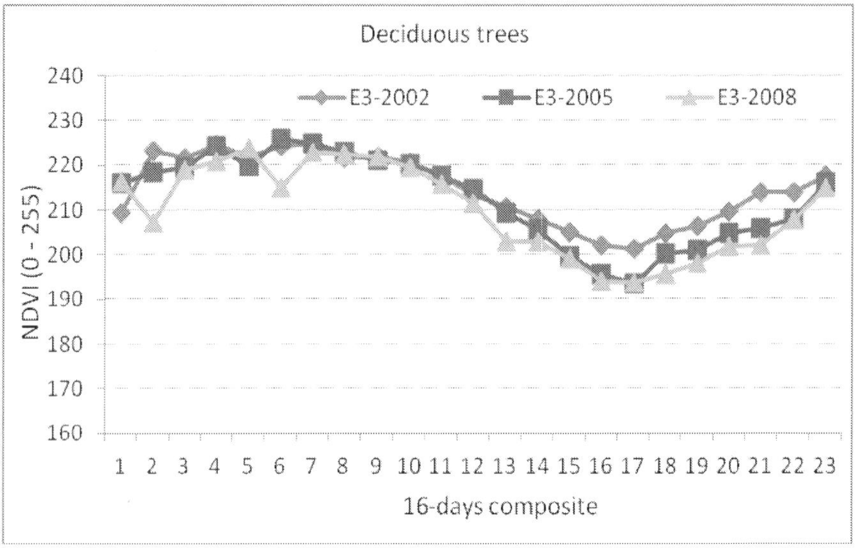

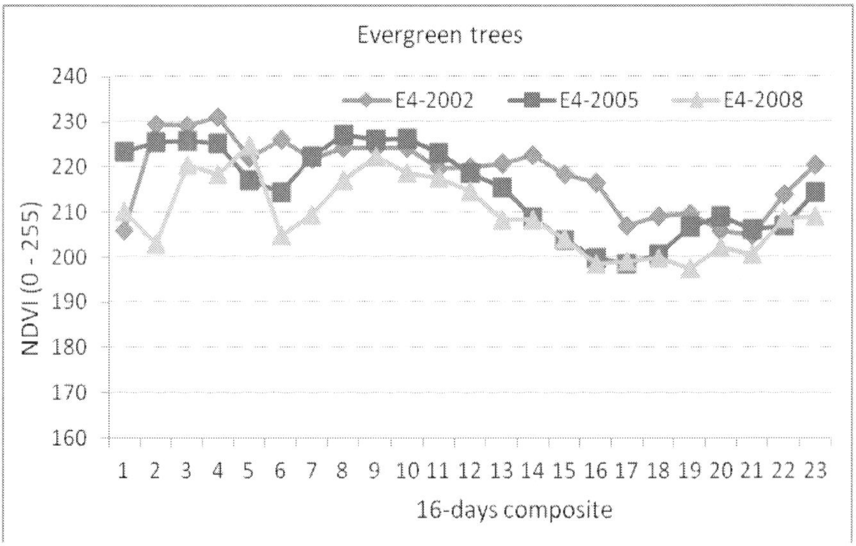

E1: herbaceous; E2: shrubs; E3: deciduous trees; and E4: evergreen trees.

Fig. 4. Annual seasonality of vegetation derived from NDVI data for the years 2002, 2005 and 2008 in the four vegetation type analyzed. Each year consists of 23 16-days composite periods.

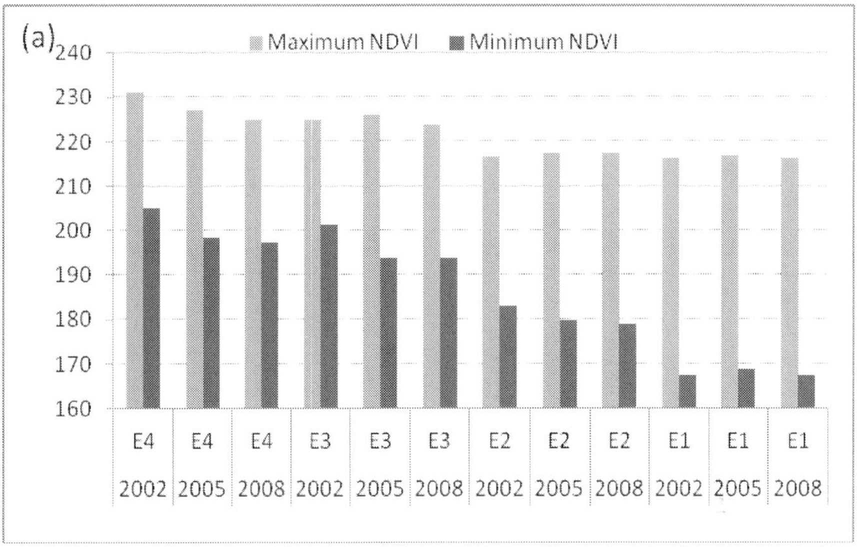

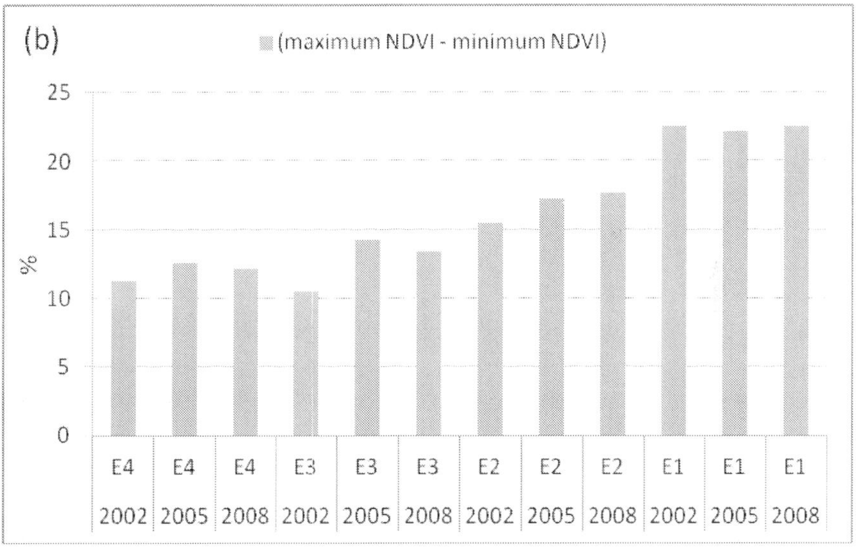

E1: herbaceous; E2: shrubs; E3: deciduous trees; and E4: evergreen trees

Fig. 5. Metrics of vegetation phenology derived from NDVI data used in Fig. 4. Maximum and minimum NDVI values indicate periods of higher and lower plant productivity (left) respectively, and the difference of both, as a percentage, indicates the degree of seasonality of each vegetational type in the three years analyzed.

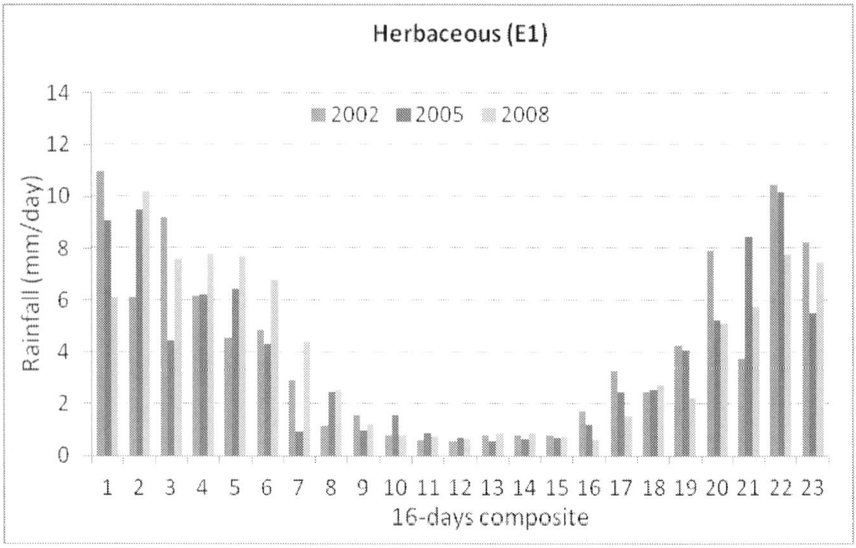

Herbaceous (E1)
2002
2005
2008
Rainfall (mm/day)
0
2
4
6
8
10
12
14
1 2 3 4 5 6 7 8 9 10 11 12 13 14 15 16 17 18 19 20 21 22 23
16-days composite

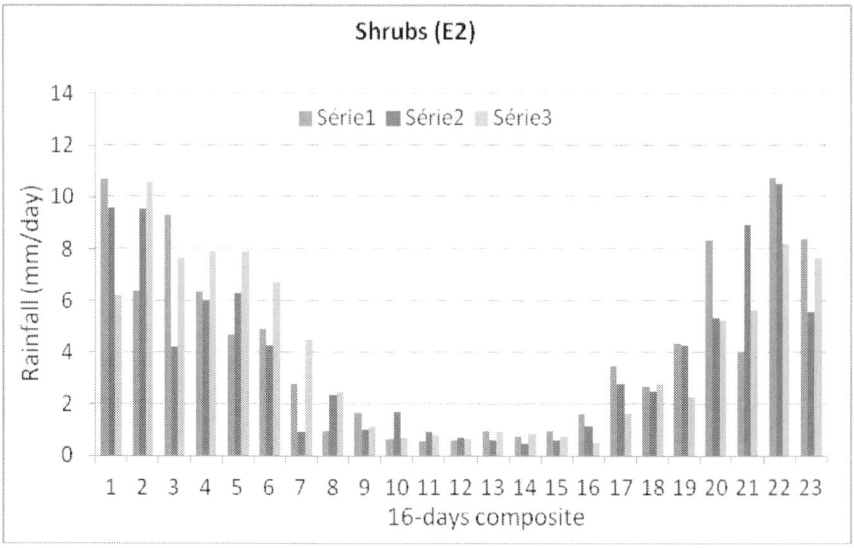

Shrubs (E2)
Série1
Série2
Série3
Rainfall (mm/day)
0
2
4
6
8
10
12
14
1 2 3 4 5 6 7 8 9 10 11 12 13 14 15 16 17 18 19 20 21 22 23
16-days composite

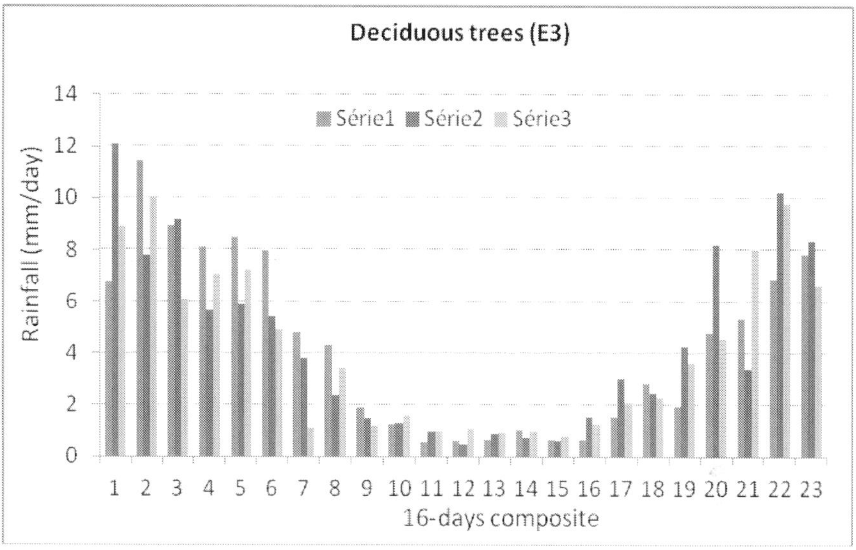

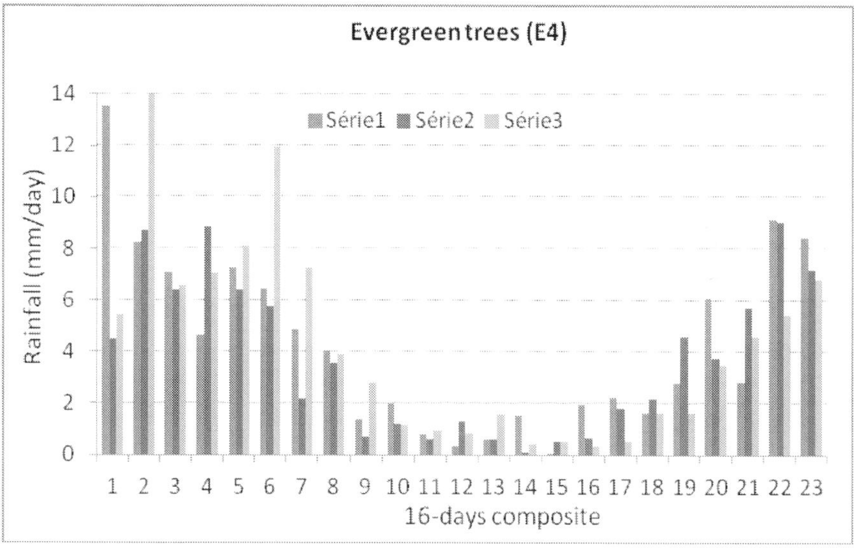

E1: herbaceous; E2: shrubs; E3: deciduous trees; and E4: evergreen trees.

Fig. 6. Annual seasonality of precipitation for the years 2002, 2005 and 2008, in places where we sampled the four vegetation types analyzed. Each year consists of 23 16-days composite periods. Precipitation in mm is the daily mean rainfall values for a 16-days composite period.

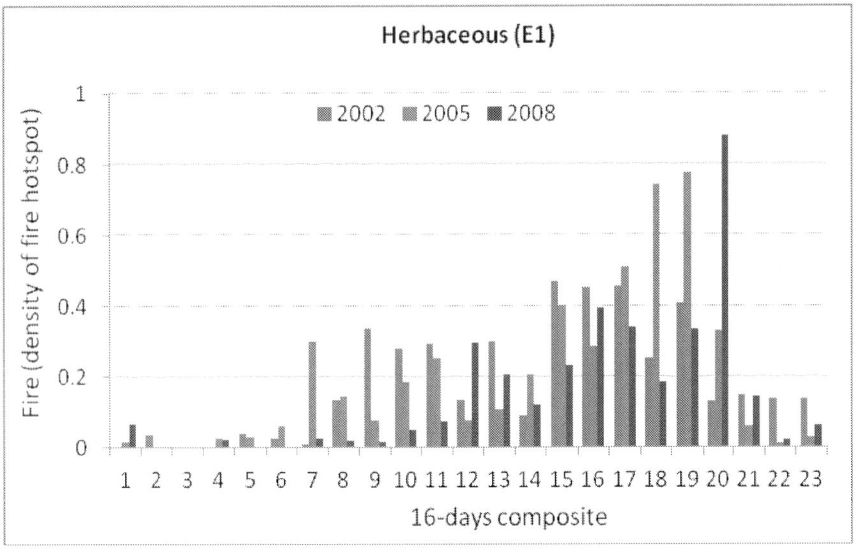
Herbaceous (E1)
Fire (density of fire hotspot)
1
0.8
0.6
0.4
0.2
0
2002 2005 2008
1 2 3 4 5 6 7 8 9 10 11 12 13 14 15 16 17 18 19 20 21 22 23
16-days composite

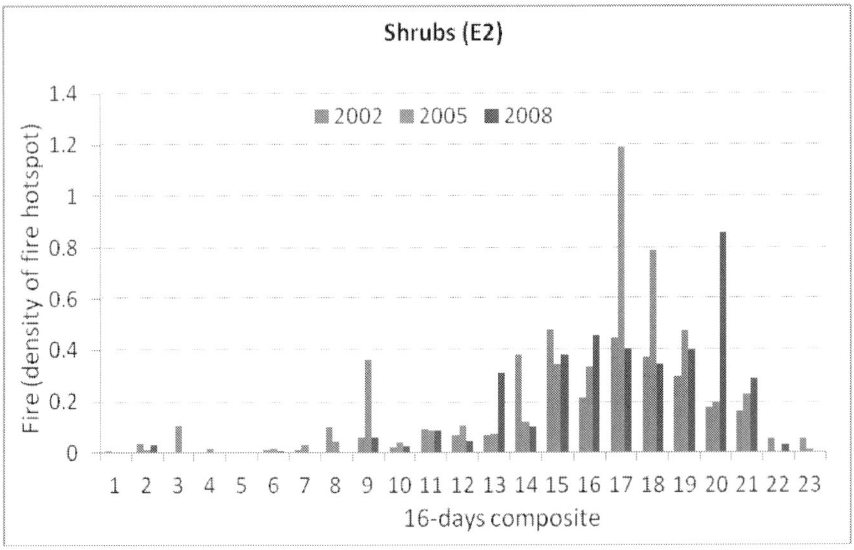
Shrubs (E2)
Fire (density of fire hotspot)
1.4
1.2
1
0.8
0.6
0.4
0.2
0
2002 2005 2008
1 2 3 4 5 6 7 8 9 10 11 12 13 14 15 16 17 18 19 20 21 22 23
16-days composite

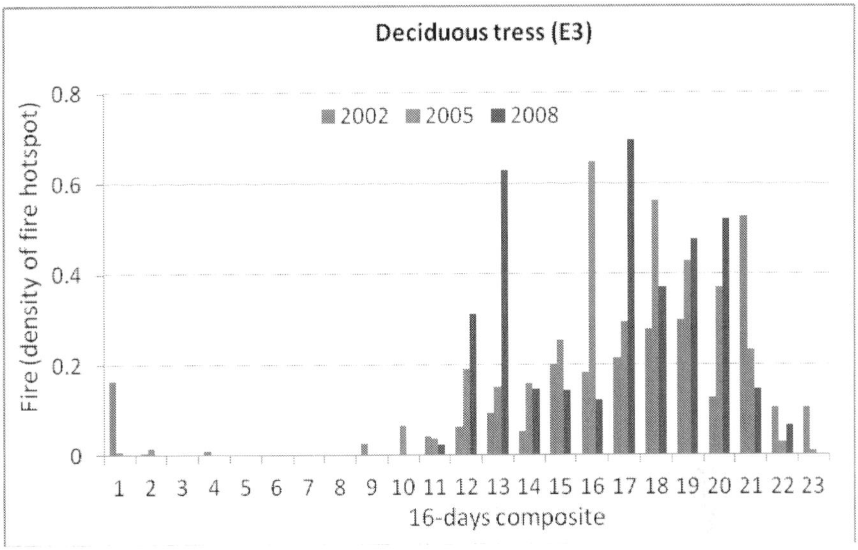

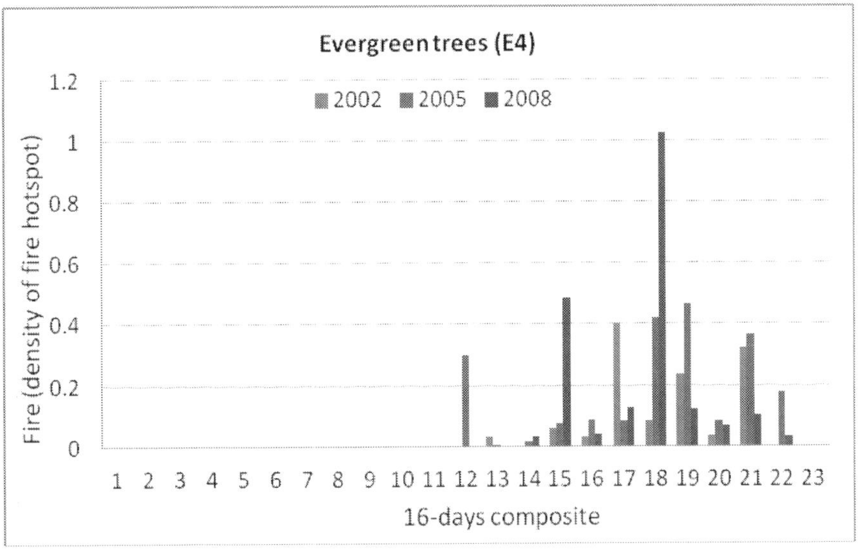

E1: herbaceous; E2: shrubs; E3: deciduous trees; and E4: evergreen trees.

Fig. 7. Annual seasonality of fire for the years 2002, 2005 and 2008, in places where we sampled the four vegetation types analyzed. Each year consists of 23 16-days composite periods. Fire is the daily mean value of the density of hotspot within a 10km radius for a 16-days composite period.

3.2.4 Relationship between vegetation (NDVI) and environmental variables (precipitation and fire)

The results showed in Table 1 indicate significant positive correlation between NDVI and precipitation in herbaceous, shrubs and deciduous trees, and negative correlation between NDVI and fire in the same three vegetation types. In the case of evergreen trees, the correlation between NDVI and precipitation is positive but not significant, and between NDVI and fire is negative, but also not significant. These results are corroborated in subsequent analysis.

	E1 Prec	E2 Prec	E3 Prec	E4 Prec	E1 Fire	E2 Fire	E3 Fire	E4 Fire
E1-NDVI	**0.60**	0.58	0.67	0.70	**-0.69**	-0.68	-0.66	-0.41
E2-NDVI	0.52	**0.51**	0.60	0.65	-0.68	**-0.72**	-0.70	-0.43
E3-NDVI	0.20	0.19	**0.31**	0.36	-0.61	-0.73	**-0.75**	-0.51
E4-NDVI	0.00	-0.01	0.16	**0.09**	-0.43	-0.57	-0.66	**-0.47**
E1-Fire	**-0.40**	-0.39	-0.43	-0.49	1.00	0.77	0.65	0.35
E2-Fire	-0.31	**-0.30**	-0.34	-0.42	0.77	1.00	0.65	0.41
E3-Fire	-0.25	-0.24	**-0.31**	-0.38	0.65	0.65	1.00	0.47
E4-Fire	-0.09	-0.08	-0.19	**-0.20**	0.35	0.41	0.47	1.00

Table 1. Correlation matrix of vegetation, rainfall and fire variables, highlighting the significant correlations between the following couple of variables: NDVI and rainfall, NDVI and fire, and rainfall and fire; which taking into account the four types of vegetation analyzed (E1: herbaceous; E2: shrubs; E3: deciduous trees; and E4: evergreen trees).

Figure 8 shows the result of the linear regression analysis between vegetation and precipitation for each vegetation type. Each line in this figure with a specific color shows the degree of fit between the points distributed for both variables by type of vegetation. Although this degree of fit between both variables is low, the results indicate that there is a gradient of fit between precipitation and vegetation, here named as precipitation gradient, which ranges from high to low coefficient of correlation (R2) following the sequence: herbaceous-E1 (high R2), shrubs-E2, deciduous trees-E3 and evergreen trees-E4 (low R2).

Thus, as the R2 value increases the influence of precipitation on vegetation increases, so herbaceous is more dependent on rainfall, in the annual cycle, than the other types of vegetation analyzed. That means, herbaceous are strongly dependent on rainfall in order to increase its vegetation cover. In the dry season, these kinds of species lose their leaves or even die.

At the opposite end of the precipitation gradient, where the evergreen trees-E4 are positioned, precipitation has weak influence on the vegetation cover, which means that in

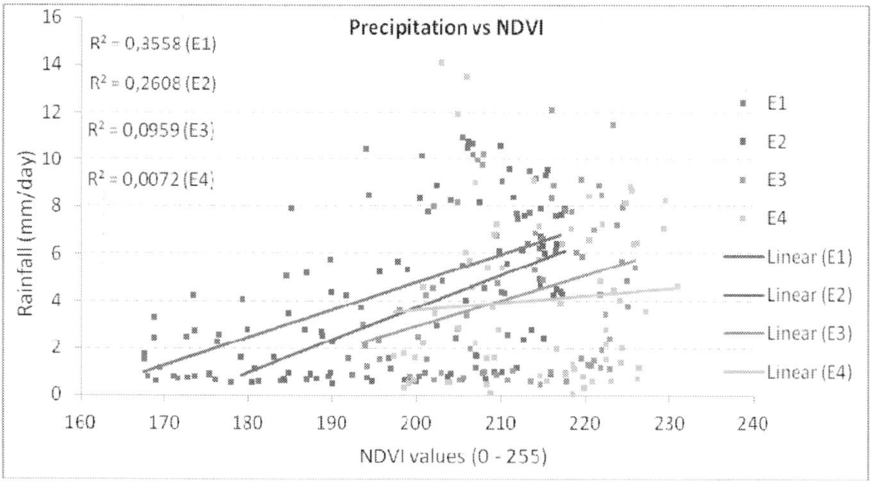

E1: herbaceous; E2: shrubs; E3: deciduous trees; and E4: evergreen trees.

Fig. 8. Regression of precipitation (independent variable) and NDVI (dependent) for each vegetation type analyzed. NDVI values range from 0 to 255 (x-axis). Precipitation in mm is the daily mean rainfall values for a 16-days composite period (y-axis). N = 69.

the dry season, evergreen trees are able to capture water from the vicinity of river courses, as occurs in gallery forests, or from deep soil, where the length of tree roots reach deep and moist soil layers, allowing these trees to replace their leaves throughout the year, which gives them their evergreen nature.

An analyses of variance (ANOVA) performed to evaluate these regressions is shown in table 2. Results indicate that, except for the regression between NDVI and Precipitation for the evergreen trees (E4) class, all regressions are significant at the 0.99 confidence level. Moreover, the relationships between NDVI and Fire were significant for all classes.

	NDVI x Prec			NDVI x Fire		
	R^2	F	p	R^2	F	p
E1	0.3558	37.01	<0.01	0.4795	61.72	<0.01
E2	0.2608	23.64	<0.01	0.5172	71.77	<0.01
E3	0.0959	7.11	<0.01	0.5575	84.40	<0.01
E4	0.0072	**0.49**	**0.49**	0.2218	19.10	<0.01

Table 2. Analysis of Variance (ANOVA) of linear regression NDVI x Precipitation (Prec) and NDVI x Fire. Bold values indicate the case where regression was not significant.

Figure 9 shows the result of the linear regression analysis between vegetation and fire for each vegetation type. Each line in this figure with a specific color shows the degree of fit between the points distributed for both variables by type of vegetation. The results indicate that there is a gradient of fit between fire and vegetation, here named as fire gradient, which ranges from high to low coefficient of correlation (R2) following the sequence: deciduous trees-E3 (high R2), shrubs-E2, herbaceous-E1, and evergreen trees-E4 (low R2).

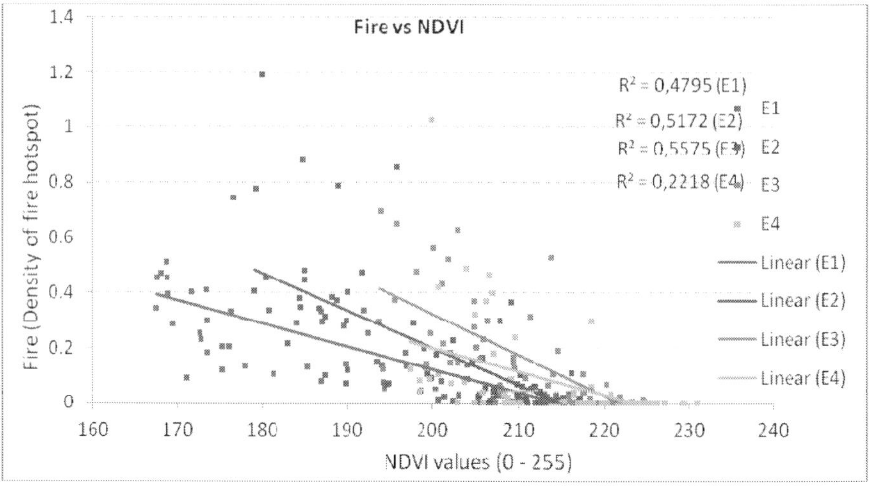

E1: herbaceous; E2: shrubs; E3: deciduous trees; and E4: evergreen trees.

Fig. 9. Regression of fire (independent variable) and NDVI (dependent) for each vegetation type analyzed. NDVI values range from 0 to 255 (x-axis). Fire is the daily mean value of the density of hotspot within a 10km radius for a 16-days composite period. N = 69.

The fire gradient identified above indicates that there is direct relationship between NDVI of the main vegetation types (herbaceous, shrubs and deciduous trees), which make up the Cerrado vegetation, and fire, indicating the role of fire in the maintenance of these vegetation types.

Fire occurs with greater intensity at the end of dry season. First of all, fire consumes part of the burk and organic matter of the plant, after the first rains, in the beginning of the rainy season, these partially burned plant sprouts new shoots with greater vigor.

At the opposite end of the fire gradient, where the evergreen trees-E4 are positioned, the fire occurs in lower proportion in these trees, however, unlike what happens with other types of vegetation, the effect of fire is pernicious, it can damage or even eliminate some species in this vegetation type according to the intensity level.

The multiple regression analysis indicates that there is a direct relationship between precipitation and fire, and vegetation index (NDVI) in the four vegetation types of the savanna vegetation. The multiple coefficients of determinations (R^2) show that the environmental variables as a whole (precipitation an fire) follow a gradient of high influence in vegetation types with low vegetation cover (herbaceous R^2=0.67 and shrubs R^2=0.65) to low influence in that with high vegetation cover (deciduous trees R^2= 0.55 and evergreen trees R^2=0.27). Results from the ANOVA of the multiple regression presented in Table 3 indicate that, when the analysis is performed considering both independent variables, the multiple regression gives statistically significant parameters, for all classes of vegetation. However, an univariate test of significance performed for each independent variable show that precipitation alone is not significant correlated to the vegetation index for both tree classes (E3 and E4).

	Whole model R			Univariate test of significance			
	R^2	F	p	F_prec	F_fire	p_prec	p_fire
E1	0.6001	49.52	<0.01	19.90	40.31	<0.01	<0.01
E2	0.6141	52.52	<0.01	16.59	60.43	<0.01	<0.01
E3	0.5648	42.83	<0.01	**1.12**	71.11	**0.29**	<0.01
E4	0.2220	9.42	<0.01	**0.01**	18.22	**0.92**	<0.01

Table 3. Analysis of Variance (ANOVA) of the multiple regression between NDVI (dependent variable) and precipitation and fire (independent variables). Bold values indicate the cases where regression was not significant.

4. Conclusions

The response of vegetation NDVI is more related to the variation of fire than to variations in precipitation in Cerrado region. Vegetation NDVI responds to variation of precipitation with a time lag ranging from 16 to 48 days, while vegetation NDVI responds to variation of fire with a time lag ranging from 0 to 48 days.

The relationship between vegetation types, derived from NDVI, and precipitation, derived from TRMM, shows a gradient of positive correlations in vegetation types with low vegetation cover, herbaceous (r= 0.60) and shrubs (r= 0.51), to very little or none with high vegetation cover, deciduous trees (r= 0.31) and evergreen trees (r= 0.09). On the other hand, the relationship between vegetation and fire hotspot shows a gradient of negative correlation, which is stronger in herbaceous (r= 0.72), shrubs (r= 0.74) and deciduous trees (r= -0.73) than in evergreen trees (r= -0.52).

Our analyses show that vegetation cover increases are related to increases in precipitation and decreased in density of fire hotspots. We also found high density of fire hotspot in the dry season in deciduous trees, shrubs and herbaceous which suggesting the high removal of CO_2 (greenhouse gas) of the land cover to the atmosphere somehow influencing the dynamic equilibrium of this (atmosphere) in the region of the Brazilian tropical savanna.

5. References

Allan, G.; Johnson, A.; Cridland, S. & Fitzgerald, N. (2003). Application of NDVI for predicting fuel curing at landscape scales in northern Australia: can remotely sensed data help schedule fire management operations? *International Journal of Wildland Fire* 12, 299–308.

Aquino, F.G.; Miranda, G.H.B. (2008). Consequencias ambientais da fragmentação de habitats no cerrado. *In* Sano, S.M.; Almeida, S.P.; Ribeiro, J.F. (Eds.). *Cerrado: ecologia e flora*. Brasília: Editora Embrapa, p.385-398.

Barnes, S. L. (1973). Mesoscale objective analysis using weighted time-series observations, NOAA Tech. Memo. ERL NSSL-62 National Severe Storms Laboratory, Norman, OK 73069, 60 pp. [NTIS COM-73-10781.], 1973. 2359

Bartlett, H.H. (1955, 1957, 1961). Fire in relation to primitive agriculture and grazing in the tropics: annotated bibliography, Vol. 1-3. Mimeo. Publ. Univ. Michigan Bot. Gardens, Ann Arbor, USA.

Borlaug, N.E. (2002). Feeding a world of 10 billion people: the miracle ahead. In: R. Bailey (ed.). *Global warming and other eco-myths*. pp. 29-60. Competitive Enterprise Institute, Roseville, EUA.

Bowman, D.M.J.S. & Murphy, B.P. (2010). Fire and Biodiversity. In Sodhi, N.S.; Ehrlich, PR., eds 2010. *Conservation Biology for all* , pp. 163-180.

Bowman, D.M.J.S., Zhang, Y., Walsh, A. & Williams, R.J. (2003). Experimental comparison of four remote sensing techniques to map tropical savanna fire-scars using Landsat-TM imagery. *International Journal of Wildland Fire* 12, 341–48.

Brandis, K. & Jacobson, C. (2003). Estimation of vegetative fuel loads using LandsatTM imagery in New South Wales, Australia. *International Journal of Wildland Fire,* Vol. 12, 185–94.

Castro, L.H.R.; Moreira, A.M. & Assad, E.D. (1994). Definição e regionalização dos padrões pluviométricos dos Cerrados brasileiros. In: ASSAD, E.D. *Chuvas nos Cerrados: análise e espacialização*. Brasília, Embrapa-CPAC/Embrapa-SPI, 1994. 423p.

Coutinho, L.M. (1978). O conceito de cerrado. *Revista Brasileira de Botânica* 1(1):17-23.

Coutinho, L. M. (2000). O bioma do cerrado, in: Klein, A. L. Eugen Warming e o cerrado brasileiro: um século depois. UNESP, 156 p.

Diaz-Delgado, R.; Lloret, F. & Pons, X. (2004). Statistical analysis of fire frequency models for Catalonia (NE Spain, 1975–1998) based on fire scar maps from Landsat MSS data. *International Journal of Wildland Fire* 13, 89–99.

Goldammer, J.G. (1988). Rural land use and fires in the tropics. *Agroforestry Systems*, Vol.6: 235-253

Justice, C.O.; Smith, R.; Gill, A.M. & Csiszar, I. (2003). A review of current space-based fire monitoring in Australia and the GOFC/GOLD program for international coordination. *International Journal of Wildland* Fire, 12, 247–258.

Klink, C.A. & Machado, R.B. (2005). A conservação do cerrado brasileiro. *Megadiversidade,* Vol.1: 147-155.

Koch S. E.; desJardins, M. & Kocin, P. J. (1983). An interactive Barnes objective map analysis scheme for use with satellite and conventional data. *J. Climate Appl. Meteor.*, Vol.22, 1487-1503.

Law, R. M.; Kowlaczyk, E. A. & Wang, Y.P. (2006). Using atmospheric CO2 data to assess a simplified carbon-climate simulation for the 20th century. *Tellus*, Vol.58B, 427-437.

Miller, J.D.; Danzer, S.R.; Watts, J.M.; Stone, S. & Yool, S.R. (2003). Cluster analysis of structural stage classes to map wildland fuels in a Madrean ecosystem. *Journal of Environmental Management*, Vol. 68, 239-52.

Myers, N. et al. 2000. Biodiversity hotspots for conservation priorities. *Nature* 403: 853-858.

Nimer, E. (1989). *Climatologia do Brasil*. Secretaria de Planejamento e Coordenacao da Presidencia da Republica e IBGE, Rio de Janeiro, 421p.

Nimer, E. 1977. Clima. Pp 47-48 in: IBGE. *Geografia do Brasil: região Nordeste*. Rio de Janeiro.

Parker, A. (2000). Wet-Dry Tropical Climate (Aw), In: Climate Savanna, 01.06.2000, Available from
http://www.blueplanetbiomes.org/savanna_climate_page.htm

PROBIO-MMA (Projeto de Conservação e Utilização Sustentável da Diversidade Biológica Brasileira) do Ministério do Meio Ambiente. (2007). *Mapeamento de Cobertura Vegetal do Bioma Cerrado*. Technical Report. Brasilia, Brasil.

Rao, V.B. & Hada, K. (1990). Characteristics of rainfall over Brazil: annual variations and connections with the southern oscillations. *Theoretical and Applied Climatology*, Vol.2, p.81-91.

Ratter, J.A.; Ribeiro, J.F. & Bridgewater, S. (1997). The brazilian Cerrado vegetation and threats to its biodiversity. *Annals of Botany*, Vol.*80*: 223-230.

Reed, B.C.; Brown, J.F.; Vanderzee, D.; Loveland, T.R.; Merchant, J.W. & Ohlen, D.O. (1994). Measuring phenological variability from satellite imagery. *J. Veg. Sci.* Vol.5, 703-714.

Ritter, M.E. (2006). *The Physical Environment: an Introduction to Physical Geography*. Available from
http://www.uwsp.edu/geo/faculty/ritter/geog101/textbook/ title_page. html

Rollins, M.G.; Keane, R.E. & Parsons, R.A. (2004). Mapping fuels and fire regimes using remote sensing, ecosystem simulation, and gradient modeling. *Ecological Applications*, Vol.14, 75-95.

Roy, P.S. (2004). Forest fire and degradation assessment using satellite remote sensing and geographic information system. In: Sivakumar, M.V.K.; Roy, P.S.; Harmesen, K.; Saha, S.K., eds, 2010. *Satellite Remote Sensing and GIS Applications in Agricultural Meteorology* pp. 361-400.

Satyamurty, P.; Nobre, C. A. & Silva Dias, P. L. (1998). South America. *Meteorological Monographs*, Vol.27, n. 49 (Southern Hemisphere Meteorology, cap. 3C), p. 119-139.

Silverman, B.W. (1986). Density estimation for statistics and data analysis. London; New York: Chapman & Hall, 1986. 175 p.

Schwartz, M.D. (1994). Monitoring global change with phenology: the case of the spring green wave. *Int. J. Biometeorol.* Vol.38, 18-22.

Vila, D.A.; de Goncalves, L.G.G.; Toll, D.L. & Rozante, J.R. (2009). Statistical Evaluation of Combined Daily Gauge Observations and Rainfall Satellite Estimates over Continental South America. *J. Hydrometeor.* Vol.10, 533–543.

4

Land Cover Change Detection in Southern Brazil Through Orbital Imagery Classification Methods

José Maria Filippini Alba, Victor Faria Schroder
and Mauro Ricardo R. Nóbrega
Brazilian Agricultural Reasearch Corporation (Embrapa)
Brazil

1. Introduction

Remote sensing has been considered a promising technology as support for agriculture since its beginnings, due to its contribution for a climatic perspective or for understanding of processes related to land. However, significant applications occurred only in the late twentieth century, as result of the creation of best orbital systems, with higher spatial resolution, more bands and stereoscopic capture. Several orbital platforms, as AQUA/TERRA, Quickbird and Ikonos are examples in that sense (Moreira, 2005; Embrapa, 2009).

Engineering innovations, new sensors and methods of digital image processing must be performed simultaneously so that the advances in remote sensing will be achieved. Anyway, the incorporation of orbital images on geographic information systems (GIS) and their post-processing appear as significant application since a daily life perspective, specially when classification methods are involved, because of their relation to land use, land cover and easy interpretation.

This chapter considers classification methods applied on orbital imagery in Southern Brazil, in the coastal plain of Rio Grande do Sul state (Fig. 1), where a sequence of lagoons and lakes of different sizes occurs in the context of subtropical to temperate climate with cold winters and hot summers, being organized according to the following four sections:

- About classification methods.
- Evaluation of rice planting area in the vicinity of Caiuba lagoon (1981 - 2009).
- Analysis of land cover evolution in the municipality of Montenegro (1993 - 2008).
- Comparison and evaluation of errors.

All the exposed data are related to research projects of the Embrapa Temperate Climate Research Center, Pelotas, Rio Grande do Sul state, one of the 45 research centers of the Brazilian Agricultural Research Corporation (Embrapa) spread on the national territory.

2. About classification methods

Classification methods were created in the statistical context, when a collection of objects or samples could be characterized and separated in different classes (Davis, 1986). The method

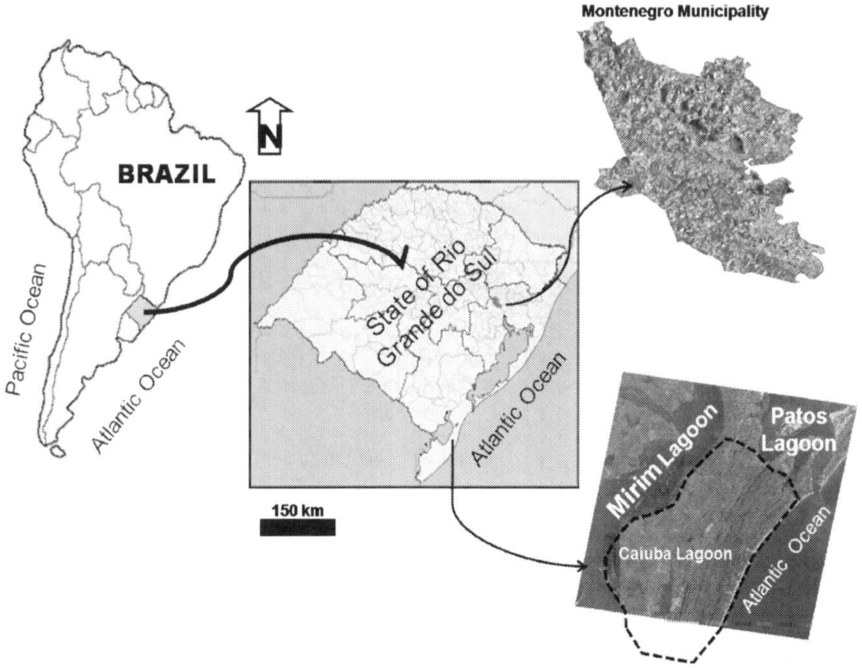

Fig. 1. Location of the areas of study in the context of South America, Brazil and the State of Rio Grande do Sul. Montenegro municipality as gray levels composite of the bands 1, 4 and 5 TM/Landsat 5 (Miranda, 2002, 2005) and Caiuba lagoon in gray levels composite of the bands 3, 2 and 1 ASTER (NASA, 2004) on september 26th, 2000.

was extended for processing of digital images considering the pixels as objects to be classified (Crosta, 1993, Lillesand & Kiefer, 1994, Jensen, 1996). The application of the classification methods on satelite imagery is affected by two main factors:

i. Intervention of user.
ii. Criteria of definition for the groups.

The "supervised classification" is included in topic (i), when the user defines the groups through digitalization of uniform spectral answer. Statistics are calculated for each group, so the classification is performed for all the image (pixel by pixel). The unsupervised classification eliminates the intervention of the user; then, the software defines the groups by means of scattergrams "band versus band", where isolines related to distribution density of the pixels are analysed (Crosta, 1993).

Different options for case (ii) are posible, by instance, criteria can use the standard deviation for the parallelepiped method or a minimum distance when an eliptical form is defined for each group or a combination of the later with statistical probability, that is, the maximum likelihood method. Jensen (1996) presented other criteria of classification, as Isodata method and the Fuzzy method.

Spectrometric methods measure the response of target materials in the laboratory or field. Then the spectral patterns are simulated for a specific sensor through a specialized software, so that a sequence of orbital images is classified according to the pattern generated by that software (Lillesand & Kiefer, 1994; Pontara, 1998; Moreira, 2005).

Classification of remote sensing images appear as useful tool in terms of land use, whether in local scale or in regional scale. Filippini-Alba and Siqueira (1999) classified land use in the municipality of Pelotas, Rio Grande do Sul state, Brazil, according to nine classes: agriculture, clay soils, forestry, natural forest, pastures, soil without vegetal cover, urban, water and wetlands. Natural forest and pastures occupied 23% and 30% of the territory respectively, with intense interference between the classes "urban" and "soil without vegetal cover". Similar classes were considered by Bolfe et al. (2009) for land use in Rio Grande do Sul state, but with different results. Agriculture and pastures occurred 32% and 50% of the territory respectively with only 3% for natural forest. The differences between both studies are easy explained in term of scale, because in the two occasions Landsat images were used and a municipality was considered at the former and a state at the latter, with territory difference of 1 to 155 times in size.

Lu et al. (2004) discriminated seven categories of change detection techniques: (i) algebra; (ii) transformation; (iii) classification; (iv) advanced models; (v) Strategies with geographic information systems (GIS); (vi) visual analysis and (vii) miscellanea. The classification methods are detached, with six different modalities. One of them, the "Post-classification comparison", is predominantly used in this chapter. That is, multi-temporal images are classified separately into thematic maps, then the classified images are compared pixel by pixel. The "Post-classification comparison" minimizes the atmospheric impacts, the environmental differences among multi-temporal imagery, as well as differences related to the sensor kind, providing a complete matrix of change information. However, some disadvantages can be appointed, because a great amount of time and expertise is required and, by other side, the final accuracy depends on the quality of the classified image due to the weather condition on that date.

Guild et al. (2004) quantified the areas of deforestation in the Amazonian forest, state of Rondonia, Brazil. The tasselled cup transformation (Crist & Kauth, 1986) was applied with the Landsat imagery from the years 1984, 1986 and 1992. The variables brightness, greenness and wetness were evaluated for each year, then, a file integrated the nine levels of information (three variables by three years). These data were processed through principal components and classification methods with overall accuracy of 79.3, 68.4% and 71.4%, for tasselled cap land cover change classification, tasselled cap with principal components land cover change classification and tasselled cap image differencing, respectivelly. Final classes were a combination between land cover and time, so change detection was quantified.

The two applications present in this chapter consider the Supervised classification method with maximum likelihood as criteria for definition of the classes. The proximity of the study areas and knowledge of the territory justify this option to take advantage of available information. Unsupervised classification is a fast proccess, good for unknown or outlying areas, when truth of field is unavailable and most time-consuming after processing, due to the need of class identification. Maximum likelihood criteira is restricted by software and time-consuming but it represents a improvement in relation simple criteria as the parallelepiped or the minimum distance.

Acording to Lu et al. (2004), methods (iii) and (v) were considered in this chapter. Classification (iii) was applied in both conditions, Caiuba lagoon and Montenegro municipality. The extraction of the poligon corresponding to the "potential area for agriculture" in the vicinity of Caiuba lagoon represents a tipical strategy of GIS (v). Softwares of digital images processing and GIS are very similar. Both can execute multilayer processing, including raster/vector files and logic/mathematical algorithms, but digital images processing is more specific for raster format and GIS for vector format.

3. Evaluation of rice planting area in the vicinity of Caiuba lagoon (1981-2009)

The Caiuba lagoon is part of the litoral lacunar complex of Rio Grande do Sul state, southern Brazil and it extends by 3300 hectares, in the municipality of Rio Grande, 15 kilometers to north extreme from the Taim Ecological Reserve and 45 kilometers to south from Patos lagoon (Fig. 1). This significant source of water is used mainly for irrigation of rice, specially when the Merin lagoon is further. Accordingly, the Foundation for Research Support in State of Rio Grande do Sul (FAPERGS) funded a research project leadered by the Federal University of Rio Grande (FURG) attempting to study the sustainability of the productive system, as well as the effects on local biodiversity. The Embrapa Temperate Climate Research Center collaborated to the evaluation of the agricultural area in the period 1973 to 2009 by satellite images. Imagery of Landsat satellite of different years was considered for similar times (Table 1), for the scenes corresponding to orbit 237 points 82 and 83 of the worldwide reference system 1 (MSS sensor) and for the scenes corresponding to orbit 221 points 82 and 83 of the worldwide reference system 2 (TM sensor). Thus, the atmospheric conditions were more or less equivalent, deriving in comparable image quality. Each image was evaluated for the various land uses and the areas occupied for the different classes were calculated in order to study the historical evolution of the process during the above period.

The first satellite of Landsat series was launched in 1972 with the multispectral scanner (MSS), with four bands in visible - near infrared and one in thermal infrared and

Sensor	Date	Range of wavelengh and IFOV
MSS	Sep. 6th, 1973	500 - 600nm, 600 - 700nm and 700 - 800nm, IFOV = 79m
MSS	Mar. 13th, 1981	
TM	Jan. 22nd, 1991	630 - 690nm, 760 - 900nm and 1550 - 1750nm, IFOV = 30m
TM	Dec. 21st, 1996	
TM	Dec. 19th, 2001	
TM	Jan. 20th, 2002	
TM	Jan. 28th, 2005	
TM	Jan. 2nd, 2007	
TM	Jan. 7th, 2009	

Table 1. Description of basical parameters of the images of Landsat series used for evaluation of the planting area of rice in the vicinity of tha Caiuba lagoon.
IFOV = instantaneous field of view. Source: INPE, 2010b.

instantaneous field of vision (IFOV) of 79 meters and 240 meters respectively. Improvements of the system included more bands (short-medium infrared) and reduction of IFOV to 30 meters and 120 meters respectively, for the thematic mapper (TM) in 1982 (Jensen, 1996). A panchromatic band was developed for the Landsat 7 satellite, with the TM plus sensor, but, the series reached to the end. The Landsat 5 satellite was an engineering success, the platform was launched in 1984 and is still on orbit. Anyway, the Landsat series represents the greatest collection of terrestrial images for environmental applications, specially, since a historical point of view.

Composites of three bands were used, with green band (500 - 600 nm), red band (600 - 700 nm) and near infarred band (700 - 800 nm) for the MSS sensor and the red band (630 - 690 nm), the nearinfarred band (760 - 900 nm) and the shortmedium infrarred band (1550 - 1750 nm) for TM sensor. These games of bands are not equivalent, then similar patterns of colour were adjusted by visual observation.

Digital imagery was registered for the Universal Transverse of Mercator projection(UTM), zone 22 South with the datum WGS84, after that, a mosaic of pairs of scenes was composed, by instance, scene 237/82 and 237/83 for MSS sensor. So the mosaic was cutted evolving the study area and a file with the mentioned three bands was created for each date. Initially, data were processed by the supervised classification according to the maximum likelyhood criteria. Eigth poligons of homogeneous features were digitalized with the software ER-Mapper (1995), deriving in the test areas, then each pixel of the corresponding image was classified according to its similarity with the parameters of each test area (beach/dunes, forestry, rice crops, pastures, sandy fields, soil without vegetal cover, water and wetlands). A second strategy was developed to improve results, so the "potential area for agriculture", that is rice crops, pastures and soil without vegetal cover, was isolated and classified by similar way.

Results of the preliminar process of classification considered a rectangle of 30 km wide and 65 km long for the images of 2001, 2002 and 2005 (Fig. 2). The "potential zone for agriculture" is represented by a "central zone" in direction south - north to the East of Merin Lagoon, where agricultural areas are discriminated. A confusion between rice crop class and wetlands class is observed in the west - north sector of the study area. Sandy fields are long structures related to old movements of the sea (Atlantic Ocean), where a low charge of livestock is a common use and forestry is developed eventually, as observed in the images. The area occupied by water bodies was almost constant, that is 19 - 21% (Table 2), but, the wetlands were reduced in area in 2005, a year of drought probably, then, there was an increment in the area occupied by the class "Soil without vegetal cover" and a reduction of the area occupied by the class "Pastures".

When the "potential zone for agriculture" was isolated, the precison of evaluation of the area occupied by pastures, rice crops and soil without vegetal cover (SWVC) was improved. The kind of sensor, the date of the image and the meteorological conditions induced diferences among the imagery of different dates (Fig. 3). The images of 1973 and 1981 present a different characteristics due to captation with the MSS sensor. The first image corresponds to september, when the culture had not been implanted yet. Some agricultural areas showed different pattern in 2001 and 2002 (same harvest) related to waterlogged soils, probably, due to intense rain in that time. A differencial answer of the vegetation in the agricultural areas was observed since 2005, what suggests a evolution of the vegetal development of the rice

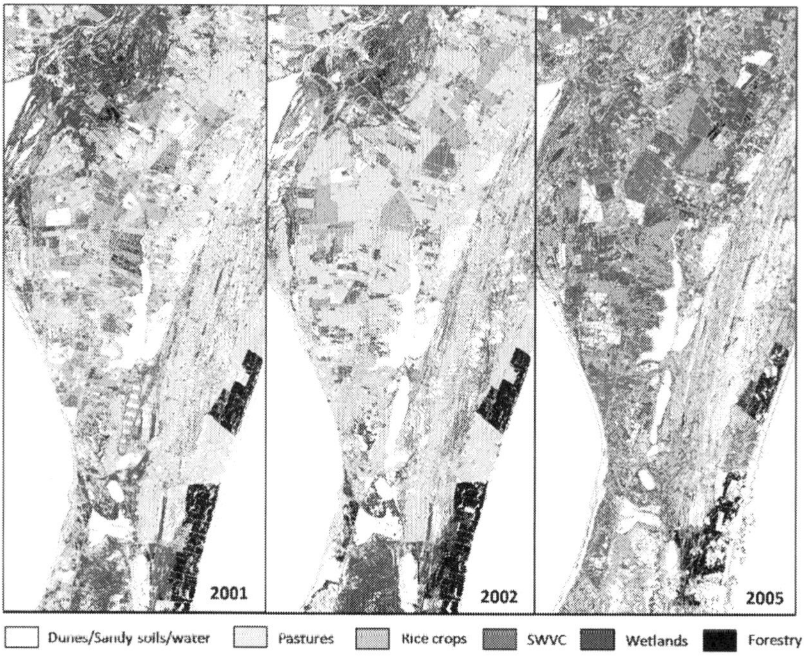

Fig. 2. Preliminary classification in the Caiuba region.

Image date	Dec. 19th, 2001	Jan. 20th, 2002	Jan. 28th, 2005
Water, %	19.2	21.3	20.0
Wetland, %	17.8	18.5	11.3
Pastures, %	20.2	12.4	8.9
Rice crops, %	10.2	9.1	10.4
Sandy fields, %	17.7	13.6	15.1
Beach/dunes, %	6.7	6.3	10.5
Florestry, %	4.3	3.6	4.6
Clouds, %		1.5	
SWVC,%	3.9	13.8	19.3
Total area, ha	202,777	204,088	208,907

Table 2. Preliminary areas of land cover calculated by classification methods in the vicinity of Caiubá region through Landsat-TM images in the period 2001 - 2005. SWVC = Soil without vegetal cover.

varieties or, perhaps, the introduction of a new crop. All the images show a intense rotation among pastures, rice and fallow lands, what lets a reduction of inputs, rest of the soil and improvement of productivity.

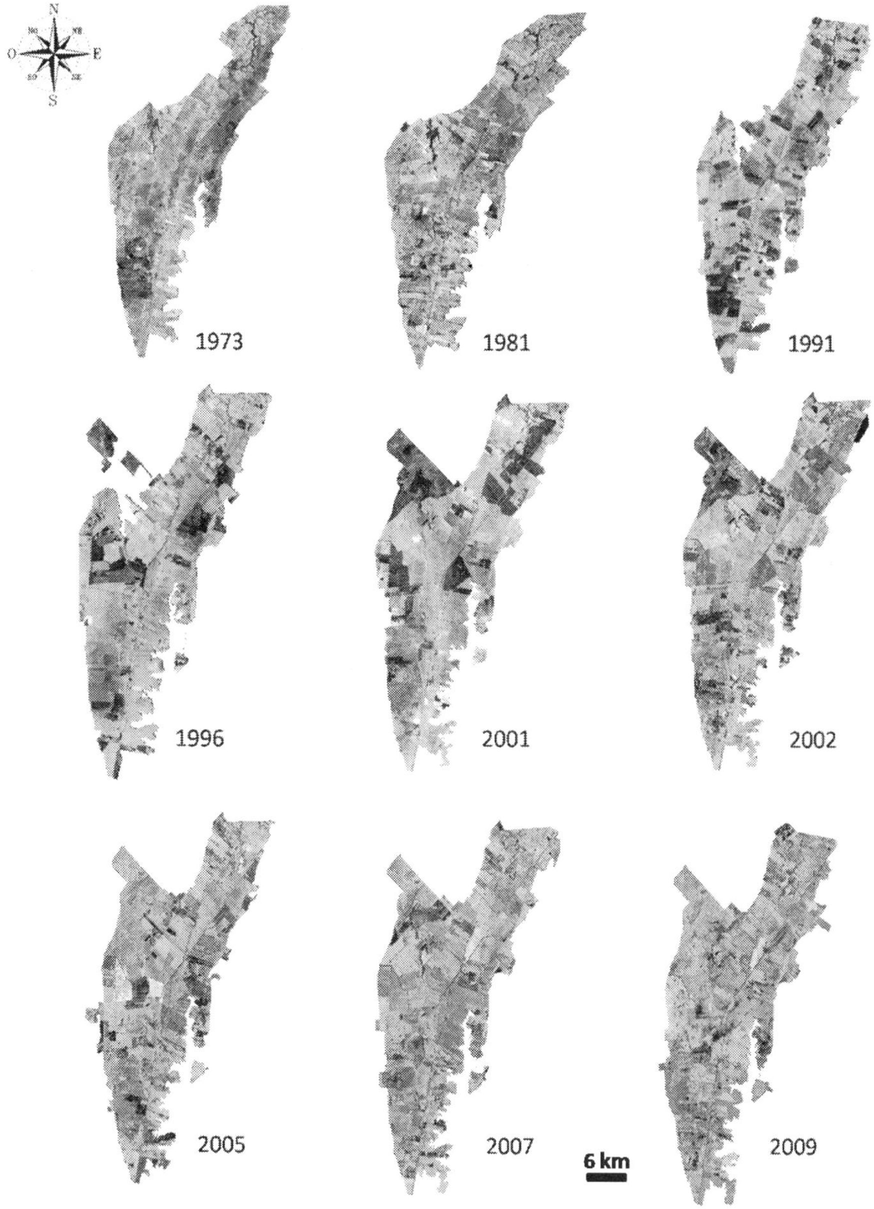

Fig. 3. Images Landsat corresponding to the "Potential zone for agriculture" for different dates.

The poligonal area was classified according to five classes: (1) Undefined; (2) Pastures; (3) Rice crops; (4) Soil without vegetal cover; (5) Water. The class "Undefined" represents rice crops or pastures depending on the year, thus it was incorporated to class "Pastures" in 1973, 1981 e 2001 and to class "Rice crops" in 1991, 1996, 2002, 2005, 2007 e 2009, accordingly the interpretation of the images. So, the classes "Pastures", "Rice crops", "SWVC" and "Water" were evaluated for occupied area (Fig. 4; Table 3).

The occurrence of water is almost insignificant inside the "potencial zone for agriculture", because the irrigation is performed through the water of the lagoons Caiuba and Mirim. The area occupied by the class "Rice crops" seems to depend on the vegetal developping, with restricted values when months previous to January are evolved. This fact was checked with the images of 2001 and 2002, corresponding to the same harvest, Dezember and January respectively.

By this reason, only the data corresponding to the months of january and march, when the vegetal developping of rice is reached, were consider in the graphic of "occupied area" as a function of time (Fig. 5).

Year	Water	Pastures	Rice crops	SWVC	Total
06/09/1973	173	19057	12856	13565	45652
13/03/1981	85	28593	13915	6592	49185
22/01/1991	52	17110	18751	15886	51798
21/12/1996	825	22534	12042	22534	57935
19/12/2001	56	30523	14404	8516	53498
20/01/2002	120	17579	20090	16299	54087
28/01/2005	49	13144	21963	20246	55402
02/01/2007	57	25062	21054	9467	55640
07/01/2009	80	5302	21029	31124	57535

Table 3. Area evaluation of land cover classes for the "potential zone for agriculture". The class "Undefined" was incorporated to the class "Pastures" or the class "Rice crops" according to the year. SWVC = Soils without vegetal cover. Data in hectares.

The area of the "Potential zone for agriculture" was delimited by digitalization, but a soft and constant increment is evident during the period 1981 to 2009. By other side, the area occupied by the class "Rice crops" was evaluated by classification methods; after a period of increment, the class reached a maximum in 2005 with 22 thousand hectares, then there was a stabilization in 2007 - 2009 with about 21 thousand hectares. The classes "Pastures" and "SWVC" showed oscillation in complementary way, because the sum of both classes was almost constant. As classes "Potential zone for agriculture" and "Rice crops" presented linear behavior in the graphic Area against time, thus, linear regression models were adjusted (Table 4).

The parameter R^2 is the correlation coefficient between the real variable and the adjusted variable by the model. So, a value near zero indicates bad adjust of the model and a value near one indicates a good adjust of the model. The parameter A indicates the annual growing rate for of the occupation area of the respective class. The area occupied by the

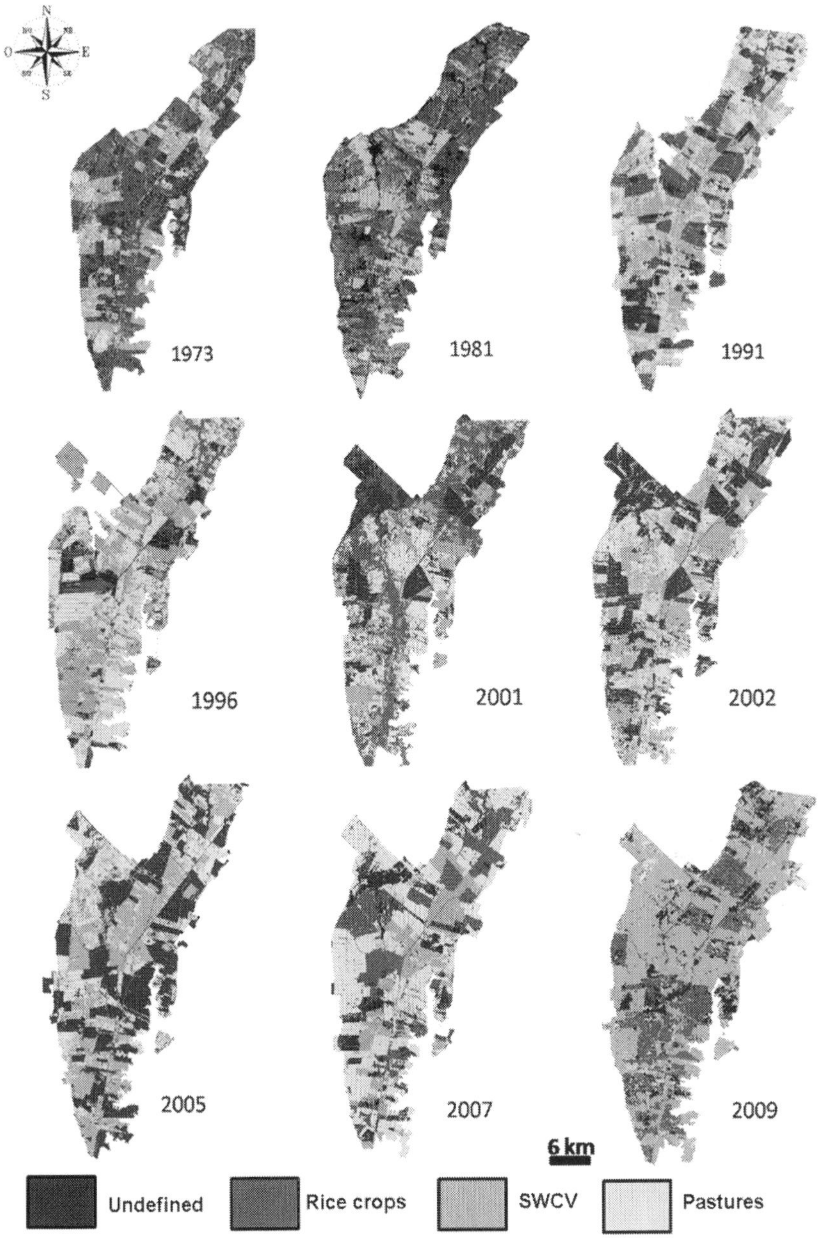

Fig. 4. Images Landsat post-classified corresponding to the "Potential zone for agriculture" for different dates.

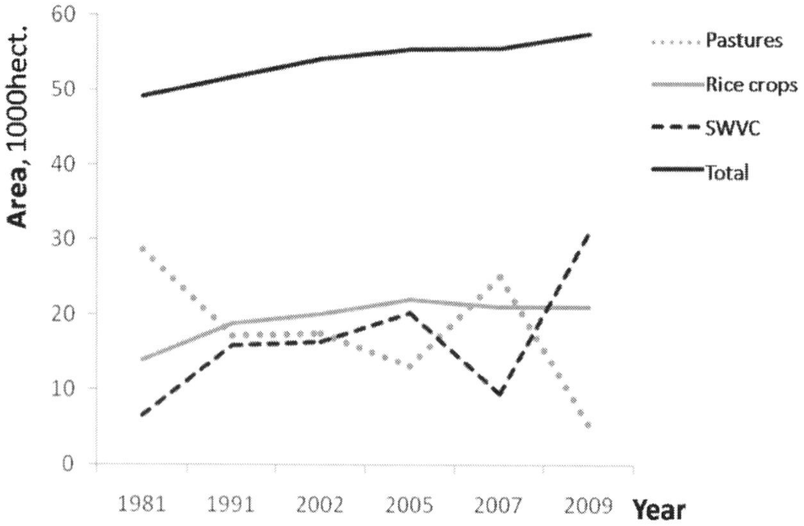

Fig. 5. Areas of land cover as function of the year in the "Potential area for agriculture" (Total), Caiuba region. SWVC = Soils without vegetal cover.

Class of land cover	A	B	R^2	Period
Potential zone for agriculture	271	486996	0.97	1981 - 2009
Rice crops	302	584736	0.93	1981 - 2005

Table 4. Parameters of the linear regression models for the area of classes "Potential zone for agriculture" and "Rice crops" as a function of time (Area = A*year - B, in hectares).

"Potential area for agriculture" grew with a rate of 271 hectares by year, little inferior than the growing rate for the area occupied by the class "Rice crops", that is 302 hectares by year. Parameter B is the value of area in the year zero without real significance in this case.

Data of the municipality of Pelotas (Filippini Alba & Siqueira, 1999) and data for the state of Rio Grande do Sul (Bolfe et al., 2009) were compared to data presented here, after legend conversion. The correlation coefficient of the data discussed here was 0.54 with data of the first paper and 0.77 with data of the second one. Some classes showed significant differences, by instances Bolfe et al. (2009) evaluated 50% of area occupied by "pastures" in the state, but the value was about 30% for the other works. The area occupied by water was 19-20% in the Caiuba region, due to the occurrence of the lagoons. This value was 1% in the municipality context and 3% in the regional one.

4. Analisys of land cover evolution in the municipality of Montenegro, Estado do Rio Grande do Sul (1993-2008)

The municipality of Montenegro is located 55 kilometers South from Porto Alegre (state capital), with a territorial area of 420 square kilometers and population about 59,557 inhabitants. Thirty-three municipalities, including Montenegro, integrate the vegetal carbon productive pole. The production of black acacia for the manufacture of tannin is an important activity for the economy of the municipality since 1948, when the first factory of tannin derived from the bark of acacia was installed (TANAC, 2010). Recently, the fruit production is becoming increasingly important in the context of local economy. The intense forest exploitation, the occurrence of new uses of land and a moderate urban occupation oriented the choice of the municipality of Montenegro for this research, focusing on the detection of temporal changes in the territorial organization, during the period 1993 to 2008, in the context of the project "Development and evaluation of products and co-products of the vegetal carbon productive chain in the State of Rio Grande do Sul, aiming for sustentability", with coordination of Embrapa Temperate Climate Reasearch Center.

The topography of the municipality is complex when compared to the previous case, while in the southeastern region occurs a flat terrain changing for slightly wavy; in the north sector occur a basalt plateau with a rugged relief.

Imagery of the Landsat 5 satellite were used, corresponding to the scene of orbit 221, points 80 and 81 for WRS-2 (INPE, 2010b), for three different dates: September 8th, 1993; August 8th, 1999 and October 3th, 2008. The initial data processing was performed with the software Marlin (INPE, 2010a), after that, the software ER-Mapper (1995) was used for classification according to isoclass likelyhood criteria. The images were registered with known ground control points, considering terrestrial features of easy identification, so that, the coordinates systems were uniformized and small errors eliminated. The projection used was the Universal Transector of Mercaptor (UTM), region 22 South, datum WGS 84.

Eigth classes were defined by the supervised classification process according to maximum likelihood criteria. The classes "Annual crops", "Perennial crops" and "Pastures/SWVC" were mapped together in gray tones (Fig. 6). The annual crops reached a maximum area of production in 1999 (Table 5) with poor production in previous and posterior times. By another hand, the perennial crops reached a maximum in 2008, after a significant increment in the previous years, as consequence of an important citrus production. Pastures and SWVC were mapped together due to the dinamic process of changes evolving both classes. A little reduction of the area occupied for both classes was observed.

Class of land cover	1993		1999		2008	
Annual crops	13.9	5835	21.2	8884	12.1	5084
Forestry	19.2	8076	5.5	2299	18.3	7686
Native forest	18.9	7948	23.0	9675	21.3	8957
Pastures/SWVC	27.0	11340	26.8	11246	23.2	9733
Perennial crops	8.9	3737	18.2	7655	19.6	8226
Unevaluated	3.3	1370	1.9	810	0.4	173
Urban	6.5	2729	2.1	878	3.3	1377
Water	2.3	968	1.3	554	1.8	767
Units	%	hectares	%	hectares	%	hectares

Table 5. Areas calculated with TM/Landsat 5 imagens for the period 1993 - 2008 through classification methods for Montenegro municipality (Schroder & Filippini-Alba, 2010a).

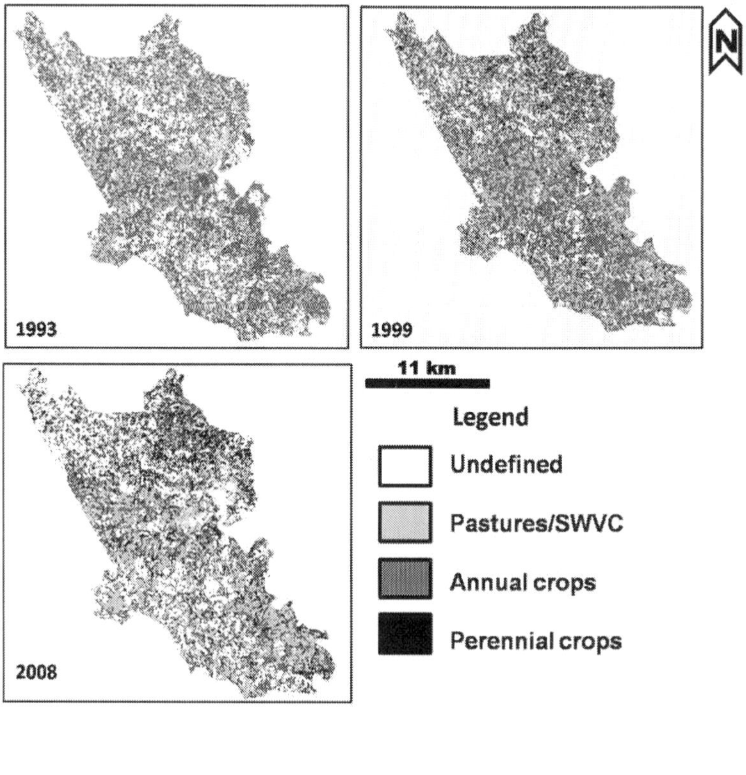

Fig. 6. Evolution of land cover related to agriculture and pastures/SWVC in the municipality of Montenegro based on Landsat 5 imagery (1993 - 2008).

The class "Forestry" showed a minimum of planting area in 1999, what is evident in the map of spatial distribution (Fig. 7.), but the class "Native forest" showed a maximum that year (Fig. 8). The class "Urban" includes other features besides the urban regions, by instance outcrops, which explains its high value in 1993. The density of the central spot in the image of 2008 suggest a real increment of urban population that year.

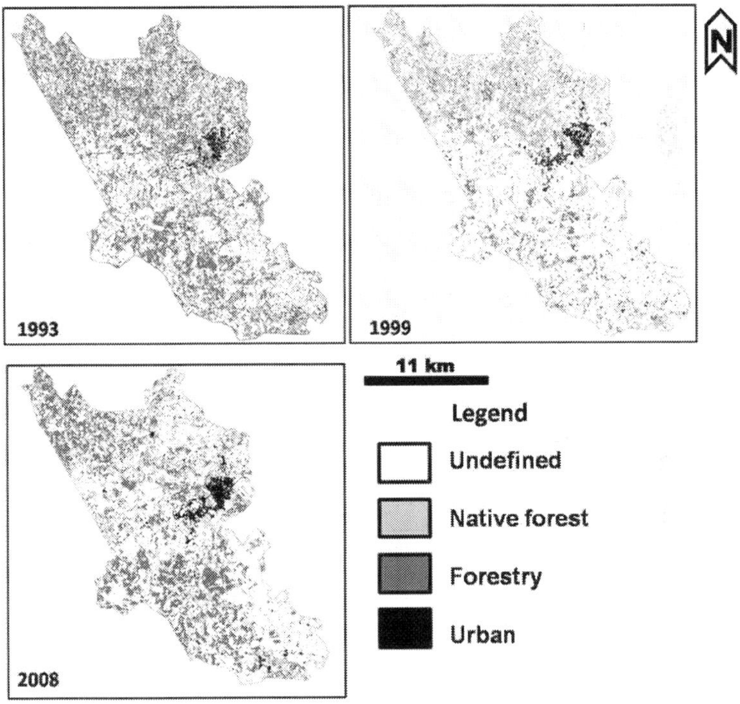

Fig. 7. Evolution of land cover related to the classes "Native forest", "Forestry" and "Urban" in the municipality of Montenegro based on Landsat 5 imagery (1993 - 2008).

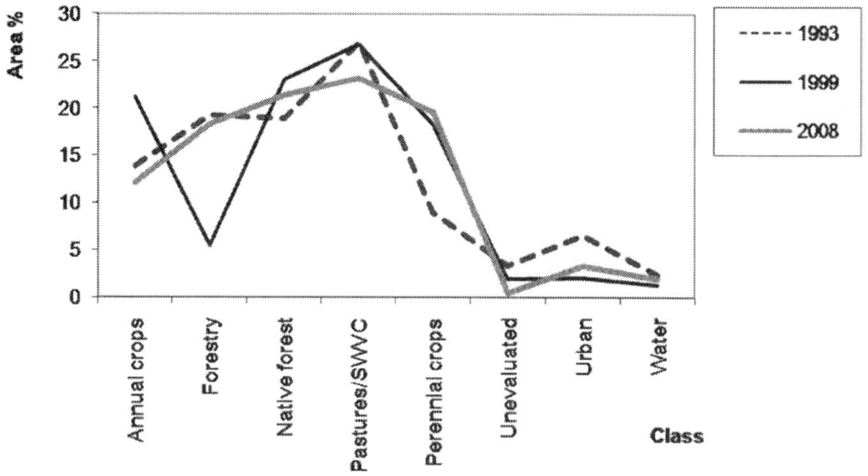

Fig. 8. Evolution of the land cover classes considered in this chapter in the municipality of Montenegro, Rio Grande do Sul state, Brazil.

5. Comparison and evaluation of errors

Two strategies were used to analize the errors of the classification methods: (a) Duplication of process with new test areas for the same classes in Caiuba region. (b) Confusion matrixs by truth of field for Montenegro municipality. Each strategy is related to a different error condition, that is, error of processing and error of the method respectively.

The maximum error for case (a) was for the class "Soils without vegetal cover", more or less 2% when the overall area was evaluated (Table 6). The interference of the clouds was of the same order (value insered with the class "Beach/dunes/clouds"). A confusion between dunes and water (sediments) occurred in the Caiuba region (central part of Fig. 9). Other sectors appear very similar for both images.

Class	Image A	Image B
Water, %	20.3	21.3
Wetland, %	18.9	18.5
Pastures, %	12.2	12.4
Rice crops, %	7.7	9.1
Sandy fields, %	13.4	13.6
Beach/dunes/clouds, %	8.4	7.8
Forestry, %	3.6	3.6
SWVC,%	15.6	13.8

Table 6. Errors derived from classification with new test areas in the Caiuba region for the image of Jan. 20th, 2002 in a total area of 204088 hectares. SWVC = Soil without vegetal cover.

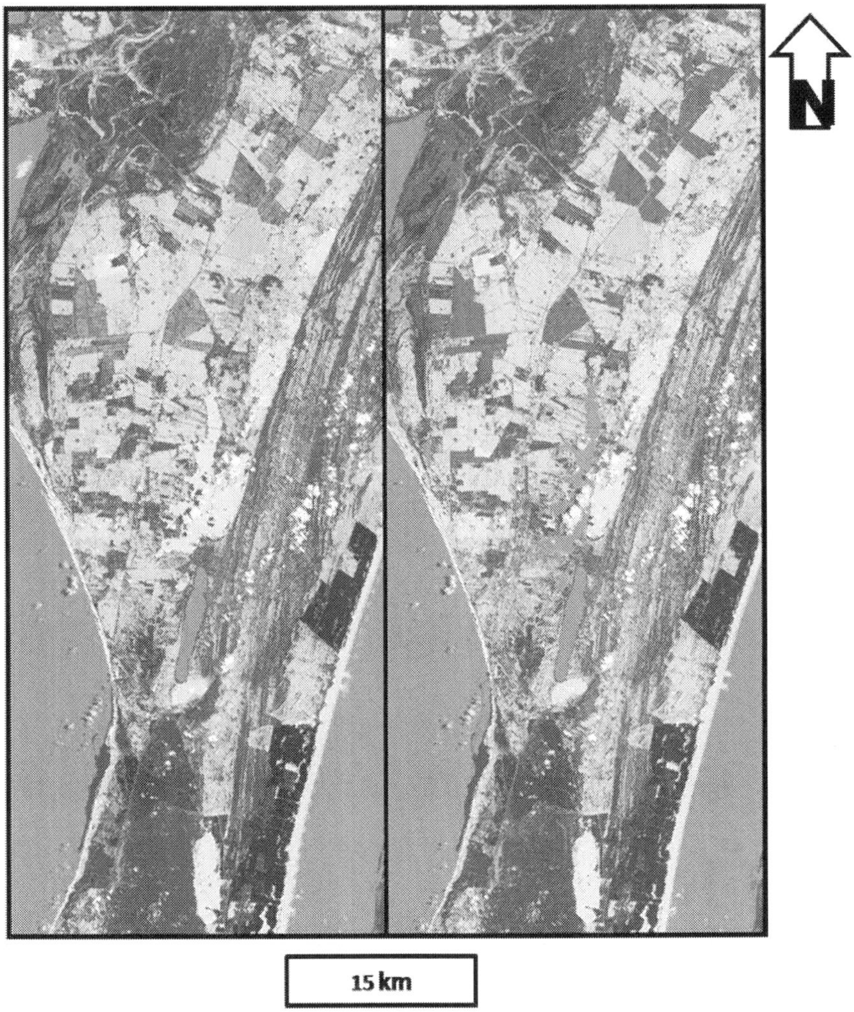

Fig. 9. Comparison of the classification in Caiuba region with Landsat image of 2002 for different test areas.

Data from Montenegro municipality considered the confusion matrix constructed with the truth of field for the Landsat 5 image of april 13, 2009. Thus, 48 control points were selected in the image, trying a "randomly - homogeneous" distribution on the territory of the municipality. Each point was verified at field in september-november 2009 and historical informations were collected with the local farmer when posible.

The accuracy of the method was moderate, that is, 42% for the full process (Table 7). Forestry, Pastures/SWVC, Perennial crops and Urban/outcrops showed the better results, with values greater or equal than 50%. The rest of the classes presented low accuracy with values in the interval 0 - 25%. The correlation coefficient of the quantity of control points and the accuracy was 0.43, suggesting few dependence between both variables. Anyway, a critical case occurred with the class "Native forest" with 11 control points and only two hits. An explanation for the low accuracy of the classification process and some specific classes is the shadow derived from the steep topography, causing confusion among classes and inconsistent results. A improvement of the results is obtained when principal components are considerer before classification, with a potential increment of accuracy of 10% (Schroder & Filippini-Alba, 2010b).

	Control points	SCC	Accuracy
Forestry	9	5	56%
Pastures/SWVC	11	6	55%
Water	2	0	0%
Perennial crops	7	4	57%
Native forest	11	2	18%
Urban/outcrops	4	2	50%
Anual crops	4	1	25%
Total	**48**	**20**	**42%**

Table 7. Results of the confusion matrix for the process of classification in the municipality of Montenegro, Rio Grande do Sul state, Brazil (Schroder & Filippini-Alba, 2010b). SCC = samples correctly classified.

6. Conclusion

Two categories of change detection techniques (Lu et al., 2004) were considered in this chapter, all of them including classification methods: Post-classification comparison and strategy with GIS.

The strategy with GIS isolated the poligon corresponding to the "Potential area for agriculture", then, the interference between some pair of classes was eliminated, by instance, wetlands and rice crops. The post-classification comparison allowed a rapid approach about the region with minor accuracy (preliminary results). Definition of the method used depends on the ratio between cost and efficiency according to the designed objectives.

Errors associated to classification methods are mainly due to the spectral answer, by undefinition of classes or occurrence of pixels of transition, because the errors derived from digitalization were insignificant. Atmospheric conditions and the regional topography also influence the process of classification.

Land cover changes in a dynamic way, sometimes with significant transformation rates of one class to another, as the discussed cases confirm. Truth of field appears as an optimal method to improve results, but the cost of process, in time, financial and human resources is incremented.

7. Acknowledgments

The authors are grateful to Dr. Angela Campos Diniz (Embrapa) and Dr. Cleber Palma (Federal University of Rio Grande), leaders of the projects that funded research activities. Partial support was obtained through the Foundation for Research Support in the State of Rio Grande do Sul (FAPERGS).

8. References

Bolfe, L.E., Siqueira, J.O.W. de, Pereira, R.S., Filippini -Alba, J.M. & Miura, A.K. (2009). Uso, ocupação das terras e banco de dados geográficos da metade sul do Rio Grande do Sul. *Ciência Rural*, Vol. 39, No. 6, (September 2009), pp. 1729 - 1737, ISSN 0103-8478.

Crist, E.P. & Kauth, R.J. (1986). The Tasseled Cap De-Mystified. *Photogrammetric Engineering and Remote Sensing*, Vol. 52, No. 1, (January 1986), pp. 81-86, ISSN 5201-0081.

Crosta, A. P. (1993). *Processamento digital de imagens de sensoriamento remoto*. State University of Campinas, ISBN 85-85369-02-7, Campinas, Brazil.

Davis, J. (2002). *Statistics and data analysis in geology*. Wiley, ISBN 04-71172-75-8, New York.

Embrapa (2009). Sistemas orbitais de monitoramento e gestão territorial. In: *Embrapa Monitoramento por Satélite*, 27/06/2011, Available from http://www.sat.cnpm.embrapa.br.

ER-Mapper (1995). *ER Mapper Reference*. Earth Resource Mapping, West Perth, Australia.

Filippini-Alba, J.M. & Siqueira, O.W. de. (1999). Caracterização da ocupação das terras em escala municipal, através do processamento de imagens Landsat. Área piloto, Pelotas, RS. *Agropecuária Clima Temperado*,Vol. 2, No. 2, (December 1999), pp. 223 - 231, ISSN 1415-6822.

Guild, L.S.; Cohen, W.B.; Kauffman, J.B. (2004). Detection of deforestation and land conversion in Rondônia, Brazil, using change detection techniques. *International Journal of Remote Sensing*, Vol. 25, No. 4, (February 2004), pp. 731-750, ISSN 0143-1161.

Inpe (n.d.). Catálogo de imagens. In: *National Institute of Spatial Research (Inpe)*, 10/jun/2010b. Available from <http://www.dgi.inpe.br/CDSR/>.

Inpe (n.d.). Marlin. In: *National Institute of Spatial Research (Inpe)*, 01/03/2010a. Available from <http://www.dgi.inpe.br/CDSR/marlin_PT.php>

Jensen, J.R. (1996). *Introductory digital image processing: a remote sensing perspective*. Prentice Hall , ISBN 0-13-205840-5, Upper Saddle River, USA.

Lillesand, T. M.; Kiefer, R.W. *Remote sensing and image interpretation*. Wiley, ISBN 0-471-57783-9, New York: 1994. 750p.

Lu, D.; Mausel, P.; Brondízio, E.; Moran, E. (2004). Change detection techniques. *International Journal of Remote Sensing*, Vol. 25, No. 12, pp. 2365-2407, ISSN 0143-1161.

Miranda, E. E. (2002). Brasil visto do espaço: Rio Grande do Sul. In: *Embrapa Monitoramento por Satélite*, 1 CD-rom.

Miranda, E. E. (2005). Brasil em Relevo. In: *Embrapa Monitoramento por Satélite*, 02/08/2010. Available from <http://www.relevobr.cnpm.embrapa.br>.

Moreira, M. A. (2005). *Fundamentos de sensoriamento remoto e metodologias de aplicação*. Universidade Federal de Viçosa, ISBN, Viçosa, Brazil.

Nasa (2004). ASTER: Advanced spaceborne thermal emission and reflection radiometer. In: *Nasa/JPL*, 27/06/2011. Available from < http://asterweb.jpl.nasa.com>.

Pontara, R.C.P. (1998). *Análise do comportamento espectral dos filitos carbonosos para interpretação de imagens*. Brasilia University, Academic dissertation, Brasília, Brazil.

Schroder, V.F.; Filippini-Alba, J.M. (2010). Classificaçção de imagens orbitais com auxílio da análise por componentes principais no município de Montenegro - RS, 2009. In: *Infoteca/Embrapa*, 11/10/2011. Available from http://www.infoteca.cnptia.embrapa.br/handle/doc/884278

Schroder, V.F.; Filippini-Alba, J.M. (2010). Potencialidade do uso de imagens orbitais para detecção de mudanças temporais: estudo de caso no município de Montenegro - RS, 1993 - 2008. In: *Infoteca/Embrapa*, 11/10/2011. Available from http://www.infoteca.cnptia.embrapa.br/handle/doc/884267

Tanac (n.d.). Histórico. In: *Tanac Cia.* 08/07/2010, Available from <http://www.tanac.com.br/PT/institucional.php?codCategoriaMenu=148&nomArea=Hist%F3rico&codDado=2&menu=138>.

5

Mapping Soil Salinization of Agricultural Coastal Areas in Southeast Spain

Ignacio Melendez-Pastor, Encarni I. Hernández,
Jose Navarro-Pedreño and Ignacio Gómez
Department of Agrochemistry and Environment,
University Miguel Hernández of Elche
Spain

1. Introduction

Soil salt content is a key factor that determines soil chemical quality together with soil reaction, charge properties and nutrient reserves (Lal et al., 1999). An adequate salt supply is essential for an optimum development of photosynthetic mechanism and other biochemical processes in plants (Sitte et al., 1994). Soil salt content constitutes an environmental problem when salt accumulation generates drastic changes in soil physical and chemical properties, adversely affecting soil productivity and plant growth (Richards, 1954; Qadir et al., 2000).

Salinization affects about 30% of the irrigated land of the world, decreasing this area approximately 1-2% per year due to salt-affected land surfaces (FAO, 2002). In Europe, about 1-3 million hectares of the land are affected by salinization (European Commission, 2003), and most of these areas are situated in the Mediterranean basin. In Spain, about 18% of the 3.5 million hectares of irrigated land are severely affected or at serious risk of soil salinization (European Commission, 2002). Soil salinization is a frequent problem in arid and semiarid regions like Southeast Spain (Hernández Bastida et al., 2004). In these areas, agriculture with a great water requirement combined with high water tables and an adverse climate (increased occurrence of extreme drought events) have forced irrigation with poor quality water, causing processes of soil degradation and salinization, limiting crop growth and the production capacity (Pérez-Sirvent et al., 2003; Acosta et al., 2011).

Evaluating the spatial variability of basic soil properties in saline soils, and mapping spatial distribution patterns of these soil properties helps to make effective site-specific management decisions (Ardahanlioglu et al., 2003). Accordingly, remote sensing techniques and geographic information systems (GIS) have introduced a new era for soil resources assessment and monitoring in terms of information quality (Mermut and Eswaran, 2001). *A priori* knowledge of spectral characteristics of remotely sensed materials is fundamental to any valuable quantitative analysis (Ben-Dor et al., 1997). The variety of absorption processes occurring in the soil and their wavelength dependence allow us to derive information about the chemistry of the minerals composing it from the reflected or emitted light (Clark, 1999). Reflectance spectra of soils are attributed to numerous soil properties. There are no narrow absorption bands linked to soil salinity status, since it is determined by soil properties such as pH, electrical conductivity, salt content and exchangeable sodium percentage (Csillag et al., 1993; Farifteh et al., 2008). In this sense, soil reflectance is derived from the particular spectral behaviour of the heterogeneous combination of minerals, organic matter and soil water (Ben-Dor and Banin, 1994). Salt-affected soils cations (Na^+, Mg^{2+}, K^+, and Ca^{2+}) and anions (Cl^-, SO_4^{2-}, CO_3^{2-} and HCO_3^-) can be detected by optical spectrometers since salt minerals have diagnostic spectral features occurring in the visible and near infrared (VNIR) and short-wave infrared (SWIR) spectral regions (Farifteh et al., 2008). Saline soils usually have evaporate minerals, which spectral features that can be explained by vibrational absorption due to water molecules chemically bound as part of the crystal structure (Howari et al., 2000). In this sense, the spectral differences of evaporates of single salt compounds are determinant of the type and mineralogy of the soils (Howari et al., 2000).

Remote sensing has been extensively employed in soil salinity studies. Data from aerial photography, videography, and optical, thermal, microwave or geophysical sensors has been used in soil salinity mapping (Metternich and Zinck, 2003). Perhaps, the most widely used remote sensing data in recent decades have been provided by multispectral (Landsat, SPOT, IRS, ASTER) or hyperspectral (DAIS, HyMap, AVIRIS, Hyperion) sensors in the spectral range approximately between 400 and 2500 nm. Researchers have frequently employed remote sensing data to map soil salinity with multispectral (Metternich and Zinck, 1997; Dwivedi et al., 2001; Melendez-Pastor et al., 2010a) and hyperspectral images (Dehaan and Taylor, 2002, 2003; Schmid et al., 2009, Ghrefat and Goodell, 2011). Pioneering studies in the 1970s employed air-borne and satellite-borne multispectral scanners to detect soil salinity, indicating the better capability of infrared bands over visible bands to locate saline soils and the low contribution of thermal bands to improve the delineation of saline areas (Richardson et al., 1976; Dalsted et al., 1979). Nowadays, imaging spectroscopy techniques are employed for the automatic detection of soil salinization with airborne or satellite sensor (Dehaan and Taylor, 2002, 2003; Dutkiewicz et al., 2009; Schmid et al., 2009; Weng et al., 2009; Melendez-Pastor et al., 2010a; Ghrefat and Goodell, 2011). Imaging spectroscopy deals with the mapping of ground materials by detecting and analysing reflectance/absorbance features in hyperspectral (or multispectral) images (Clark, 1999). Imaging spectroscopy adds a new dimension of remote sensing by expanding point spectrometry into a spatial domain and under field conditions, which is a very good approach for the study of soil properties (Ben-Dor et al., 2009).

The aim of this chapter is the application of remote sensing for the study of soil salinity of an agricultural area in southeast coast of the Iberian Peninsula. Different digital image processing techniques were applied to satellite multispectral images (Landsat TM). 'Conventional' hard classification techniques were combined with spectral mixture analysis and soil properties to achieve a better understanding of the soil salinization process in the study area.

Multispectral satellite images such as those obtained by the Landsat program provide low or free cost worldwide coverage for four decades. Moreover, salinization problems are concentrated in arid and semi-arid regions, often in developing countries with few economic resources. Although there are more advanced sensors that can provide a more precise quantification of the extent of soil salinity (e.g. hyperspectral), their high cost difficult its extensive use. Therefore, it is necessary to continue investigating the application of multispectral image repositories as a tool to assist in the monitoring and management of saline soils.

2. Material and methods

This study will evaluate the applicability of various remote sensing techniques for studying salinization processes in an agricultural coastal area. One of the greatest difficulties in the application of remote sensing techniques to the study area is the fragmentation of the territory by the existence of small plots and buildings that create a dispersed mixture of spectral signals to the scale of a moderate spatial resolution multispectral remote sensing image as those acquired by the Landsat Thematic Mapper sensor. This difficulty motivates the need to evaluate various techniques and methodological approaches to carry out this study as necessary to help monitoring the processes of salinization.

Representative soils of the area were sampled and their properties were characterized at the laboratory by standard methods. Predominant land cover classes at the soil sampling plots and at additional land cover validation points were identified. Land cover is a fundamental variable that impacts on and links many parts of the human and physical environments (Foody, 2002) with a great influence on soil properties (Caravaca et al., 2002; Majaliwa et al., 2010; Biro et al., 2011). Both kinds of information in a GIS database were included. In this sense, the effect of land cover on soil properties was statistically evaluated. Then, multispectral images were employed for a hard land cover mapping with a supervised approach using the k-nearest-neighbour classifier. Accuracy assessment methods highlighted the need to employ a mixed pixel focus to deal with the particularities of the study area. Spectral unmixing techniques allowed the identification of representative spectral endmembers and the obtainment of their corresponding fraction images. Finally, fraction endmembers were employed to characterize land cover classes and to predict soil properties with various statistical methods.

2.1 Description of the study area

The study area is located in a coastal zone of Southeast Spain, in the province of Alicante. It is located around 38.14°N and 0.73°W, at the south of the cities of Elche and Alicante. The study area (Figure 1) comprises alluvial plains resulting from the accumulation of sediments from the Segura and Vinalopó rivers. During most of the Holocene (~10,000 years ago to present) the study area was a large lagoon (Blázquez, 2003). In the last centuries, the ancient lagoon was transformed into an irrigated agricultural land draining the wetland. Nowadays, this area is a mixture of small-size cities, coastal urban areas, scattered residential houses, irrigated crops and isolated and scattered wetlands. The perimeter of the study area was delimited according to natural or man-made features in order to enclose a large coastal plain area primarily occupied by irrigated agricultural activities. The study area lies in the north with the natural parks of *El Hondo* and the *Salinas de Santa Pola*. Both natural areas are wetlands included in the RAMSAR list of wetlands of international importance. The east and south boundaries are the *Sierra del Molar* and the *Segura River* respectively. Urban areas and sclerophyllous vegetation mainly occupy the Sierra del Molar, while the Segura River is the most important watercourse in southeast of Iberian Peninsula providing water for irrigation agriculture and to fill the reservoirs that currently comprise

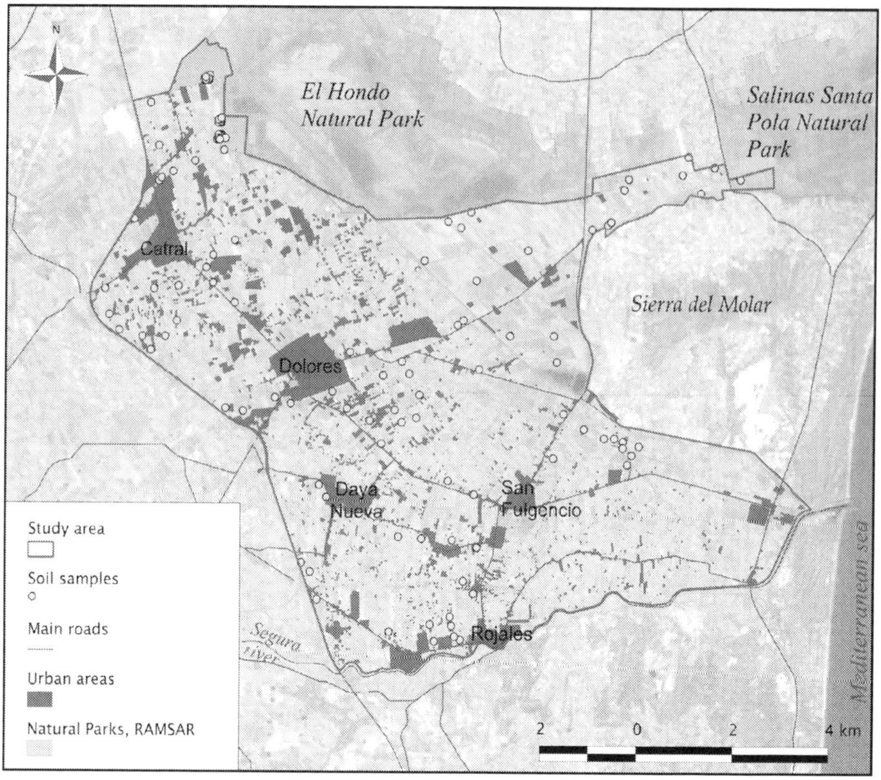

Fig. 1. Study area with the Landsat scene (false-colour composite RGB:742) and superimposed cartographic information (soil samples, urban areas, natural parks and roads).

the wetland of *El Hondo*. The western boundary of the study area is a motorway that cuts north to south the floodplain.

This coastal region has a semiarid Mediterranean climate, with a mean annual rainfall of less than 300 mm and a mean annual temperature of 17 °C and defined by the Köppen climate classification system as *Bsk* class (dry climate with a dry season in summer and a mean annual temperature about 18 °C). The climate is arid or semiarid according to the aridity index of Martonne (De Martonne, 1926) and the aridity index of UNEP (1997) respectively. Figure 2 shows the daily climatic diagram of mean temperature, precipitation and evapotranspiration (by the Penman-Monteith method) for the hydrological year 2010-2011 (from October to September) at Catral meteorological station. Mean daily temperature (blue line) varies from approximately 9°C in winter to more than 25°C in summer. Rain events (red bars) mainly occurred from December 2010 to May 2011 with total accumulated precipitation of 182 mm. This very scarce precipitation joint with an accumulated evapotranspiration of 1115 mm implied that the hydrological year was very dry.

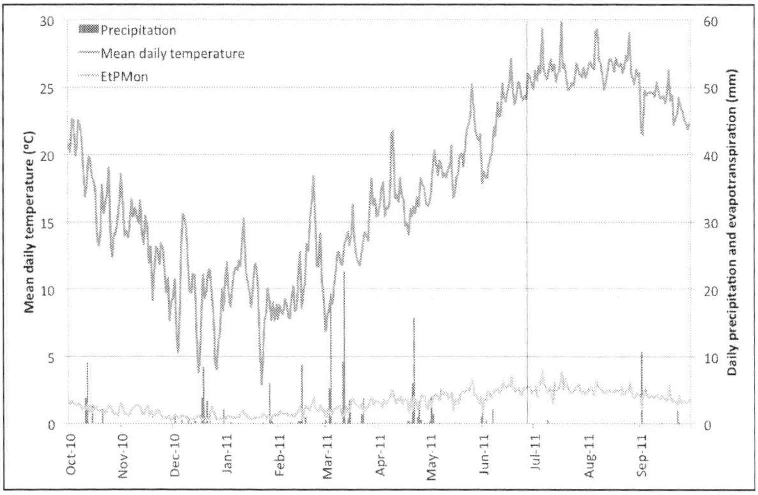

Fig. 2. Daily values of precipitation, mean temperature and evapotranspiration for the hydrological year 2010-11 at Catral station. Source data from the Spanish Ministry of Environment and Rural and Marine Affairs (MARM).

Predominant soil classes are Entisols according to the Soil Taxonomy (Soil Survey Staff, 2006) but affected by agriculture practices along years. They are characterized by a massive presence of carbonates and soluble salt content. In the studied area, irrigation is essential to support agriculture. The water deficit during several months requires irrigation while low quality water is used in the poorly drained soils of these coastal plains, being soil salinization an environmental problem. Thus, the study area soils are subjected to severe risk of physical, chemical and biological degradation (De Paz et al., 2006) that endanger agriculture sustainability.

2.2 Field survey

Field survey was done in the late spring and summer months of the hydrological year 2010-2011 to collect soil samples and identify land cover classes. An extensive soil sampling was done, and 116 samples were collected and geographically referenced. Samples were obtained from the upper 5 cm as solar radiation in VNIR spectral range has limited penetration capabilities. Soil samples were dried at room temperature and a 2 mm sieve was used to separate the fine fraction to be analysed. Analysed soil characteristics included in the study were electrical conductivity (EC) (1:5 w/v water extraction), pH and organic carbon (OC) by wet chemical oxidation (Walkley and Black, 1934) with potassium dichromate oxidation (Nelson and Sommers, 1982).

A land cover validation campaign was also conducted along with the soil survey in order to allow accuracy assessment of generated land cover maps. Land cover validation points were randomly generated in a GIS and a database with the land cover category generated. A total of 205 land cover validation points were identified, combining field observation and recent aerial orto-photography (0.5 m of spatial resolution). Land cover classes identified in the study area were: water bodies, seasonal or permanent crops, saltmarshes and misused agricultural field that tends to be saltmarshes, palm groves, marshes with almost permanent inundation, and anthropic areas (Table 1).

Land cover	ID	Features
Water	1	Wetlands water tables and irrigation ponds
Arable land	2	Herbaceous (e.g. alfalfa, barley) and horticultural (e.g. melon, broccoli) crops
Permanent crops	3	Fruit trees (e.g. orange, lemon, pomegranate)
Fallow/abandoned	4	Fallow or recently abandoned agricultural land.
Saltmarsh	5	Halophyte vegetation (e.g. *Salicornia sp.*, *Suaeda sp.*, *Limonium sp.*, *Halocnemum sp.*)
Palm groves	6	Palm trees plantations and nurseries, mainly from *Phoenix dactilifera*
Marsh	7	*Phragmites australis* dominated wetland vegetation
Man-made/urban	8	Urban areas, roads, farms or industrial areas

Table 1. Descriptions of land cover classes identified in the study area.

Land cover categories at soil sampling points were also identified and included along with soil properties in a GIS database for the land cover classification training stage and for further spatial analyses. Note that land cover (i.e. biophysical materials found on the land) and land use (i.e. how the land is being used by human beings) (Jensen, 2007) are different terms but often used together or interchangeably. In this chapter, we adopt the term land cover because we are interested in knowing about the biophysical characteristics of the study area, but the knowledge of both land use and land cover are important for land planning and land management activities (Lillesand et al., 2003).

2.3 Satellite imagery preprocessing

Remote sensing data were acquired by the Thematic Mapper (TM) sensor on-board the Landsat 5 satellite. Meteorological conditions and the satellite pass over the study area conditioned the date of image acquisition. A scene acquired on 28th June 2011 (path 199 row 33) was employed for analyses. A vertical black line on Figure 2 indicates the time of acquisition of the scene. No rain events happened 16 days prior to the scene acquisition date. Typically summer meteorological conditions without cloud coverage and high temperature were registered on the date of image acquisition, and thus the image quality was optimal.

Satellite image preprocessing included geometric and atmospheric corrections with the aim to ensure the spatial comparability with other data sources and to obtain at-ground reflectance pixel spectra, respectively. Various georeferenced data types were used for the geometric correction: aerial orthophotos (0.5 m of pixel resolution) and digital cartography (scale = 1:10000). The Landsat 5 TM scene was geometrically corrected using Ground Control Points (GCP) identified on the orthophotos and cartographic maps. A quadratic mapping function of polynomial fit and the nearest neighbour resampling method were used for the correction. The nearest neighbour resampling method was selected because it ensures that the original (raw) pixel values are retained in the resulting output image, which is an important requirement in any change detection analysis (Mather 2004). The maximum allowable root mean square error (RMSE) of the geometric correction was less than half a pixel, a reference value frequently cited (Townsend and Walsh 2001; Jensen, 2005).

Atmospheric correction involves the estimation of the atmospheric optical characteristics at the time of image acquisition before applying the correction to the data (Kaufman, 1989). This type of correction is a pre-requisite in many remote sensing applications such as in classification and change detection procedures (Song et al., 2001). Radiometric calibration was applied prior to the atmospheric correction. The conversion of raw digital numbers (DN_{raw}) of Landsat level 1 (L1) image products to at-satellite radiance values (L_{sat}) required the application of current re-scaling values (Chander et al., 2010) by applying the following expression (Chander and Markham, 2003; Chander et al., 2010):

$$L_{sat} = \left(\frac{L_{MAX\,\lambda} - L_{MIN\,\lambda}}{255} \right) (DN) + L_{MIN\,\lambda} \tag{1}$$

Where L_{sat} is at-satellite radiance [W/(m^2 sr μm)]; $L_{MIN\,\lambda}$ is the spectral radiance that is scaled to Q_{calmin} [W/(m^2 sr μm)] (Q_{calmin} is the minimum quantized calibrated pixel value, i.e. DN=0, corresponding to $L_{MIN\,\lambda}$); $L_{MAX\,\lambda}$ is the spectral radiance that is scaled to Q_{calmax} [W/(m^2 sr μm)] (Q_{calmax} is the maximum quantized calibrated pixel value, i.e. DN=255, corresponding to $L_{MAX\,\lambda}$); and DN are digital numbers of the L1 image product. Surface reflectance values (ρ) were computed by using the image based COST method (Chavez Jr, 1996). Path radiance (L_p) values were computed by using the equation reported in Song et al. (2001) that assumes 1% surface reflectance for dark objects (Chavez Jr, 1989, 1996; Moran et al., 1992). The optical thickness for Rayleigh scattering (τ_r) was estimated according to the equation given in Kaufman (1989).

2.4 Land cover classification

Image classification procedures aim to automatically categorize all pixels in an image into land cover classes or themes (Lillesand et al., 2003). Thematic mapping from remotely sensed data can be defined as grouping together cases (pixels) by their relative spectral similarity (unsupervised component) with the aim of allocating cases based on their similarity to a set of predefined classes that have been characterized spectrally (supervised component) (Foody 2002). Multispectral images (like Landsat TM scenes) are frequently used to perform the classification based on spectral pattern recognition methods that exploits the pixel-by-pixel spectral information as the basis for automated classification (Lillesand et al., 2003). In this study, a supervised land cover classification of the Landsat TM image was performed with a *k*-nearest-neighbour clustering algorithm to obtain a discrete or 'hard' categorical land cover map for the study area. *K*-nearest-neighbour (KNN) classifier searches away from the pixel to be classified in all directions of the spectral space until it encounters *k* user-specified training pixels and then assigns the pixel to the class with the majority of pixels encountered (Jensen, 2005). KNN algorithm has been successfully applied for land cover classification with remote sensing data (Franco-Lopez et al., 2001; Haapanen et al., 2004; Blanzieri and Melgani, 2008). Land cover classes assigned to the soil plots were employed in the training stage of the algorithm. Major urban areas were digitized with a GIS and masked-out of the supervised classification procedure since urban areas induce a great spectral confusion. Water areas training points were also included in the training dataset.

The land cover validation database was employed to evaluate the performance of the classification. Land cover map accuracy assessment was quantified with statistical methods such as the error matrix and the kappa statistic. The error matrix is a square array of numbers organized in rows and columns that express the number of sample units (i.e. pixels) assigned to a particular category relative to the actual category as indicated by the reference data (Congalton, 2004). Reference data are in the columns while the rows indicate the map categories to be assessed. This form of expressing accuracy as an error matrix is an effective way to evaluate both errors of inclusion (commission errors) and errors of exclusion (omission errors) present in the classification as well as the overall accuracy (Congalton et al., 1983). In addition to the error matrix, the Kappa coefficient developed by Cohen (1960) was employed to quantify the accuracy of the land cover map. Cohen's Kappa (or KHAT) is a measure of agreement for nominal scales based on the difference between the actual agreement of the classification (i.e., agreement between computer classification and reference data as indicated by the diagonal elements) and the chance agreement, which is indicated by the product of the row and column marginal (Congalton et al., 1993).

2.5 Spectral unmixing

A mixed pixel results when a sensor's Instantaneous Field of View (IFOV) includes more than one land cover type on the ground (Lillesand et al., 2003). The spectrum of a single pixel is a complex measurement that integrates the radiant flux from all the spatially unresolved materials in the IFOV, regardless of whether or not we know their identities (Adams and Gillespie, 2006). Spectral mixture analysis (SMA) has been developed as a method to transform the reflectance in the bands of multispectral images to fractions of reference endmembers, which are reflectance spectra of well-characterized materials that mix to produce spectra equivalent to those of pixels of interest in the image (Adams et al., 1995). As part of SMA techniques, linear spectral unmixing (LSU) models tread the radiation recorded by a sensor as the result of a linear mixture of spectrally pure endmember radiances (Small and Lu 2006). This method is based on the assumptions that: 1) the recorded radiation by the sensor for each pixel is limited to the sensor's IFOV, and assumes no influences by reflected radiation from neighbouring pixels (Settle and Drake 1993), 2) the overall global radiance is proportional to the surface occupied by each land cover type, and 3) the spectrally pure endmembers are valid for the whole study area (Quarmby et al. 1992). LSU models describe radiation reflected by an individual pixel (i,j) of a band k as the result of the product of reflectance for each land cover type by their respective mixture fraction plus an additional associated error for each pixel. The general expression of the model is presented in the following equation:

$$\rho_{i,j,k} = \sum_{m=1,p} F_{i,j,m}\, \rho_{m,k} + e_{i,j} \tag{2}$$

Where $\rho_{i,j,k}$ is the observed reflectance of a pixel for row i, column j, and band k ; $F_{i,j,m}$ is the proportion of component m of a pixel for row i, column j, for each one of the pure components; $\rho_{m,k}$ is the characteristic reflectance for component m in band k ; and $e_{i,j}$ is the error associated to the estimation of proportions for each pixel i, j. The Least Square Mixing Model proposed by Shimabukuro and Smith (1991) is commonly used to resolve linear spectral mixture models. The method proposed by Shimabukuro and Smith (1991) assumes two initial restrictions for the computation of the proportions of spectrally pure endmembers. The first one implies that pure endmember proportions must range between 0 and 1. This means that the proportions of the components are normalized to a common range of potential values. The following expression summarizes this first restriction:

$$0 \leq F_{i,j,m} \leq 1 \tag{3}$$

The second restriction is that the sum of the fractions for every component is equal to the total pixel surface. In this way, it is quite simple to express the individual contribution or fraction of an endmember in relation to the total reflectance of the pixel.

$$\sum\nolimits_{m=1,p} F_{i,j,m} = 1 \tag{4}$$

The choice of a LSU model must consider both the landscape of the test site and the ability of the model to depict the structure, shape and distribution of the basic landscape components (Ferreira et al. 2007). Well-chosen endmembers not only represent materials found in the scene, but provide an intuitive basis for understanding and describing the information in the image (Adams and Gillespie 2006). Endmembers were obtained after applying a spatial and spectral remote sensing data dimensionality reduction with the minimum noise fraction (MNF) and pixel purity index (PPI) techniques, respectively. The MNF is used to detect the inherent dimensionality of image data, segregating noise from the signal in the data and reducing computational requirements for subsequent processing tasks (Boardman and Kruse, 1994). The MNF as modified from Green et al. (1988) consists in two steps: 1) applying a transformation, based on an estimated noise covariance matrix to decorrelate and rescale the noise in the data (noise has unit variance and no band-to-band correlations); and 2) performing a standard principal component transformation of the noise-whitened data. A final dataset of coherent and almost noise-free bands are selected from the MNF output and can be used for subsequent processing steps. Pixel Purity Index (PPI) is a procedure for finding the most spectrally pure (extreme) pixels that typically correspond to mixing endmembers in multispectral and hyperspectral images (ITT VIS, 2008). PPI is computed by repeatedly projecting n-dimensional scatterplots onto a random unit vector; the extreme pixels in each projection (those pixels that fall onto the ends of the unit vector) are recorded and the total number of times each pixel is marked as extreme is noted. The selection of extreme pixels corresponding to analogous surface features is complex due to the great number of pixels typically found in remote sensing image data. The n-dimensional visualizer implemented in ENVI software (ITT Visual Information Solutions) is a tool to locate, identify, and cluster the purest pixels and most extreme spectral responses in a data set. The distribution of these points in n-space can be used to estimate the number of spectral endmembers and their pure spectral signatures (Boardman, 1993). Three endmembers were used in the LSU model of the study area, namely green vegetation (GV), non-photosynthetic vegetation (NPV) and shade (S). The GV endmember represents the signature of green dense vegetation, the NPV endmember is the signature of bare soil or sparse non-photosynthetic vegetation, and the shade endmember represents the signature of dark pixels and water bodies.

2.6 Statistical methods

The possible existence of differences in soil properties based on the land cover classes was determined by the use of the analysis of variance (ANOVA). ANOVA is used to evaluate significant differences between means of independent variables. The observed variance of independent variables is partitioned into components by several explanatory variables (factors). Land cover class was the factor employed in the analysis. Post-hoc analysis was performed using Tukey method.

Relationships between fraction endmembers and soil properties were studied by the principal component analysis (PCA). PCA is a technique of data dimensionality reduction that performs an orthogonal transformation to convert potentially correlated input variables into uncorrelated variables or principal components. The first components accumulate most of the variance and therefore, the most useful information about the variables.

Regression analyses between soil properties and fraction endmembers were performed for quantitative estimation of soil properties. A linear regression analysis applying a stepwise method for variable entry and removal was the selected statistical technique. Model selection was based on the lower typical error of the estimation and minimum collinearity.

3. Results and discussion

The relationship between various soil properties and land cover classes was analysed. Two approaches to the study of land cover are presented. A first categorization based on discrete land cover classes and another based on mixture fractions. Finally, statistical models for predicting soil properties of interest in the study of soil salinity through the use of mixture fractions are presented.

3.1 Soil properties

The study was focused on soil electrical conductivity (EC), pH and organic carbon (OC) (Table 2). These properties are important in chemical and biological quality of soils (Lal et al., 1999). Previous studies in semiarid areas combining remote sensing and soil analyses have indicated significant differences in these properties in different land cover classes (Biro et al.,

Land covers	EC (mS/cm)	pH	OC (%)
Arable land	1.38 ± 1.01 *ab*	8.23 ± 0.26 *ab*	1.92 ± 0.85 *a*
Permanent crops	0.70 ± 1.00 *b*	8.40 ± 0.26 *b*	1.29 ± 0.24 *a*
Fallow/abandoned	3.52 ± 2.43 *ac*	8.14 ± 0.25 *ab*	1.46 ± 0.36 *a*
Saltmarsh	3.98 ± 2.86 *c*	8.05 ± 0.07 *ab*	3.61 ± 0.81 *b*
Palm groves	3.82 ± 2.42 *ac*	8.20 ± 0.20 *ab*	2.11 ± 1.00 *a*
Marsh	4.71 ± 2.04 *c*	7.82 ± 0.09 *a*	5.24 ± 1.39 *c*
P-value	<0.001***	<0.001***	<0.001***

Table 2. Descriptive statistics (mean ± standard deviation) of soil properties based on land cover classes. The *p*-value and homogeneous subgroups (lower case letters; Tukey test, $P < 0.05$) resulting from the ANOVA test are included.

2011). Cultivated areas (i.e. arable land and permanent crops, homogenous subgroup *b*) have lower electrical conductivity values than natural or semi-natural vegetation. The construction of drainage systems at agricultural areas to encourage the leaching for salinity control has been a traditional amelioration strategy (Qadir et al., 2000). This fact explains that marshes and saltmarshes soils have higher EC values than the other land cover classes, since they are areas with poor drainage and temporally flooded. Salinity increases when farming finishes (i.e. fallow/abandoned) because irrigation water is not available to promote salts leaching.

The pH values were slightly alkaline but significantly different for marshes. Wetland soils are characterized by the permanent or seasonal inundation of the land, promoting anaerobic conditions and thus reduced redox conditions (high concentration of H^+ which implies low pH)(Reddy et al., 2000). The organic carbon content was also different depending on the type of land cover. Arable land and permanent crops soils have organic carbon content ranging from 1.46 to 2.11% that is not very high (Pérez-Sirvent et al., 2003). Opposite, wetland soils (i.e. stable saltmarshes and saltmarshes) exhibited the highest organic carbon contents. Compared to upland areas, most wetland soils show an accumulation of organic matter by the higher rates of photosynthesis in wetlands than other ecosystems and the lower rates of decomposition due to anaerobic conditions (Reddy et al., 2000).

All soil properties are significantly correlated ($P<0.01$) according to the Pearson bivariate correlation test applied to the full dataset. Figure 3 shows two scatterplots of the land cover classes average values (error bars represent the standard deviation) of pH and organics carbon versus electrical conductivity (EC). EC is negatively correlated with pH ($R=-0.61$) and positively correlated with organic carbon ($R=0.34$), while pH and organic carbon are negatively correlated ($R=-0.32$). Two sets of distinct land uses mainly dependent on the EC values are distinguished: 1) active cultures: with low EC and OC values, and 2) natural vegetation and crops of low requirements (palm groves): with high EC and OC values, increasing as the land cover is more similar to the wetland. Palm groves are the most halotolerant crop and require little tillage.

3.2 Land cover classification

A land cover map was obtained with the *k*-nearest-neighbours algorithm (Figure 4). Optimum results were obtained with $k=4$. The area occupied by the land cover classes (hectares and percentage of the total area) was quantified (Table 3).

The study area is mainly agricultural but largely occupied by urban areas. Urban/man-made areas represent 14.4% of the study area. There is a clear distinction between the northeast portion (area between the two natural parks and the *Sierra del Molar*) with large fields and less presence of buildings, and the rest of the study area, with numerous buildings scattered, villages, small-size towns and smaller parcels. Dominant land cover classes are also different at these two sectors. Close to the natural parks, there are many saltmarshes (7.09% of the study area), marshes (1.95% of the study area), palm groves (2.21% of the study area) and arable land (45.76% of the area and mainly forage, barley and melons). The other sector has a massive presence of permanent crops (16.3% of the study area and mainly citrus trees such as orange and lemon trees).

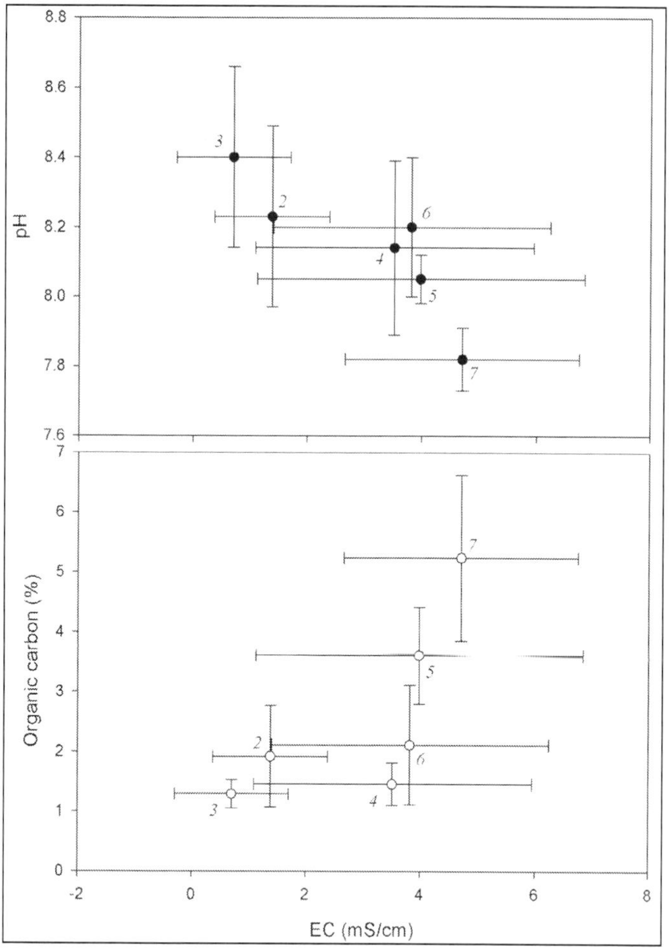

Fig. 3. Scatterplots with the average values pH and organic carbon versus electrical conductivity for land cover classes (numbers in italics are the ID number of the class). X and Y bars represent one standard deviation.

The distribution of land cover classes can be explained by the characteristics of the soils. Generally, the closest soils to the wetland areas of the natural parks are more saline. These soils have a poor drainage due to its lower altitude and very high-water tables, largely due to the horizontal flow of water and salts from the nearby water bodies. Permanent crops class dominates in areas that are close to the towns, being better drained and less saline. Fallow/abandoned areas (12.11% of the study area) are spread throughout the study area as a result of the abandonment of farming on individual fields. However, abandoned land is more present in the proximal portion of the natural parks since the conditions of salinization of soils led to their abandonment.

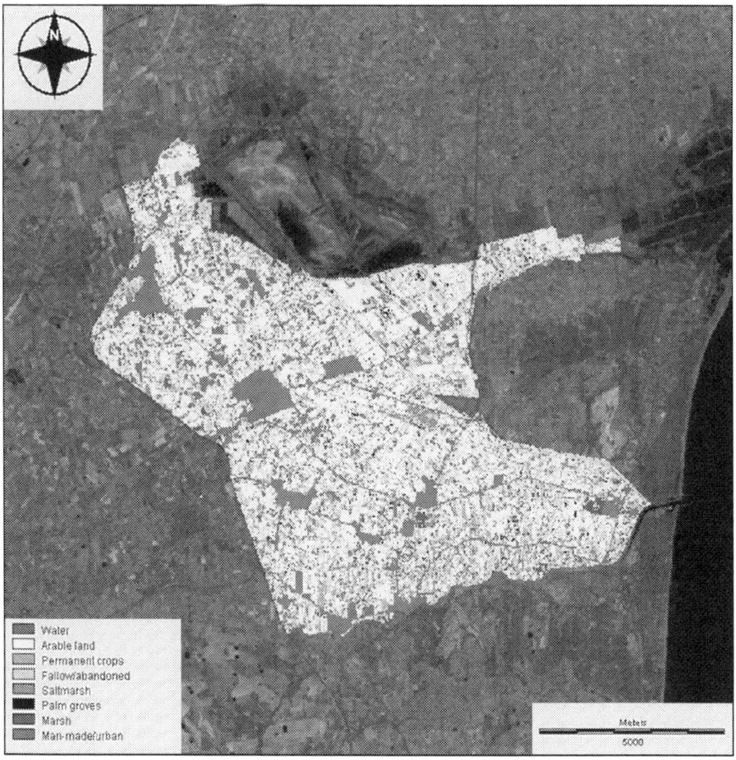

Fig. 4. Land cover map of the study area.

Land cover classes	Area (ha)	Area (%)
Water	17.37	0.20
Arable land	4061.52	45.76
Permanent crops	1446.93	16.30
Fallow/abandoned	1074.51	12.11
Saltmarsh	629.01	7.09
Palm groves	196.02	2.21
Marsh	172.71	1.95
Man-made/urban	1277.73	14.40
TOTAL	*8875.80*	*100*

Table 3. Area occupied by land cover classes according to the map obtained by k-nearest neighbour.

The land cover map accuracy was evaluated with the data set of validation points. Overall accuracy was a 68%, and KHAT value was 0.56. According to Landis and Koch (1977),

KHAT values ranging from 0.4 to 0.8 exhibit a moderate agreement. Inter-class confusion was detected analysing the error matrix. A portion of arable land (78% of producer's accuracy, 65% of user's accuracy) was wrongly classified as permanent crops, fallow/abandoned land or saltmarshes. A great portion of palm groves (21% of producer's accuracy and 75% of user's accuracy) was classified as arable land. The performance of the automatic classification for marshes (90% of producer's accuracy and 75% of user's accuracy) and water areas (100% of producer's accuracy and 80% of user's accuracy) was highly satisfactory. The performance of the KNN algorithm for our land cover classification approach was enough good and comparable with the accuracy obtained by Franco-Lopez et al. (2001) classifying a forest stand (52% of overall accuracy with k=10), and the results of the experiment carried on by Samaniego and Schulz (2009) classifying crop types (47% of overall accuracy with k=5).

3.3 Spectral unmixing and land covers

Spectral mixture analysis was applied to obtain fraction images of green vegetation (GV), non-photosynthetic vegetation (NPV) and shade endmembers. Spectral signatures of selected endmembers are highly distinctive (Figure 5). These endmembers had optimal spectral separability as measured with the transformed divergence method. GV endmember is associated with vigorous vegetation, NPV endmember is associated with bare soil and dry halophytic vegetation, and shade endmember is associated with water bodies and low illuminated areas.

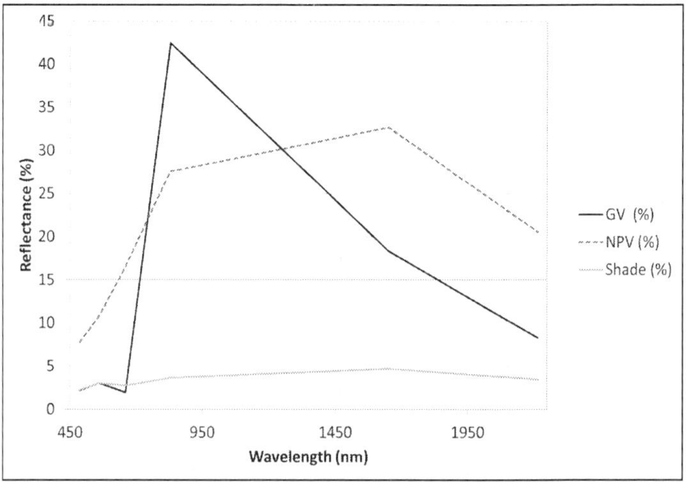

Fig. 5. Plot showing the spectral signatures of selected green vegetation (GV), non-photosynthetic vegetation (NPV) and shade endmembers.

Fraction images of the three endmembers and the residual fraction of the spectral mixture analysis were obtained (Figure 6). Values range from 0 for low high membership to the image fraction (black colour) to 1 for high membership (white colour). Fractions images are continuous variables that are graphically represented with a greyscale colour ramp. High values of the shade fraction image are present in the wetland areas of the natural parks and in a triangular area in the middle-right boundary of the study area that corresponds with a small wetland. A white area in the right of the image corresponds with the Mediterranean Sea. High green vegetation fraction values area scattered through the study area. They correspond with active crops at the time of the image acquisition. Indeed, the white areas in the NPV image fraction correspond with bare soil and saltmarshes which vegetation is quite dry in summer and have a great spectral confusion with background soil. Urban areas were also associated with this fraction image. Finally, high values of the residual fraction are located in industrial areas, whose spectral signature was notably different respect to the three endmembers of the unmixing model.

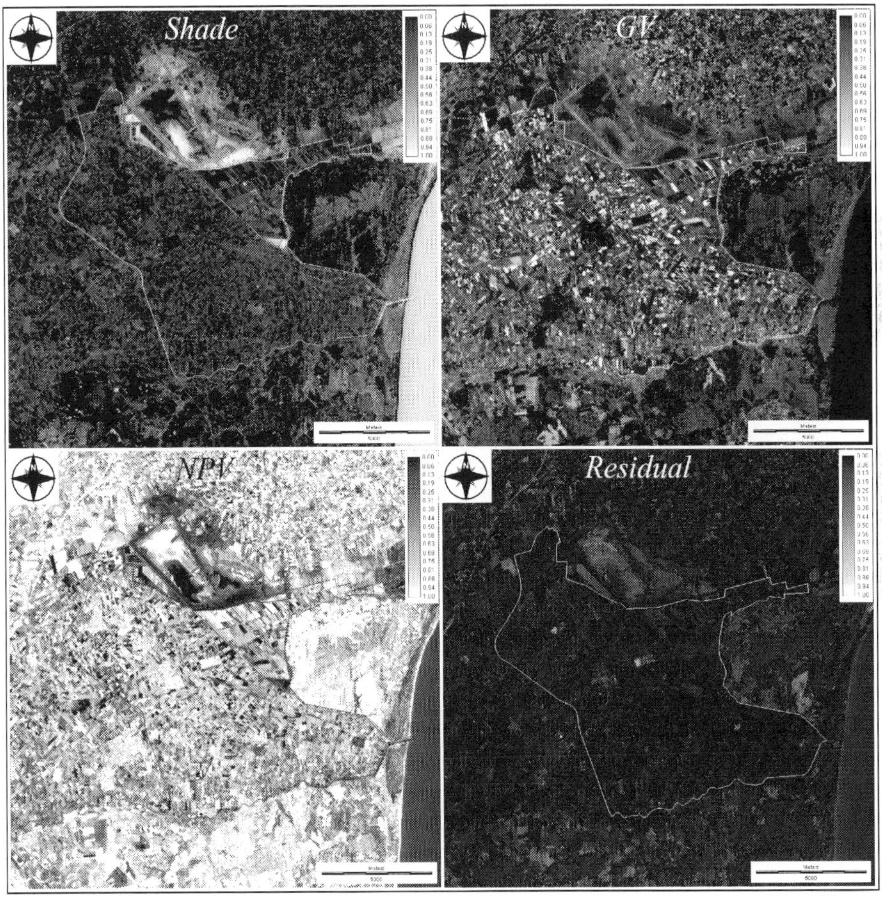

Fig. 6. Fraction images of shade/water, green vegetation (GV), non-photosynthetic vegetation (NPV) and the residual component of the linear spectral mixture analysis. White/black polygon represents the boundary of the study area.

Average values of the three fraction images for the land covers were computed and represented in a ternary diagram (Figure 7). Water land cover has high shade fraction values (>90%) and very low values for the GV and NPV fractions. Marshes have an important fraction of shade (>55%) and around 30% of the GV fraction. This mixture composition is highly indicative of the marshes structure with green *Phragmites australis* stands, growing on flooded or water-saturated soils. Shade fraction has a low contribution in the other land covers (<30%). Saltmarshes, permanent crops, arable land, fallow/abandoned and palm groves land cover classes have GV fraction values between 20-40% and NPV fraction values between 50-70%. This relative homogeneity in the mixture fractions values for different land cover classes could be attributed to the lower water availability in summer, that promotes a drying and browning of the vegetation and promotes spectral confusion. Melendez-Pastor et al. (2010b) previously observed this phenomenon in the study area. They also employed ternary diagrams, combining mixture fraction and land cover classes for a drought year and an average year. Soil or NPV fractions increase their contribution in a dry weather scenario (i.e. drought or summer) and the water and GV fractions have a lower contribution.

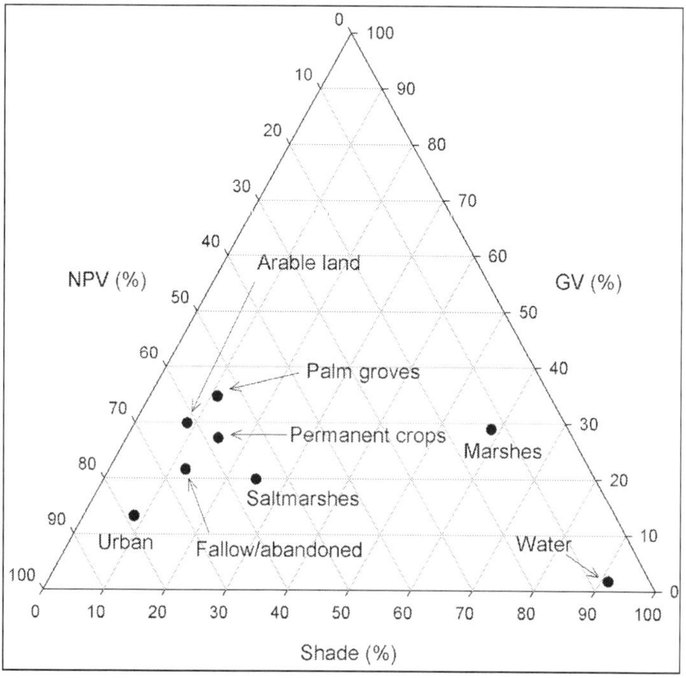

Fig. 7. Ternary diagram of the average mixture fraction values for the land cover classes.

3.4 Fraction endmembers to predict soil properties

Mixture fraction values were statistically related to soil salinity. Principal component analysis provided valuable information about the relationship among soil properties and spectral mixture analysis fractions. The first three principal components accumulated 75.8%

of total variance. PC1 was positively correlated with NPV fraction (factor loading = 0.988) and negatively correlated with GV fraction (factor loading = -0.902). PC1 might be used to separate vegetated from non-vegetated pixels. PC2 was positively correlated with electrical conductivity (factor loading = 0.778) and the shade fraction (factor loading = 0.604) and negatively correlated with the pH (factor loading = -0.785). PC2 might be used to differentiate soil salinity status.

Salinization status seems to be related to the abundance of the shade fraction. This result could be explained by the presence of water at the soil profile, which is an evidence of poor drainage that could lead to salt accumulation. Thus, monitoring shade fraction values along a year could be an indirect method to detect the evolution of soil electrical conductivity with remote sensing. PC3 was positively correlated with the residual fraction of the spectral unmixing (factor loading = 0.853) and organic carbon content (factor loading = 0.711). Evident negative correlations with the PC3 were not found.

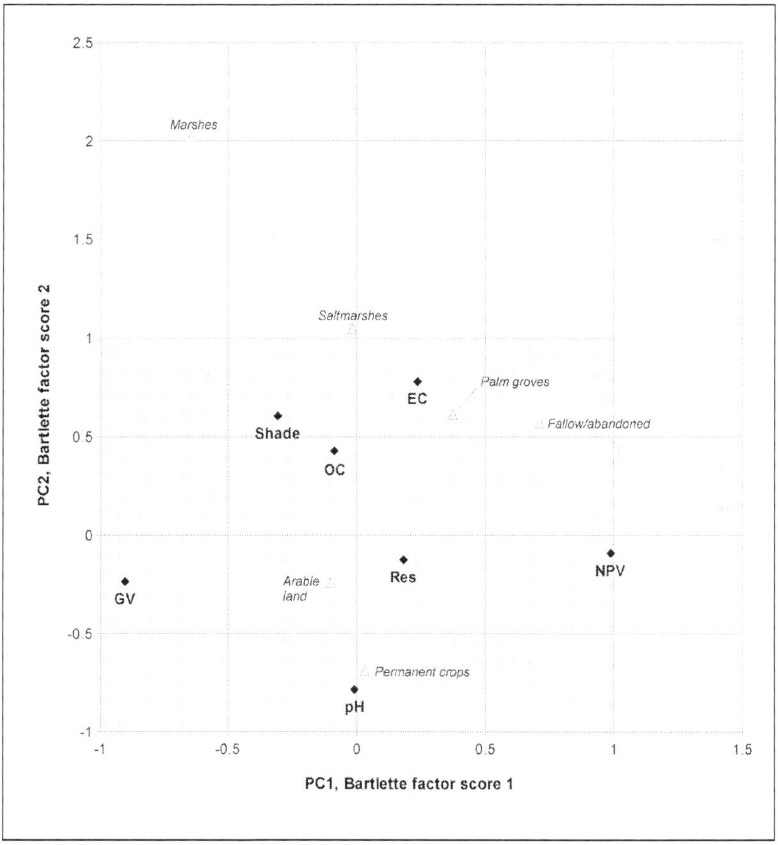

Fig. 8. Factor loadings plot for the measured soil properties and mixture fractions, and average values factor scores plot for the land cover classes.

Previous studies assessed the relationships between PC factor loadings of soil properties and PC factor scores of land cover classes (Biro et al., 2011). We also included the mixture fraction values for soil plots in the principal component analysis. Soil properties and mixture fractions factor loadings and average factor scores for land cover classes of the first two components were plotted to explore their relationship (Figure 8). Factor loading values range from -1 to 1. Factor scores of land cover classes were also included in the plot. Land cover classes were differentiated from each other along PC1, mainly because of the high positive factor loading of the NPV fraction and high negative factor loading of the GV fraction. Also, land cover classes were differentiated from each other along PC2, mainly because of the high positive loadings of the shade fraction, electrical conductivity and organic carbon content, and high negative loading value of pH.

Finally, we tested the usefulness of mixture fractions to predict EC. Stepwise linear regression was employed to model relationship among EC and mixture fractions. EC variable was normalised with the natural logarithm, while the other variables were normal. Table 4 summarizes the main parameters of two linear regression models. Moderate adjustment was obtained from the regression with R=0.338 for the first model and R=0.408 for the second model. The ANOVA test (data not included in the table) indicated the usefulness of the models with a *p*-value <0.001 for both cases. Model 1 included GV fraction as the unique mixed fraction predictor variable, while model 2 included GV and also the shade fraction. In both models, the coefficient B has a negative value for the variable GV, suggesting an inverse relationship between the green covers and EC. By contrast, the regression coefficient B was positive for the shade fraction, indicating a direct relationship between the presence of shadows/water and EC. This latter observation corroborates the interpretation given in the principal component analysis on the usefulness of the shade fraction to predict soil salinity. Collinearity statistics revealed the absence of collinearity problems as we obtained tolerance values much greater than 0 and VIF values much lower than 15 (SPSS, 2009).

					Unstandardized Coefficients			Collinearity Statistics	
Model	**R**	**R^2**	**Adjusted R^2**	**Predictors**	**B**	**Std. Error**	**Sig.**	**Tolerance**	**VIF**
1	0.338	0.115	0.107	(Constant)	7.483	0.106	0.000		
				GV	-1.081	0.279	0.000	1.000	1.000
2	0.408	0.167	0.152	(Constant)	7.302	0.123	0.000		
				GV	-1.005	0.273	0.000	0.989	1.011
				Shade	1.320	0.492	0.008	0.989	1.011

VIF: variable inflation factor

Table 4. Summary of results of linear regression models.

Regression models with mixture fractions did not show collinearity that could lead us to false predictions. A major constraint using proximal and remote sensing data for mapping salinity is related to the fact that there is a strong vertical, spatial and temporal variability of salinity in the soil profile (Mulder et al., 2011). Direct and precise estimation of the salt quantities is difficult by using satellite data with a low spectral resolution because these fail to detect specific absorption bands of some salt types, and the spectra interfere with other soil chromophores (Mougenot et al., 1993). More research will help to improve the prediction of soil properties with remote sensing data for a fast assessment of soil status over large areas and at low cost.

4. Conclusions

This chapter provides an interesting case of study on the application of remote sensing to soil salinity. The land use and management greatly affect soil salinity and land cover mapping helps to delineate areas with different severity of salinization. The use of spectral mixture analysis in combination with land cover maps and soil properties data is a more advanced technique. Mixture fractions help to know the spectral behaviour of land cover and their constituents using a simple three endmembers model. In addition, mixture fractions can be used as predictors in regression models to predict the electrical conductivity of soils. The results of regression models were encouraging but require further research to improve them. Since mixture fractions are sensitive to spectral changes due to changes in ground surface, they may be particularly useful for mapping the severity of soil salinization processes over time with low coast satellite images. The combined use of soil properties analytics, land cover maps and spectral mixture analysis is feasible for monitoring saline soils and land management over large areas with a reduced cost.

5. Acknowledgment

The authors acknowledge the Spanish Ministry of Science and Innovation for the financial support of the project '*Efectos de la aplicación de un compost de biosólido sobre la calidad de suelos con distinto grado de salinidad*' with the reference CGL2009-11194 that allowed this research.

6. References

Acosta, J.A.; Faz, A.; Jansen, B.; Kalbitz, K. & Martínez-Martínez, S. (2011). Assessment of salinity status in intensively cultivated soils under semiarid climate, Murcia, SE Spain. *Journal of Arid Environments*, Vol. 75, No. 11 (November 2011), pp. 1056-1066, ISSN 0140-1963

Adams, J.B. & Gillespie, A.R. (2006). *Remote Sensing of Landscapes with Spectral Images. A Physical Modeling Approach.* Cambridge University Press, ISBN 0521662214, Cambridge, UK

Adams, J.B.; Sabol, D.E.; Kapos, V.; Filho, R.A.; Roberts, D.A.; Smith, M.O. & Gillespie, A.R. (1995). Classification of Multispectral Images Based on Fractions of Endmembers: Application to Land-Cover Change in the Brazilian Amazon. *Remote Sensing of Environment*, Vol. 52, No. 2 (May 1995), pp. 137-154, ISSN 0034-4257

Ardahanlioglu, O.; Oztas, T.; Evren, S.; Yilmaz, H. & Yildirim, Z.N. (2003). Spatial variability of exchangeable sodium, electrical conductivity, soil pH and boron content in salt- and sodium-affected areas of the Igdir plain (Turkey). *Journal of Arid Environments*, Vol. 54, No. 3 (July 2003), pp. 495-503, ISSN 0140-1963

Ben-Dor, E. & Banin, A. (1994). Visible and near-infrared (0.4-1.1 μm) analysis of arid and semiarid soils. *Remote Sensing of Environment*, Vol. 48, No. 3 (June 1994), pp. 261-274, ISSN 0034-4257

Ben-Dor, E.; Inbar, Y. & Chen, Y. (1997). The reflectance spectra of organic matter in the visible near-infrared and short wave infrared region (400-2500 nm) during a controlled decomposition process. *Remote Sensing of Environment*, Vol. 61, No. 1 (July 1997), pp. 1-15, ISSN 0034-4257

Ben-Dor, E.; Chabrillat, S.; Demattê, J.A.M.; Taylor, G.R.; Hill, J.; Whiting, M.L. & Sommer, S. (2009). Using Imaging Spectroscopy to study soil properties. *Remote Sensing of Environment*, Vol. 113, No. SUPPL. 1 (September 2009), pp. S38-S55, ISSN 0034-4257

Biro, K.; Pradhan, B.; Buchroithner, M. & Makeschin, F. (2011). Land use/land cover change analysis and its impact on soil properties in the Northern Part of Gadarif Region, Sudan. *Land Degradation & Development*, Vol. in press DOI: 10.1002/ldr.1116, ISSN 1085-3278

Blanzieri, E. & Melgani, F. (2008). Nearest Neighbor Classification of Remote Sensing Images With the Maximal Margin Principle. *IEEE Transactions on Geoscience and Remote Sensing*, Vol. 46, No. 6 (June 2008), pp. 1804-1811, ISSN 0196-2892

Blázquez, A.M. (2003). *L'Albufera d'Elx: evolución cuaternaria y reconstrucción paleoambiental a partir del estudio de los foraminíferos fósiles*. PhD dissertation. Universitat de Valencia, Valencia, Spain.

Boardman, J.W. (1993). Automating Spectral Unmixing of AVIRIS Data Using Convex Geometry Concepts, *Summaries of the Fourth Annual JPL Airborne Geosciences Workshop, Volume 1: AVIRIS Workshop, JPL Publication 93-26*, pp. 11-14. Washington D.C., USA.

Boardman, J.W. & Kruse, F.A. (1994). Automated spectral analysis: A geological example using AVIRIS data, northern Grapevine Mountains, Nevada. *Proceedings, Environmental Research Institute of Michigan (ERIM) Tenth Thematic Conference on Geologic Remote Sensing*, pp. I-407 - I-418. Ann Arbor (MI), USA.

Caravaca, F.; Masciandaro, G. & Ceccanti, B. (2002). Land use in relation to soil chemical and biochemical properties in a semiarid Mediterranean environment. *Soil and Tillage Research*, Vol. 68, No. 1 (October 2002), pp. 23-30, ISSN 01671987

Chander, G. & Markham, B. (2003). Revised Landsat-5 TM Radiometric Calibration Procedures and Postcalibration Dynamic Ranges. *IEEE Transactions on Geoscience and Remote Sensing*, Vol. 41, No. 11 PART II, (November 2003), pp. 2674-2677, ISSN 0196-2892

Chander, G.; Haque, M.O.; Micijevic, E. & Barsi, J.A. (2010). A Procedure for Radiometric Recalibration of Landsat 5 TM Reflective-Band Data. *IEEE Transactions on Geoscience and Remote Sensing*, Vol. 48, No. 1 (January 2010), pp. 556-574, ISSN 0196-2892

Chavez Jr, P.S. (1989). Radiometric calibration of Landsat Thematic Mapper multispectral images. *Photogrammetric Engineering & Remote Sensing*, Vol. 55, No. 9 (September 1989), pp. 1285 -1294, ISSN 0099-1112

Chavez Jr, P.S. (1996). Image-based atmospheric corrections - Revisited and improved. *Photogrammetric Engineering & Remote Sensing*, Vol. 62, No. 9 (September 1996), pp. 1025-1036, ISSN 0099-1112

Clark, R.N. (1999). Chapter 1: Spectroscopy of Rocks and Minerals, and Principles of Spectroscopy. In: *Manual of Remote Sensing, Volume 3, Remote Sensing for the Earth Sciences*, A.N. Rencz, (Ed.), pp. 3-58. John Wiley and Sons, ISBN 0471294055, New York, USA

Cohen, J. (1960). A Coefficient of Agreement for Nominal Scales. *Educational and Psychological Measurement*, Vol. 20, No. 1 (April 1960), pp. 37-46. ISSN 0013-1644

Congalton, R.G. (2004). Putting the map back in map accuracy assessment. In: *Remote Sensing and GIS accuracy assessment*, R.S. Lunetta & J.G. Lyon, (Eds.), pp. 1-11. CRC Press, ISBN 156670443X, Boca Raton (FL), USA

Congalton, R.G.; Oderwald, R.G. & Mean, R.A. (1983). Assessing Landsat Classification Accuracy Using Discrete Multivariate Analysis Statistical Techniques. *Photogrammetric Engineering & Remote Sensing*, Vol. 49, No. 12 (December 1983), pp. 1671-1678, ISSN 0099-1112

Csillag, F.; Pasztor, L. & Biehl, L.L. (1993). Spectral band selection for the characterization of salinity status of soils. *Remote Sensing of Environment*, Vol. 43, No. 3 (March 1993), pp. 231-242, ISSN 0034-4257

Dalsted, K.J.; Worcester, B.K. & Brun, L.J. (1979). Detection of Saline Seeps by Remote Sensing Techniques. *Photogrammetric Engineering & Remote Sensing*, Vol. 45, No. 3 (March 1979), pp. 285-291, ISSN 0099-1112

Dehaan, R.L. & Taylor, G. R. (2002). Field-derived spectra of salinized soils and vegetation as indicators of irrigation-induced soil salinization. *Remote Sensing of Environment*, Vol. 80, No. 3 (June 2002), pp. 406-417, ISSN 0034-4257

Dehaan, R.L. & Taylor, G.R. (2003). Image-derived spectral endmembers as indicators of salinisation. *International Journal of Remote Sensing*, Vol. 24, No. 4 (February 2003), pp. 775-794, ISSN 0143-1161

De Martonne, E. (1926). L'indice d'aridité. *Bulletin de l'Association des geógraphes français*, Vol. 9, pp. 3-5, ISSN 0004-5322

De Paz, J.M.; Sánchez, J. & Visconti, F. (2006). Combined use of GIS and environmental indicators for assessment of chemical, physical and biological soil degradation in a Spanish Mediterranean region. *Journal of Environmental Management*, Vol. 79, No. 2 (April 2006), pp. 150-62, ISSN 03014797

Dutkiewicz, A.; Lewis, M. & Ostendorf, B. (2009). Evaluation and comparison of hyperspectral imagery for mapping surface symptoms of dryland salinity. *International Journal of Remote Sensing*, Vol. 30, No. 3 (February 2009), pp. 693-719, ISSN 0143-1161

Dwivedi, R.S.; Ramana, K.V.; Thammappa, S.S. & Singh, A. N. (2001). The Utility of IRS-1C LISS-Ill and PAN-Merged Data for Mapping Salt-Affected Soils. *Photogrammetric Engineering & Remote Sensing*, Vol. 67, No. 10 (October 2001), pp. 1167-1175, ISSN 0099-1112

European Commission. (2002). *Towards a Thematic Strategy for Soil Protection. COM(2002) 179 final*. European Commission, Brussels, Belgium.

European Commission. (2003). *Extent, causes, pressures, strategies and actions that should be adopted to prevent and to combat salinization and sodification in Europe*. Directorate General Environment, Directorate B, Erosion Working Group (Task 5; Topic: Salinization and Sodification), Brussels, Belgium.

FAO. (2002). *Crops and Drops, making the best use of water for agriculture*, Food and Agriculture Organization of the United Nations, Rome, Italy.

Farifteh, J.; Van Der Meer, F.; Van Der Meijde, M. & Atzberger, C. (2008). Spectral characteristics of salt-affected soils: A laboratory experiment. *Geoderma*, Vol. 145, No. 3-4 (June 2008), pp. 196-206, ISSN 0016-7061

Ferreira, M.E.; Ferreira, L.G.; Sano, E.E. & Shimabukuro, Y.E. (2007). Spectral linear mixture modelling approaches for land cover mapping of tropical savanna areas in Brazil. *International Journal of Remote Sensing*, Vol. 28, No. 2 (January 2007), pp. 413-429, ISSN 0143-1161

Foody, G.M. (2002). Status of land cover classification accuracy assessment. *Remote Sensing of Environment*, Vol. 80, No. 1 (April 2002), pp. 185-201, ISSN 0034-4257

Franco-Lopez, H.; Ek, A.R., & Bauer, M.E. (2001). Estimation and mapping of forest stand density, volume, and cover type using the k-nearest neighbors method. *Remote Sensing of Environment*, Vol. 77, No. 3 (September 2001), pp. 251-274, ISSN 0034-4257

Ghrefat, H.A. & Goodell, P.C. (2011). Land cover mapping at Alkali Flat and Lake Lucero, White Sands, New Mexico, USA using multi-temporal and multi-spectral remote sensing data. *International Journal of Applied Earth Observations and Geoinformation*, Vol. 13, No. 4 (August 2011), pp. 616-625, ISSN 0303-2434

Green, A.A.; Berman, M.; Switzer, P. & Craig, M.D. (1988). A transformation for ordering multispectral data in terms of image quality with implications for noise removal. *IEEE Transactions on Geoscience and Remote Sensing*, Vol. 26, No. 1 (January 1988), pp. 65-74, ISSN 0196-2892

Haapanen, R.; Ek, A.R.; Bauer, M.E. & Finley, A.O. (2004). Delineation of forest/nonforest land use classes using nearest neighbor methods. *Remote Sensing of Environment*, Vol. 89, No. 3 (February 2004), pp. 265-271, ISSN 0034-4257

Hernández Bastida, J.A.; Vela de Oro, N. & Ortiz Silla, R. (2004). Electrolytic Conductivity of Semiarid Soils (Southeastern Spain) in Relation to Ion Composition. *Arid Land Research and Management*, Vol. 18, No. 3 (July 2004), pp. 265-281, ISSN 15324982

Howari, F.M., Goodell, P.C. & Miyamoto, S. (2000). Spectral properties of salt crusts formed on saline soils. *Journal of Environmental Quality*, Vol. 31, No. 5 (September 2002), pp. 1453-61, ISSN 00472425

ITT VIS. (2008). *ENVI 4.5 User's Guide*. ITT Visual Information Solutions Boulder (CO), USA.

Jensen, J.R. (2005). *Introductory Digital Image Processing* (3rd Edition). Prentice Hall, ISBN 0131453610, Upper Saddle River (NJ), USA

Jensen, J. R. (2007). *Remote Sensing of the Environment: An Earth Resource Perspective* (2nd edition), Pentice Hall, ISBN 0131889508, Upper Saddle River (NJ), USA

Kaufman, Y.J. (1989). The atmospheric effect on remote sensing and its corrections. In: *Theory and Applications of Optical Remote Sensing*, G. Asrar, (Ed.), pp. 336-428, Wiley-Interscience, ISBN 0471628956, New York, USA

Lal, R., Mokma, D. & Lowery, B. (1999). Relation between soil quality and erosion. In: *Soil quality and soil erosion*, R. Lal, (Ed.), pp. 237-258, CRC Press, ISBN 1574441000, Boca Raton (FL), USA

Landis, J.R. & Koch, G. G. (1977). The measurement of observer agreement for categorical data. *Biometrics*, Vol. 33, No. 1 (March 1977), pp. 159-174, ISSN 0006341X

Lillesand, T.M.; Kiefer, R.W. & Chipman, J.W. (2003). *Remote Sensing and Image Interpretation* (5th edition). John Wiley and Sons, ISBN 9812530797, Hoboken (NJ), USA

Majaliwa, J.G.M.; Twongyirwe, R.; Nyenje, R.; Oluka, M.; Ongom, B.; Sirike, J.; Mfitumukiza, D.; Azanga, E.; Natumanya, R.; Mwerera, R. & Barasa, B. (2010). The Effect of Land Cover Change on Soil Properties around Kibale National Park in South Western Uganda. *Applied and Environmental Soil Science*, Article ID 185689, ISSN 1687-7667

Mather, P.M. (2004). *Computer Processing of Remotely-Sensed Images: An Introduction* (3rd edition). Wiley, ISBN 0470849185, Chichester, UK

Melendez-Pastor, I.; Navarro-Pedreño, J.; Koch, M., & Gómez, I. (2010). Applying imaging spectroscopy techniques to map saline soils with ASTER images. *Geoderma*, Vol. 158, No. 1-2 (August 2010), pp. 55-65, ISSN 0016-7061

Melendez-Pastor, I.; Navarro-Pedreño, J.; Koch, M., & Gómez, I. (2010). Multi-resolution and temporal characterization of land use classes in a Mediterranean wetland with land cover fractions. *International Journal of Remote Sensing*, Vol. 31, No. 20 (October 2010), pp. 5365-5389, ISSN 0143-1161

Mermut, A.R., & Eswaran, H. (2001). Some major developments in soil science since the mid-1960s. *Geoderma*, Vol. 100, No. 3-4 (May 2001), pp. 403 -426, ISSN 0016-7061

Metternicht, G.I. & Zinck, J.A. (2003). Remote sensing of soil salinity: Potentials and constraints. *Remote Sensing of Environment*, Vol. 85, No. 1 (April 2003), pp. 1-20, ISSN 0034-4257

Metternicht, G. & Zinck, J.A. (1997). Spatial discrimination of salt- and sodium-affected soil surfaces. *International Journal of Remote Sensing*, Vol. 18, No. 12 (June 1997), pp. 2571-2586, ISSN 0143-1161

Moran, M.S.; Jackson, R.D.; Slater, P.N. & Teillet, P.M. (1992). Evaluation of simplified procedures for retrieval of land surface reflectance factors from satellite sensor output. *Remote Sensing of Environment*, Vol. 41, No. 1-2 (August-September 1992), pp. 169-184. ISSN 0034-4257

Mougenot, B.; Pouget, M. & Epema, G.F. (1993). Remote sensing of salt affected soils. *Remote Sensing Reviews*, Vol. 7, No. 3-4, pp. 241-259, ISSN 02757257

Mulder, V.L.; De Bruin, S.; Schaepman, M.E. & Mayr, T. R. (2011). The use of remote sensing in soil and terrain mapping—A review. *Geoderma*, Vol. 162, No. 1-2 (April 2011), pp. 1-19, ISSN 0016-7061

Nelson, D.W. & Sommers, L. E. (1982). Total carbon, organic carbon, and organic matter. In: *Methods of Soil Analysis*. A.L. Page; R.H. Miller & D.R. Keeney (Eds.), pp. 539-579, American Society of Agronomy (ASA),Madison (WI), USA

Pérez-Sirvent, C.E.; Martínez-Sánchez, M.J.; Vidal, J. & Sánchez, A. (2003). The role of low-quality irrigation water in the desertification of semi-arid zones in Murcia, SE Spain. *Geoderma*, Vol. 113, No. 1-2 (April 2003), pp. 109-125, ISSN 0016-7061

Qadir, M.; Ghafoor, A., & Murtaza, G. (2000). Amelioration strategies for saline soils: A review. *Land Degradation & Development*, Vol. 11, No. 6 (November-December 2000), pp. 501-521, ISSN 1085-3278

Quarmby, N.A.; Townshend, J.R.G.; Settle, J.J.; White, K.H.; Milnes, M.; Hindle, T.L. & Silleos, N. (1992). Linear mixture modelling applied to AVHRR data for crop area estimation. *International Journal of Remote Sensing*, Vol. 13, No. 3 (February 1992), pp. 415-425, ISSN 0143-1161

Reddy, K.R.; D'Angelo, E.M. & Harris, W.G. (2000). Biogeochemistry of Wetlands. In: *Handbook of Soil Science*, M.E. Sumner, (Ed.), p. G-89-G-119, CRC Press, ISBN 0849331366, Boca Raton (FL), USA

Richards, L.A. (1954). *Diagnosis and improvement of saline and alkali soils. Agriculture Handbook No. 60*. United States Department of Agriculture (USDA), Washington D.C., USA.

Richardson, A.J.; Gerbermann, A.H.; Gausman, H.W. & Cuellar, J.A. (1976). Detection of Saline Soils with Skylab Multispectral Scanner Data. *Photogrammetric Engineering & Remote Sensing*, Vol. 42, No. 5 (May 1976), pp. 679-684, ISSN 0099-1112

SPSS (2009). *PSAW 18.0 for Windows Help System*. IBM Corpotation, Armonk (NY), USA

Samaniego, L. & Schulz, K. (2009). Supervised Classification of Agricultural Land Cover Using a Modified k-NN Technique (MNN) and Landsat Remote Sensing Imagery. *Remote Sensing*, Vol. 1, No. 4 (December 2009), pp. 875-895, ISSN 2072-4292

Schmid, T.; Koch, M. & Gumuzzio, J. (2009). Applications of hyperspectral imagery to soil salinity mapping. In: *Remote Sensing of Soil Salinization: Impact and Land Management*, G. Metternicht & J.A. Zinck, (Eds.), pp. 113-140. CRC Press, ISBN 1420065025, Boca Raton (FL), USA

Settle, J.J. & Drake, N.A. (1993). Linear mixing and the estimation of ground cover proportions. *International Journal of Remote Sensing*, Vol. 14, No. 6 (March 1993), pp. 1159-1177, ISSN 0143-1161

Shimabukuro, Y.E. & Smith, J.A. (1991). The least-squares mixing models to generate fraction images derived from remote sensing multispectral data. *IEEE Transactions on Geoscience and Remote Sensing*, Vol. 29, No. 1 (January 1991), pp. 16-20, ISSN 0196-2892

Sitte, P.; Ziegler, H.; Ehrendorfer, F. & Bresinsky, A. (1994). *Strasburger. Tratado de Botánica* (8th Spanish edition), Omega, ISBN 8428209790, Barcelona, Spain

Small, C. & Lu, J.W.T. (2006). Estimation and vicarious validation of urban vegetation abundance by spectral mixture analysis. *Remote Sensing of Environment*, Vol. 100, No. 4 (February 2006), pp. 441-456, ISSN 0034-4257

Soil Survey Staff. (2006). *Keys to Soil Taxonomy* (10th Edition).United States Department of Agriculture (USDA), Washington D.C., USA

Song, C.; Woodcock, C.E.; Seto, K.C.; Lenney, M.P. & Macomber, S.A. (2001). Classification and Change Detection Using Landsat TM Data: When and How to Correct Atmospheric Effects? *Remote Sensing of Environment*, Vol. 75, No. 2 (February 2001), pp. 230-244, ISSN 0034-4257

Townsend, P.A. & Walsh, S.J. (2001). Remote sensing of forested wetlands: application of multitemporal and multispectral satellite imagery to determine plant community composition and structure in southeastern USA. *Plant Ecology*, Vol. 157, No. 2 (December 2001), pp. 129-151, ISSN 1385-0237

UNEP. (1997). *World Atlas of Desertification* (2nd edition). United Nations Environment Programme (UNEP), ISBN 0340691662, Nairobi, Kenya

Walkley, A. & Black, I.A. (1934). An Examination of the Degtjareff Method for Determining Soil Organic Matter, and A Proposed Modification of the Chromic Acid Titration Method. *Soil Science*, Vol. 37, No. 1 (January 1934), pp. 29-38, ISSN 0038-075X

Weng, Y.; Gong, P. & Zhu, Z. (2008). Reflectance spectroscopy for the assessment of soil salt content in soils of the Yellow River Delta of China. *International Journal of Remote Sensing*, Vol. 29, No. 19 (October 2008), pp. 5511-5531, ISSN 0143-1161

6

Remote Sensing Based Modeling of Water and Heat Regimes in a Vast Agricultural Region

A. Gelfan[1], E. Muzylev[1], A. Uspensky[2], Z. Startseva[1] and P. Romanov[3]
[1]*Water Problem Institute of Russian Academy of Sciences, Moscow*
[2]*State Research Center of Space Hydrometeorology Planeta, Moscow*
[3]*City College of City University of New York, New York, NY*
[1,2]*Russia*
[3]*USA*

1. Introduction

In many parts of the world, solution of the problem of water security is closely associated with possibility and predictability of access to soil water which accounts, globally, for approximately 60% of total rainfall. Soil water, so-called "green water", is fundamentally significant for maintaining sustainability of ecosystems and providing water for agriculture (in average, agriculture demands about 70% of global water use, and over 90% in some regions (Global Water Security, 2010). In addition, changes in the store of soil water due to evapotranspiration, infiltration of surface water and other water cycle processes strongly affect land-atmosphere interactions on intraseasonal to interannual timescales (Yang, 2004) and understanding physical mechanisms of these processes is central for effective modeling mesoscale atmospheric circulation.

Taking into account a vital role of "green water" both for ecosystems and agriculture, there is a need for physically based models allowing to describe land-surface processes in their interaction with atmosphere and to predict soil water availability under changing environment and climate. Processes of mass and energy exchange at the land surface, linking the atmosphere and soil water are of interest to several geophysical disciplines; particularly, the possibility of developing adequate models of these processes was studied intensively by hydrological and meteorological communities for a few decades. Recently, a huge contribution to our understanding these processes has been made by remote sensing community. Combination of these efforts results in intensive developing so called land-surface models (LSMs), considering as a key tool to predict successfully the likely future states of the terrestrial systems under anthropogenic pressure and climate change. There are a lot of reviews providing a detailed and comprehensive discussion of LSMs (e.g. Sellers et al., 1997; Pitman, 2003; Overgaard et al., 2006). Most of them are concentrated on the development of LSMs designed for use in weather and climate models. Below a review of LSMs will be done with emphasis on their availability to capture processes controlling soil water dynamics, and primarly, evapotranspiration. Additionally, focus will be on utilization of remotely sensed data in LSMs.

According to Sellers et al. (1997), the LSMs could be broadly divided onto three generations distinguishing largely on the level of complexity of the evapotranspiration process description. The first generation, developed in the late 1960s and 1970s, used simple bulk aerodynamic transfer formulations. Land surface considers as homogeneous one in the first-generation LSMs and spatially uniform surface parameters (water-holding capacity, albedo, roughness length etc.) are used, i.e. these models do not discriminate soil evaporation and transpiration. Examples are the "bucket" model by Manabe (1969) (probably, the earliest first-generation LSM) and more recent TOPUP model by Schultz et al. (1998), PROMET model by Mauser & Schädlich (1998). Use of the single value of the aerodynamic resistance regardless of the surface type is one of the common simplifications and major conceptual limitations of the first-generation LSMs (Pitman, 2003). Another one is that these models include a single soil layer for soil moisture and temperature simulations. According to the findings of the Project for the Intercomparison of Landsurface Parameterisation Schemes (PILPS; Henderson-Sellers et al., 1995), soil moisture dynamics simulated by the first-generation LSMs can not be often reproduced adequately by these models because of aforementioned conceptual limitations.

The second generation of the LSMs, developed in the early 1980s, explicitly represented, in contrast to the first-generation models, the influence of vegetation on the interaction between land surface and atmosphere. Taking into account difference between soil and vegetation provided an opportunity to begin integrating satellite data into LSMs (Pitman, 2003). Also, these models improve soil moisture representation replacing the simple conceptual scheme of Manabe (1969) by more sophisticated parametrizations (particularly, the vertically distributed Richards equation is often used). Such a new direction in land surface modeling took roots from a pioneering work of Deardorff (1978) developed a model for simulating soil moisture and heat transfer in two layers and vegetation as a single layer controlling heat balance on a land surface, as well as from the works of Dickinson (1986) and Sellers et al. (1986) developed the Biosphere Atmosphere Transfer Scheme (BATS) and Simple Biosphere Model (SiB) based on the ideas of Deardorff (1978). During the subsequent decades, a huge number of the second-generation LSMs have been developed; examples include SVAT (Kuchment & Startseva, 1991), VIC (Wood et al., 1992), BASE (Desborough & Pitman, 1998), CLM (Dai et al., 2002), SWAP (Gusev & Nasonova, 2002), LaD (Milly & Shmakin, 2002).

One of the principal advantages of the second-generation LSMs is their ability to consider snow processes affecting a major part of terrestrial water balance in cold regions. Snow sub-models of different vertical discretization have been implemented in LSMs and an intensive work on their evaluation and intercomparison (e.g. the recent SnowMIP2 Project (Rutter et al., 2009)) has been made. Additionally, some LSMs include parametrization of hydrothermal processes in a frozen soil (Wood et al., 1992; Gusev & Nasonova, 2002 among others) but an adequate description of these processes strongly affected soil moisture content before vegetation season still remains problematic (Pitman, 2003).

A step forward of the third-generation LSMs is in explicit description of a canopy physiology including biophysical mechanisms of stomatal conductance, photosynthesis, plant growth, etc. This ability opens up new opportunities not only to improve reproduction of evapotranspiration but, importantly, to address carbon exchange by plant. Description of the other land surface processes (soil moisture and temperature dynamics, snow processes, etc.) is similar to one utilized in the second-generation LSMs. One of the first LSMs of third generation were developed by Collatz et al. (1991) and Sellers et al. (1992), some recent examples representing improvements of the aforementioned second-generation LSMs include modifications of the SVAT-model (Kuchment et al., 2006), the CLM-model (Oleson et al., 2008), the SiBcrop model (Lokupitiya et al., 2009).

One of the major concerns restricting availability of LSMs, particularly on a regional scale, is the issue of spatial heterogeneity of land surface characteristics (soil, vegetation, topography) required for assigning the model parameters. Regional (mesoscale) heterogeneity is not captured by the existing ground-based observational network that leads to an excessive aggregation of the parameters and, as a rule, to decrease in accuracy of reproducing spatial distribution of the desired processes. Significant improvement of performance of LSMs can be reached by assimilation of information that is additional to the ground observations, first of all, information on land surface provided by satellite remote sensing. Remote sensing allows substituting the missed ground observations by measurements of the incoming and outgoing land surface radiation fluxes conversed into physical distributed parameters. No exaggeration to say that spatial data on land surface derived from remote sensing is the only source of the distributed parameters for LSMs at regional scale. Additionally, these data can be used for model evaluation purposes. Applicability of satellite remote sensing for improvement of the land surface modeling is reviewed, e.g. by Overgaard et al. (2006), including results of a number of field experiments (FIFE, HAPEX, KUREX and others). However, in spite of the fact that quantity and quality of satellite products have largely increased for the last decade and they have recognized as a potentially valuable source of distributed information, the majority of satellite products still needs considerable improvement and applicability of these products, even being improved, should be verified both for the specific region and for the used LSM.

This paper has three major objectives. First, the existing satellite-derived data of land surface and snow cover characteristics will be overviewed in brief and the specific developed technologies of the satellite data thematic processing will be presented for the study region located within the agricultural Black Earth area of the European Russia. Secondly, structure of the physically based distributed Remote Sensing Based Land Surface Model (RSBLSM) developed for simulation of vertical water-and-heat transfer in vegetation, unfrozen and frozen soil, snow cover will be described and the results of its testing against the available ground-based observations will be shown. Finally, the results of utilizing satellite-derived land surface and snow characteristics as the parameters and input variables of RSBLSM, as well as for the model evaluation, will be demonstrated and discussed.

2. Case study

The case study has been carried out for agricultural Black Earth region of the European Russia of 227,300 km^2 located in the steppe-forest physiographic zone (Fig. 1)

Relief of the region is low middle-hilly plain dissected by broad river valleys, ravines and gullies. Dominant slopes are ~1-4°. Absolute elevation marks of surface in the region are in the range of 150-260 m. Annual net radiation is 27-32 kcal/cm^2, and the sum of incoming solar radiation during the summer months is 41-44 kcal/cm^2. Annual precipitation is 519 mm, over 40% of which falls as snow. Snow water equivalent significantly varies over the area. At the northern part of the region snow cover deceases in the middle of April, southern parts are snow-free in March. Maximum rainfall (60-70 mm) falls as a rule in July. Ground water level lies at the depth of 15-30 m between rivers and of 3-5 m in river valleys.

Soils are mainly chernozem (podzolized, leached and typical), small part of the territory is occupied by floodplain meadow and gray soil. In texture, soils mostly relate to the loam. Sandy loam and sand are found rarer. In the north-western part of the region gray forest-steppe soils with patches of degraded chernozem are located. When moving to the south gray soils give place to leached chernozem often occurring in combination with powerful chernozem. The southeastern part of the region is occupied by ordinary chernozem. Most of the region territory is under cultivation, the natural vegetation is preserved mainly in river valleys and on slopes of gullies and ravines. Plough-lands compose the most part of the region (78 %), forests occupy about 5%; pastures take up about 16%; urbanized lands occupy less than 1%. The main crops are cereals (spring wheat and barley, winter wheat, less corn, buckwheat and rye) as well as sugar beets, potatoes and forage grasses. Ratio of crops in different households differs substantially. Region-averaged grain wedge is about 60% of all plough-lands. In the region there are 48 agricultural meteorological stations at which observations on meteorological characteristics and soil moisture under different crops, as well as snow cover characteristics are conducted. At several of them measurements of the evaporation pans are also made.

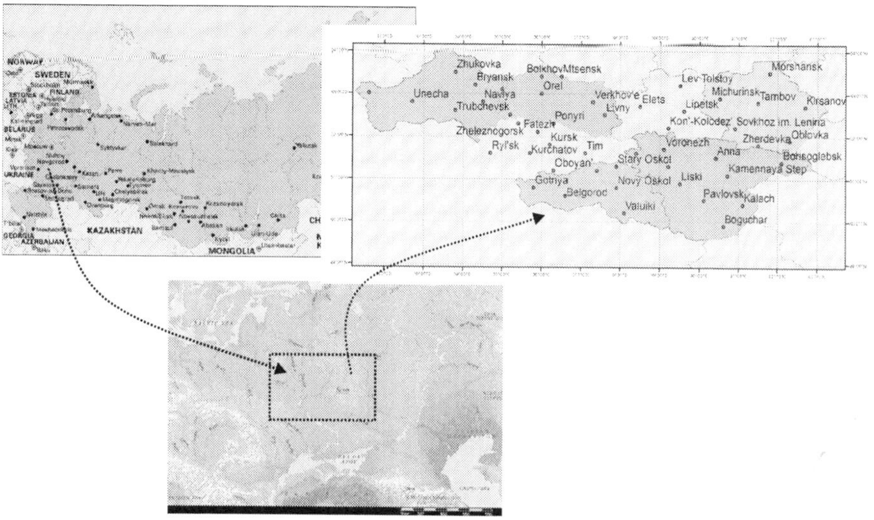

Fig. 1. Study region (points at the right upper map represent location of meteorological stations)

To assign the model parameters, most of which are the measured soil, vegetation and snow cover characteristics as well as to calibrate and validate the model the data have been attracted of above observations at agricultural meteorological stations for 21 years from 1971 to 2010.

3. Remote sensing of agricultural regions: products and algorithms of data processing

3.1 Remote sensing of land surface and vegetation characteristics

This section contains a brief overview of satellite instruments and methods for remote sensing of various land surface parameters, valuable in particular for hydrological applications. During the past decades, a series of sensors have been developed and launched, such as the Advanced Very High Resolution Radiometer (AVHRR) and the MODerate resolution Imaging Spectroradiometer (MODIS), which are respectively onboard the polar-orbiting satellites NOAA and NASA EOS Terra and Aqua. The AVHRR instrument has 6 VIS and IR channels with spatial resolution of 1 km (NOAA KLM Users Guide, 2005) that are informative with respect to different land surface and vegetation parameters. The MODIS instrument has 36 channels in VIS and IR (Justice et al., 1998) with 1 km spatial resolution, which provides information on even more geophysical parameters of land surface, vegetation, atmosphere, etc. In recent years new geostationary satellites have been developed, such as European Meteosat Second Generation (MSG), i.e. Meteosat-8 and -9. Meteosat main payload is the optical imaging radiometer, the so called Spinning Enhanced Visible and Infrared Imager (SEVIRI). SEVIRI has 12 spectral channels from VIS to IR (Schmetz et al., 2002). It has channels similar to AVHRR, and the benefit is that SEVIRI provides measurements of the Earth-disc every 15 minutes comparing to the coverage of the polar-orbiting satellites (2 times a day for the same territory).

Among the above mentioned geophysical parameters we will focus on: land surface temperature (LST or T_{ls}), air foliage temperature (T_a), emissivity (E), as well as normalized difference vegetation index ($NDVI$), vegetation cover fraction (B) and leaf area index (LAI). LST is one of the important geophysical parameters. Together with the land surface spectral emissivity (LSE) the LST affect the heat and water transport between the surface and the atmosphere. There is a strong need in the remote sensing LST, since the conventional surface temperature observations are rather sparse (in space and time).

The possibilities of extracting LST and LSE information from thermal IR multichannel measurements in the "atmospheric window" spectral range (3.7-4.0, 10.5-12.5 μm) has been the subject of numerous investigations during last 20 years, see (Becker & Li, 1995).

Because of the land surface heterogeneity, the satellite measurements usually come from mixed pixels. At satellite pixel scale, LSE refers to the area-weighted and channel-averaged emissivity (E) and LST refers to the radiometric surface temperature corresponding to the Field Of View (FOV) of a radiometer (Becker & Li, 1995).

3.1.1 AVHRR- and MODIS-based remote sensing products

The methods have been developed and tested for AVHRR/NOAA data processing (Muzylev et al, 2002, 2005) that provide the retrieval of two types of LST (efficient radiation temperature $T_{s.eff}$ and land skin temperature T_g) and emissivity E as well as the derivation of air-vegetation temperature T_a and three vegetation characteristics, namely, vegetation index $NDVI$, fraction B and leaf area index LAI. The algorithms for AVHRR-based estimation of $T_{s.eff}$, T_g, T_a utilize cloud-free measurements in the split window channels 4 and 5 and linear statistical regression similar to well-known split window technique (Becker et al., 1995; PUM

LST, 2008; Wan et al., 1996). The values of emissivity E_4 and E_5 are specified *a priori* using one of alternate approaches. The required ancillary information is extracted from the classification-based emissivity model (Snyder et al., 1998) as well as from the empirical relationships between the emissivity and/or $NDVI$ / B, see (Muzylev et al., 2002, 2005). To estimate LAI, the empirical relationships between LAI and $NDVI$ (established for different land covers) have been applied (see Section 5.1, formulas (27) and (28)).

The threshold technique of cloud detection in the AVHRR FOV has been used that has allowed increasing the reliability of cloud-free fragment detection (Volkova & Uspenskii, 2007). The developed software package has been applied for AVHRR/NOAA cloud-free data thematic processing to generate named remote sensing products and cloud/precipitation parameters for various dates of the 1999-2010 vegetation periods. The error statistics of T_a, T_{sg} and $T_{s.eff}$ derivation has been investigated for various samples using comparison with synchronous collocated in-situ measurements that has given root-mean square (RMS) errors in the range of 1.5-2.0, 3.5-4.5, and 2.5-3.5°C respectively (Muzylev et al., 2005, 2006, 2010). The archive of synchronous AVHRR/NOAA measurements, remote sounding data, and in-situ hydro-meteorological observations has been compiled for the study area and its separate parts for 1999-2010 vegetation seasons.

The dataset of MODIS-based remote sensing products has also been compiled on the base of special technology using LP DAAC web-site https://lpdaac.usgs.gov, that includes estimates of land surface temperature (LST) T_{ls}, E, $NDVI$, LAI for the region of interest and 2003-2010 vegetation seasons. Two types of MODIS-based T_{ls} and E estimates have been extracted (for separate dates of the named time period): LST/E Daily L3 product (MOD11B1) with spatial resolution ~ 4.8 km and LST/E 5-Min L2 product (MOD11_L2) with spatial resolution ~ 1 km. The verification of T_{ls} estimates has been performed by the comparison against analogous and collocated AVHRR-based ones (Muzylev et al, 2010).

3.1.2 SEVIRI-based land surface and land air surface temperature estimates

In the recent years there were a lot of studies on LST derivations from SEVIRI/MSG data, see (PUM LST, 2008; Solovjev et al., 2009, 2010). In the State Research Center of Space Hydrometeorology "Planeta" (Moscow, Russia) the new methodology has been developed for the derivation of LST and LSE from cloud-free brightness temperatures measured in the SEVIRI channels 9 (10.8 μm) and 10 (12.0 μm) at three different times based on combination of two well-known techniques, i.e. split-window method (Wan et al., 1996) and two temperature method, see (Faysash et al., 2000) with additional hypothesis that the emissivity values E_9 and E_{10} remain constant during the time interval between the first and the last image cycles used. The detailed description of the proposed technique can be found in (Uspensky et al., 2009; Solovjev et al., 2010). It is important to mention that unlike the technique from (PUM LST, 2008), this method does not require the accurate knowledge of emissivity (E) in the split-window channels.

The method described has been used for the period of 2009-2011 to produce LST estimates over Europe and, in particular, for Black Earth zone of Russia. An example image of this LST mapping is presented in Fig. 2.

The comparison with synchronous and collocated LST products from Satellite Application Facility on Land Surface Analysis, SAF LSA (Lisbon, Portugal) (PUM LST, 2008) has been

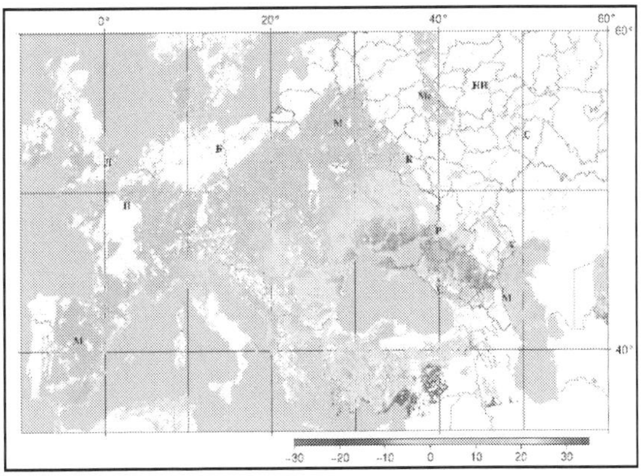

Fig. 2. *LST* map, 27.09.2011, 07:00 UTC

performed to validate the results over the central Europe. Good level of correlation has been reached, which could be treated as indirect proof of the method's efficiency. RMS deviation between the above mentioned LST estimates is in the range of 0.9-3.0°C. The upper limit could be shifted down to 2.4°C by subtracting systematic biases.

An additional validation has been performed through the inter-comparison with MODIS-based *LST* for the study region, see Fig. 3.

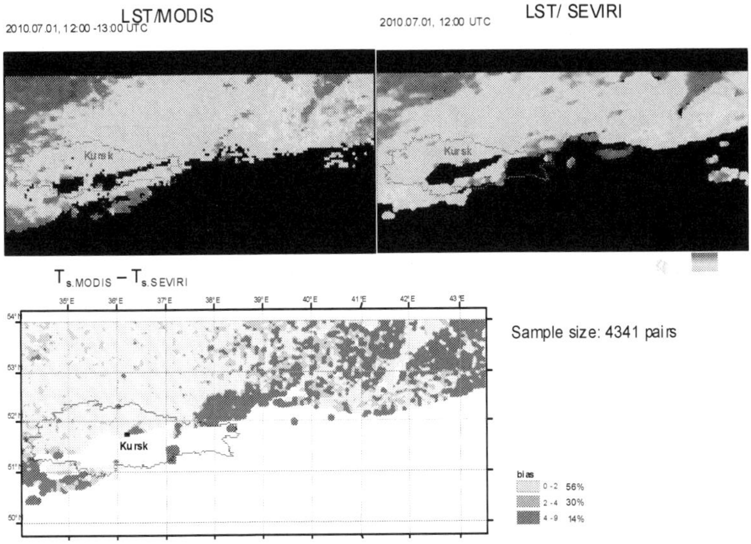

Fig. 3 Inter-comparison results for MODIS- and SEVIRI-based *LST* estimates

It is obvious that both *LST* maps are similar to each other. The discrepancy is rather small (0.0-2.0°C) for the most of the territory.

Along with this, a very important parameter such as land surface-air temperature T_a can also be derived from SEVIRI data. A new method of T_a derivation has been developed in (Uspensky et al., 2011). A multiple linear regression model has been constructed that estimates air temperature from satellite-observed LST, solar zenith angle (or related day of the year number), and land elevation. Land surface temperatures estimates from above were used in this scheme to calculate regression coefficients, as well as time-synchronous collocated in-situ measurements from the ground observation network. The development and validation experiments were carried out for the study area for the vegetation seasons of 2006-2009. Data from the above 48 agricultural meteorological stations were used for 8 standard synoptic times a day. All the data were subdivided on separate training and validation data sets. For vegetation season of 2009, an overall bias and standard deviation of calculations are approximately 0 and 1.9-2.1°C, respectively. The performance of the method is similar to the one presented in (Good, 2009).

3.2 Remote sensing of snow cover

Seasonal snow cover is among the most important factors for mid-latitude agriculture. Water accumulated in the snow pack in winter and released through the snowmelt is critical for crop development in early spring. For winter grain crops that are widely grown in Canada, Russia and Ukraine, snow pack presents an insulation material protecting the plants from freezing temperatures. Accurate information on the snow cover extent and variations during the winter season helps to identify areas of potential winterkill and to predict adverse conditions for crops development (e.g. Romanov, 2011).

Although most information on the snow cover distribution and properties for agriculture is traditionally obtained from in-situ observations at ground-based meteorological stations, the use of satellite-based snow products is becoming increasingly popular. First, this section presents a brief overview of current satellite-based snow mapping techniques and of snow products that can be used in agricultural applications. Then, a new technique used in this study and based on synergy of optical and microwave snow cover products available from MODIS and Advanced Microwave Sounding Radiometer (AMSR-E) instruments onboard EOS satellites Terra and Aqua is described.

3.2.1 Existing snow mapping techniques and products

To derive information on snow cover from satellite data a number of different techniques, both interactive and automated has been developed and is actively used. The most popular interactive snow product based on visual inspection of satellite optical imagery is Northern Hemipshere snow charts that have been generated by NOAA since 1972. Interactive maps of snow cover are currently produced within a computer-based Interactive Multisensor Snow and Ice Mapping System (IMS) that provides software tools and access to various datasets to facilitate the image analysis and map drawing by human analysts (Helfrich et al, 2007). Daily IMS snow maps are generated at 4 km resolution and are available from the National Snow and Ice Data Center (NSIDC) at http://nsidc.org/data/g02156.html. Despite some weaknesses associated primarily with subjectivity in the image analysis and interpretation NOAA IMS snow maps present a robust and consistent product. The overall accuracy of daily interactive snow maps is quite high with the yearly mean rate of agreement with surface observations data exceeding 90% (e.g., Brubaker et al., 2005). Most errors occur during fast and large-scale snow advance or retreat or when persistent cloudiness obscures the land surface.

In contrast to interactive techniques the interest to automated algorithms is attracted due to their low exploitation costs and ability to better utilize potentials of satellite data, particularly their high spatial resolution and multi-spectral capability. The two major techniques for mapping and monitoring snow cover from satellites are based, correspondingly on passive observations in the microwave and in the optical (visible to infrared) spectral range.

The advantage of using microwave sensors consists in their ability to "see" through most clouds and to provide information (although quite limited) on snow depth and snow water equivalent (SWE). The primary limitations of microwave measurements are associated with their coarse spatial resolution of 25-50 km, poor sensitivity to shallow and melting snow (Walker and Goodison, 1993) and difficulty to distinguish between snow and frozen rocks and soil (Grody and Basist, 1996). Snow depth and SWE estimates from observations in the microwave are limited only to dry snow packs while corresponding retrieval errors range typically between 50 and 100% (e.g. Kelly et al., 2003). Global monitoring of snow with microwave sensors data started in 1978 with the launch of Nimbus-7 with Scanning Multichannel Microwave Radiometer (SMMR) onboard and continued with a number of other sensors, including in particular, Special Sensor Microwave Imager (SSM/I) on Defense Meteorological Satellite Program (DMSP) satellites since 1987 (Armstrong and Brodzik 2005) and AMSR-E onboard EOS Aqua satellite since 2002 (Kelly et al, 2003).

As compared to satellite passive microwave measurements, observations in the optical spectral range allow for more accurate mapping of snow cover at higher spatial resolution. The mean accuracy of snow identification in optical bands usually exceeds 90%, but drops to 80-90% over dense coniferous forests (Simic, et al, 2004, Hall & Riggs, 2007). Daily global snow cover maps routinely generated with data from MODIS onboard NASA EOS Terra and Aqua satellites at 500 m spatial resolution (Hall et al, 2002) and from AVHRR onboard NOAA satellites at 4 km resolution (http://www.star.nesdis.noaa.gov/smcd/emb/snow/HTML/snow.htm). It is important that snow retrievals in the optical spectral bands are possible only under clear sky conditions. Partial improvement in the map area coverage can be achieved with geostationary satellites which provide multiple observations per day and hence increase the chance to see the land surface cloud clear (e.g, De Wildt et al., 2007). With geostationary satellites, however, the map coverage is only regional and is limited to the area within ~ 65° N and S.

Because of physical limitations of both principal snow remote sensing techniques, snow products generated with single sensor data lack either continuity or sufficient accuracy and spatial resolution and thus are hard to use in numerical model applications. In an attempt to improve satellite-based snow cover characterization several techniques have been proposed that combine snow cover observations in the optical and microwave spectral bands (e.g., Romanov et al. (2000), Brodzik et al. (2007), Foster et al. (2011)). The objective of these techniques is to maximize advantages offered by optical and microwave observations, to compensate for their weaknesses and to generate continuous snow maps at the highest possible spatial and temporal resolution. Most often in these algorithms optical snow retrievals are used in clear sky conditions, whereas microwave retrievals complement the optical data when cloudy.

3.2.2 Snow cover mapping through synergy of optical and microwave products from EOS satellites

A new technique used in this study is based on synergy of optical and microwave snow cover products available from MODIS and AMSR-E instruments onboard EOS satellites Terra and Aqua. The objective was to generate an advanced product providing continuous (gap free) characterization of the global snow cover distribution at 5 km spatial resolution at daily time step.

The algorithm utilizes two NASA daily snow products, MODIS snow cover map on a latitude-longitude grid at 5 km resolution (labeled by NASA as MOD10C1 and MYD10C1 correspondingly for MODIS Terra and Aqua) and Aqua AMSR-E-based snow water equivalent product AE_DySno. In the developed blending technique we took a cautious approach to the microwave data: microwave retrievals indicating no snow as well as retrievals over mountains were disregarded due to frequent omission of melting snow and shallow snow and frequent overestimates of snow cover in the mountains by microwave algorithms. The remaining microwave retrievals were used to complement snow cover distribution mapped with MODIS data in clear sky conditions. Within this approach some pixels in the daily map may remain undetermined. To eliminate these gaps in the coverage and to achieve continuity in the derived snow cover distribution pixels that remain undetermined in the current day snow map were filled in with the data from the previous day's blended snow map.

All available MODIS and AMSR-E snow products since 2002 have been reprocessed to derive almost 10-years-long time series of daily global snow cover maps. Snow maps in binary format are available at ftp://www.orbit.nesdis.noaa.gov/ pub/ smcd/ emb/ snow/ eos/. An example of a snow cover map generated through synergy of MODIS and AMSR-E data is presented in Fig. 4. To estimate the accuracy of the new snow product we compared it with available surface observations of snow cover. The comparison made for 9 consecutive winter seasons from 2002-2003 to 2010-2011 has shown that the yearly mean agreement of the blended MODIS and AMSR-E snow product to surface observations was 87%. This is only about 3% less than the accuracy of NOAA interactive snow maps estimated using the same method. The accuracy of the EOS blended product dropped to 80-85% in the middle of the winter season and increased to close to 100% in late spring, summer and early fall.

In order to assess consistency of satellite snow retrievals derived from the different satellites, we compared estimates of the derived snow covered area (SCA) from daily snow cover maps of MODIS Terra with ones of MODIS Aqua. The analysis has shown that the SCA derived from MODIS Terra and from MODIS Aqua changes synchronously.

Additionally, we compared dates of snow melt off as determined from satellite data with the dates of snow melt off as determined from the available ground-based observations. The comparison was performed for 48 meteorological stations. The results show that in most cases these dates differ by 1 to 10 days, however in some cases the difference exceeded two weeks (results of 2003 are shown in Fig. 5 as an example). Most probably the primary reason for the difference between satellite and surface estimates of the snow melt-off date consists in the difficulty of detecting shallow wet snow from satellites. However, the observed dates of snow cover decease averaged over the whole area are appeared to be very close to satellite-derived estimates of these dates.

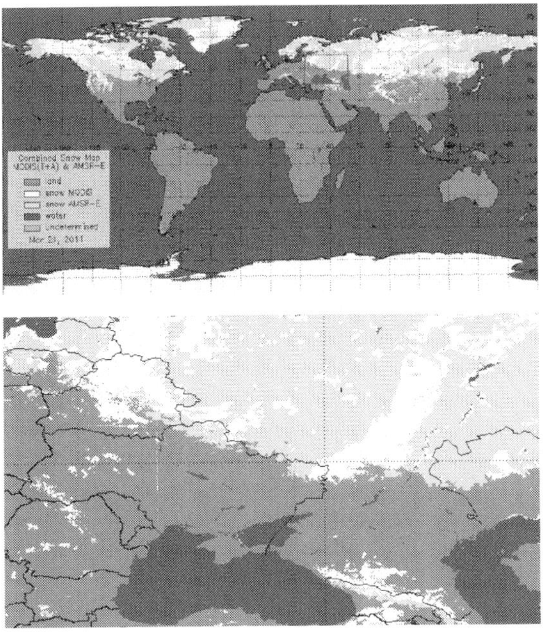

Fig. 4. Example of a blended global daily snow cover map derived from combined data of MODIS and AMSR-E instruments onboard Aqua and Terra satellites (upper) and zoomed in portion of the map covering Eastern Europe. Light blue color and white color represents snow cover identified correspondingly with microwave and optical satellite data (lower).

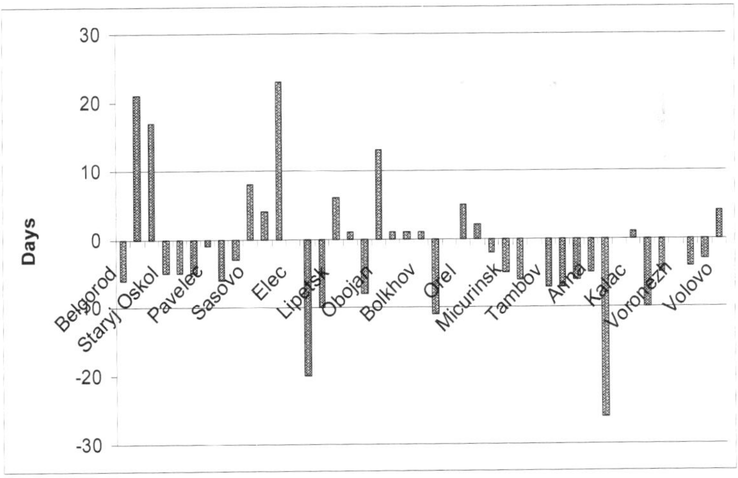

Fig. 5. Difference of snow melt off dates determined from satellite data and from the available ground-based observations

We consider the blended automated snow products as the most advanced ones providing a reliable and robust characterization of snow cover distribution at the satisfactory accuracy. Potential of using these products in conjunction with land surface model for reproducing snow cover distribution over the agricultural area will be demonstrated in section 5.2.

List of the satellite-derived data, both on land surface characteristics and snow, used in this study is shown in Table 1.

Land Surface Characteristic	**Name of Product**	**Sensor/ Satellite**	**Latitude-Longitude Resolution**	**Time Resolution**	**Time Period**
Land Surface Temperature	NASA_MOD11_B1	MODIS/ TERRA&AQUA	0.05°x0.05°	Twice per day	1 Apr to 31 Oct 2003-2010
	NASA_MOD11_L2		0.01°x0.01°	Twice per day	1 Mar to 30 May 2002, 1 Mar to 31 Oct 2003-2010
	Thematic processing from	AVHRR/NOAA	1′x1.5′	Twice per day	1 Apr to 31 Oct 1999-2010
		SEVIRI/ METEOSAT-9	0.06°x0.06°	Every 15min (at cloud-free condition)	1 Apr to 31 Oct 2009-2011
Land Surface Albedo	NASA_MOD043_C1	MODIS/TERRA	0.05°x0.05°	16-day product	1 Mar to 30 May 2003-2010
Snow Water Equivalent	NASA_AE-DySno	MODIS/AQUA	0.20°x0.20°	Once per day	20 Jan to 30 May 2003-2010
Snow Covered Area	NASA_MOD10_L2	MODIS/TERRA	0.01°x0.01°	Once per day	20 Jan to 30 May 2002-2010
Vegetation Cover Fraction	Thematic processing from	AVHRR/NOAA	1′x1.5′	Twice per day	1 Apr to 31 Oct 1999-2010
Leaf Area Index	Thematic processing from	AVHRR/NOAA	1′x1.5′	Twice per day	1 Apr to 31 Oct 1999-2010
	NASA_MOD15_A2	MODIS/ TERRA&AQUA	0.01°x0.01°	8-day product	1 Apr to 31 Oct 2003-2010
Land Cover Classification	Land Cover Type	AVHRR/NOAA	0.01°x0.01°	Static data generated from AVHRR data	
Tree Cover Fraction	Tree Cover Fraction				
Fraction of Evergreen Tree Cover	Evergreen Tree Cover				

Table 1. Satellite-derived products used in this study

4. Remote sensing based land surface model: Structure, calibration and validation by the ground-based data

The developed comprehensive Remote Sensing Based Land Surface Model (RSBLSM) contains as major components the model of vertical water and heat transfer in the "Soil-Vegetation-Atmosphere" system (SVAT) for vegetation season as well as the model of vertical water and heat transfer in the "Soil-SNow-Atmosphere" system (SSNAT) for cold season. The first versions of these models were developed in the 1990s (Kuchment et al., 1989; Kuchment & Startseva, 1991; Kuchment & Gelfan, 1996). Later, conceptualization of the model was improved through accounting for additional processes (Gelfan et al., 2004; Kuchment et al., 2006), and, importantly, the methods of the model adaption to remote sensing data on land surface were developed (Muzylev et al., 2002; 2005; 2010; Kuchment et al., 2010). Below RSBLSM components used in this paper are described in brief.

4.1 SVAT component of the RSBLSM system

The SVAT model is intended for simulating evaporation from bare soil, transpiration by vegetation, vertical latent and sensible heat fluxes, vertically distributed soil water and heat content, soil surface and foliage temperatures, land surface radiation temperature as well as other variables characterized water and heat regimes of soil-vegetation system during a warm season. The model flowchart is shown in Fig. 6.

Land surface is considered as a two-component soil-vegetation system. Water/heat fluxes incoming to and outgoing from bare soil and vegetation cover are accounted for separately. So evapotranspiration (Ev) is described as sum of two fluxes: bare soil evaporation (E_g) and transpiration by vegetation (E_f) as

$$E_g = \rho_a \cdot (r \cdot q^*(T_g) - q_a) / r_{ag} \quad (1a)$$

$$E_f = \rho_a \cdot (q^*(T_f) - q_a) \cdot LAI / (r_a + r_s) \quad (1b)$$

Sensible heat fluxes from surface of bare soil H_g and from vegetation cover H_f are calculated as:

$$H_g = \rho_a \cdot c_p \cdot (T_g - T) / r_{ag}, \quad (2a)$$

$$H_f = \rho_a \cdot c_p \cdot (T_f - T) / r_a \quad (2b)$$

where T_g and T_f are the soil surface and the foliage temperatures, respectively, T and q_a are the air temperature and specific air humidity at 2 m height, respectively; $q^*(T_g)$ and $q^*(T_f)$ are the specific air saturation humidity at the temperatures T_g and T_f, correspondingly, r is the relative air humidity near the soil surface, r_{ag} and r_a are the aerodynamic resistance between soil surfaces and foliage and between foliage and atmosphere, respectively,

$$r_{ag} = (C_g \cdot U)^{-1} \quad (3)$$

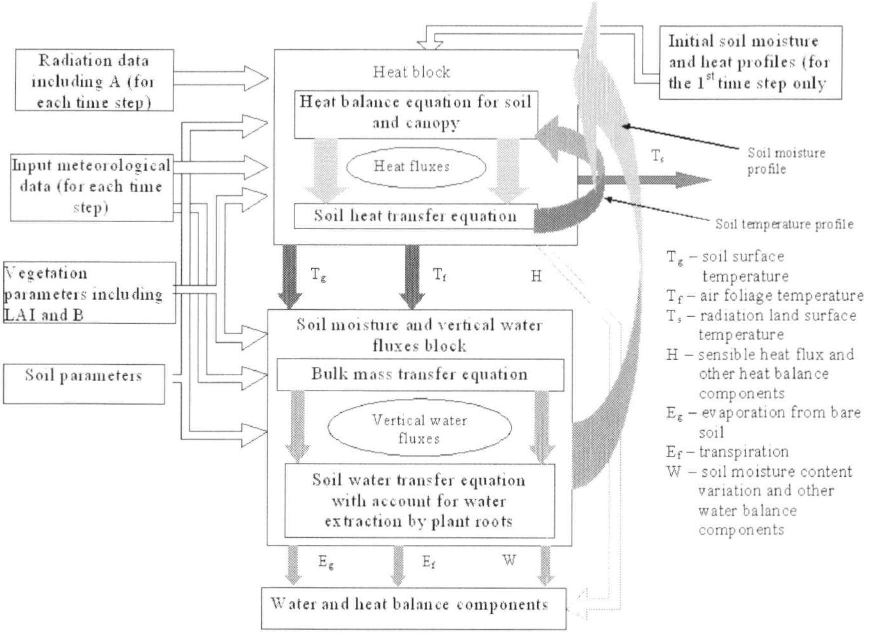

Fig. 6. The SVAT model flowchart

$$r_a = (C_e \cdot U)^{-1}, \tag{4}$$

where C_e and C_g are the vapour transfer coefficients at the canopy level and at the ground one, respectively, depending on the surface roughness; U is the wind velocity at 10 m height, r_s is the stomatal resistance calculated by formula

$$r_s = r_0 \frac{\psi(\theta_{wp}) - \varsigma\psi_f}{\psi(\theta_{wp}) - \psi_f} \tag{5}$$

r_0 is the minimum stomatal resistance, $\psi(\theta_{wp})$ is the soil matrix potential corresponding to wilting point θ_{wp}, ς is the coefficient, LAI is the leaf area index, ρ_a is the air density, c_p is the specific heat at constant pressure.

The relative air humidity r near the soil surface in (1a) is defined by formula

$$r = \exp \frac{Mg\psi(\theta)}{R(T_g + 273)} \tag{6}$$

where θ is the volumetric soil moisture content, $\psi(\theta)$ is the soil matrix potential, M is the molecular mass of water, g is the acceleration of gravity, R is the universal gas constant.

To describe water transfer in the soil, the equation of soil moisture diffusion is applied taking into consideration water extraction by plant roots

$$\frac{\partial\theta}{\partial t}=\frac{\partial}{\partial z}\left[D(\theta)\frac{\partial\theta}{\partial z}-K(\theta)\right]-R(\theta,z) \quad (7)$$

where $K(\theta)$ is the hydraulic conductivity of soil, $D=K(\partial\psi/\partial\theta)$ is the soil moisture diffusion coefficient, $R(\theta,z)$ is the intensity of water extraction by plant roots

$$R(\theta,z)=-K(\theta)\cdot[\psi_r-\psi(\theta)]\cdot b_r\cdot\rho_r(z) \quad (8)$$

ψ_r and $\rho_r(z)$ are the rootage water potential and the density, respectively, b_r is the coefficient.

The soil matrix potential $\psi(\theta)$ and the soil hydraulic conductivity $K(\theta)$ can be assigned using different parametrizations. We compared some of them (Brooks & Corey, 1964; Clapp & Hornberger, 1978; van Genuhten, 1980) and found that they give similar simulation results. Hereafter, van Genuhten's formulas are used:

$$\psi(\theta)=-\frac{\left(S^{-1/m}-1\right)^{1/n}}{\alpha} \quad (9)$$

$$K=K_0S^{0.5}\left[1-\left(1-S^{1/m}\right)^m\right]^2 \quad (10)$$

where $S=(\theta-\theta_r)(\theta_s-\theta_r)^{-1}$ is the relative saturation; θ_s and θ_r are the residual and the saturated water contents respectively; K_0 is the saturated hydraulic conductivity; $\alpha>0$ is the parameter, which is related to the inverse of the air entry pressure; $m=1-n^{-1}$, $n>1$ is the parameter, which is a measure of the pore-size distribution.

The foliage water potential ψ_f is assumed to be expressed in terms of ψ_r using the relationship

$$\psi_f=\psi_r-r_r\rho_w\int_0^{Z_{\max}}R(\theta,z)dz \quad (11)$$

where r_r is the rootage resistance, $Z_{\max}$ is the maximum length of roots, and ρ_w is the water density.

Heat transfer within a soil layer is described by

$$C_{eff}(\theta)\frac{\partial T}{\partial t}=\frac{\partial}{\partial z}(\lambda(\theta)\frac{\partial T}{\partial z}) \quad (12)$$

where $C_{eff}(\theta)$ and $\lambda(\theta)$ are the effective soil heat capacity and the soil heat conductivity.

The soil surface temperature T_g is calculated from Eq. (12). The foliage temperature T_f is obtained from the heat balance equation for the vegetation cover neglecting its heat content. Both T_g and T_f are used for assessing latent and sensible heat fluxes from bare soil and vegetation surfaces (Eqs. (1a) – (2b)).

The land surface radiation temperature T_s is calculated from the long-wave radiation balance equation

$$\sigma \cdot T_s^4 = R_a - R_{lf} - R_{lg} \tag{13}$$

where R_a is the atmosphere counterradiation, R_{lf} and R_{lg} are long-wave components of radiation balance for vegetation and bare soil, respectively, calculated as functions of the measured meteorological variables, σ is the Stefan-Boltzmann constant.

The input variables of the SVAT model are incoming radiation, air temperature, humidity and pressure, cloudiness, precipitation, wind speed assigned from the standard meteorological observations. Initial conditions for Eqs. (7) and (12), namely soil moisture and temperature profiles at the beginning of the vegetation season, are calculated by Eqs. (25), (26) (see the following section) through spin-up simulations of soil moisture and heat regimes during a winter previous to vegetation season.

Eqs. (7), (12) are numerically integrated by an implicit, four-point finite difference scheme with the time and spatial steps of 3 hours and 10 cm, respectively.

Spatial heterogeneity of land surface characteristics is taking into account in the model by the mosaic approach, i.e. by selection of plots on the studied territory with different soils, land-use and vegetation types which correspond to specific parameter values. Soils are characterized by bulk density, maximum hygroscopicity, porosity, field capacity, and saturated hydraulic conductivity. The heterogeneity of vegetation is represented by the minimum stomatal resistance, the leaf area index LAI, vegetation cover fraction *B*, and the aerodynamic resistance r_a .

The principal model parameters have been adjusted by calibration against ground-measured soil water content W, evapotranspiration Ev, vertical soil moisture profiles, and soil surface temperature. Also, the values of several parameters have been retrieved from the specific measurements at agricultural meteorological stations, some of them have been estimated using satellite data, and the values of certain parameters have been derived from literature sources. The spatial distributions of the most meteorological parameter values (being input model variables) have been built using interpolation procedures. The water and heat balance components for the entire considered territory have been calculated as weight-averages accounting for size of the area occupied by specific soil and vegetation.

In developing version of the model designed for utilizing satellite estimates of the land surface characteristics (built in a quasi-regular grid nodes) the uniform grid with 3x3 AVHRR/NOAA pixel cells (~ 7x5 km^2) has been superimposed on the entire investigated territory divided into plots with different soils and vegetation. Grid size for other sensors was assigned as close to one of AVHRR/NOAA. (Note, that pixel size for the IR channels of the AVHRR radiometer is 1' in latitude and 1.5' in longitude, for similar MODIS channels resolution is equal to 1 and 4.8 km, and for the same channels of SEVIRI it is 0.05° in latitude and 0.06° in longitude) . For nodes of the grids there have been built AVHRR-derived estimates of T_a, T_{sg}, $T_{s.eff}$, $NDVI$, E, B, and LAI, MODIS-derived estimates of T_{ls}, E, $NDVI$, B and LAI, and SEVIRI-derived estimates of T_{ls} . Maps of some AVHRR-derived land surface characteristics for part of the study area are shown in Fig. 7. There have been also defined model parameters and input variables as well as there have been calculated the values of Ev, W and other water and heat balance components together with T_f, T_g and T_s .

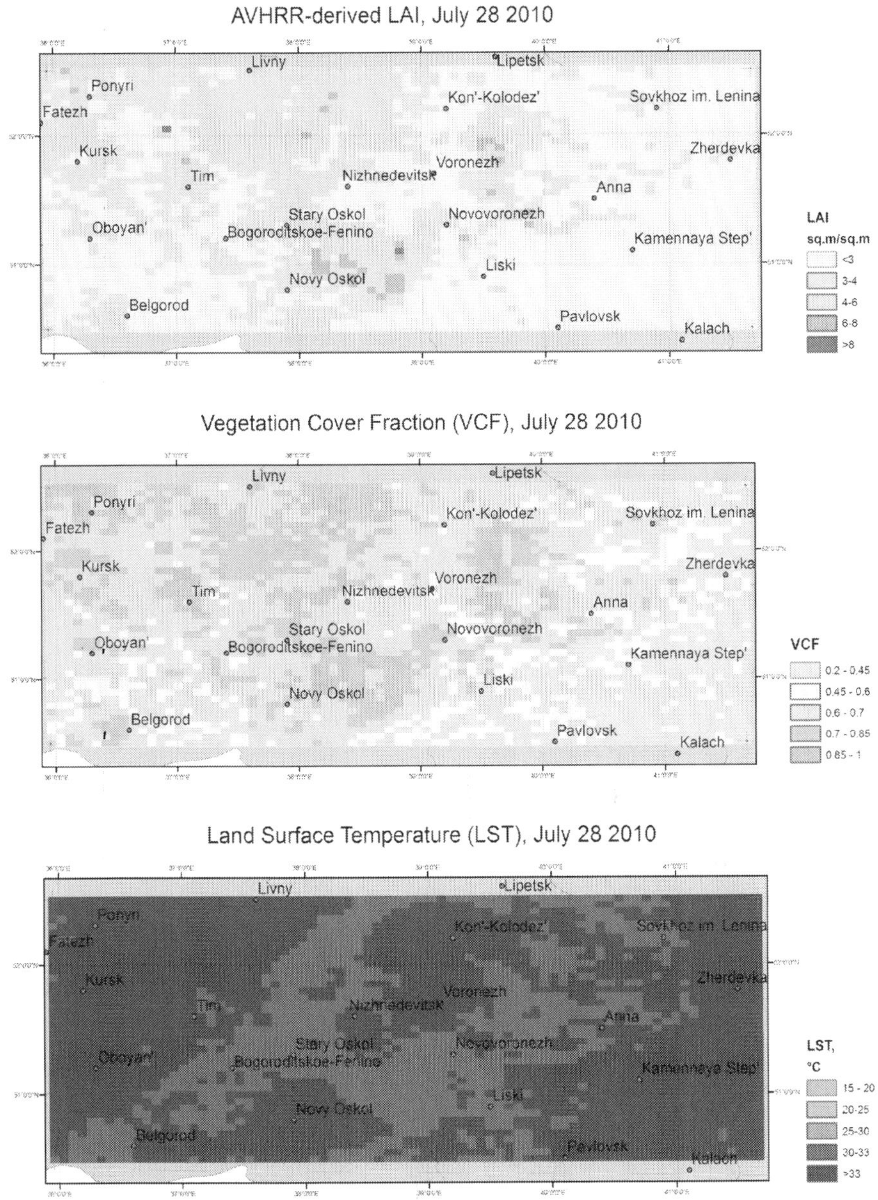

Fig. 7. Maps of some AVHRR-derived land surface characteristics

4.2 SSNAT component of the RSBLSM system

SSNAT model describes hydrothermal processes in soil and snow cover during a cold season, including processes of snow accumulation and melt, changes of soil moisture and temperature during soil freezing and thawing, as well as snowmelt water infiltration.

To simulate temporal changes of the snow depth and density, heat content of snow, water phase transformation and other processes within a snowpack during snow accumulation and melt, one-layer snow model proposed in (Kuchment, Gelfan, 1996) has been applied. The main equations of the model are written as:

$$\frac{dH}{dt} = \rho_w \left[X_s \rho_0^{-1} - (S + E_s)(\rho_i i)^{-1} \right] - V \tag{14}$$

$$\frac{d}{dt}(iH) = \frac{\rho_w}{\rho_i}\left(X_s - S - E_s\right) + S_i \tag{15}$$

$$\frac{d}{dt}(wH) = X_l + S - E_l - R - \frac{\rho_i}{\rho_w} S_i \tag{16}$$

$$c_s \frac{d}{dt}(T_s H) = Q_a - Q_g - \rho_w LS + \rho_i LS_i \tag{17}$$

where H is the snow depth; i and w are the vertically averaged volumetric contents of ice and liquid water, respectively; T_s is the vertically averaged temperature of snowpack; S is the melt rate; S_i is the rate of freezing of liquid water in snow, E_l is the rate of liquid water evaporation from snow; E_s is the rate of snow sublimation, Q_a is the net heat flux at the snow surface; Q_g is the ground heat flux; X_s and X_l are the snowfall and rainfall rates, respectively (partitioning of the total precipitation, X, into solid and liquid phase is a function of the air temperature); V is the snowpack compression rate; R is the snowmelt outflow from snowpack; c_s is the specific heat capacity of snow; ρ_w, ρ_i, and ρ_0 are the densities of water, ice, and fresh-fallen snow, respectively; L is the latent heat of ice fusion.

The melt rate S is found from the energy balance of the snowpack at zero snow temperature as:

$$S = \begin{cases} \left(Q_a - Q_g\right)\left(\rho_w L\right)^{-1} = \left(Q_{sw} + Q_{lw} - Q_{ls} + Q_T + Q_E + Q_P - Q_g\right)\left(\rho_w L\right)^{-1}, Q_a - Q_g > 0 \\ 0, Q_a - Q_g < 0 \end{cases} \tag{18}$$

where Q_{sw} is the net short wave radiation; Q_{lw} is the downward long wave radiation; Q_{ls} is the upward long wave radiation from snow; Q_T is the sensible heat exchange; Q_E is the latent heat exchange; Q_P is the heat content of liquid precipitation.

The heat flux components of Q_a for an open agricultural site are calculated by the empirical relationships using the observed meteorological variables (air temperature, air humidity, wind speed, precipitation, and cloudiness) as the inputs.

The rate S_i of freezing of liquid water in snowpack is calculated as:

$$S_i = \begin{cases} H\dfrac{dw}{dt}, \; T_s = 0^\circ C \wedge Q_a - Q_g < 0 \wedge \dfrac{|Q_a - Q_g|}{\rho_i L} \geq H\dfrac{dw}{dt} \\ \dfrac{|Q_a - Q_g|}{\rho_i L}, \; T_s = 0^\circ C \wedge Q_a - Q_g < 0 \wedge \dfrac{|Q_a - Q_g|}{\rho_i L} < H\dfrac{dw}{dt} \\ 0, T_s = 0^\circ C \wedge Q_a - Q_g \geq 0 \\ X_l, T_s < 0^\circ C \end{cases} \quad (19)$$

The snowpack compression rate V (in cm s-1) is found from:

$$V = \frac{v_1 \rho_s}{\exp(v_2 T_s + v_3 \rho_s)} \frac{H^2}{2} \quad (20)$$

where ρ_s is the density of snowpack (in g cm^{-3}) equal to $\rho_s = \rho_i i + \rho_w w$; v_1, v_2, and v_3 are the coefficients equal to 2.8×10^{-6} cm^2 s^{-1} g^{-1}; -0.08 $^{\circ}$C^{-1}; 21 cm^3 g^{-1}, respectively.

The outflow of liquid water from snow is calculated as:

$$R = \begin{cases} X_l + S - E_l - w_{\max} \dfrac{dH}{dt}, w = w_{\max} \\ 0, w < w_{\max} \end{cases} \quad (21)$$

where $w_{\max}$ is the holding capacity of snowpack related to its density ρ_s as

$$w_{\max} = 0.11 - 0.11 \frac{\rho_s}{\rho_w} \quad (22)$$

Numerical integration of the Eqs. (14) – (17) is carried out by an explicit finite-difference scheme with the time-step of 3 hours.

To calibrate the snow pack model for the study area, the meteorological data from 48 stations for the period from 1 November 2001 to 31 May 2002 were used. Then the model was validated against snow depth observations during the winter-spring seasons of 2002-2003 and 2003-2004. The comparison of snow modeling results with the observed snow depth at the ground-based stations has demonstrated a good correspondence between the two datasets. The standard error of the simulated snow depth was 5.8 cm during the calibration stage and 7.7 cm at the validation stage.

Water and heat transfer in a soil during the processes of soil freezing, thawing and infiltration of water are described by the following equations (Gelfan, 2006):

$$\frac{\partial W}{\partial t} = \frac{\partial}{\partial z}\left(D_\theta \frac{\partial \theta}{\partial z} + D_I \frac{\partial I}{\partial z} - K \right) \quad (23)$$

$$c_T \frac{\partial T}{\partial t} - \rho_w L \frac{\partial W}{\partial t} = \frac{\partial}{\partial z}\left(\lambda \frac{\partial T}{\partial z} \right) + \rho_w c_w \left(D_\theta \frac{\partial \theta}{\partial z} + D_I \frac{\partial I}{\partial z} - K \right) \frac{\partial T}{\partial z} \quad (24)$$

where W and I are the total water content and ice content of soil, respectively ($W = \theta + \rho_i I / \rho_w$); $c_T = c_{eff} + \rho_w L(\partial\theta / \partial T)$; c_{eff} is the effective heat capacity of soil equals $c_{eff} = \rho_g c_g (1 - P) + \rho_w c_w \theta + \rho_i c_i I$; ρ and c are the soil density and the specific heat capacity, respectively (indexes w, i and g refer to water, ice and soil matrix, respectively); P is the soil porosity; $D_\theta = K(\partial\psi / \partial\theta)_I$; $D_I = K(\partial\psi / \partial I)_\theta$; $\psi = \psi(\theta, I)$ is the matrix potential of soil.

If soil is frozen ($I(z,t) \neq 0$), than hydraulic and thermal characteristics of soil are functions of ice content, i.e. $K = K(\theta, I)$, $\lambda = \lambda(\theta, I)$. One can see that for an unfrozen condition ($I(z,t) = 0$), Eq. (23) reduces to Eq. (7) (neglecting water extraction by plant roots) and Eq. (24) reduces to Eq. (12).

The matrix potential, $\psi = \psi(\theta, I)$, and the hydraulic conductivity, $K=K(\theta,I)$ are determined from (Gelfan, 2006):

$$\psi(\theta, I) = -\frac{\left(S_f^{-1/m} - 1\right)^{1/n}}{\alpha} \times \left[\frac{\theta_0 - \theta_r}{\theta_0 - I - \theta_r} + \frac{\theta_r}{\theta}\left(1 - \frac{\theta_0 - \theta_r}{\theta_0 - I - \theta_r}\right)\right](1 + 8I)^2 \tag{25}$$

$$K(\theta, I) = K_0 S_f^{0.5} \left[\frac{1 - \left(1 - S_f^{1/m}\right)^m}{(1 + 8I)}\right]^2 \tag{26}$$

where $S_f = (\theta - \theta_r)(\theta_0 - I - \theta_r)^{-1}$ is the relative saturation of frozen soil.

Note, that under $I(z,t) = 0$ formulas (25), (26) reduce to the van Genuchten's formulas (9), (10) for an unfrozen condition.

The values of θ_0 and θ_r are assumed to be equal to the measured soil porosity and maximum hydroscopicity, respectively. The values of a, n, c_T, and λ are calculated from the measured soil characteristics, such as bulk density, field capacity, and wilting point by the formulas presented in Gelfan (2006).

Equations (23) and (24) are numerically integrated by an implicit, four-point finite difference scheme with the time and spatial steps of the finite difference scheme of 1 hour and 10 cm, respectively.

5. Results

5.1 Satellite based modeling water and heat regimes of the study region during a vegetation season

The model has been verified by comparing the calculated and measured vertical soil moisture and temperature profiles, values of land surface temperature, radiation balance, soil water content of one-meter soil layer, and evapotranspiration. Results of such a collation for several characteristics above are shown in the Figs. 8-10 by the example of some meteorological stations located in the study area.

Moreover, there has been executed comparison of modeled values of T_f, T_g and T_s with their satellite-derived analogues T_a, T_{sg} and $T_{s.eff}$ for AVHRR and T_{ls} for MODIS and

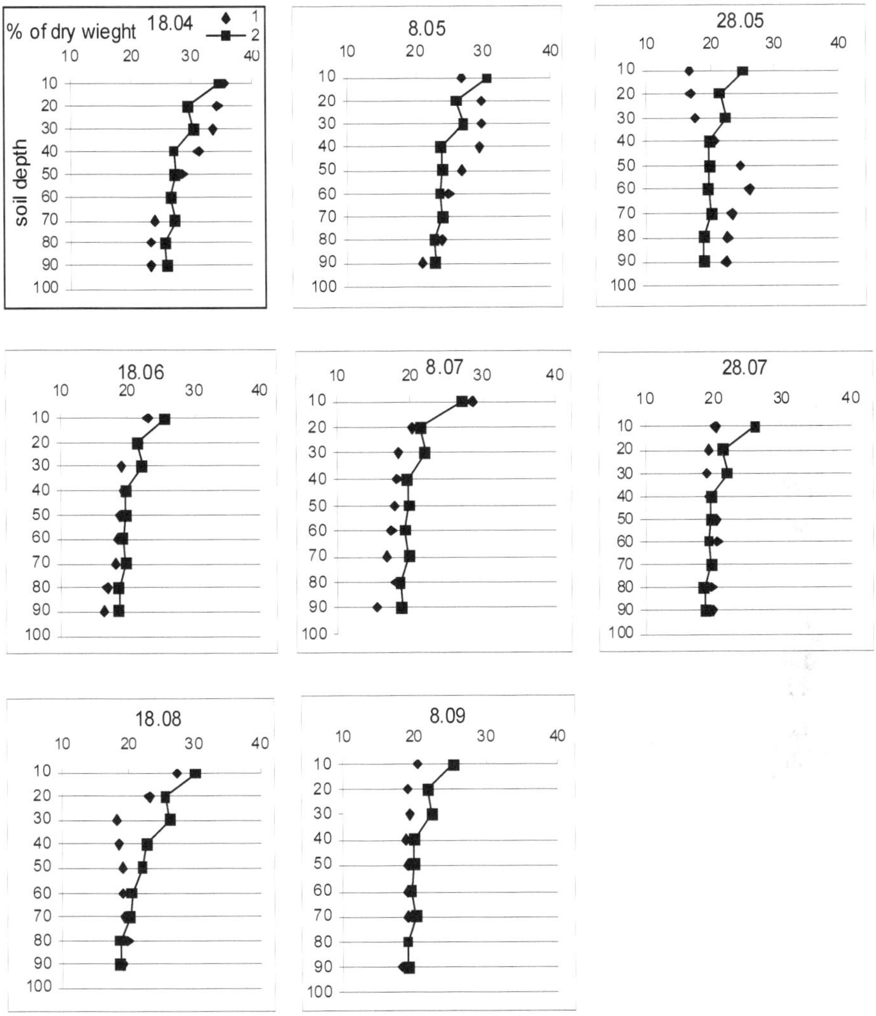

Fig. 8. Modeled (1) and measured (2) vertical soil moisture profiles for perennial grasses at water balance station Nizhnedevitsk for 2003 vegetation season

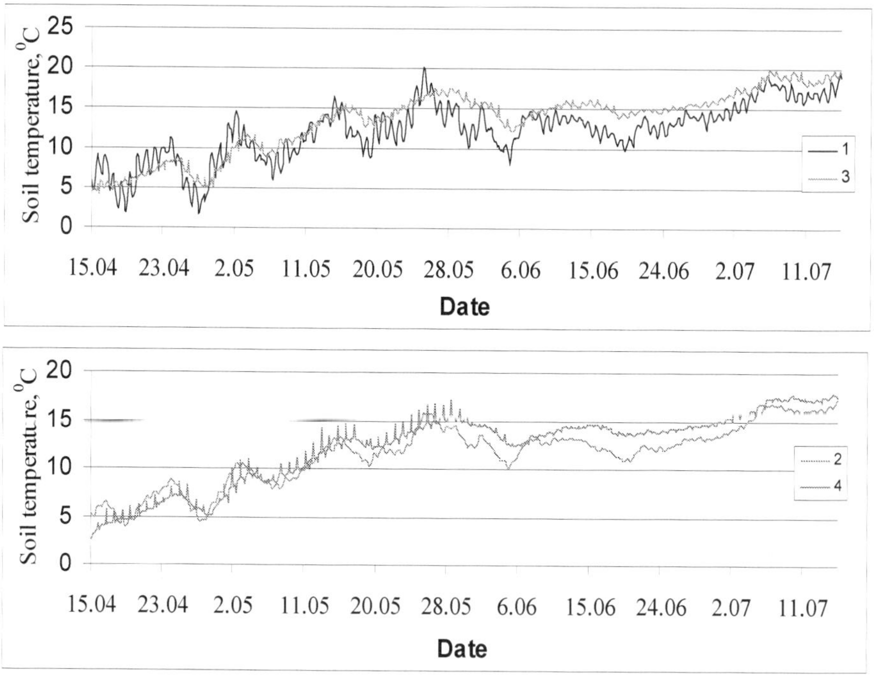

Fig. 9. Soil temperature under vegetation cover modeled for the depth 20 (1) and 40 (2) cm and measured at the same depths for perennial grasses at Nizhnedevitsk water balance station in 2003

SEVIRI. For most terms of each considered vegetation season the differences $T_{s.eff} - T_s$, $T_a - T_f$, and $T_{sg} - T_g$ have not exceed the standard errors of AVHRR-derived estimates of $T_{s.eff}$, T_a and T_{sg} as one can see from Fig. 11 illustrating this result for the part of the study area. Separate local spots with a difference of 20°C on these figures correspond to the clouds above given plots.

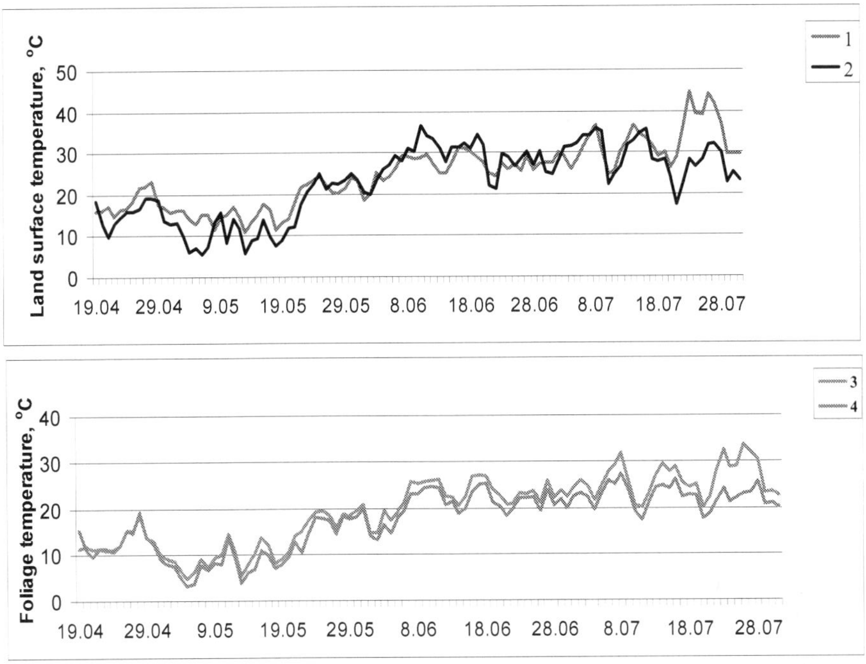

Fig. 10. Modeled (1) and measured (2) soil surface temperatures, modeled foliage temperature (3) and measured air temperature (4) for winter wheat at agricultural meteorological station Petrinka for vegetation season 1999

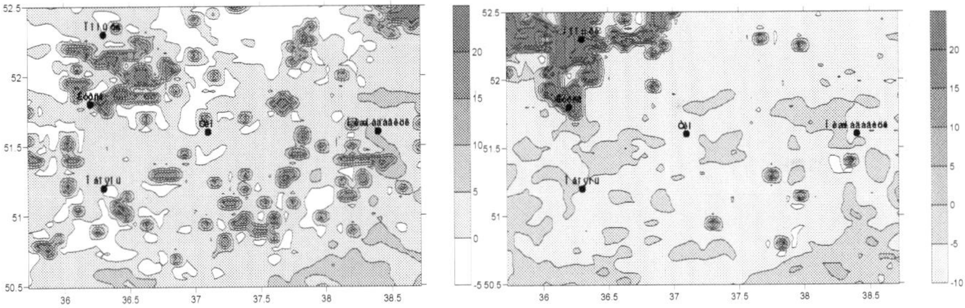

Fig. 11. LST difference map ($\Delta T = T_{s.eff}$ (AVHRR) - T_s(MODEL)) for part of the study region of 23,500 km^2 with spacing of ~ 5 km, 11.00 GMT 27 May (left) and 12 August (right) 2009

Similar distributions have been also built for the difference of $T_{ls}(MODIS) - T_s(MODEL)$. Results of comparing SEVIRI-derived temperature T_{ls} defined by continuous measurements during 6-day interval of the vegetation season 2009 with three-hour ground-based observation data on air temperature T at agricultural meteorological station Livny and with modeled temperature T_f calculated using the same ground-based data are shown in Fig. 12. As seen from this figure, the temporal behavior of these variables is quite close.

To discover relevance of the model parameters (stomatal resistance r_0, leaf area index LAI, soil and vegetation albedos (A_g, A_f) and emissivities (E_s, E_p) for describing the water and heat exchange processes, sensitivity of the model (in particular, of the quantities Ev and T_s) to these parameter variations has been investigated. In (Kuchment & Startseva, 1991) it has been shown that Ev is strongly influenced by r_0. The present study has confirmed this effect. Particularly, changing the values of r_0 for winter wheat and perennial grasses by 20 and 60 % results in changing the values of Ev by 4-5 and 8-10 mm per decade, correspondingly. Strong sensitivity of quantities Ev and T_s to variations of LAI was also found. It was particularly notable in periods of rapid plant growth, their yellowing, mowing, as well as in the beginning of vegetation season.

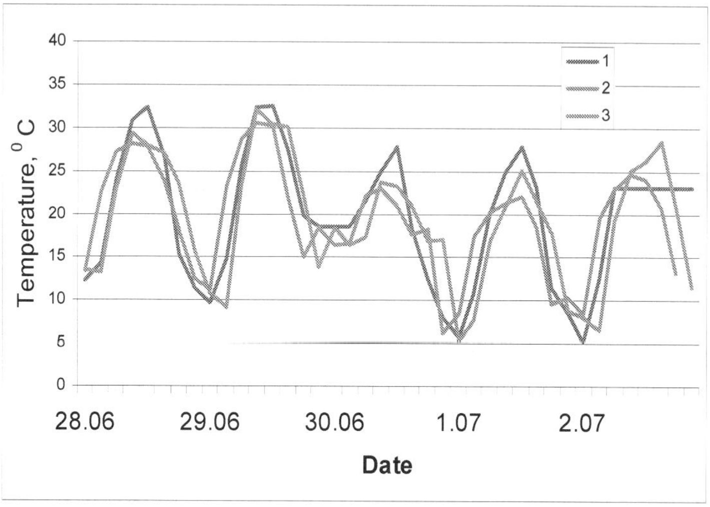

Fig. 12. SEVIRI-derived LST (1), air temperature (2), and air vegetation temperature modeled using three-hour ground observation data (3) on 28.06.-03.07.2009 at agricultural meteorological station Livny

Changing the relative values of LAI by 0.1 and 0.2 resulted in changing the values of Ev by 2-2.5 and 3-3.5 mm per decade and T_s by 1.0-1.2 and 1.4-1.6°C, respectively, and changing the maximum values of LAI from 3 to 5 leads to increase of Ev 7-8 mm per decade and decrease of T_s ~ 2°C. Numerical experiments with a shift of time mowing led to changing values of Ev for perennial grasses to 15-17 mm per decade and values of T_s to more than 3°C (fig. 13).

Such sensitivity makes it possible to select appropriately the values of LAI at the specific site, that, in turn, leads to close to actual assessment of vegetation cover fraction B that is shown below. The influence of the integral soil and vegetation albedos A_g and A_f on the values of Ev and T_s was different. Evapotranspiration was varied more when changing A_f whereas temperature T_s was changed significantly by variation of A_g. Particularly, change of A_g by 0.2 led to changes of Ev for several crops by 1.5-2.5 mm per decade and corresponding

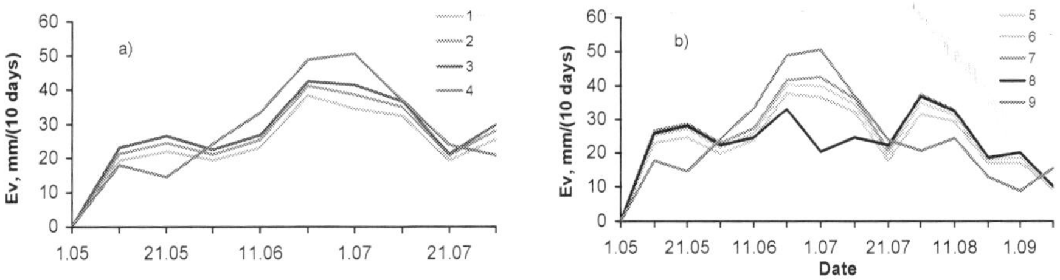

Fig. 13. Modeled values of evapotranspiration Ev for 1997 vegetation season at agricultural meteorological station Petrinka for winter wheat (a); perennial grasses (b) and different values of LAI_{max}: 2.5 (1); 3.5 (2); 4.5 (3); 3.0 (5); 4.0 (6); 5.0 (7); (8) 4.5 with shift of mowing day for 3 weeks. (4) and (9) corresponds to measured values of Ev for perennial grasses.

variations of T_s reached 3°C and more. Similar variations of A_f led to changes of Ev by 4-5 mm and more per decade and also of T_s by 1.5-2°C. Direct effect of soil and vegetation emissivities E_s and E_p on Ev and T_s was negligible.

The main conclusion from all experiments described above is that the key parameters affecting evapotranspiration and soil water content under wet soil conditions are r_0 and shading parameters LAI and B. At the same time the influence of the latter two is often more substantial than of the first one. Under dry soil conditions (usually occurring when increasing land surface temperature), this effect becomes less noticeable and the value of evapotranspiration is mainly determined by soil water content of the upper one meter soil layer.

High sensitivity of Ev and T_s to LAI variations as well as possibility to control current values of LAI by comparing modeled T_f, T_g and T_s with satellite-derived T_a, T_{sg} and $T_{s.eff}$ allowed specifying time behavior of LAI for several crops in the absence of phytometry data. Satellite-based values of LAI were estimated using empirical relationships between LAI and $NDVI$ for grasslands (Biospheric Aspects, 1993) (27) and for agricultural crops (Biftu & Gan, 2001) (28):

$$LAI = NDVI \cdot 1.71 + 0.48 \tag{27}$$

$$LAI = -2.5 \cdot \ln(1.2 - 2 \cdot NDVI) \tag{28}$$

Specified in that way time behavior of LAI is presented at Fig. 14. Here sudden changes correspond to time intervals of plant mowing.

In initial versions of the model the fraction B was calculated as follow

$$B = 1 - \exp(-p \cdot LAI) \tag{29}$$

Here p is empirical coefficient that was adjusted by comparing modeled magnitudes of T_f, T_g, and T_s with satellite-derived values of T_a, T_{sg}, and $T_{s.eff}$ using LAI estimates determined from both satellite and ground data. Numerical experiments were carried out under three scenarios of estimating LAI and B: 1) values of B were calculated by (29) using LAI determined by ground observation data on phenological stage changes and plant heights for different land-use; 2) values of B were also calculated by (29) using satellite-

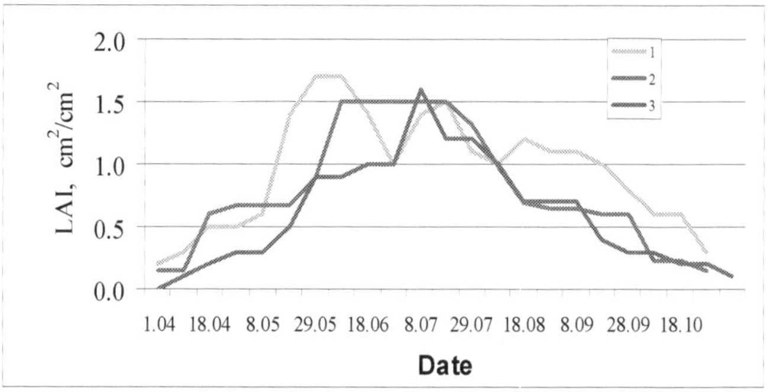

Fig. 14. Temporal behavior of LAI for vegetation season 2004 built by data of AVHRR/3 (1), MODIS (3) and by ground observation data at water balance station Nizhnedevitsk for perennial grasses (1, 2) and without allocation of cultures (3)

derived LAI; 3) estimates of B and LAI were generated from satellite data. Values of T_a and T_f under these scenarios are close to each other: for the most of observation times their differences do not exceed 2.5 and 3.5°C respectively (i.e. they are close to RMSE for satellite-derived T_a). These results are confirmed by rather high (0.65-0.75 at different seasons) T_f and T_a correlation coefficient. Similar results were also obtained when comparing T_s with $T_{s.eff}$ and T_{sg} with T_g under all the scenarios. Examples of comparing AVHRR-derived and ground-based estimates of B for perennial grasses are presented in Fig. 15.

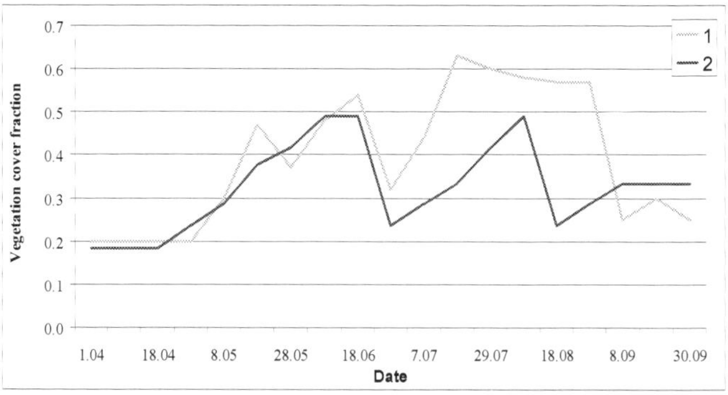

Fig. 15. Comparing AVHRR-derived (1) and ground-based estimates (2) of vegetation cover fraction B for perennial grasses at the station Kursk for vegetation season 1999

Described results gave impetus to research possibility of direct use of satellite-based estimates of vegetation cover and LSTs in the model. High desirability of such use while model simulating water and heat balance components for vast territories is due to the necessity of distributed estimates of the land surface characteristics, especially under the lack of ground-based observation data. To assimilate satellite-based estimates of vegetation and meteorological characteristics the updating of the SVAT model has been performed including:

1. Replacing the ground point-wise estimates of the model parameters *LAI* and *B* by their AVHRR- or MODIS-based analogues. The efficiency of such approach has been proved through comparisons: between satellite-derived and ground-based data on *LAI* and *B* behavior during vegetation season; between satellite-derived, modeled, and in-situ measured temperatures; between modeled and actual values of evapotranspiration *Ev* (Fig. 16) and of soil water content *W* for one-meter soil layer (Fig. 17). The

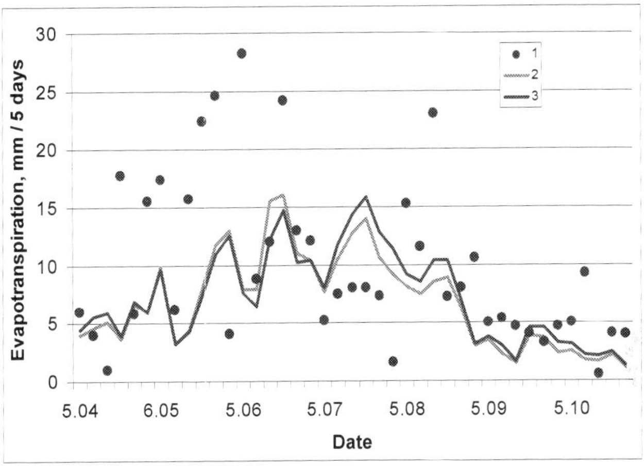

Fig. 16. Evapotranspiration Ev for vegetation season 2008 measured on grassland at water balance station Nizhnedevitsk (1), modeled using AVHRR-derived LAI for perennial grasses (3) and MODIS-derived LAI (2)

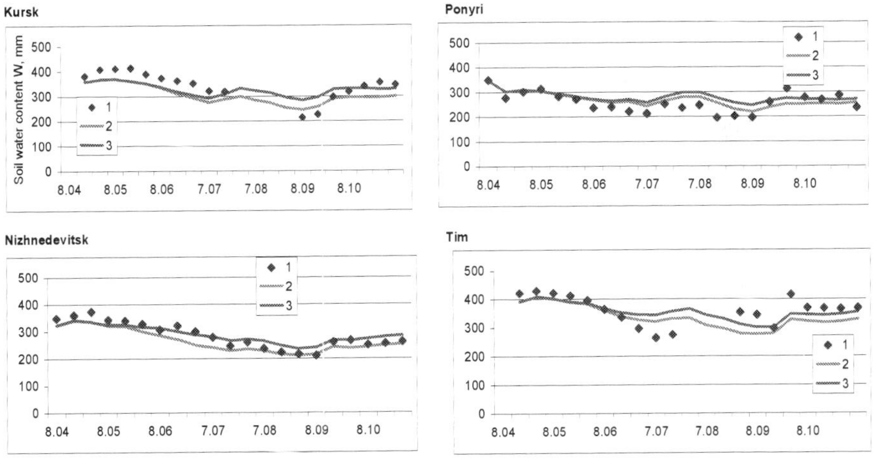

Fig. 17. Soil water content: ground measured for corn at four agricultural meteorological stations (1), modeled using AVHRR-derived LAI and B for corn (2) and MODIS-derived LAI (3) for vegetation season 2008.

discrepancies between T_a, T_s, $T_{s.eff}$ and T_f, T_g, T_s as well as between T_{ls} and T_s and ground-measured ones do not exceed standard errors of satellite-derived estimates T_a, T_{sg} and $T_{s.eff}$ respectively while the modeled and measured values of Ev and W are found close to each other within a standard error of their estimation.

2. Entering the AVHRR- or MODIS-based LST estimates as the input SVAT model variables instead of their standard ground-based estimates if the time-matching of satellite and ground-based observations takes place. The SEVIRI-derived T_{ls} estimates can also play the same role. Permissibility of such replacement has been verified while comparing remote sensed, modeled and ground-based temperatures as well as calculated and measured values of W and Ev (Fig. 18) The SEVIRI-based T_{ls} estimates are found to be very informative and useful due to their high (up to 15 min) temporal resolution.
3. Inputting AVHRR- and MODIS-derived LAI and B, AVHRR-, MODIS-, and SEVIRI-based LSTs in each grid cell of the model in order to account for the space variability of vegetation cover parameters and meteorological characteristics. Ground-based data on precipitation, air temperature and humidity prepared by Inverse Distance Squared (IDS) interpolation method are also inserted into the model in each grid node. The calculations of vertical water and heat fluxes, soil water and heat contents and other water and heat balance components for above region of interest have been carried out using the described updated SVAT model and the fields of AVHRR/3- or MODIS-derived LAI and B estimates together with AVHRR/3- or MODIS- or SEVIRI-derived T_{ls} retrievals for 1999-2010 vegetation seasons. Fig.19 shows the results of calculation of latent heat flux and soil water content at the same date of 2010 as shown in Fig. 7. The acceptable accuracy levels for above values assessment have been achieved under all scenarios of parameter and input model variable specification using satellite- and ground-based data.

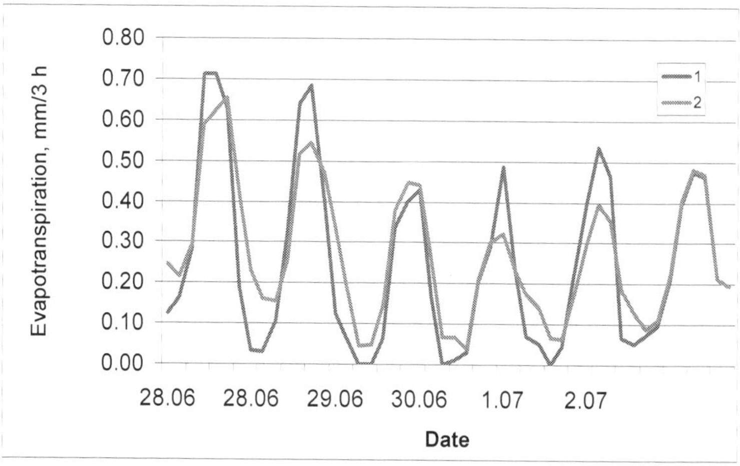

Fig. 18. Three-hour values of evapotranspiration modeled using SEVIRI-derived data (1) and three-hour observation data at the agricultural meteorological station Livny (2) on 28.06.-03.07.2009

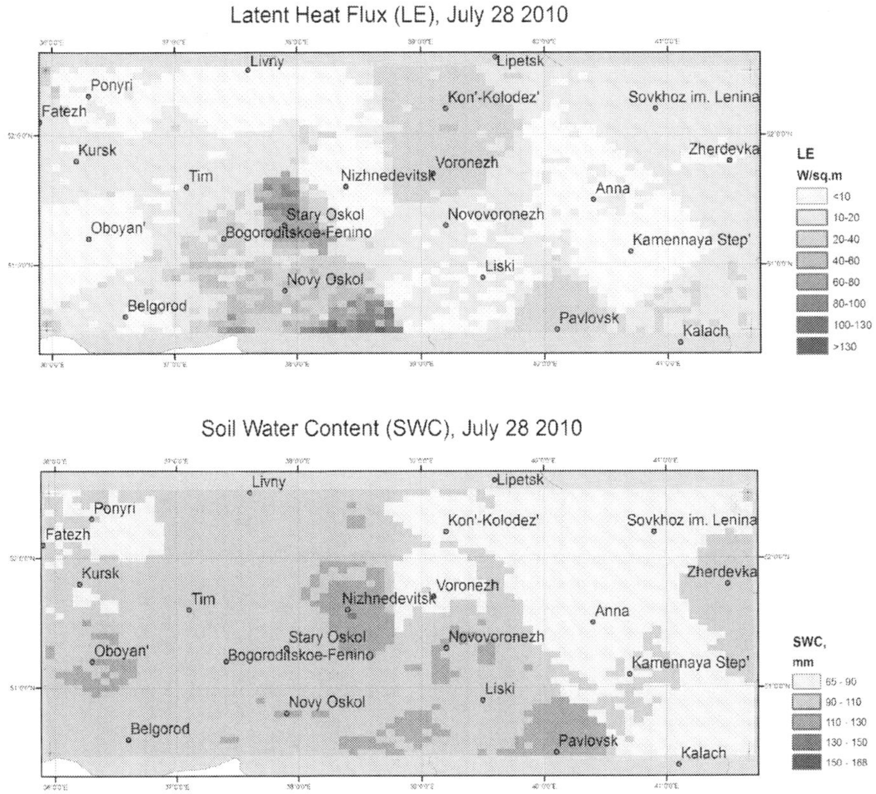

Fig. 19. Calculated latent heat flux and soil water content distributed over the study region

As follows from the above, the presented approach can be used for vast territories under the lack or absence of ground observations. The most promising in this case is the utilization of SEVIRI data due to their frequent occurrence.

5.2 Satellite based modeling of snow cover

The calibrated snow model (14) – (22) was applied to calculate snow cover characteristics in each 0.01° pixel of the spatial grid within the study areas. The meteorological data interpolated from the available meteorological stations to each pixel by the IDS method were used as the input to the snow cover models. In pixels where MODIS data were available the simulated land surface temperature and albedo were replaced by corresponding satellite-derived variables. Fields of snow cover characteristics based on satellite observations and the snow pack model were generated for the study area for the time period from January 1st to May 31th of the years 2002 – 2005. The initial snow cover distribution in the model was assigned following AMSR-derived SWE data on 1 January for all pixels. For the pixels where the initial SWE values were unavailable because of the lack of coverage or for some other reasons, these values were interpolated from neighboring pixels by the IDS interpolation method.

Maps of the simulated distributions of these snow characteristics were compared with the corresponding satellite maps. The simulated changes of SCA were appeared to be in satisfactory correspondence with the satellite-derived SCA. As an example of this correspondence, simulated and satellite-derived dynamics SCA are shown in Fig. 20 for several sub-regions adjoining the meteorological stations of the study area.

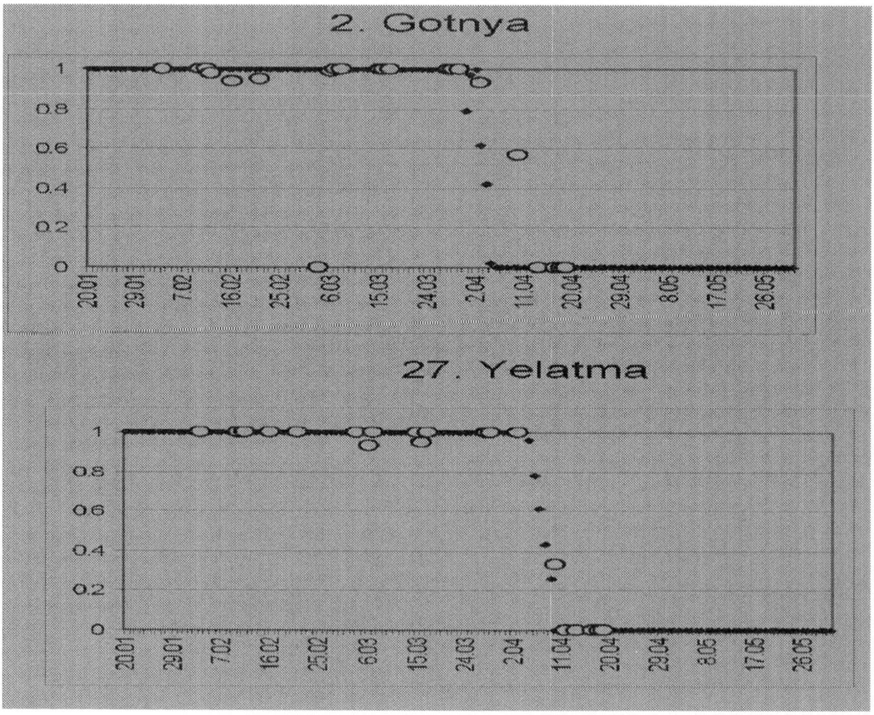

Fig. 20. Simulated (solid dots) and MODIS-derived (circles) dynamics of SCA for some polygons within the study area (winter-spring season of 2004)

However, the simulated maps of SWE substantially differ from the corresponding AMSR-derived SWE maps as it is illustrated by Fig. 21. We assume that this difference is resulted from by the fact that the accuracy of the SWE estimated from the radiometric satellite measurements noticeably decreases during melt period when snowpack is saturated by melted water so the AMSR-derived SWE maps may not adequately represent the real SWE distribution for this period of the year.

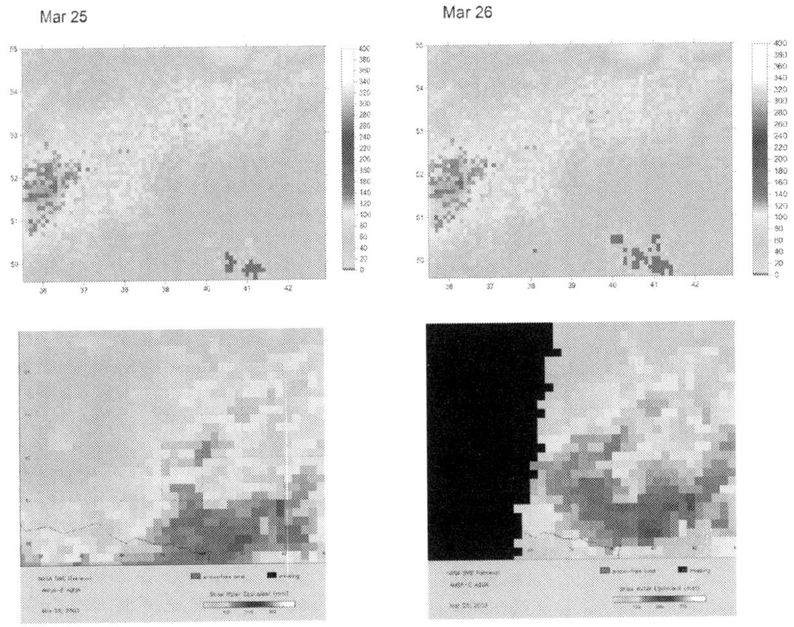

Fig. 21. Simulated (upper raw) and AMSR-derived (lower raw) distribution of SWE (mm) on 25-26 March, 2003.

6. Conclusion

The paper highlights the importance of satellite remote sensing data used in conjunction with a land surface model in describing water and heat regimes of vast agricultural regions. Accuracy and robustness of the corresponding quantified assessments by LSM is strongly restricted by insufficiency, both in space and time, of ground observations of highly heterogenic soil, vegetation and snow characteristics, soil temperature, etc., which can be used as the model parameters or input variables, as well as for calibration and validation of the model. Satellite remote sensing is not only an additional source of land surface data allowing substitution of the missed ground observations into the model, but, for majority of regions, it is a unique data source. From the other side, in spite of the diversity of satellite data on land surface characteristics and technologies of the data thematic processing, opportunity of using these data in LSMs is often problematic because of many reasons, e.g. irregularity of data, uncertain data accuracy, etc.

Moreover, there is no general approach allowing reliable choice of appropriate satellite data and/or processing technique; this choice depends on both specific features of the region of interest and the used model. In this study, we tried, firstly, to demonstrate opportunities of utilizing land-surface/snow remote sensing products obtained from the different satellites (NOAA, EOS Terra and Aqua, METEOSAT-9) and sensors (SEVIRI, AMSR-E, AVHRR, MODIS) in the developed LSM. Secondly, we analyzed sensitivity of the simulation results to different satellite remote sensing data. It is important that most of the used products were derived by the originally developed processing techniques, especially new technique for processing SEVIRI measurement and the technology based on synergy of optical and microwave snow products. Other main emphasis of the study is to develop a new physically based distributed Remote Sensing Based Land Surface Model (RSBLSM). The model is aimed to simulation of vertical water-and-heat transfer and, importantly, it takes into account hydrothermal processes in the "frozen soil-snow-atmosphere" system. These processes are critical for cold region agriculture, as they define crop development in early spring before the vegetation season beginning.

The obtained results allow us to conclude that utilization of the differently derived satellite-based estimates in the developed physically based model, intensively calibrated and validated against the available ground observations, provides an opportunity for reproducing spatial fields of evapotranspiration, soil moisture and temperature at different soil depths, temperature of soil/vegetation surfaces, snow and other water and heat characteristics for the vast agricultural region.

7. Acknowledgment

We thank E.Volkova, A.Kukharsky, S.Uspensky, V.Solovjev from SRC "Planeta" for their participation in the development of satellite data processing methods, as well as M.Alexadrovich from Water Problem Institute of RAS for technical assistance in modeling and data processing. The presented study was carried out under support of the Russian Foundation of Basic Researches (grant № 10-05-00807).

8. References

Armstrong, R. L., Brodzik, M. J. (2005) Northern Hemisphere EASE-Grid Weekly Snow Cover and Sea Ice Extent Version 3. *Digital media, http://nsidc.org/data/nsidc-0046.html,* National Snow and Ice Data Center, Boulder, Colorado, USA.

Becker F, Li, Z.-L. (1995) Surface temperature and emissivity at various scales: definition, measurement and related problems. *Rem. Sens. Rev.*,Vol. 12, pp. 225-253.

Biospheric Aspects of the Hydrological Cycle (BAHC). (1993) Report № 27, 103 p. Ed. by BAHC Core Project Office, Institut fűr Meteorologie, Freie Universitat Berlin, Germany.

Biftu, G.F., Gan, T.Y. (2001) Semi-distributed, physically based, hydrologic modeling of the Paddle River basin, Alberta, using remotely sensed data. *J. Hydrol.*, Vol. 244, pp. 137-156.

Brodzik, M. J., Armstrong, R., Savoie, M. (2007) Global EASE-Grid 8-day Blended SSM/I and MODIS Snow Cover. *Digital media. http://nsidc.org/data/docs/daac/nsidc0321_8day_ssmi_modis_blend/index.html,* National Snow and Ice Data Center, Boulder, Colorado, USA.

Brooks, R.H., Corey, A.T. (1964) *Hydraulic properties of porous media.* Hydrol. Pap. Colorado State Univ. № 3. 27 p.

Brubaker, K. L., Pinker, R. T., Deviatova, E. (2005) Evaluation and Comparison of MODIS and IMS Snow-Cover Estimates for the Continental United States Using Station Data. *J. Hydrometeor.*, Vol. 6, pp. 1002–1017. doi: 10.1175/JHM447.1

Clupp, R.B., Hornberger, G.M. (1978) Empirical equations for some hydraulic properties. *Water Resources Research*, Vol. 14, № 4, pp. 601-604.

Collatz, G.J., Ball. J.T., Grivet, C., Berry, J.A. (1991) Physiological and environmental regulation of stomatal conductance, photosynthesis and transpiration: a model that includes a laminar boundary layer. *Agricultural and Forest Meteorology*, Vol. 54, pp. 107–136.

Dai, Y., et al. (2003) The Common Land Model (CLM) version 1.0. *Bull. Am. Meteorol. Soc.*, Vol. 84, pp. 1013– 1023.

de Wildt, M. D., Gabriela, S., Gruen, A. (2007) Operational snow mapping using multitemporal Meteosat SEVIRI imagery. *Remote Sensing of Environment*, Vol. 109, pp. 29-41.

Deardorff ,J.W. (1978) Efficient prediction of ground surface temperature and moisture with inclusion of a layer of vegetation. *Journal of Geophysical Research*, Vol. 83, pp. 1889–1903.

Desborough, C.E., Pitman, A.J. (1998) The BASE land surface model. *Global and Planetary Change*, Vol. 19, pp. 3-18.

Dickinson, R.E., Henderson-Sellers, A., Kennedy, P.J., Wilson, M.F. (1986) *Biosphere Atmosphere Transfer Scheme (BATS) for the NCAR Community Climate Model*, NCAR Tech. Note, TN-275 + STR, 69 pp.

Faysash, A., Smith, E.A. (2000) Simultaneous Retrieval of Diurnal to Seasonal Surface Temperatures and Emissivities over SGP ARM-CART Site Using GOES Split Window. *J. Appl. Meteor.*, Vol. 39, pp. 971-982.

Foster, J.L., Hall, D. K., Eylander, J.B., Riggs, G.A., Nghiem, S.V., Tedesco, M., Kim, E., Montesano, P.M., Kelly, R. E.J., Casey, K.A., Choudhury, B. (2011) A blended global snow product using visible, passive microwave and scatterometer satellite data. *International Journal of Remote Sensing*, Vol. 32, pp. 1371-1395.

Gelfan, A. N., Pomeroy, J. W., Kuchment, L.S. (2004) Modelling forest cover influences on snow accumulation, sublimation, and melt. *J. Hydrometeorology*, Vol. 5, No. 5, pp. 785–803.

Gelfan, A. N. (2006) Physically based model of heat and water transfer in frozen soil and its parametrization by basic soil data. In: *Predictions in Ungauged Basins: Promises and Progress. Proceedings of symposium S7 held during the Seventh IAHS Scientific Assembly at Foz do Iguazu, Brazil, April 2005*, Eds: M. Sivapalan,. IAHS Publ., Vol. 303, pp. 293-304.

Global Water Security - an engineering perspective. *Report of the The Royal Academy of Engineering*, Publ. by The Royal Academy of Engineering, London, UK, April 2010.

Good, E. (2009) Blending in situ and satellite data for monitoring land air temperatures. *Proc. of the 2009 Eumetsat Meteorol. Sat. Conf, Bath, UK, 21 - 25 September 2009*, 5 p.

Grody, N.C., Basist, A.N. (1996) Global identification of snow cover using SSM/I Measurements. *IEEE Trans. Geosci. Remote Sens.*, Vol. 34, N 1, pp. 237-249.

Gusev, E. M., Nasonova, O.N. (2002) The simulation of heat and water exchange at the land-atmosphere interface for the boreal grassland by the land-surface model SWAP. *Hydrol. Processes*, Vol. 16, pp. 1893-1919.

Hall, D.K., Riggs, G., Salomonson, V., DiGirolamo, N.E., Bayr, K.J. (2002) MODIS snow cover products. *Remote Sensing of Environment*, V. 83, pp. 181-194.

Hall, D.K, Riggs, G.A. (2007) Accuracy assessment of the MODIS snow-cover products. *Hydrological Processes*, Vol. 21, pp. 1534–1547.

Helfrich, S.R., McNamara, D., Ramsay, B.H., Baldwin, T., Kasheta, T. (2007) Enhancements to, and forthcoming developments in the Interactive Multisensor Snow and Ice Mapping System (IMS). *Hydrological Processes*, Vol. 21, pp. 1576–1586.

Henderson-Sellers, A, Pitman, A.J., Love, P.K., Irannejad, P., Chen, T. (1995) The project for intercomparison of land surface parameterization schemes (PILPS) phases 2 and 3. *Bulletin of the American Meteorological Society*, Vol. 76, pp. 489–503.

Justice, C. O., Vermote, E., Townshend, J. R. G., et al. (1998) The Moderate Resolution Imaging Spectroradiometer (MODIS): land remote sensing for global change research. *IEEE Trans. Geosci. Remote Sens.*, Vol. 36, pp. 228-1249.

Kelly, R. E., Chang, A. T., Tsang, L., Foster, J.L. (2003) A prototype AMSR-E global snow area and snow depth algorithm. *IEEE Transactions on Geoscience and Remote Sensing*, Vol. 1, pp. 230-242.

Kuchment, L.S., Motovilov, Yu., G., Startseva, Z.P. (1990) Modeling of moisture transport in the soil-vegetation-surface boundary layer system for hydrological problems. *Water Resources*, Vol. 16, N 2, pp. 121-128. (Translated from Russian, *Vodnye Resursy* (1989), N 2, pp.32-39, Plenum Publishing Corporation, Consultants Bureau, New York, USA).

Kuchment, L.S., Startseva, Z.P. (1991) Sensitivity of evapotranspiration and soil moisture in wheat fields to changes in climate and direct effects of carbon dioxide. *Hydrol. Sci. J.*, Vol. 36, N 6, pp. 631-643.

Kuchment, L. S., Gelfan, A.N. (1996) The determination of the snowmelt rate and the meltwater outflow from a snowpack for modeling river runoff generation. *J. Hydrology*, Vol.179, pp. 23-36.

Kuchment, L.S., Demidov, V.N., Startseva, Z.P. (2006) Coupled modeling of the hydrological and carbon cycles in the soil-vegetation–atmosphere system. *J. Hydrology*, Vol.323, pp.4-21

Kuchment, L.S., Romanov, P., Gelfan, A.N., Demidov, V.N. (2010) Use of satellite-derived data for characterization of snow cover and simulation of snowmelt runoff through a distributed physicall basedmodel of runoff generation *Hydrol. Earth Syst. Sci.*, Vol. 14, pp. 339–350.

Lokupitiya, E., et al. (2009) Incorporation of crop phenology in Simple Biosphere Model (SiBcrop) to improve land-atmosphere carbon exchanges from croplands. *Biogeosciences*, Vol.6, pp.969–986.

Manabe, S. (1969) Climate and the ocean circulation. The atmospheric circulation and hydrology of the Earths surface. *Mon.Wea. Rev.*, Vol. 97, pp. 739–774.

Mauser, W., Schädlich, S. (1998) Modelling the spatial distribution of evapotranspiration on different scales using remote sensing. *J. Hydrol.*, Vol. 212–213, pp. 250–267.

Milly, P.C.D., Shmakin, A.B. (2002) Global modeling of land water and energy balances. Part I: the Land Dynamics (LaD) model. *Journal of Hydrometeorology*, Vol. 3, pp. 283–299.

Muzylev, E.L., Uspensky, A.B., Volkova, E.V., Startseva, Z.P. (2002) Simulation of hydrological cycle of river basins using synchronous high resolution satellite data. *Russian Meteorology and Hydrology*, N 5, pp. 52-63.

Muzylev, E.L., Uspensky, A.B., Volkova, E.V., Startseva, Z.P. (2005) Using satellite information for modeling heat and moisture transfer in river watersheds. *Earth Research from Space*, N 4, pp. 35-44. (In Russ.)

Muzylev, E., Uspensky, A., Startseva, Z., Volkova, E. (2006) Modeling vertical heat and water fluxes from river basin area with AVHRR/NOAA-based information on land surface characteristics. *Proc. of the 7th International Conference on Hydroinformatics, Nice, France, 4-8 September 2006*. Ed. P.Gourbesville, J.Cunge, V.Guinot, S.-Y.Liong, Vol. 2, pp. 1163-1170, Research Publishing, Chennai, India.

Muzylev, E.L., Uspenskii, A.B., Startseva, Z.P., Volkova, E.V., Kukharskii, A.V. (2010) Modeling water and heat balance components for the river basin using remote sensing data on underlying surface characteristics. *Russian Meteorology and Hydrology*, Vol. 35, N 3, pp. 225-235, DOI: 10.3103/S1068373910030106

NOAA KLM User's Guide. (2005) *http://www.ncdc.noaa.gov/oa/pod-guide/ncdc/docs/klm/html/c1/sec1-2.htm.*

Oleson, K.W., Niu, G.-Y., Yang, Z.-L., Lawrence, D. M., Thornton, P. E., Lawrence, P. J., Stöckli, R., Dickinson, R. E., Bonan, G. B., Levis, S., Dai, A., Qian, T. (2008) Improvements to the Community Land Model and their impact on the hydrological cycle. *J. Geophys. Res.*, Vol. 113, G 01021, doi:10.1029/2007JG000563.

Overgaard, J., Rosbjerg, D., Butts, M. B. (2006) Land-surface modelling in hydrological perspective - a review. *Biogeosciences*, Vol. 3, pp. 229–241.

Pitman, A.J. (2003) The evolution of, and revolution in, land surface schemes designed for climate models. *Int. J. Climatol.*, Vol. 3, pp. 479–510.

PUM LST (Product User Manual. Land Surface Temperature) (2008) *SAF/LAND/IM/PUM_LST/2.1*, 49p.

Romanov, P., Gutman. G., Csiszar, I. (2000) Automated monitoring of snow over North America with multispectral satellite data. *Journal of Applied Meteorology*, Vol. 39, pp. 1866–1880.

Romanov, P. (2011) Satellite-Derived Information on Snow Cover for Agriculture Applications in Ukraine, in Use of Satellite and In-Situ Data to Improve Sustainability. *NATO Science for Peace and Security Series C: Environmental Security*. Part 2, pp.81-91, DOI: 10.1007/978-90-481-9618-0_9

Rutter, N. et al. (2009), Evaluation of forest snow processes models (SnowMIP2), J. Geophys. Res., 114, D06111, doi:10.1029/2008JD011063

Schmetz, J., Pili, P., Tjemkes, S., Just, D. et al., (2002) An introduction to Meteosat Second Generation (MSG). *Bull. Amer. Meteor. Soc.*, Vol. 83, pp. 977-992.

Schultz, K., Franks, S., Beven, K. (1998) *TOPUP - A TOPMODEL based SVAT model to calculate evaporative fluxes between the land surface and the atmosphere, Version 1.1, Program documentation*, Department of Environmental Science, Lancaster University, UK.

Sellers, P.J, Mintz, Y., Sud, Y.C., Dalcher, A. (1986) A Simple Biosphere model (SiB) for use within general circulation models. *Journal of the Atmospheric Sciences*, Vol. 43, pp. 505–531.

Sellers, P.J., Berry, J.A., Collatz, G.J., Field, C.B., Hall, F.G. (1992) Canopy reflectance, photosynthesis and transpiration. III. A reanalysis using improved leaf models and a new canopy integration scheme. *Remote Sensing of the Environment*, Vol. 42, pp. 187–216.

Sellers, P. J., Dickinson, R. E., Randall, D. A., Betts, A. K., Hall, F. G., Berry, J. A., Collatz, G. J., Denning, A. S., Mooney, H. A., Nobre, C. A., Sato, N., Field, C. B., Henderson-Sellers, A. (1997) Modeling the exchanges of energy, water, and carbon between continents and the atmosphere. *Science,* Vol. 275, pp. 502-509.

Simic, A., Fernandes, R., Brown, R., Romanov, P., Park, W. (2004). Validation of VEGETATION, MODIS, and GOES C SSM/I snow-cover products over Canada based on surface snow depth observations. *Hydrological Processes,* Vol. 18, No. 6, pp. 1089–1104.

Snyder, W.C., Wan, Z., Zhang, Y., Feng, Y.-Z. (1998) Classification-based emissivity for land surface temperature measurement from space . *Int. J. Rem. Sens.,* Vol. 19, No. 14, pp. 2753-2774.

Solovjev, V.I., Uspensky, S.A. (2009) Monitoring of land surface temperatures based on second generation geostationary meteorological satellites, *Earth Studies from Space,* No. 3, pp.17-25, (in Russian).

Solovjev, V.I., Uspensky, A.B., Uspensky, S.A. (2010) Derivation of land surface temperature using measurements of IR radiances from geostationary meteorological satellites, *Russian Meteorology and Hydrology,* Vol. 35, No. 3, pp. 159-167, DOI 10.3103/ S1068373910030015.

Uspensky, S., Solovjiev, V., Uspensky, A.,(2009) Monitoring of land surface temperatures based on SEVIRI/METEOSAT-9 measurements. *Proceedings of 2009 EUMETSAT Meteorological Satellite Conference, 2009, EUMETSAT,* pp 55-60.

Uspensky, S., Uspensky, A., Rublev, A. (2011) Land surface air temperature retrieval capabilities with geostationary meteorological satellite data. *Proc. "Intern. Symp. on Atmospheric Radiation and Dynamics", St.Petersburg, June 21-24, 2011,*pp. 37-38.

van Genuchten, M.Th. (1980) A closed form equation for predicting the hydraulic conductivity of unsaturated soils. *Soil Sci. Soc. Am. J.,* No. 44, pp.892-898.

Volkova, E.V., Uspenskii, A.B. (2007). Detection of clouds and identification of their parameters from the satellite data in the daytime. *Russian Meteorology and Hydrology.* Vol. 32, No. 12, pp. 723-732, DOI: 10.3103/S1068373907120011.

Walker, A.E., Goodison, B.E. (1993) Discrimination of a wet snow cover using passive microwave satellite data. *Annals of Glaciology,* Vol. 17, pp. 301–311

Wan, Z., Dozier, J. (1996) A generalised split-window algorithm for retrieving land surface temperature from space. *IEEE Trans. Geosci. Rem. Sens.,* Vol. 34, No. 4, pp. 892-905.

Wood, E.F, Lettenmaier, D.P, Zartarian, V.G. (1992) A land-surface hydrology parameterization with subgrid variability for general circulation models. *Journal of Geophysical Research* Vol. 97, pp. 2717–2728.

Yang, Z.-L. (2004) Modeling land surface processes in short term weather and climate studies, In: *Observation, Theory and Modeling of Atmospheric Variability,* Ed. X. Zhu, X. Li, M. Cai, S. Zhou, Y. Zhu, F.-F. Jin, X. Zou, and M. Zhang, pp. 288-313, World Scientific Series on Meteorology of East Asia, World Scientific, New Jersey.

7

High Resolution Remote Sensing Images Based Catastrophe Assessment Method

Qi Wen et al.*
National Disaster Reduction Center of China
China

1. Introduction

In recent years, there seems to be more and more occurrences of natural disasters happening around the world due to abnormal climate change. To deal with natural disasters, disaster assessment technology will provide technical support and help facilitate decision making for disaster relief, disaster prevention and reduction, post-disaster recovery and reconstruction (The six editing room of Press of China Standards, 2010). Airborne remote sensing and satellite remote sensing, which feature no time limitation, no geographical restriction, wide coverage and high accuracy, are widely used in disaster assessment scenario, because they can provide prompt and accurate information (Xie & Zhang, 2000). After several earthquakes happened around China (Xingtai, Haicheng, Tangshan, Longling, Datong) during 1970s and 1980s, China had widely implemented airborne remote sensing photography and seismic damage interpretation (Zhang, 1993). Remote sensing had also played an important role in disaster assessment of recent occurring disasters, including Wenchuan Earthquake, Yushu Earthquake, Zhouqu Debris flow and Yingjiang Earthquake (Chen et al., 2008; Shi et al., 2010). This chapter will mainly discuss the catastrophe assessment method and technical flow used by National Disaster Reduction Center of China (NDRCC) during Wenchuan Earthquake(2008), Yushu Earthquake(2010), Zhouqu Debris Flow(2010). Further discussion and advises are also given.

2. The flow of catastrophe assessment method

Traditionally, catastrophe assessment process can be divided into three major steps: disaster scope assessment, physical quantity assessment and direct economic losses assessment. In addition, two assessment processes: rapid disaster assessment and ground investigation process are needed to supplement the major steps. Rapid disaster assessment is usually carried out to preliminarily judge the disaster condition after the disaster occurred. The reported data from Rapid Disaster Assessment, combined with that from following Disaster Scope Assessment, are integrated to have an overview of disaster scope and extent. Another supplementary process, Ground Investigation Process, is usually carried out to cross-

* Yida Fan, Siquan Yang, Shirong Chen, Haixia He, Sanchao Liu, Wei Wu, Lei Wang, Juan Nie, Wei Wang, Baojun Zhang, Feng Xu, Tong Tang, Zhiqiang Lin, Ping Wang and Wei Zhang
National Disaster Reduction Center of China, China

validate disaster condition after Rapid Disaster Assessment process. The rapid disaster assessment result, ground investigation data, combined with high resolution remote sensing data will provide a detailed comprehensive assessment report of disaster condition. Figure 1 illustrates the flow of catastrophe assessment method.

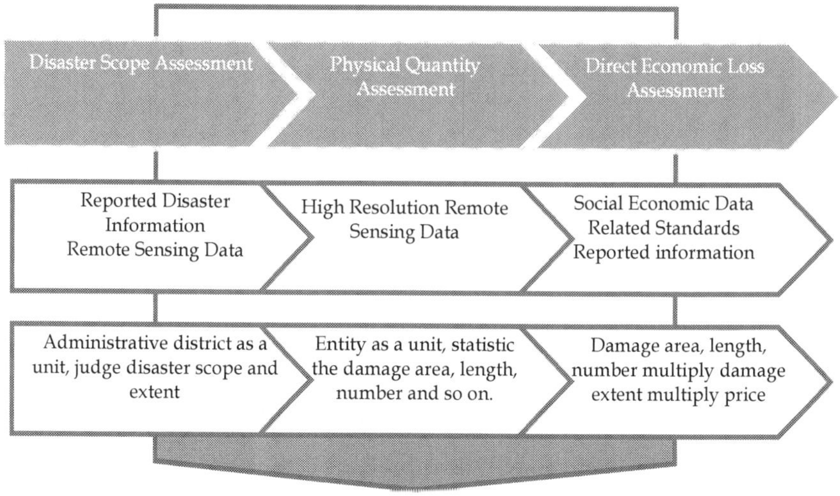

Fig. 1. The flow of catastrophe assessment method

2.1 Preliminary judgment of disaster condition

When the disaster happens, the remote sensing system will response immediately. By comparing pre-disaster and post-disaster high resolution remote sensing images, data on damage of transportation lines, houses, variation of terrain will be generated, and preliminarily overview of damage and disaster trend will be concluded based on these data.

2.2 Ground investigation

When the disaster situation is stable, according to preliminary judgment result, an expert investigation group will be sent to investigate in-field the disaster situation of sample region. Prior to ground investigation, the disaster region should be partitioned as multi-level grids in high-resolution remote sensing images. The first level partitioning is to partition the remote sensing images to functional zone according to administrative district, based on which the second level partitioning is to partition each functional zone to smaller unit, named "Entity", according to street in order to facilitate the ground investigation. The third level partitioning is to partition the "Entity" to even smaller grid unit according to house and parcel of rice field. All the grids are indexed, so that the investigation team members can fill out the corresponding forms. Investigation team uses Ground Disaster Information Collection System (Figure 2) to retrieve disaster information of 3rd level grid unit, including number of houses and their floors, estimated construction cost, and general information of 1st level unit, including affected population, damage extent and field photos. This information is collected and sent to the assessment team in NDRCC.

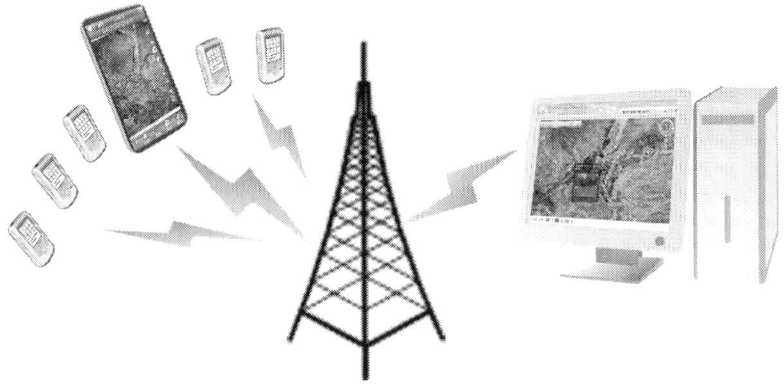

Fig. 2. Ground Disaster Information Collection System

Ground Disaster Information Collection System is a GIS software with the client-server architecture. The client side implemented in mobile phone is used to collect text and picture disaster information, label the position and send information to the sever side. The server side receives information from client side, interprets it and stores it to database.

2.3 Disaster scope assessment

When the disaster situation is stable, data reported by local government, assessment results generated by investigation expert group, interpreted data from remote sensing images are integrated and analyzed together to implement disaster scope assessment for sample regions. Regression relations between the sample regions and the entire disaster area are then established. Also, damage assessment data in adjacent regions are interpolated so as to generalize assessment of the whole area based on that of sample region.

In the case of an earthquake, disaster loss assessment is graded using Comprehensive Disaster Index. Disaster scope and extent of loss are assessed as a unit of administrative district(such as "county" in China). The Comprehensive Disaster Index is usually calculated as a weighted sum of the average seismic intensity, the toll of death and missing, the rate of number of death and missing per ten thousands persons, the number of collapsed houses, the rate of number of collapsed houses per ten thousands people, the geological disaster risk, and the rate of number of relocated person per ten thousand people.

$$DI = \sum (f_k \times DI_k) \tag{1}$$

DI_k is the kth normalized single indicator:

$$DI_k = [DI_k - \min(DI_k)] / [\max(DI_k) - \min(DI_k)] \tag{2}$$

f_k is the kth weight, and should be adjusted according to the specific condition of the disaster.

Average seismic intensity can be calculated by weighted sum of rates of different seismic intensity level for different area:

$$I = \sum (I_i \times S_i / S) \tag{3}$$

I_i is the intensity level. S_i / S is the rate of different seismic intensity level for different area; the toll of death and missing is the number of dead and missing persons in each county; number of collapsed houses is counted separately for different counties; geological disaster risk is calculated as a weighted sum of number of endangered residential areas, broken roads, blocked rivers, collapsed bridges and reservoirs. The rate of number of relocated per ten thousands of persons is calculated based on the base household population of each county.

2.4 Physical quantity assessment

The physical quantity assessment is carried out based on the previous data gathered in the first step of Disaster Scope Assessment. Based on extensive research and practical work experiences, NDRCC has established NNDDCA (the National Natural Disaster Damage Comprehensive Assessment Framework). Disaster loss is classified as five dimensions, and further divided to two hundred and twenty nine indicators. The five dimensions include: population loss, economic losses, residents' housing and property damage, national economy and industry loss, government offices and social undertakings loss. Several indicators, which are suitable to be assessed in remote sensing images, are first selected from the two hundred and twenty nine indicators and the damage of which is then assessed using artificial visual interpretation method and automatically machine recognition technology. The remote sensing index framework is then established.

One thing to notice is that, the affected population is not included in the remote sensing assessment index framework, and the economic losses can be statistically calculated from residents' housing and property damage, national economy and industry loss, government offices and social undertakings loss. The losses of residents' property are also not included in the remote sensing assessment index framework.

According to the characteristics of remote sensing technology, housing of three categories of statistical objects is classified as housing type; select specific object from agriculture, industry and services of state-owned enterprises; select water conservancy, environment, public facilities and other departments from government offices and social undertakings. Specific objects in the three categories can be divided into three new categories of remote sensing index of physical quantity damage assessment: housing and buildings, important infrastructure and natural resources. The first category, housing and buildings, includes building, bungalow, chimney and water tower. In the second category, important infrastructures include transportation, municipal, power, water, telecommunication and radio communication, pipeline facilities, etc., natural resources include land resources and water resources. Table 1 below shows a selection of assessment indicators.

In our case, data extracted from high-resolution remote sensing images in the step of Physical Quantity Assessment is usually brought to the decision-making chain as major information resource. By using pre-disaster and post-disaster UAV and airborne high-resolution remote sensing images collected from members of NDRCC, we have

<table>
<tr><td rowspan="26">Remote sensing indexes of physical quantity damage assessment</td><td colspan="2" rowspan="3">Housing and buildings</td><td>Building</td></tr>
<tr><td>Bungalow</td></tr>
<tr><td>Chimney and water tower</td></tr>
<tr><td rowspan="18">Important infrastructures</td><td rowspan="6">Transportation infrastructure</td><td>National road</td></tr>
<tr><td>Provincial road</td></tr>
<tr><td>County road</td></tr>
<tr><td>Bridge</td></tr>
<tr><td>Railway</td></tr>
<tr><td>airport pavement</td></tr>
<tr><td rowspan="4">Municipal infrastructure</td><td>Municipal road</td></tr>
<tr><td>Public traffic facility</td></tr>
<tr><td>Urban green land</td></tr>
<tr><td>Street lamp</td></tr>
<tr><td rowspan="2">Power facility</td><td>Power generation, substation facility</td></tr>
<tr><td>Overhead power line</td></tr>
<tr><td rowspan="4">Water conservancy facility</td><td>reservoir</td></tr>
<tr><td>channel</td></tr>
<tr><td>dam</td></tr>
<tr><td>embankment</td></tr>
<tr><td rowspan="2">telecommunication and radio communication facility</td><td>emission station</td></tr>
<tr><td>transmission line</td></tr>
<tr><td>Pipeline resources</td><td>Overhead and ground pipeline</td></tr>
<tr><td rowspan="5">Natural resources</td><td rowspan="3">Land resources</td><td>Farmland</td></tr>
<tr><td>Woodland</td></tr>
<tr><td>Building land</td></tr>
<tr><td rowspan="2">Water area resources</td><td>River</td></tr>
<tr><td>Lake</td></tr>
</table>

Table 1. Remote sensing indexes of physical quantity damage assessment

implemented disaster targets recognition and change detection, through which disaster data is extracted and integrated. The data includes number and area index of houses, number and length of roads, number and area index of power facilities, water conservation facilities, and communication stations, municipal utilities, and land resources.

2.5 Direct economic losses assessment

Based on physical quantity assessment results and related standards, Direct Economic Loss is assessed for houses, infrastructure of government, industry, and personal property within disaster areas. For example, the economic loss of a building can be calculated by multiplying build-up area, damage rate, and unit price. The damage rate is usually calculated using damage extent grading method. Other indicators follow the same method in calculating that of building.

3. Case studies of remote sensing catastrophe assessment

Catastrophe afflicted areas usually have complex terrain and frequently changing weather in a short time after the disaster, which makes earth observation technology one of the best methods for disaster emergency monitoring and disaster assessment due to the acquisition capability of high resolution images. In this section, three cases in China will be discussed in detail to illustrate the Disaster Assessment Technology NDRCC had used: Wenchuan Earthqueake, Yushu Earthquake, and Zhouqu Debris Flow Disaster. During Wenchuan Earthquake in May 2008, we used high resolution airborne remote sensing images, reported data collected by local government and ground investigation data in combination for Disaster Scope Assessment (Fan et al., 2008a); during Yushu Earthquake in April 2010, we implemented housing damage assessment (Yang et al., 2011); during Zhouqu Debris Flow Disaster in August 2010, we implemented Physical Quantity Assessment and Direct Economic Loss Assessment (Qi et al., 2011). The Physical Quantity Assessment is implemented based on disaster scope assessment result with indicators like houses and building, important infrastructure, land resources, and the Direct Economic Loss Assessment is implemented based on physical quantity assessment result.

3.1 Wenchuan earthquake

On May 12th, 2008, an earthquake of magnitude 8.0 Richter occurred in the countryside of Wenchuan, Sichuan province, China. The maximum seismic intensity was 11 degree which is higher than that of Tangshan Earthquake (7.8 Richter). Aftershocks frequently occurred and the largest one was of magnitude 6.4 Richter. Four hundred and seventeen counties out of ten provinces (Sichuan, Gansu, Shanxi, Chongqing, Guizhou, Yunnan, Henan, Hubei, Shanxi, Hunan) suffered from this disaster, leading to 69197 deaths, 18289 missing and 374000 injuries by July 13th 2010. The highly affected areas are most in mountainous valley where traffic was not convenient. The earthquake and following aftershock wrecked the fragile transportation and communication facilities, making disaster relief a very difficult job since rescue personnel, supplies, vehicles and large-scale disaster rescue equipments couldn't access the disaster scene in time.

3.1.1 Preliminary judgment of disaster scope

A seismic intensity distribution map was immediately made after the earthquake using empirical data from USGS (Figure 3). Results showed that the maximum intensity was beyond 9 degrees. Then, data of population and areas of affected towns and counties in the affected regions is calculated according to the map.

3.1.2 Remote sensing assessment of housing damage

Short after the earthquake in Wenchuan, we rapidly collected 1277 remote sensing images captured by 24 satellites from 12 countries using the International Charter "SPACE AND MAJOR DISASTERS" and domestic satellites data sharing mechanism. Several airplanes of

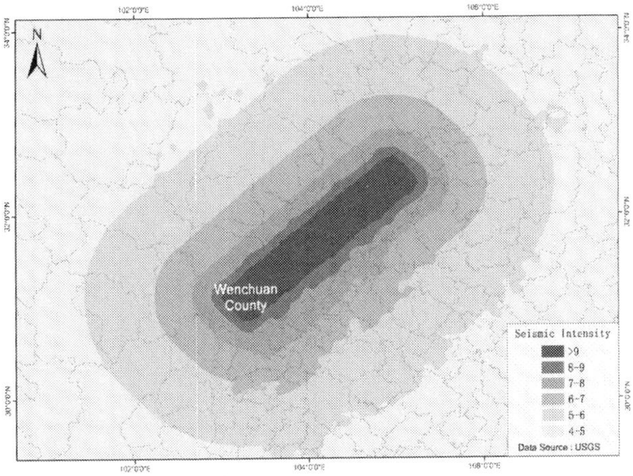

Fig. 3. Seismic intensity distribution map

National Administration of Surveying and Chinese Academy of Sciences flew to affected area to take airborne remote sensing image, in a resolution of 0.5m and 2m.

Take the indicator of houses for example. With the help of remote sensing images, damage condition of houses in satellite-covered regions was estimated. For uncovered regions, damage condition of houses was estimated by proportional spatial extrapolation and interpolation based on data from satellite-covered regions in a unit of village or town. Based on destruction probability of general fortified buildings, combined with the seismic intensity data, the relationship between collapse rate and damage rate was calculated. Damage condition of housing in affected region was then estimated. Figure 4 illustrates the flow of damage assessment process for the indicator of houses during Wenchuan Earthquake (Fan

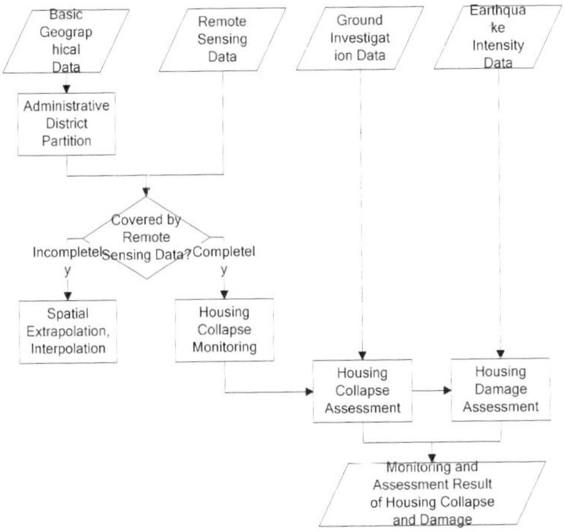

Fig. 4. The flow of collapse and damage assessment process for houses

et al., 2008b). Figure 5, 6, and 7 show the damage assessment map of sample regions: Maoxian County, Qingchuan County and the city of Dujiangyan. Figure 8 demonstrates the comprehensive assessment result of damage of the whole disaster area for single indicator of houses.

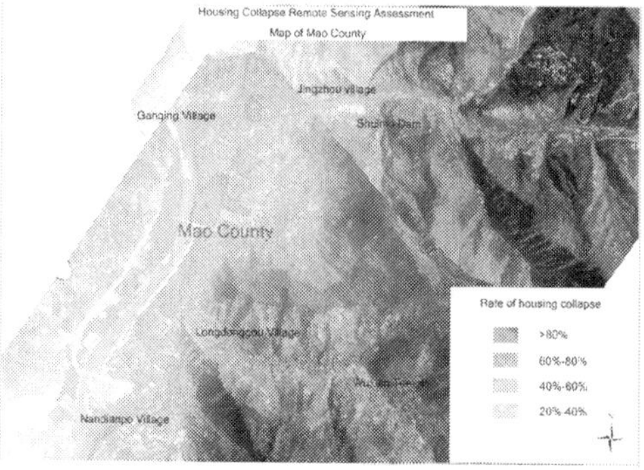

Fig. 5. The damage assessment result of Maoxian County for houses.

Fig. 6. The damage assessment result of Qingchuan County for houses.

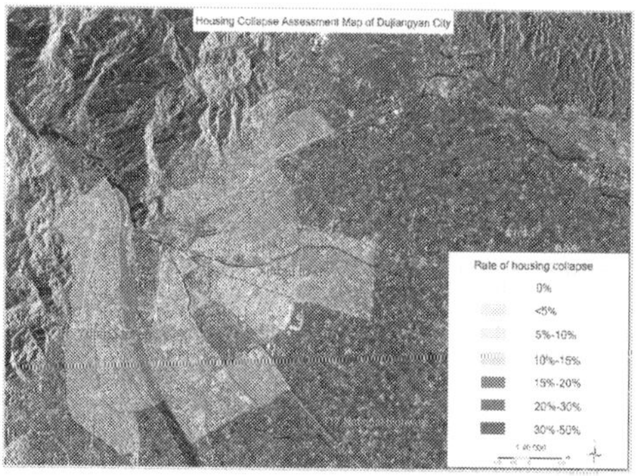

Fig. 7. The damage assessment result of Dujiangyan City for houses.

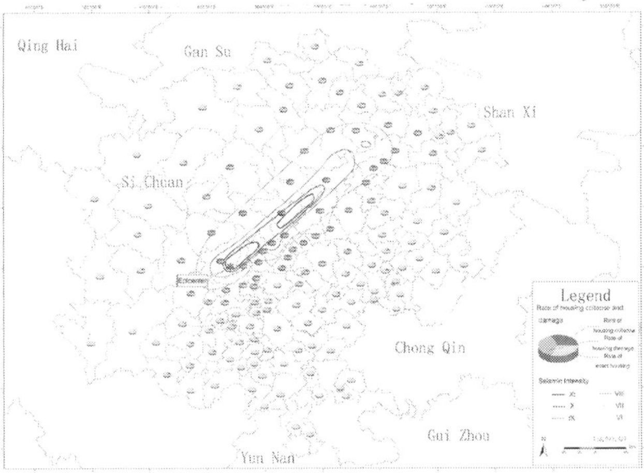

Fig. 8. Comprehensive assessment map of damage for houses

3.1.3 Disaster monitoring

In the following days after the quake, we continually and extensively monitored traffic, main national roads and settlements in disaster area to fast and accurately identify and locate traffic jam and landslide with the help of high-resolution remote sensing images.

EROS-B satellite image on May 15th illustrates several traffic jams occurred along 213 and 317 national highway in suburban area around Wenchuan due to secondary disaster landslide (Figure 9).

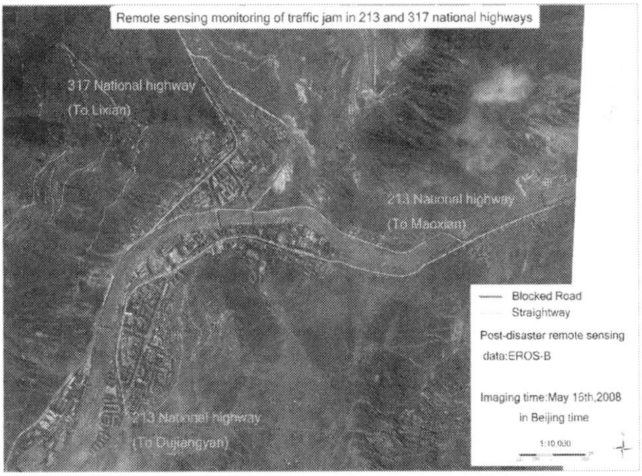

Fig. 9. Remote sensing image of the area along 213 and 317 national highways in the countryside of Wenchuan. Be noticed traffic jams are spotted with red lines

Quickbird satellite image on May 16th illustrates 5 landslides happened in a 2-kilometre diameter area around Yongan County, which seriously interrupted traffic (Figure 10).

Fig. 10. Remote sensing image from Quickbird satellite about landslide around Yongan County

Comparing EROS-B satellite image of 0.7m on May 27th and airborne remote sensing image of 0.5m resolution on May 18th, about 10 large refugee settlements distributed along both sides of urban main road in Qingchuan County (Figure 11), which increased significantly in ten days.

Fig. 11. Remote sensing images about quantity increasing of refugee settlements

3.1.4 Disaster scope assessment results

Combined with remote sensing interpretation results, ground investigation information and reported data from local government, we built Comprehensive Disaster Index (CDI) for Disaster Scope Assessment. The index will be demonstrated in detail as follows:

The weight of average seismic intensity was 0.3; the weight of rate of death and missing per ten thousands persons was 0.15, and the total weight was 0.3; the weight of houses collapsed rate per ten thousands was 0.1, and the total weight was 0.2; the weight for geological disaster risk was 0.1; the weight of relocated rate per ten thousands was 0.1.

Disaster area was partitioned into three sub region (Figure 12): serious region, severe region, and general region. The area and CDI of serious region was around 26000 km^2, and 0.4 above; the area and CDI of severe region was around 90000 km^2, and between 0.15 and 0.4; the area and CDI of general region was around 384000 km^2, and between 0.01 and 0.15.

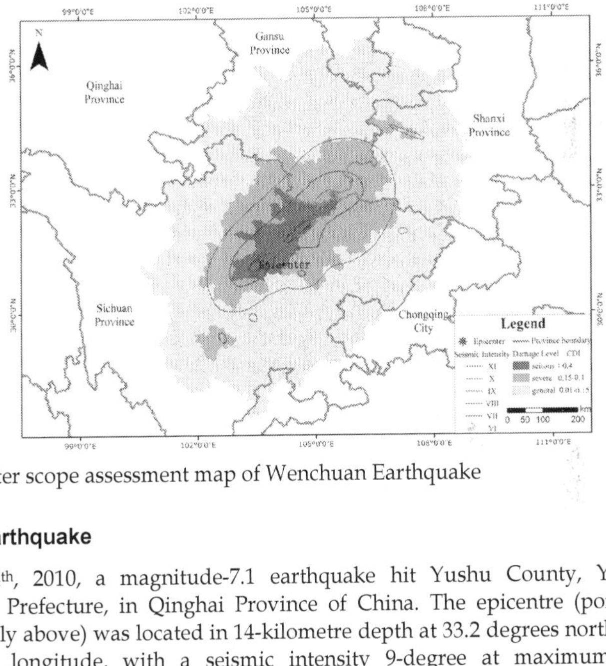

Fig. 12. Disaster scope assessment map of Wenchuan Earthquake

3.2 Yushu earthquake

On April 14th, 2010, a magnitude-7.1 earthquake hit Yushu County, Yushu Tibetan Autonomous Prefecture, in Qinghai Province of China. The epicentre (pointer on earth surface directly above) was located in 14-kilometre depth at 33.2 degrees north latitude, 99.6 degrees east longitude, with a seismic intensity 9-degree at maximum. Aftershocks frequently occurred with the largest one of magnitude-6.3. The earthquake wrecked Jiegu town and its surrounding area, resulted in 2220 deaths, 70 missing and 12135 injuries by p.m. 5, April 14th 2010. Many factors, such as high latitude, inconvenient traffic and cold weather, make disaster relief very difficult.

3.2.1 Preliminary assessment

Firstly, a seismic intensity distribution map was made by earthquake intensity empirical model according to earthquake magnitude, depth of epicentre and regional fracture zone distribution. Jiegu Town, Yushu County was in the hard-hit area with IX degree. Figure 13 illustrates contour of seismic intensity and terrain distribution.

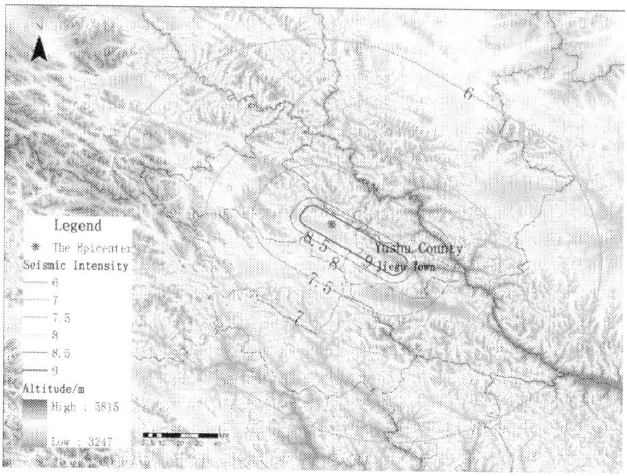

Fig. 13. Seismic intensity and terrain distribution map

By comparing pre-disaster airborne remote sensing images and post-disaster EROS B high resolution satellite images, we analyzed the damage condition of housings, and estimated disaster scope and damage extent rapidly. Figure 14 illustrates the assessment result.

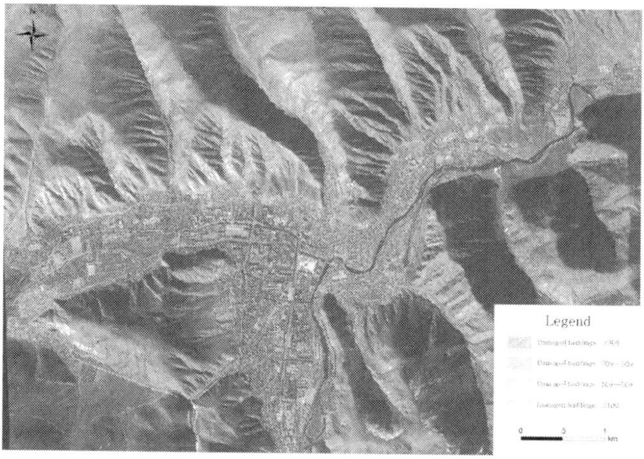

Fig. 14. Rapid assessment of disaster scope and extent of houses damage

3.2.2 Ground investigation and comprehensive assessment

Immediately after the quake, an airplane of Chinese Academy of Sciences (CAS) was sent to Yushu to take airborne remote sensing images. Images with a total storage of 409G and 0.4

meter spatial resolution were acquired, covering 4500 km^2 hard-hit area. Then combined with the result from preliminary assessment process, we partitioned residential region of Jiegu Town into 3-level grids according to street, function, and structure. About 685 girds of functional level were indexed.

First level (Figure 15) includes 4 categories: government office, community, industry area, and restricted area; second level (Figure 16) includes 6 categories: government office, school, general office, community, industry area, and restricted area; third level (Figure 17) includes 5 categories: building, bungalow, greenhouse, square, and stadium.

Fig. 15. Grids of first level

Fig. 16. Grids of second level

Fig. 17. Grids of third level

Hazard assessment expert group was then sent for field investigation. Damage extent, building size, architecture style, and actual pictures of each third level grid were collected and sent back to experts in NDRCC for interpretation and verification.

3.2.3 Disaster monitoring

At the same time, some key areas were monitored 24 hours a day using satellite and UAV images, including landslide region in Road 214, traffic jams in Road 308, refugee settlements area, terrain.

According to UAV monitoring images of 0.2m in Figure 18, 9861 refugee tents were placed in Jiegu urban region by April 19th, 5 days after the quake.

Fig. 18. Tent settlement monitoring

26 landslide regions were spotted, among which 5 were along Road 214 in the south of Jiegu town, which resulted in traffic jam. The other 21 were mainly located in mountain region and had little affect to traffic. Figure 19 illustrates traffic jam caused by landslide along Road 214.

Fig. 19. Traffic jam monitoring and assessment in landslide regions

3.2.4 Disaster scope assessment results

Figure 20 illustrates CDI about the disaster scope and extent of loss. CDI for this case is a little different compared to that of previous Wenchuan Earthquake. For serious regions covering around 992 km^2, CDI was 1; for severe regions of around 7030 km^2, CDI was [0.2,1]; for general regions of 27840 km^2, CDI was [0.01,0.2].

3.2.5 Housing damage assessment results

The housing damage scope and extent of loss were defined again by integrating field investigation data, data reported by local government (Figure 21). This is the first time satellite-airplane-ground integration assessment mode used in national disaster assessment. The damage scope was partitioned into three parts: collapsed building area, about 1,314,000m^2; serious damaged building area, about 2,332,000 m^2; and minor damaged building area, about 680,000 m^2. Collapsed building is defined as houses collapsed into ruins; serious damaged building as structures of houses been destroyed, thus it must be reconstructed; minor damaged building as structure been slightly affected, and it can still be used after reinforcement. Figure 8 and table 1 show housing damage assessment results. Figure 22 illustrates the remote sensing hazard monitoring and assessment flow in Yushu Earthquake.

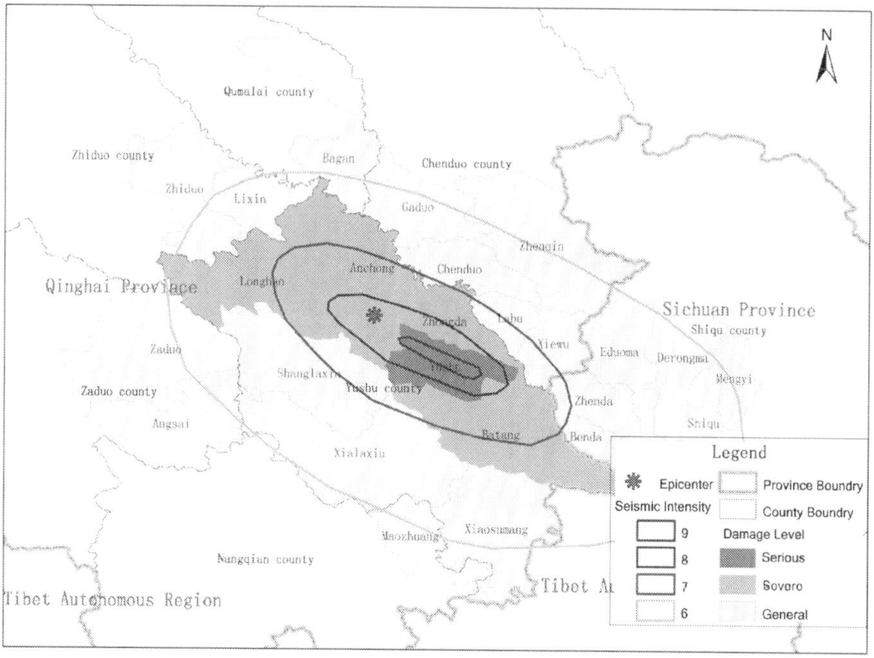

Fig. 20. Disaster scope assessment map of Yushu Earthquake

Fig. 21. Housing damage assessment results

Collapse & damage type	Completely collapsed (ten thousand square meters, %)				Serious damage (ten thousand square meters, %)				Minor damage (ten thousand square meters, %)			
	Bungalow	Building	Total	Ratio	Bungalow	Building	Total	Ratio	Bungalow	Building	Total	Ratio
Residential	118.5	1.6	120.1	27.8	116.6	31.2	147.8	34.2	27.7	8.5	36.2	8.4
Office	0.8	4.9	5.7	1.3	3.8	62.7	66.5	15.3	4.2	17.3	21.5	4.9
School	0.7	0	0.7	0.2	0.2	1.4	1.6	0.4	0	0.9	0.9	0.2
Industrial and mining enterprises	0.1	0.5	0.6	0.1	2.2	4.8	7	1.6	0.6	2.9	3.5	0.8
Special	4	0.2	4.2	1.0	8.9	1.4	10.3	2.4	5.4	0.5	5.9	1.4
Total	124.1	7.3	131.4	30.4	131.7	101.5	233.2	53.9	37.9	30.1	68	15.7

Table 2. Housing damage construction area statistic table of Jiegu town urban area.

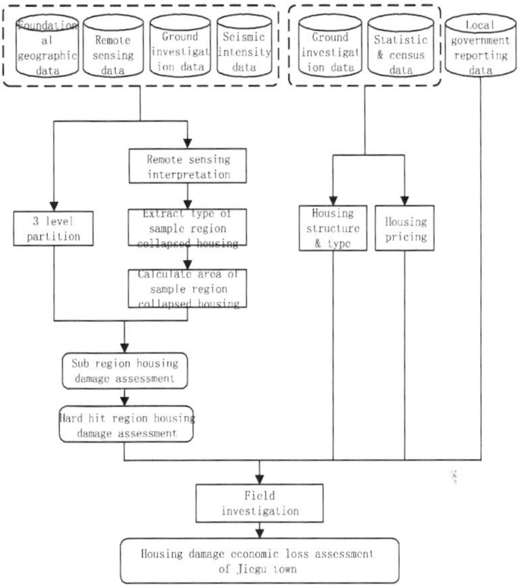

Fig. 22. Remote sensing hazard monitoring and assessment flow

3.3 Zhouqu debris flow

On August 8th 2010, a serious debris flow disaster occurred in Zhouqu County, Gannan Tibetan Autonomous Prefecture, Gansu Province, and China. The debris flow breaks many records since 1949, including number of death, extent of damage, and relief difficulty. The casualty consists of 1447 deaths, 318 missing, and about 21000 residents were forced to evacuate by p.m.4, August 24th. The disaster lasted for a long time, and repeatedly occurred in some regions, resulting in several barrier lakes, and part of town under water. After 8th, the disaster area had suffering from several heavy rains, which, triggered more debris flow, and interrupted the traffic. The hard-hit region was in remote mountain area with narrow road, making disaster relief a very difficult task.

3.3.1 Preliminary assessment

An UAV and airplane were sent to Zhouqu one day after the debris to acquire affected area images, including UAV image of 0.2m resolution and aerial remote sensing images of 1m resolution. Once the images were sent back, we compared pre-disaster and post-disaster remote sensing images, continually monitored disaster scope, damage extent of housing, residents , variation of river water level, newly happened landslide, and preliminary analyzed and estimated the damage condition and disaster trend.

Figure 23 illustrates preliminary assessment result. Collapsed area was 0.14 km^2, among which more than 90% were bungalows; serious damaged area was 0.15 km^2, among which more than 79% were buildings; minor damaged area was 1.12 km^2, among which more than 40% were buildings. Affected crop area was about 0.45 km^2, and affected forest area was about 0.27 km^2.

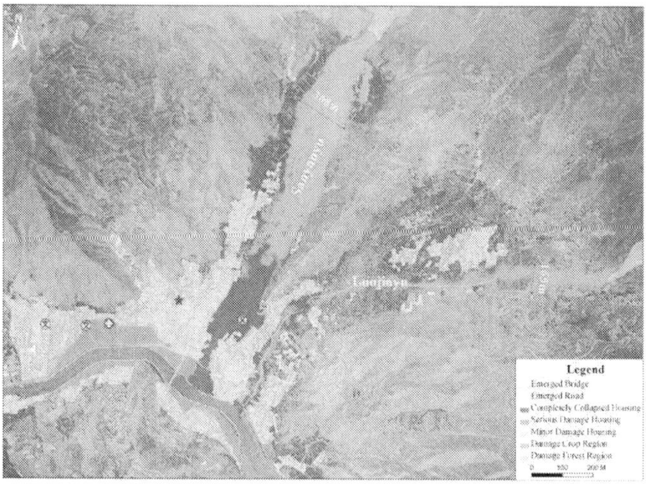

Fig. 23. Preliminary assessment results of Zhouqu Debris Flow disaster

According to debris flow and landslide monitoring result (Figure 24), Sanyanyu debris flow area was 0.66 km^2; Luojiayu debris flow area was 0.2 km^2. The 5 landslide regions in the mountain area of Zhouqu had a total area of 17272 km^2.

Figure 25 and Table 3 shows refugee settlement monitoring result. Tents number increased a lot from August 15th to August 8th.

Settlement	Area (m^2)	Tents number in 8th August	Tents number in 15th August	Increasing number
No.1 high school	2079	13	55	42
No.3 high school	3590	36	98	62
total	5669	49	153	104

Table 3. Comparison of tents number

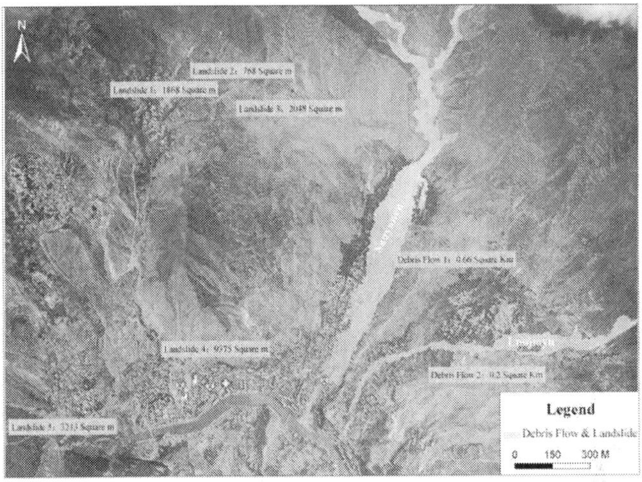

Fig. 24. Debris flow and landslide monitoring map

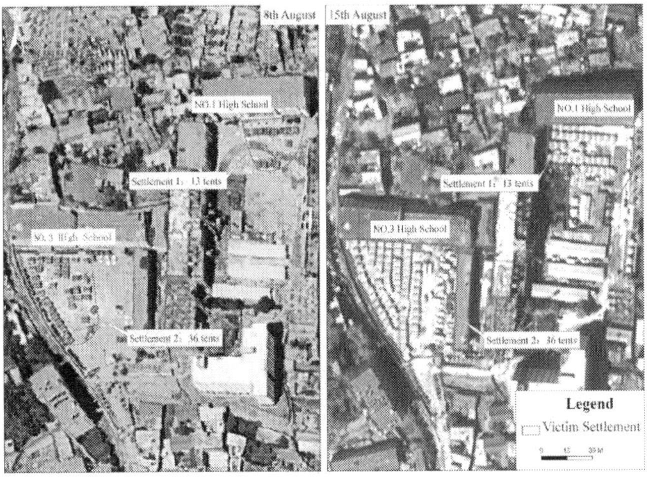

Fig. 25. Settlement monitoring map

The flooded area expanded obviously (Figure 26), with about 226,000 m^2 in August 15th, 23,000 m^2 more compared with that in August 10th, and 102,000 m^2 more compared with that in May 5th.

Fig. 26. Water area monitoring map

3.3.2 Ground investigation

Then according to the preliminary analysis results, we partitioned affection area into 2-level grids similar with previous case, using house as unit in high-resolution UAV and airborne remote sensing images. A total of 2457 girds were indexed (Figure 27). Hazard assessment experts were sent for field investigation. The damage extent, building size, architectural style, and field pictures of each grid were collected and sent back to experts in NDRCC for interpretation and verification.

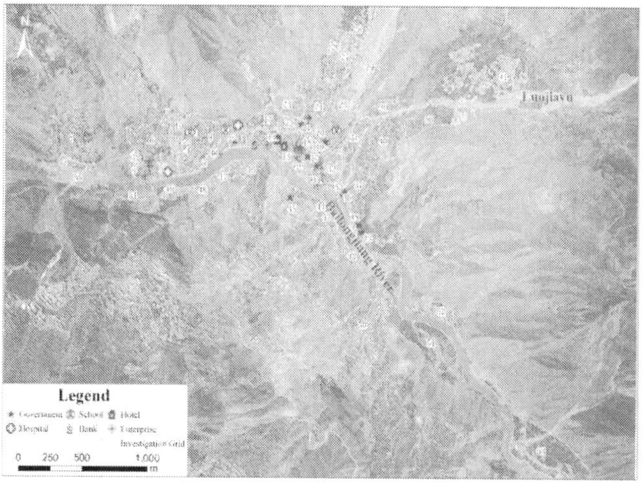

Fig. 27. Field investigation grids partition map

First level includes 3 categories of 59 grids: seriously damaged area, seriously flooded area, and minor affect area; Second level includes 2457 subgrids.

3.3.3 Disaster scope and housing damage assessment

The disaster scope and damage extent were defined again by integrating pre-disaster, post-disaster remote sensing information, ground investigation data and data reported by local government (Figure 28).

Disaster area was partitioned into three parts: serious affected region refers to the regions which bore main force of landslide, resulting in serious damage to houses, infrastructure and farmland; severe affected area refers the region which suffered from a long time soaked in water and mud, causing damage to houses; minor affected region refers to the region where houses and infrastructures are mildly damaged by floods and landslides.

The serious affected area was 1.2 km^2, severe affected area was 0.2 km^2 and minor affected area was 1.0 km^2.

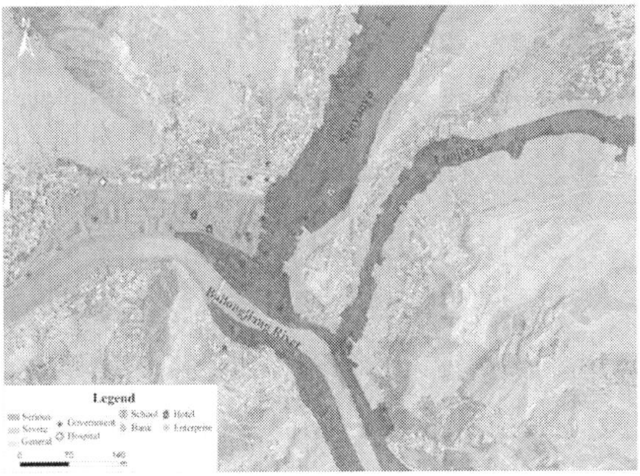

Fig. 28. Disaster scope assessment map of Zhouqu debris flow disaster

3.3.4 Physical quantity assessment

In the following Physical Quantity Assessment process, with the help of data reported by local government, we carefully assessed the physical quantity of houses, roads, power facilities, communication facilities, water conservation facilities, municipal utilities and land resources in UAV and airborne images to estimate approximate direct economic loss. It is worth to mention that some damage quantity ignored in field investigation was detected in remote sensing images, such as submerse street trees, telegraph poles, smart street pavilions.

According to remote sensing images and ground investigation verification result (Figure 29), in rush destroyed region, collapsed residential building area was about 125,000 m^2, seriously damaged area was about 102,000 m^2, minor damaged area was about 147,000 m^2. In flooded region, collapsed building area was around 419,827 m^2.

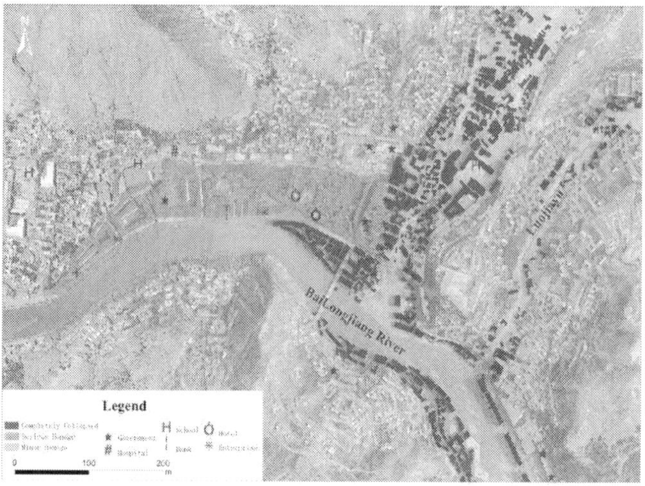

Fig. 29. Housing damage assessment map

Roads were divided into 3 classes (Figure 30): provincial road, urban road and rural road. According to remote sensing assessment result, 2 provincial roads were damaged, with a length of 2 km; 25 urban roads were damaged, with a length of 7.5 km; 5 rural roads were damaged, with a length of 4.8 km. Three bridges of a total length of 242 meters were out of function.

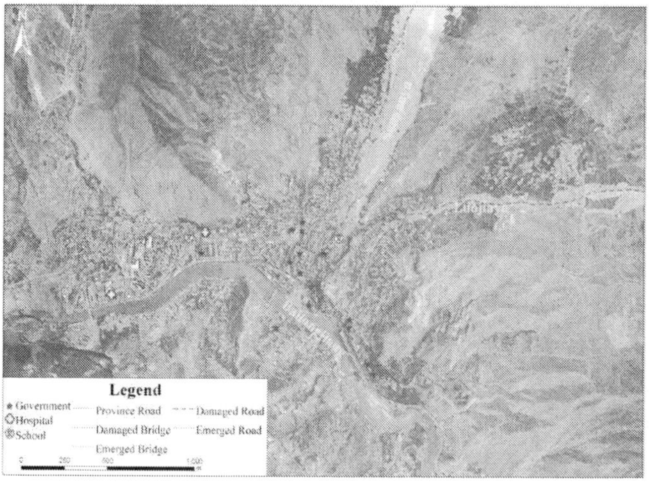

Fig. 30. Road and bridge damage assessment map

The physical quantity assessment results were used for direct economic losses estimation. It is for the first time for China that remote sensing images are integrated into the national catastrophe assessment flow as a major data source.

4. Conclusion

Throughout the three cases discussed in this chapter, one interesting thing is that the importance of remote sensing images is increasing in the whole process of disaster assessment: preliminary assessment step, disaster scope and damage assessment step, and physical quantity assessment. Also, the resolution of remote sensing images is improved, from 0.5m to 0.2m, providing more accurate data for assessment and decision making.

Although the whole process of integration of remote sensing images and data from ground investigation has been growing mature, the sample region choosing, physical quantity assessment still need further improvement.

Another thing to notice is that, automatic operation system is needed to connect the whole disaster assessment flow to provide efficient and accurate data for decision making. The satellite-airplane-ground disaster monitoring and assessment system is the trend we are and will realize in the near future.

5. Acknowledgment

This chapter is sponsored by High Resolution Earth Observation System Project of China.

6. References

Chen, S.; Ma, H. & Fan, Y. et al. (2008). Road Damage Assessment from High Resolution Satellite Remote Sensing Imagery in Wenchuan Earthquake. *Journal of Remote Sensing*, Vol.12, No.6, pp. 949-955, ISSN 1007-4619.

Fan, Y.; Yang, S. & Wang, L. et al. (2008a). Study on Urgent Monitoring and Assessment in Wenchuan Earthquake. *Journal of Remote Sensing*, Vol.12, No.6, pp. 858-864, ISSN 1007-4619.

Fan, Y.; Yang, S. & Wang, W. et al. (2008b). Comprehensive Assessment of Disaster Scope, In:*Comprehensive Analysis and Assessment of Wenchuan Earthquake*, Zhu, H., pp. (94-103), Science Press, ISBN 978-7-03-023855-9, Beijing, China.

Shi, F.; He, H.; Zhang, Y. (2010). Remote Sensing Interpretation of the MS7.1 Yushu Earthquake Surface Ruptures, *Technology for Earthquake Disaster Prevention*, Vol.5, No.2, pp. 220-227, ISSN 1673-5722.

The sixth editing room of Press of China Standards (2010). *Compilation of Earthquake and Related Standards*, Press of China Standards, ISBN 978-7-5066-5651-1, Beijing, China.

Wen, Q.; He H. & Wang X. et al. (2011). UAV remote sensing hazard assessment in Zhouqu Debris Flow Disaster, Proceedings of SPIE European Remote Sensing, ISBN 9780819488022, Prague, Czech Republic, September, 2011.

Xie, L.; Zhang, J. (2000). Application of Satellite Remote Sensing Technology in Earthquake Disaster Reduction, *Journal of Natural Disasters*, Vol.9, No.4, pp. 1-8, ISSN 1004-4574.

Yang, S.; Liu, S. & Wu, W. et al. (2011). Remote Sensing Applications in Qinghai Yushu Earthquake Monitoring and Assessment, *Spacecraft Engineering*, Vol.20, No.6, pp. 90-96, ISSN 1673-8748.

Zhang, D. (1993). Preliminary Study on Visual Interpretation Marks of Building Damages Caused by Earthquakes on Aero Photograph, *Earthquake*, Vol.13, No.1, pp. 26-30, ISSN 1000-3274.

8

Automatic Mapping of the Lava Flows at Piton de la Fournaise Volcano, by Combining Thermal Data in Near and Visible Infrared

Z. Servadio[1,2], N. Villeneuve[1] and P. Bachèlery[3,4]
[1]*Laboratoire Géosciences Réunion, Université de la Réunion, Institut de Physique du Globe de Paris, CNRS, UMR 7154, Géologie des Systèmes Volcaniques, Saint Denis*
[2]*Institut de Recherche pour le Développement, US 140, BP172, 97492 Sainte-Clotilde Cedex*
[3]*Clermont Université, Université Blaise Pascal, Laboratoire Magmas et Volcans, CNRS, UMR 6524, Observatoire de Physique du Globe de Clermont-Ferrand, BP 10448, F-63000 Clermont-Ferrand*
[4]*IRD, R 163, LMV, F-63038 Clermont-Ferrand*
France

1. Introduction

Knowing the eruptive history of a volcano is an essential key to the understanding of its functioning, and therefore of the evolution of the character of dangerousness of its eruptions. For an essentially effusive basaltic volcano such as the Piton de la Fournaise, the spatial and temporal distribution of lava flows allows to deduct numerous parameters of its activity, on a magmatic and a structural point of view. Satellite imaging brings more advantages than the methods used in aerial pictures studies, especially by supplying bigger temporal and spectral series. The revisiting of satellites over a region can allow the generation of dynamic mappings of the implementation of the lava flow, and also bring information on the phenomenology of the eruptions: Surface, volume, flow, spatial distribution…

Furthermore, satellite images have the advantage of supplying data that grant a global visualization of the study area, and information on not easily accessible areas. The interpretation of these satellite data enables obtaining information on the surfaces and volumes of the lava field flows, but also on its nature and behavior. In a tropical environment such as La Reunion, where the climatological context presents a strong cloudiness, a satellite revisit is statistically necessary.

The optical satellite images have already been successfully used to realize mappings of lava flows. For example, in Nevado Subancaya in Peru (Legelay-Padovanie et al., 1997) or in Etna in Italy (Honda et al.,2002), the combination of spectral and morphological properties helped to elaborate surface lava flows mappings and also allowed to individualize the main structures. Lu et al. (2004) propose an association of OPTICAL and RADAR imaging in order to define more accurate outlines for the lava flows. In the later example, the downstream part of the lava flow presenting vegetation, the discrimination was realized by the infrared. The upstream part being a heavily snow covered zone, RADAR images properties become a precious source of information. Thanks to the use of LIDAR (Airborne Light Intensity Detection and Ranging) on volcanoes, Digital Ground Models of very high-resolution can be generated and various retro-reflecting properties of the lava's different textures can be studied (Favalli et al., 2009) .The comparison of the different DGM produced at different dates allow Favalli et al. (2009) to obtain mappings of the thickness and outlines of the lava flow.

The use of thermal satellite imaging to characterize the relative chronology of the implementation of lava flows was the aim of Kahle et al. (1998)'s work. Abrams et al. (1991), used it in association with optical satellite imaging to realize a chronological mapping of the lava flows in Hawaii. Other authors have realized mappings of lava flow's temperature (Hirn et al., 2005) and implementation maps by using a thermal camera (Harris et al., 2007; James et al., 2007; Lombardo et al., 2009).

At the Piton de la Fournaise, the first research that used photogrammetry to realize mappings of the outlines and thickness of lava flows took place on lava flows dated back to 1972 and 1976 (Lénat, 1987). Bonneville et al. (1989) mapped the main geological units by using SPOT1 images. After which, Despinoy (2000) realized a mapping of the lava flows above Les Grandes Pentes using a CASI hyperspectral sensor. Villeneuve (2000) realized outlines, volume calculations and a chronological follow up of the implementation thanks to stereophotogrammetry and the use of DGPS (Digital Global Positioning System). Lénat et al. (2001) associated RADAR and SPOT images to map the field of lava of the Enclos Fouqué's caldera. Recently, De Michele et Briole (2007) used a technic of correlation of images to extract lava flows which implemented between two series of aerial pictures. The study of the incoherencies in the interferograms helped realize a dynamic follow up and a mapping of the lava flows (Tinard, 2007 ; Froger et al., 2007 ; Froger et al., in press).

The aim of this article is to propose an original approach by combining thermal images with ones acquired in the visible and near infrared in order to extract independent outlines for each lava flow. The extraction is therefore independent of the operator's subjectivity. Only one automatic extraction is possible when associating thermal and optical images for the implementation of a lava flow. An automatic extraction of the outlines of lava flow is realized and then compared to a mapping of references realized by photo-interpretation. We can therefore estimate the precision of the automatic extraction.

2. Context

The Piton de la Fournaise (Reunion Island, Indian Ocean) is a basaltic volcano whose functioning is connected to the activity of a hot spot (Courtillot et al. 1986). The eruptions take place inside the Enclos Fouqué's caldera and have contributed to the creation of a volcanic cone whose summit is occupied by two craters: the Bory on the east side and the Dolomieu on the west side (figure 1).

During the past three decades and besides the 1986 and 1998 exceptions, the lava flows have been taking place inside de Enclos Fouqué's caldera (figure 1 and 4). They are mainly fed by

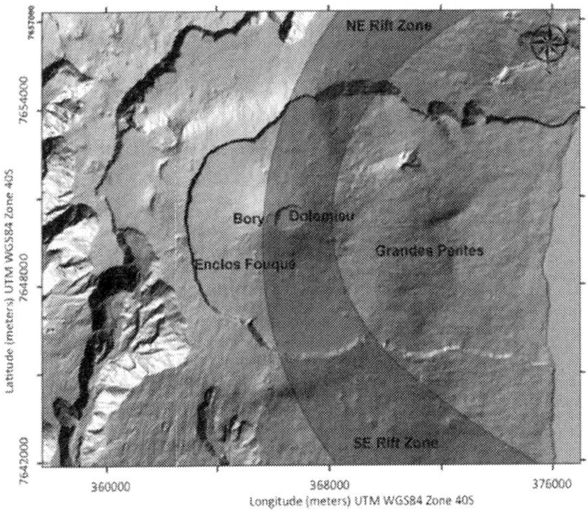

Fig. 1. Major morphological figures of the Piton de la Fournaise, the grey area is the rift zone defined by Bachèlery (1981).

intrusions situated along the rift zones (Bachèlery, 1981). The intrusions are interconnected at the same level as the central cone. Eruptions are of three kinds: summit zone eruptions, proximal eruptions, and distal eruptions (Peltier et al. 2009). It is clear that in the recent history of the Piton de La Fournaise, summit zone eruptions are the most frequent (Villeneuve et Bachèlery, 2006; Peltier et al., 2009). Most of these eruptions take place according to leveled cracks that progressively migrate by furthering from the central cone (Bachèlery, 1981).

Degassing at the event during eruptions can generate lava fountains which cause pyroclastic deposits and cones of several meters high. Two morphologies of lava flow are observed at Piton de la Fournaise: 'a'a type lava flows and pahoehoe type lava flows.

The occidental part of the Enclos Fouqué is largely recovered by a vast field of lava that Lénat et al. 2001 name the CLEF (Champ de Lave de l'Enclos Fouqué). This field of lava, essentially formed by a pahoehoe lava type flow, may have been constituted from slow emissions from the volcano's summit zone, between 1750 and 1794.

At the Piton de la Fournaise, the flow of lava emitted by eruptive cracks represent long (several meters to several kilometers), thin, (about ten meters), shallow (one meter in the slopes to 5 meters in flat zones (Letourneur et al., 2008)) lava flows, which shows the poor viscosity of the emitted magma. The juxtaposition of several individual flows during a same phase will contribute to the constitution of fields of lava, and particularly for long time eruptions (more than one month). In this case, in regards to the initial lava flow, the new income contributes essentially to the thickening and widening of the lava field flow. In this case, the lava field will be considerated here as a same unit.

The spectral properties of the lava flows differ according to the type of the surface (mainly 'a'a type at Piton de la Fournaise), but also according to the age, either because of a chemical transformation of the rock on the surface, or because of the implantation (always very fast in Reunion Island) of a vegetation cover (lichen, moss, shrub…). Lava flows which implementations were separated by several years can therefore be distinguished by their spectral properties. For the summit zone of the volcano, all the more in the Dolomieu crater, where the rocks are superimposed with only a few years or a few months of interval, the spectral properties of the diverse lava flows can then be very similar.

The oldest known eruption at the Piton de la Fournaise dates back to 1644. About 200 events have been counted since that date thanks to archives, 95% of them took place in the caldera (Lacroix, 1936; Stieltjes et al., 1989; Peltier et al. 2009). Never the less, this database is incomplete, particularly in the case of short time and low scale eruptions, before 1980. The mean average magma emission at Piton de la Fournaise estimated over a century, is 0,01 $km^3.an^{-1}$ (Lénat et Bachèlery, 1987), or 0.3 $m^3.s^{-1}$. The debit estimations show a temporal estimation. For example, Stieltjes et al. (1989), calculate a mean debit of 0.3 $m^3.s^{-1}$ over 54 years (1931-1985), but obtain 0.78 $m^3.s^{-1}$ for a period of 25 years (1960-1985). These variations are partly due to the existence of long periods of inactivity. For example, no eruption took place during 1992 and 1998; witch is to say 6 years of inactivity. Also, another inactivity as long was observed between 1966 and 1972 (Villeneuve, 2000). Peltier et al. (2009) illustrate these debits variation and show a more important activity since 1998. Between January 1990 and January 2010, 61 eruptions have been registered with a total volume of emitted lava estimated at 473 Mm^3 (figure 2), and 33 eruptions between 1998 and 2010, with a total volume of emitted lava of 313 Mm^3 (Peltier et al. 2009, OVPF 2009; 2010). From these observations, we have calculated a mean debit estimated between 0.45 $m^3.s^{-1}$ and de 0.82 $m^3.s^{-1}$, from 1980 to 2010. These estimations are superior to those obtained by Stieltjes et al. (1989), on former periods.

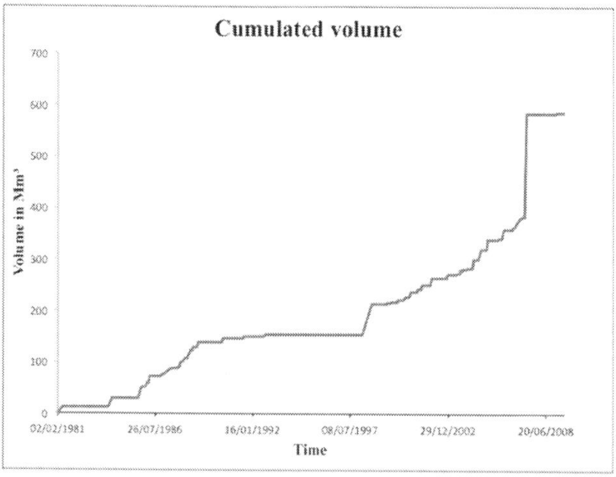

Fig. 2. Estimation of the cumulative volume of lava emitted from 1980 to 2010 by the Piton de la Fournaise.

3. Method

The originality of this research states in the use of thermal data as an analyze mask. In spite of its low spatial resolution (90m), thermal data brings essential information in our automatic mapping method. It allows determining with certitude the zone where the newly implemented lava flow is localized. The automatic extraction of the outline can be realized in this analyze mask. Also, its utilization enables treating the lava flows separately from one another because for one thermal image, only one lava flow is associated in this methodology. This is particularly adapted in the case of the constitution of a lava field flow.

3.1 Optical and thermal data

The automatic extraction method has been realized by the combination of thermal and optical data. SPOT and ASTER data have been used. SPOT data have **a** wavelength from the visible to mid infrared, and a spatial resolution from 2,5 to 20 meters. ASTER data have a spectrum from visible to thermal infrared and a spatial resolution that varies from 15m (VNIR) to 90m (TIR).

The ASTER TIR thermal data have to be acquired at the end of the eruption or very little time afterwards, in order for the thermal anomaly to be clearly visible on the entire zone. The maximum post eruption delay of acquisition is variable and depends on the thickness of the lava flow and therefore on its speed of cooling. In most of the cases, it is less than a month. ASTER VNIR and SPOT data can be acquired long after the lava flow's implementation. The principal is not having a new lava flow implementing on the same zone. The recent lava flows present low reflectances between the visible and mid infrared wavelength. The basalt spectrum, in the visible and short wavelength of the infrared (0.4-2.4 μm), is dominated by the presence of iron, which, at different levels of crescent oxidation, increases the reflectance (Despinoy, 2000). In the same way, the presence of lichen that grows on the lava flow increases the reflectance. In near and medium infrared, the presence of chlorophyll in the vegetation induces a strong signal (Kahle et al. 1995), permitting to discriminate precise outlines in zones with vegetation cover, especially near the Grandes Pentes area as for the Piton de la Fournaise.

The KALIDEOS project from the CNES (Centre National d'Etudes Spatiales, and GEO Grid (AIST/METI), and the NASA, grant free satellite data in the case of a research program. The data used in this article are from SPOT data from the KALIDEOS program, and from the GEO Grid program for the ASTER (Advanced Spaceborne Thermal Emission and Reflection Radiometer) data. The climatological conditions must be optimal because the presence of clouds masks, deforms or reduces the thermal anomaly, which in a tropical zone and especially in the volcano's zone is frequent. Only seven ASTER TIR data (the ASTER satellite was released in orbit in December 1999), acquired at the end of the eruption don't present these types of issues, and have therefore been used for the treatments (Table 1 and figure 3). Optical data acquired after the eruption have been associated to the former images.

3.2 Error calculation and outlines precision of photo-interpretation

The precision of the outlines extracted by automatic method leans entirely on their comparison with a base of outlines realized from very high-resolution satellite data' photo interpretation, and aerial pictures from IGN. This base is considerate as a reference.

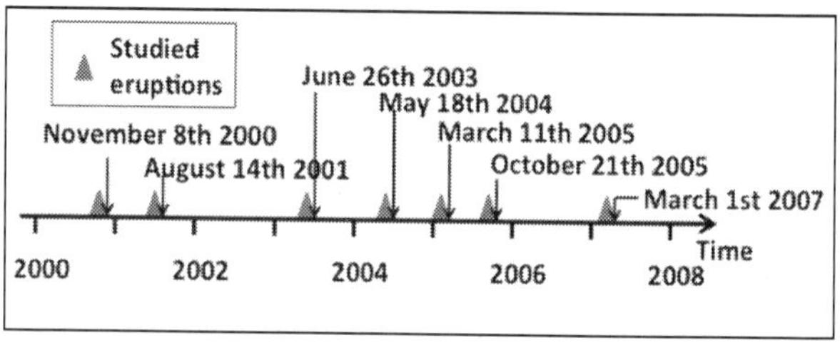

Fig. 3. Chronology of the acquisitions of the thermal data from the eruptions at the Piton de la Fournaise.

Eruption	ASTER TIR thermal Data		SPOT et ASTER optical data		
	Acquisition date (AAAAMMJJ)	Resolution	Acquisition date	Angle	Resolution
20070218	20070301	90m	20070314	R 19.1	10m
20051014	20051021	90m	20051222	R 25.9	10m
20050217	20050311	90m	20050217	6.7	15m
20040502	20040518	90m	20040518	-5.6	15m
20030530	20030626	90m	20031120	R 25.8	20m
20010611	20010814	90m	20010715-20011223	R 26.2 R 19.2	20m
20001012	20001108	90m	20001213	L 4.3	20m

Table 1. SPOT and ASTER data used for automatic treatments.

The outlines realized after photo interpretation, also known as reference, are not without errors. They contain errors linked to the operator's subjectivity and from the resolution of the data used to extract the outlines. One has to take into consideration these imprecisions before interpreting the results. This will be approached in the photo interpretation chapter, where outlines extracted by satellite data' photo interpretation are tested and compared on the base of those extracted from IGN's aerial pictures.

The method chosen lies on the constitution of an error matrix, because it permits to know the precision of all or part of the classification. It's expressed in percent or area.

3.2.1 Error matrix

The error matrix enables comparing two thermal maps. These matrixes are constructed according to a methodology developed by the Remote sensing Center of Canada (http://ccrs.nrcan.gc.ca/glossary/index_e.php?id=3124).

In our study case, an outline is compared to another that will be considered as reference (table 2).

In the matrix, numbers contained by cells correspond to areas. The total of the areas by column represent the total area of the class obtained by automatic classification. The total of the areas of each line corresponds to what has been correctly classified. The area values present in the diagonal correspond to the ones correctly classified. The other cells represent the areas which were classified wrong, either by omission, or by commission (table 2).

Areas (Km²)	Classification Data		
Reference Classification	Lava Flow	No Lava Flow	Σ
Lava Flow	Correctly classified Lava Flow	Omission Error	**Reference Lava Flow Area**
No Lava Flow	Commission Error	Correctly classified No Lava Flow	**Reference No Lava Flow Area**
Σ	**Automatic Lava Flow Area**	**Automatic No Lava Flow Area**	**Study Area**

Table 2. An error matrix for two classes (http://ccrs.nrcan.gc.ca/glossary/index_e.php?id=3124 (modifiée)).

It is therefore possible to calculate a global precision (in percent) of the classification, which is to say the area of what has been classified correctly in regards to the total area of the zone that has been classified

$$Pg = \frac{\sum \text{Well class area}}{\text{total area of the study}} * 100$$

It is also interesting to look at the mean precision of each class of classification:

$$Pm = \frac{Pmi}{i}$$

The precision of each class is different. Some pixels can be attributed to classes that don't suit them, if the spectral properties are similar. The exactitude producer (table 2) bases itself on the area of the pixels correctly classified in regards to the area of the lava flow considered as reference. As for the exactitude user (table 2), it bases on the area of the pixel correctly classified in regard to the area of the lava flow obtained by extraction.

The exactitude producer of the "lava flow" class allows knowing the accuracy of the outline obtained by automatic extraction:

$$\text{Pm (lava)} = \frac{\text{Well class area}}{\text{Real lava flow area}} * 100$$

The mean distance separating outlines vectors is calculated from the area which was not correctly classified, divided by the smallest of the two compared lava flow outline's perimeter:

$$\text{Error } \beta = \frac{\sum \text{Not correctly class area}}{\text{Perimeter}} * 100$$

3.2.2 Photo interpretation

The « referenced » outlines have been realized on a period from 1980 to 2010, by vectorization on the base of IGN aerial data from 1997 and 2003. As for the eruptions post August 2003, the research was done on a base of SPOT THR, or SPOT 4 AND 5 satellite data (Figure 4). The vectorization of the lava flows by using aerial photos is one of the most precise (Paparoditis et al. 2006), but it is also one of the most binding to realize. It could ask for several worked days.

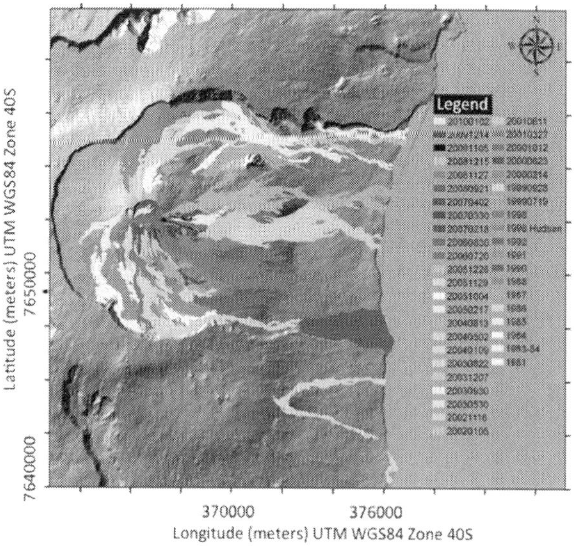

Fig. 4. Piton de la Fournaise lava flow cartography between 1980 and 2010 (Servadio et al. 2008 modified).

For the entire French territories, the temporal recurrence in the IGN's acquisitions is 5 years. It is a problem when dealing with superimposed lava flows. Satellite imaging is then a complementary tool because, even if the spatial resolution of SPOT data is lower (2,5m to 20m with optical SPOT data, or inferior or equal to 1 meter with aerial photos delivered by IGN for the BDORTHOs), the temporal resolution grants data between eruptions and defines outlines for the superimposed zones. Unfortunately, the resolution of the outlines is then less precise. The SPOT satellite was put into function in 1986, the outline data base was completed by the mappings of the OVPF and the BRGM (Stieltjes et al., 1985 et 1989 ; Billard, 1974 ; Bussière, 1967 ; Mc Dougall, 1971).

Our methodology leans on different types of satellite data, it is important to know the influence of the spatial resolution on the extracted outlines. A comparison between the outlines obtained by photo-interpretation of different data and the referenced outlines is then done by using an error matrix and a mean distance between the outlines (table 3). According to the used satellite data, the awaited error on the outline is about the same size of the pixel (table 3). The classification's precision can tell that a mapping by satellite data' photo-interpretation is 85% more reliable for satellite data with a 10m to 20 m resolution, and 95% reliable for THR SPOT data (table 3). The other error due to the referenced outlines extraction can be the consequence of the operator's subjectivity. It presents an error from 2% to 5%. The same test was run from aerial photos, and the errors didn't exceed 2%.

3.2.3 Automatic process: TIR- VNIR method

The lava flows which implement in zones where spectral properties are different, such as a vegetation zone or a soil that presents high spectral reflectances, are the easiest to identify (figure 5).

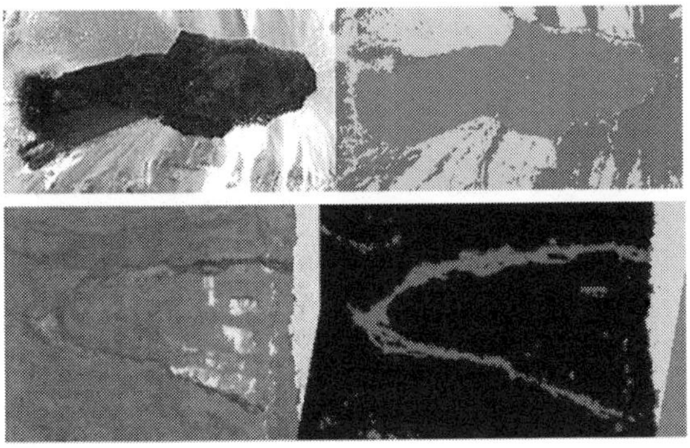

Fig. 5. Examples of automatic extraction of the lava flow outline when the reflectance of the substrate is very different from that of the lava. A and B: lava flow from the eruption of September 2009. C and D: lava flow from the eruption of 1986 off enclosures.

To automatically extract an outline, different methods can be used: classifications, threshold, or automatic detection of change (Inglada et al., 2003; Habib et al., 2007). The distance between the referenced outlines and the outlines automatically extracted is then proportional to the pixel's size, and the lava flow's classification precision is between 95% and 99%. On the other hand, lava flows which implement in low spectral reflectance zones, such as the central cone of the Piton de la Fournaise or the upperstream part of the Grandes Pentes, ask for more complex treatments. A data treatment methodology is then put together by using ENVI software (Figure 6). The visible data Principal Composant Analysis (PCA) is applied in order to maximize the data's anti-correlation. The thermal bands, near

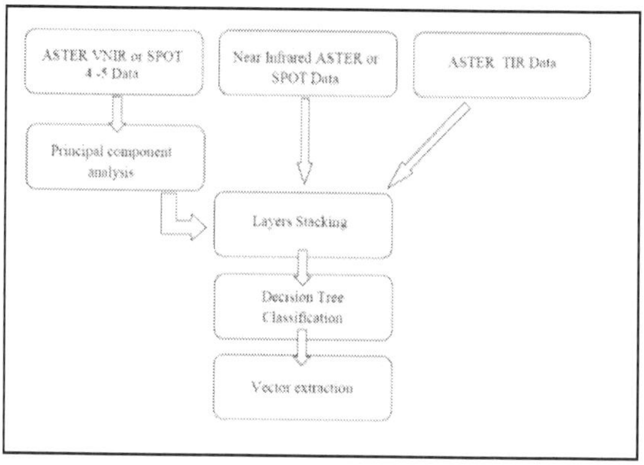

Fig. 6. TIR-VNIR Treatment sequences

infrared and the PCA are grouped in a multi-band data. A multi-level binary classification named decisional tree classification is then applied by using the following steps:

- By using thermal data, distinguish « hot » from « cold » zones.
- Extract and keep only « hot » zones.
- Discretization of the bare soil and covered vegetation zones by using the Near Infra-red band.
- Classify the pixels with low anti-correlation values.

The seven lava flows for which ASTER data were available (table 1) were tested. For half of them, the tested eruptions are on a weak slope zone, with a substrate composed of lava flows with similar spectral properties; the other half shows various substrates and slopes (figure 7).

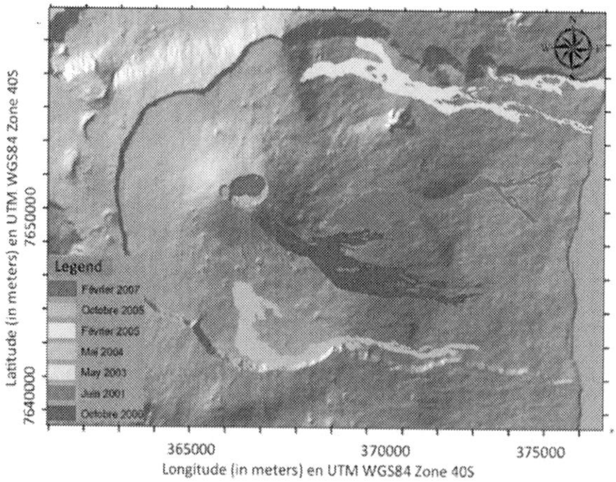

Fig. 7. Localization of the studied lava flows

Aerial Pictures	Spot 4 20m			Aerial Pictures	ASTER VNIR 15m		
1991	Lava Flow class	No lava flow class	Producer's Accuracy	20070402	Lava Flow class	No lava flow class	Producer's Accuracy
Lava flow	0.7632	0.0594	0.8226	Lava flow	3.568	0.041	3.609
No Lava flow	0.3848	3.0936	3.4784	No Lava flow	0.141	1.738	1.879
User's accuracy	1.148	3.153	4.301	User's accuracy	3.709	1.779	5.488
Global Accuracy	89.67%			Global Accuracy		96.68 %	
Mean Accuracy	90.86%			Mean Accuracy		95.68%	
Producer accuracy lava flow	92.78%			Producer accuracy lava flow		98.86%	
Producer accuracy no lava flow	88.94%			Producer accuracy no lava flow		92.49%	
Mean distance between the outlines	30.5m			Mean distance between the outlines		13.7m	
20001012	Lava Flow class	No lava flow class	Producer's Accuracy	Aerial Pictures	Spot 5 THR 5m		
Lava flow	3.354	0.727	4.081	1998	Lava Flow class	No lava flow class	Producer's Accuracy
No Lava flow	1.273	9.796	11.069	Lava flow	7.533	0.364	7.897
User's accuracy	4.627	10.523	15.15	No Lava flow	1.327	23.636	24.963
Global Accuracy	86.80%			User's accuracy	8.860	24	32.86
Mean Accuracy	85.34%			Global Accuracy		94.85%	
Producer accuracy lava flow	82.18%			Mean Accuracy		95.04%	
Producer accuracy no lava flow	88.99%			Producer accuracy lava flow		95.09%	
Mean distance between the outlines	38.86m			Producer accuracy no lava flow		96.57 %	
Global Accuracy	Spot 5 10m			Mean distance between the outlines		5m	
1986	Lava Flow class	No lava flow class	Producer's Accuracy	20000214	Lava Flow class	No lava flow class	Producer's Accuracy
Lava flow	1.017	0.165	1.182	Lava flow	1.705	0.088	1.7893
No Lava flow	0.126	11.912	12.038	No Lava flow	0.219	6.171	6.39
User's accuracy	1.143	12.077	13.22	User's accuracy	1.924	6.259	8.183

Table 3. Continued

Global Accuracy		97.95%		Global Accuracy		96.25%	
Mean Accuracy		92.5%		Mean Accuracy		95.83 %	
Producer accuracy lava flow		86.04%		Producer accuracy lava flow		95.09%	
Producer accuracy no lava flow		98.95%		Producer accuracy no lava flow		96.57 %	
Mean distance between the outlines		10.7m		Mean distance between the outlines		6.6m	
20010611	Lava Flow class	No lava flow class	Producer's Accuracy	**20010327**	Lava Flow class	No lava flow class	Producer's Accuracy
Lava flow	3.119	0.339	3.458	Lava flow	2.510	0.105	2.615
No Lava flow	0.412	13.22	13.632	No Lava flow	0.879	11.476	12.355
User's accuracy	3.531	13.559	17.09	User's accuracy	3.389	11.581	14.97
Global Accuracy		95.61%		Global Accuracy		93.43%	
Mean Accuracy		93.59%		Mean Accuracy		94.43%	
Producer accuracy lava flow		90.20%		Producer accuracy lava flow		95.98%	
Producer accuracy no lava flow		96.98%		Producer accuracy no lava flow		92.89%	
Mean distance between the outlines		8.9m		Mean distance between the outlines		3.2m	
20030530	Lava Flow class	No lava flow class	Producer's Accuracy	**20020105**	Lava Flow class	No lava flow class	Producer's Accuracy
Lava flow	0.2536	0.0295	0.2831	Lava flow	2.581	0.169	2.750
No Lava flow	0.0908	0.4277	0.5185	No Lava flow	1.057	8.103	9.16
User's accuracy	0.3444	0.4572	0.8016	User's accuracy	3.638	8.272	11.91
Global Accuracy		84.90 %		Global Accuracy		89.71 %	
Mean Accuracy		86.03%		Mean Accuracy		91.16%	
Producer accuracy lava flow		89.58%		Producer accuracy lava flow		93.85%	
Producer accuracy no lava flow		82.49%		Producer accuracy no lava flow		88.46 %	
Mean distance between the outlines		11.7m		Mean distance between the outlines		5.3m	

Table 3. Error matrices of outlines obtained by photo-interpretation, comparing those from the photo-interpretation of aerial photographs with those of the different types of satellite data.

The classification achieved, it is then possible to export the « lava flow » class as a vector that represents the outline of the lava flow (figure 8). The extraction is realized in less than an hour, once the data collected.

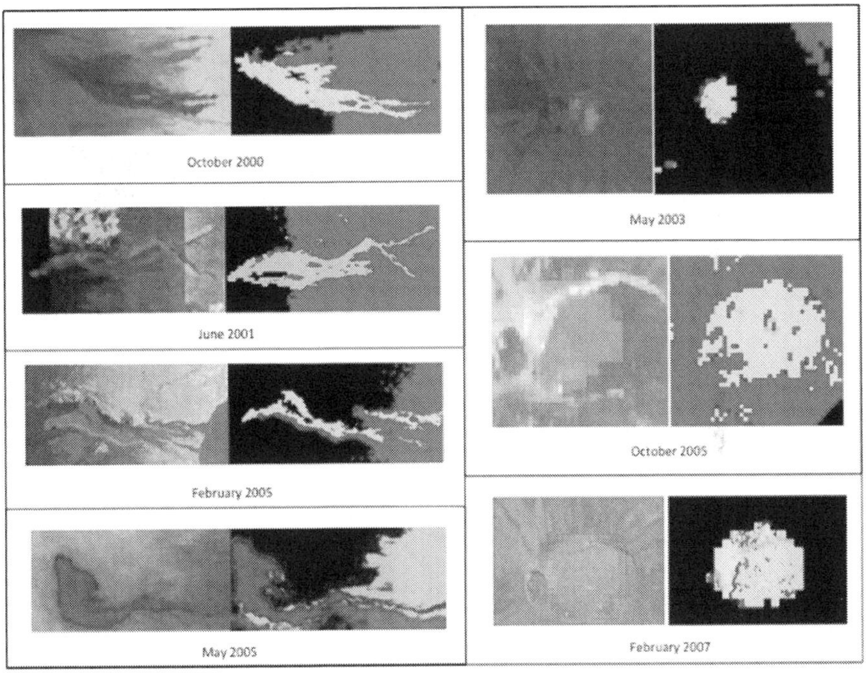

Fig. 8. Automatic extraction of lava flows outlines.

4. Results and interpretations

In order to validate the outlines automatic extraction method, each extraction result has been compared to the referenced outline. The error matrixes (table 4) represent the tests run on the seven tested objects localized on figure 7.

The error matrixes enable calculating global errors of classifications between 77% and 96% (table 4). We can observe a disparity between the summit zone eruptions and those implemented on sloppy substrates with various spectral properties. The first ones show a global precision between 77% and 88%, whereas the others variate from 91% to 96%. This is partly due to low reflectances observed for the substrate at the summit of the volcano. For example, the outlines of a lava flow newly implemented are hardly distinguished from the intra Dolomieu lava effusion zone that presents similar ages. The mean precisions show the same disparity, with values included between 77% and 84% at the summit zone, and included between 86% and 90% for the lava flows situated on the cone's flanks and on the slopes. The area of the lava flows also play a role in the classification precision's difference,

20001012	Lava Flow class	No lava flow class	
Lava Flow Aerial picture	3.422	0.659	4.081
No Lava Flow Aerial picture	0.667	11.602	12.269
Total Area of aerial picture	4.089	12.261	16.35
Global Accuracy		91.89%	
Mean Accuracy		89.205%	
Producer accuracy lava flow		83.85%	
Producer accuracy no lava flow		94.56%	
Mean distance between the outlines		28.41m	
20010611	Lava Flow class	No lava flow class	
Lava Flow Aerial picture	2.620	0.425	3.045
No Lava Flow Aerial picture	0.696	9.039	9.735
Total Area of aerial picture	3.316	9.464	12.78
Global Accuracy		91.228%	
Mean Accuracy		89.445%	
Producer accuracy lava flow		86.04%	
Producer accuracy no lava flow		92.85%	
Mean distance between the outlines		28.54m	
20030530	Lava Flow class	No lava flow class	
Lava Flow Aerial picture	0.2386	0.0445	0.2831
No Lava Flow Aerial picture	0.0784	0.3439	0.4223
Total Area of aerial picture	0.317	0.4223	0.7054
Global Accuracy		82.58%	
Mean Accuracy		82.855%	
Producer accuracy lava flow		84.28%	
Producer accuracy no lava flow		81.43%	
Mean distance between the outlines		30.45m	
20040502	Lava Flow class	No lava flow class	
Lava Flow Aerial picture	1.8389	0.6246	2.4635
No Lava Flow Aerial picture	0.6610	26.2154	26.8319
Total Area of aerial picture	2.4999	26.84	29.2949
Global Accuracy		95.31%	
Mean Accuracy		86.09%	
Producer accuracy lava flow		74.64%	
Producer accuracy no lava flow		97.70%	
Mean distance between the outlines		31.55m	
20050217	Lava Flow class	No lava flow class	
Lava Flow Aerial picture	3.291	0.901	4.192
No Lava Flow Aerial picture	0.816	17.002	17.818
Total Area of aerial picture	4.107	17.903	22.01
Global Accuracy		92.2%	
Mean Accuracy		86.965%	
Producer accuracy lava flow		78.51%	
Producer accuracy no lava flow		95.42%	
Mean distance between the outlines		29.65m	
20051004	Lava Flow class	No lava flow class	
Lava Flow Aerial picture	0.2727	0.0761	0.3488
No Lava Flow Aerial picture	0.1462	1.2272	1.3734
Total Area of aerial picture	0.4189	1.3033	1.7222
Global Accuracy		87.1%	

Mean Accuracy		83.8%	
Producer accuracy lava flow		78.2%	
Producer accuracy no lava flow		89.4%	
Mean distance between the outlines		26m	
20070218	Lava Flow class	No lava flow class	
Lava Flow Aerial picture	0.2960	0.0713	0.3673
No Lava Flow Aerial picture	0.0862	0.2617	0.3479
Total Area of aerial picture	0.3822	0.333	0.7152
Global Accuracy		77.98%	
Mean Accuracy		77.9%	
Producer accuracy lava flow		80.588%	
Producer accuracy no lava flow		75.22%	
Mean distance between the outlines		55.14m	

Table 4. Error matrices for the seven outlines obtained by automatic extraction, by comparing those from the photo-interpretation of aerial photographs with those from automatic extraction.

the summit zone lava flows have less superficy than the ones observed along the slopes. An error on the outline will echo as importantly as the aria of **the mapped** lava flow is low. As to say, the more the mapped lava flow's area is important, the less the error will echo on the precision of the classification. The exactitude of the producer of the "lava flow" class is the most important in the validation of an automatic lava flow outline extraction method. The later varies from 74% and 87% on the entire tested lava flow. There is no observed difference between the eruptions that occur in the Dolomieu crater and the ones that occur on the flanks and the Grandes Pentes of the volcano. On another hand, the morphology of the lava flows plays a role on this precision (figure 8). The more the lava flow's superficy is important and the more compact it is, like a cercle shape form or very large, the more precise the "lava flow" class will be.

The distances between the automatic extracted lava flow outlines, in the case of superimposed lava flows, and those considered as reference is about 30m (table 4 and figure 9). Now, the satellite data used in the automatic extraction have a visible, near and medium infrared spatial resolution from 10m to 20m. If the extraction was realized without any possible confusion, we should have mean distance of about the size of these pixels. Nevertheless, in some low reflectances zones, only the thermal imaging is able to bound the outline's localization zone (figure 8 and 9). It therefore considerably reduces the precision because this data has a 90m pixel. We notice that in zones where the lava regularly effuses, the reflectance's difference cannot allow a free from error extraction of the outline, and it is therefore deducted from the thermal mask. This mask is essential because it allows the definition of a zone of interest inside witch the outline can be determinated. There is no possible confusion: to one thermal data, only one lava flow can be associated.

In vegetalized zones, the extraction can be considered as a simple threshold of the infrared, the outline is then about the size of the pixel. In some zones, especially the Dolomieu crater or certain parts of the cone, the differences of reflectances of the lava flows don't allow to differentiate them. The outline is then obtained thanks to the thermal data. It can represent up to 40% of the summit zone lava flows outline, but only 5% to 0% of the proximal and distal eruption's outlines. The mean error observed on the outline's extraction can then reach twice the size of the optical data's pixel. Some light diffusion and wedging effects can interact on the precision.

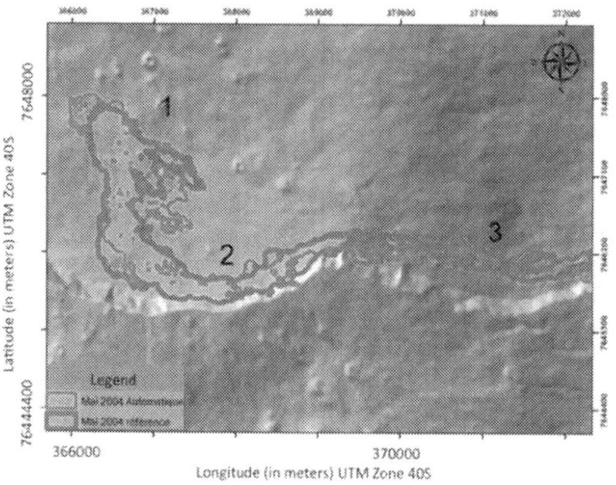

Fig. 9. Automatic lava flow outline extraction of May 2004 (blue) compared with the referenced outline (purple).

Lets considerate the example of the May 2004 lava flow (figure 9 and table 4). Three zones are sharply distinguished in the extraction: a very low reflectance and high thermal zone (1), a low reflectance with few or no thermal diffusion zone (2), and a various reflectance with few or no thermal diffusion zone (3). The weakest precision is for the first zone, for the second zone, the low reflectance is due to the substrate's nature and to luminosity issues in the zone, because the shadow projected by the rampart can interfere. A luminosity parameter is to be taken into consideration when choosing and time acquiring SPOT data. As for the extraction of the third zone, the thermal infrared band essentially obtains it, because the lava flow implemented in a vegetalized zone with a high spectral signature difference compared to the lava flow. It is the zone where the extraction is the most precise because only based on the SPOT data.

5. Discussion and conclusion

The vectors of the outlines obtained by automatic extraction properly match with the referenced outlines since we obtain a mean exactitude producer of 80% for the "lava flow" class. If we compare our automatic extraction results to those obtained by DGPS, they are less precise. The DGPS has a precision of about one centimeter at its antenna, now it's the operator who transports it. The error is due to the positioning of the antenna regards to the outlines. The error then is about one meter, for pluri-meter in our extractions. Nevertheless, the effusion zones are not all accessible, and cannot let realizing the outlines in their integrality, especially in the Grandes Pentes. Computer assisted drawing methods are on the other hand less reliable than our methodology for the data' distortion are not taken into account and can provoke hundred of meters error. The results of our classification are close to those obtained in other contexts' literature. (Azerzaq et al., 1997 ; Messar et Messar, 1997 ; Yüksel et al., 2008).

The vectorization time by photo-interpretation can take several days, whereas the automatic method enables obtaining this vector in less than one hour. There is therefore a considerable gain of time. The gaps of observations due to clouds, or zone with high thermal diffusion and low reflectances, could be filled by using RADAR imaging in treatment sequences (Weisseil et al.2004). However, by adding Bi data will increase the treatment time, especially by using coherence data.

The association of thermal and optical data has already been realized in other automatic classifications with outlines extractions contexts: glaciology (Raciviteanu et al. 2008), the canopy (Joshi et al., 2006), agriculture (Kasdan, 1979; Saito et al., 2001). For similar spectral resolution data, the error matrix results are comparable to those obtained by our methodology.

This methodology was developed in order to automate lava flow outline's extraction and therefore ensure a fast update of the Piton de la Fournaise's database. The lava flow map was updated thanks to photo-interpretation and automatic extraction (figure 4). It allowed us to test the reliability of the outlines extracted according to each methodology used in this article, and to know their precision.

The errors measured by the matrixes give us the extracted surface's error, by comparing the automatic lava flow's area and the referenced one. We saw that: 1/ the difference of interpretation between two operators for the same data can be of 2% for the aerial photos and it varies between 2% and 5% with satellite data. 2/if the outlines extracted from satellite data's photo interpretation are compared to those extracted from aerial data, the exactitude producer of the "lava flow" class can reduce to 80%, but is at an average of 90%. 3/ Those obtained with the same type of data varies from 74% to 87%, which is to say an average of less than 10%. It influences the obtained area in less than 10%, exception made for low scale summit zone eruptions which represent an error of 20%, which remains modest and without major consequences on the volume and production estimation rates for the volcano.

6. Acknowledgements

These reasearch were financed by the "Region Reunion" and the ministry of "l'Enseignement Supérieur et de la Recherche" program. The authors would like to thank the CNES (Centre National d'Etudes Spatial) for the free access to the SPOT data via the KALIDEOS program (http://kalideos.cnes.fr).

This piece of work also used free ASTER data obtained thanks to AIST GEO Grid (Ministry of Economy, Trade and Industry) programm.

Thanks to Rebecca Roger for the translation.

7. Note

All the lava flow outlines' vectors will be available online on the Laboratoire de Geosciences Reunion website.

8. References

Abrams M., Abbott E., KahleA., 1991. Combined Use of Visible, Reflected Infrared, and Thermal Infrared Images for Mapping Hawaiian Lava Flows. Journal of Geophysical Research, n°96 (B1), pp. 475-484.

Azerzaq M., Assafi M., Fahsi A., 1997. Utilisation des images HRV de SPOT pour la détection du changement périurbain à Casablanca (Maroc). Télédétection des milieux urbains et périurbains. Éd. AUPELF-UREF, pp. 275-282.

Bachèlery P., 1981. Le Piton de la Fournaise (Ile de la Réunion). Étude volcanologique, structurale et pétrologique. Thesis manuscript, Univ. Clermont-Ferrand II, 1-215.

Billard G., 1974. Carte Géologique de l'île de la Réunion. 1 :50000.Bureau de Recherche Géologique et Minières, Paris. pp. 1-4.

Bonneville A., Lanquette A.M., Pejoux R.1989. Reconnaissance des principales unités géologiques du Piton de la Fournaise, La Réunion, à partir de SPOT 1. Bulletin de la société géologique de France, n°87, pp. 1101-1110.

Bussière P., 1967. Carte Géologique du département de La Réunion, 1 :1000000. Service de la Carte Géologique de la France.

Courtillot V., Besse J., Vandamme D., Montigny R., Jaeger J-J., Cappetta, H., 1986. Deccan flood basalts at the Cretaceous/Tertiary boundary? Earth and Planetary Sciences Letters, n°80 (3-4) , pp. 361-374.

De Michele M., Briole P., 2007. Deformation between 1989 and 1997 at Piton de la Fournaise volcano retrieved from correlation of panchromatic airborne images. Geophysical Journal International, n°169, pp. 357–364.

Despinoy M., 2000. Potentiel de la télédétection haute résolution spatiale et spectrale en milieu intertropical. Une approche transdisciplinaire à partir d'expériences aéroportées à la Réunion et en Guyane française. Thesis manuscript, pp.1-244.

Favalli M., Karátson D., Mazzarini F., Pareschi M.T., Boschi E., 2009, Morphometry of scoria cones located on a volcano flank: A case study from Mt. Etna (Italy), based on high-resolution LiDAR data. Journal of Volcanology and Geothermal Research, v. 186, p. 320-330.

Froger J.-L., Fukushima Y., Tinard P., Cayol V., Souriot T., Mora O., Staudacher T., Durand Ph., Fruneau B., Villeneuve N.,2007. Multi Sensors InSAR Monitoring of Volcanic Activity: The February & April 2007 Eruptions at Piton de la Fournaise, Reunion Island, Imaged with ENVISAT-ASAR and ALOS-PALSAR data. Proc. of FRINGE 2007 Workshop, Frascati, Italy,

Habib T., Chanussot J., Inglada J., Mercier G., 2007. Abrupt change detection on multitemporal remote sensing images: a statistical overview of methodologies applied on real cases. Proceedings IEEE IGARSS, pp. 2593–2596.

Harris A.J.L., Dehn J., James M.R., Hamilton C., Herd R., Lodato L., Steffke A., 2007, Pahoehoe flow cooling, discharge, and coverage rates from thermal image chronometry. Geophysical Research Letters, n° 34, pp. 6.

Hirn B.R., Di Bartola C., Ferrucci F.,2005. Automated, Multi-Payload, High-Resolution Temperature Mapping and Instant Lava Effusion Rate Determination at Erupting Volcanoes. Geoscience and Remote Sensing Symposium, 2005. IGARSS '05. Proceedings. IEEE International. n° 7, pp. 5056 - 5059.

Honda K., Nagai M., 2002. Real-time volcano activity mapping using ground-based digital imagery. ISPRS Journal of Photogrammetry and Remote Sensing, n°57(1-2), pp.159-168.

Inglada J., Favard J-C., Yesou H., Clandillon S., Bestault C., 2003. Lava flow mapping during the Nyiragongo January, 2002 eruption over the city of Goma (D.R. Congo) in the frame of the international charter space and major disasters. Proceedings of the IEEE International Geoscience and Remote Sensing Symposium, IGARSS '03, n°3, pp.1540- 1542.

James M.R., Pinkerton,H., Robson S., 2007, Image-based measurement of flux variation in distal regions of active lava flows. Geochem. Geophys. Geosyst., n° 8, pp. Q03006.

Joshi C., De Leeuw J., Skidmore,A.K., Van Dure, I.C., Van Oosten, H., 2006. Remotely sensed estimation of forest canopy density: a comparison of the performance of four methods. International Journal of Applied Earth Obeservation and Géoinformation. n°8, pp.84-95.

Kahle A.B., Abrams M.J., Abbott E.A., Mouginis-Mark P.J., Realmuto V.J., 1995. Remote sensing of Mauna Loa. Mauna Loa Revealed: Structure, Composition, History, and Hazards. Geophys. Monogr. N°92, AGU, pp. 145–169.

Kahle A., Gillespie A., Abbott E., Abrams M, Walker R., Hoover G., Lockwood J.P., 1988. Mapping and relative dating of Hawaiian basalt flows using multispectral thermal infrared images. Journal of Geophysical Research, n°93(B12), pp. 239-251

Kasdan H., 1979. Feasibility of Using Optical Power Spectrum Analysis Techniques for Automatic Feature Classification from High Resolution Thermal, Radar, and Panchromatic Imagery. Recognition systems inc van nuys calif, pp. 1-194

Lacroix A., 1936. Le volcan actif de l'île de la Réunion et ses Produits. Paris, Gauthier et Villard ed., pp. 1-297

Legelay-Padovani A., Mering C., Guillande R., Huaman D., 1997. Mapping of lava flows through SPOT images - an example of the Sabancaya volcano (Peru), International Journal Remote Sensing, n°18(15), pp. 3111-3133.

Lénat J.-F., 1987. Structure et Dynamique internes d'un volcan basaltique intraplaque océanique : le Piton de la Fournaise (Ile de la Réunion). Thèse de doctorat d'Etat, Univ. Clermont II, France.

Lénat J.F., Bachèlery P. ,1987. Dynamics of Magma Transfer at Piton de La Fournaise Volcano (Réunion Island, Indian Ocean), Earth Evolution Sciences, Special Issue "Modeling of Volcanic Processes", Chi-Yu and Scarpa (Eds.), pp. 57-72, Friedr. Vieweg and Sohn, Brauschweig/Wiesbaden

Lénat J.-F., Bachèlery P. et Desmulier F., 2001 . Genèse du champ de lave de l'Enclos Fouqué : une éruption d'envergure exceptionnelle du Piton de la Fournaise (Réunion) au 18 ème siècle. Bull. Soc. Géol Fr. 2, pp.177-188.

Letourneur L., Peltier A., Staudacher T., Gudmundsson A., 2008, The effects of rock heterogeneities on dyke paths and asymmetric ground deformation: The example of Piton de la Fournaise (Réunion Island). Journal of Volcanology and Geothermal Research, n°173, pp. 289-302.

Lombardo V., Harris A.J.L., Calvari S., Buongiorno M.F., 2009, Spatial variations in lava flow field thermal structure and effusion rate derived from very high spatial resolution hyperspectral (MIVIS) data. Journal of Geophysical Research, n° 114, pp. B02208.

Lu Z., Rykhus R., Masterlark T., Dean K.G., 2004, Mapping recent lava flows at Westdahl Volcano, Alaska, using radar and optical satellite imagery. Remote Sensing of Environment, n°91(3-4), pp. 345-353.

Messar Y., Messar N., 1997. Apport de la télédétection à la cartographie de l'espace urbain au 1: 20000 en terme d'emprise au sol du bâti : cas de la ville d'Oran (Algérie). Télédétection des milieux urbains et périurbains. Éd. AUPELF-UREF., pp. 315-323.

Observatoire Volcanologique du Piton de La Fournaise internal annual report, 2009

Observatoire Volcanologique du Piton de La Fournaise internal report, janvier 2010

Paparoditis N., Souchon J-P., Martinoty G., Pierrot-Deseilligny M., 2006. High-end aerial digital cameras and their impact on the automation and quality of the production workflow. ISPRS Journal of Photogrammetry & Remote Sensing, n°60, pp. 400–412.

Peltier A., Bachèlery P., Staudacher T., 2009. Magma transport and storage at Piton de la Fournaise (La Réunion) between 1972 and 2007: A review of geophysical and geochemical data. Journal of Volcanology and Geothermal Research, n°184, pp. 93-108.

Raciviteanu A.E., Williams M.W., Barry R.G., 2008. Optical Remote Sensing of Glacier Characteristics: A Review with Focus on the Himalaya. Sensors , n°8, pp. 3355-3383

Saito G., Ishitsuka N., Matano Y., Kato M., 2001. Application on Terra/ASTER data on agriculture land mapping. Paper presented at the 22nd Asian Conference on Remote Sensing, Singapore, pp.1-6

Servadio Z., Villeneuve N., Gladys A., Staudacher T., Urai M., 2008. Preliminary results of lava flow mapping using remote sensing in Piton de la Fournaise, La Réunion island. Workshop on the "Use of Remote Sensing Techniques for Monitoring Volcanoes and Seismogenic Areas", IEEE Naples. Poster session and Proceeding.

Stieltjes L., Moutou P., 1989. A statistical and Probabilitic study of the historic activity of Piton de la Fournaise, Reunion Island, Indian Ocean. Journal of Volcanology and Geothermal Research, n°36, pp. 67-86.

Tinard P., 2007. Caractérisation et modélisation des déplacements du sol associés à l'activité volcanique du Piton de la Fournaise, île de La Réunion, à partir de données interférométriques. Août 2003 - Avril 2007. Ph.D. Thesis, Univ. Blaise Pascal, pp.1-334

Villeneuve N., 2000. Apports multi-sources à une meilleure compréhension de la mise en place des coulées de lave et des risques associés au Piton de la Fournaise : Géomorphologie quantitative en terrain volcanique. Thesis manuscript, pp.1-378.

Villeneuve N., Bachèlery P., 2006. Revue de la typologie des éruptions au Piton de La Fournaise, processus et risques volcaniques associés. Cybergéo : revue européenne de Géographie, n° 336

Yüksel A., Akay A.E., Gundogan R., 2008. Using ASTER Imagery in Land Use/cover classification of Eastern Mediterranean Landscapes According to CORINE Land Cover Project. Sensors. n° 8,pp. 1237-1251.

Section 2

Climate and Atmosphere

9

Coupled Terrestrial Carbon and Water Dynamics in Terrestrial Ecosystems: Contributions of Remote Sensing

Baozhang Chen
Institute of Geographic Sciences and Nature Resources Research, Chinese Academy of Sciences, Beijing P.R. China

1. Introduction

The Earth climate is a complex, interactive system, determined by a number of complex connected physical, chemical and biological processes occurring in the atmosphere, land and ocean. The terrestrial biosphere plays many pivotal roles in the coupled Earth system providing both positive and negative feedbacks to climate change (Treut et al., 2007). Terrestrial vegetation via photosynthesis converts solar energy into carbon that would otherwise reside in the atmosphere as a greenhouse gas, thereby regulating climate. Vegetation also transfers water between belowground reservoirs and the atmosphere to maintain precipitation and surface water flows.The terrestrial carbon (C) cycle is closely linked to hydrological and nutrient controls on vegetation (Betts et al., 2000; Cox et al., 2000). Understanding the coupled terrestrial C and water cycle is required to gain a comprehensive understanding of the role that terrestrial ecosystems play in the global climate change. Much progress has been made in gaining insight of the coupling processes between C and water cycles across a range of time and spatial scales (Pielke Sr, 2001; Friedlingstein et al., 2003; Seneviratne et al., 2006; Betts et al, 2007a,b; Baldocchi, 2008). Since the early 1990s, there has been an increased interest in monitoring of the CO_2, water vapor and energy exchange between the atmosphere and terrestrial ecosystems by a variety of methods, such as the eddy-covariance techniques (EC), satellite and other airborne remote sensing, CO_2 concentration and isotope measurements. Meanwhile, there are various kinds of models have been developed to better understanding of these processes and for large-scale C and water budgeting.

Remote sensing (RS) from satellite and airborne platforms, along with many other sources of land ground-based measurements (e.g., eddy covariance flux tower network, biometric plots, radar network, etc.) is playing and will continue to play a vital role in better understanding the coupled C and water cycle. Satellite RS allows the study of ecosystems from a completely new vantage point, facilitating a holistic perspective like viewing the Earth does for astronauts. Satellite-borne RS offers unique opportunities to parameterize land surface characteristics over large spatial extents at variable spatial and temporal resolutions. While there are challenges relating RS data recorded in radiance or backscatter to variables of interest, and RS has poor temporal resolution compared to ground-based measurement devices, RS and spatial analytical techniques and distributed biogeochemical modeling embedded in Geographical Information Systems (GIS) have allowed us to better understand the coupled C and hydrological dynamics across a large range of temporal and spatial scales.

The large number of papers published since the 1980s on the terrestrial and C/water cycles have resulted in the publication of several major reviews from different perspectives. For example, Running *et al.* (2004) described a blueprint for more comprehensive coordination of the various flux measurement and modeling activities into a global terrestrial monitoring network by reviewing the literature published before the middle of 1990s. Baldocchi (2008) recently provided a comprehensive review of research results associated with a global network of C flux measurements systems. The topics discussed by this review include history of the network, errors and issues related with the EC method, and a synopsis of how these data are being used by ecosystem and climate modellers and the remote-sensing community (Baldocchi, 2008). Kalma *et al.* (2008) reviewed satellite-based algorithms for estimating evepotranspiration (ET) and land surface temperatures at local, regional and continental scales, with particular emphasis on studies published since the early 1990s; while Verstraeten *et al.* (2008) provided a comprehensive review of remote sensing methods for assessing ET and soil moisture content across different scales based on the literature published after 1990s. Marquis and Tans (2008) reviewed satellite-based instruments on CO_2 concentration measurements.

In this chapter, I distil and synthesise the rapidly growing literature on C and water cycles using remote sensing in direct or indirect ways across local to global spatial scales and over a range of time scales. To give the reader a perspective of the growth of this literature, a search of Web of Science produced over 1500 papers with the key words 'ecosystem carbon, water cycles and remote sensing' published since 1990 which is indicative of the large amount of research recently being undertaken on these topics. In order to filter through this large body of literature, I concentrate on papers discussing on the coupling processes between C and water and I extract information from a database of published results that I have collated during the past decade (available on request). In terms of content, the report covers the state of knowledge, monitoring and modeling of the coupled terrestrial C and water cycles. My aim is to highlight the recent advances in this field, and propose areas of future research based on perceived current gaps in the literature.

This is a synthesis of state-of-the-art research on how RS has informed the study of coupled C and hydrology cycles. The review is divided into several inter-connected sections. First, I review the scientific background of the linkage between terrestrial ecosystems and climate, and revise the state of knowledge on terrestrial C cycling, coupling of the C and water cycles. Second, I discuss the ground-based and satellite-based monitoring methods and observation networks associated with measuring C and water fluxes, CO_2 concentration and C isotopes. Third, I report on the recent advances in modeling approaches associated with the terrestrial biochemical and hydrological studies. Fourth, I discuss research gaps in C sinks/sources estimates and finally, I discuss the current research trends and the near-future directions in this field and propose an upscaling framework for landscape and regional C and water fluxes estimates.

2. Scientific background and state of knowledge

2.1 Overview of terrestrial ecosystems and climate

The climate system is controlled by a number of complex coupled physical, chemical and biological processes (Figure 1). The terrestrial biosphere plays a crucial role in the climate system, providing both positive and negative feedbacks to climate change through biogeophysical and biogeochemical processes (Treut et al., 2007). Couplings between the climate system and biogeochemistry are mainly through tightly linked dynamics of C and water cycles. The importance of coupled C and water dynamics for the climate system has been increasingly recognized (Cox et al., 2000; Pielke Sr, 2001; Friedlingstein et al., 2003; Seneviratne et al., 2006; Betts et al, 2000, 2007a,b); however the mechanisms behind these coupled cycles are still far from well understood.

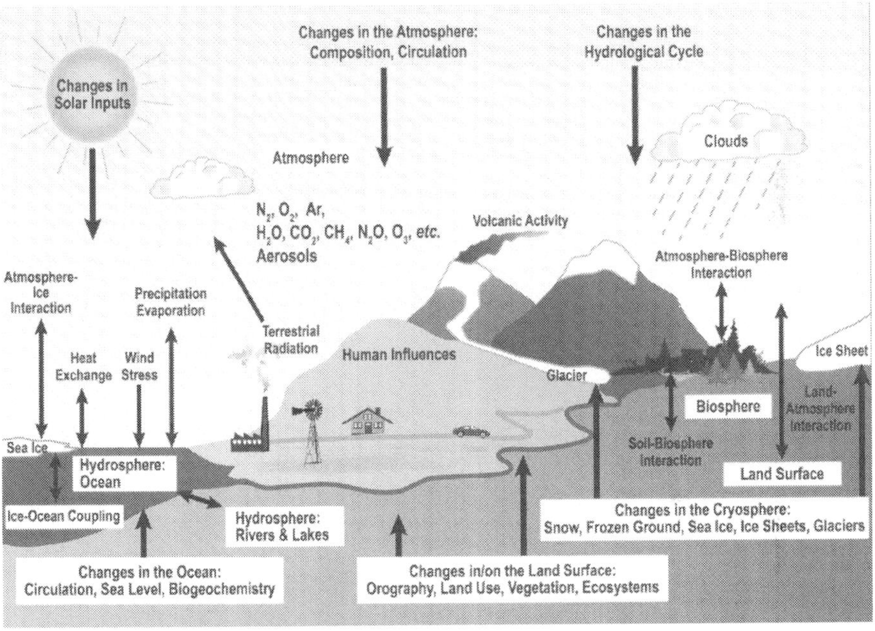

Fig. 1. Schematic view of the components of the climate system, their processes and interactions (Treut et al., 2007).

2.2 Terrestrial C cycling

One of the crucial issues in the prognosis of future climate change is the global budget of atmospheric CO_2. The growth rate of atmospheric CO_2 is increasing rapidly. Three processes contribute to this rapid increase: fossil fuel emission, land use change (deforestation), and ocean and terrestrial uptake. As shown in Figure 2, terrestrial C budgets have large uncertainties and interannual variability.

Terrestrial ecosystems mediate a large part of CO_2 flux between the Earth's surface and the atmosphere, with ~120 Pg C yr^{-1} taken up by photosynthesis and roughly the same amount

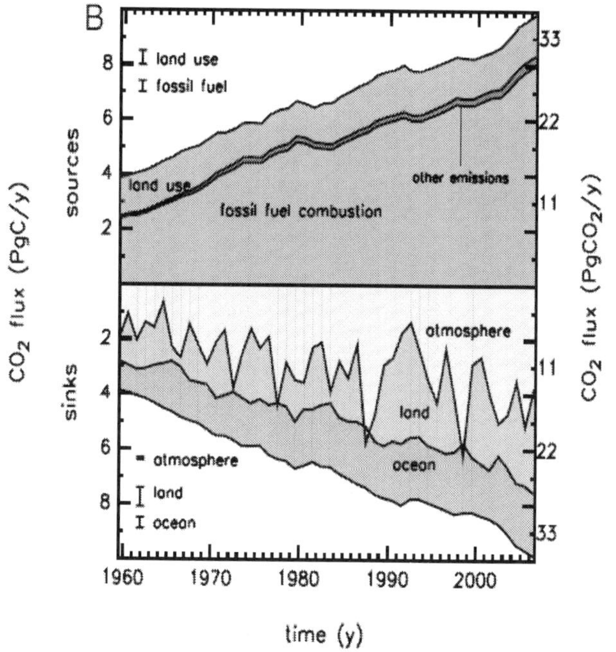

Fig. 2. Global CO_2 budget from 1959 to 2006. *Upper* panel: CO_2 emissions to the atmosphere (sources) as the sum of fossil fuel combustion, land-use change, and other emissions. *Lower* panel: The fate of the emitted CO_2, including the increase in atmospheric CO_2 plus the sinks of CO_2 on land and in the ocean (Canadell et al., 2007).

released back to the atmosphere by respiration annually (Treut et al., 2007; Prentice et al., 2001). Imbalances between gross ecosystem photosynthesis or gross primary productivity (GPP) and ecosystem respiration (R_e) lead to land surfaces being either CO_2 sinks or sources. The magnitudes of sinks and sources have fluctuated on annual and longer time scales due to variable climate, land use change, disturbance, and changes in the age distribution and species composition of ecosystems (Battle et al., 2000; Arain et al., 2002; Law et al., 2002; Morgenstern et al., 2004; Humphreys et al., 2005, 2006; Urbanski et al., 2007). Terrestrial ecosystems modify atmospheric C balance through many mechanisms. A detailed understanding of the interactive relationships in atmosphere-biosphere exchange is relevant to ecosystem-scale analysis and is needed to improve our knowledge of the global C cycle (Falk et al., 2008).

In recent years, scientists have learnt that terrestrial ecosystems' vegetation, soil (Melillo et al., 1989; Knapp et al., 1993) and animals (Naeem et al., 1995; Hattenschwiler and Bretscher, 2001) play key roles in mediating the terrestrial C cycle. Plants being the primary producers, it is from them that mass and energy gets transformed to other living organisms (Engel and Odum, 1999) within an ecosystem. The process of photosynthesis fixes atmospheric C into the biosphere. Atmospheric CO_2 enters the plant through stomatal opening that is controlled by a variety of environmental factors (Jarvis, 1976; Griffis et al., 2003). These factors include ambient temperature, atmospheric CO_2 concentration, nutrient availability, soil water availability and forest age (Schimel, 1995; Prentice et al., 2001). Changes in the atmospheric CO_2 concentration and the corresponding changes in the climate have altered the magnitudes of terrestrial C cycling. For example, a climate change induced increases in vegetation growth due to earlier springs and lengthened growing seasons were detected by the phase shift of seasonal atmospheric CO_2 cycle by Keeling et al. (1996) and satellite-based vegetation index analysis by Myneni et al. (1997). Studies indicate that an increase in atmospheric CO_2 enhances photosynthesis (e.g. Woodward and Friend, 1988) and hence increases assimilation of atmospheric CO_2 by the terrestrial vegetation. Nitrogen (N) availability to plants is another factor that can affect photosynthesis. This is because N is a primary nutrient for plant growth. In the recent years, variations in plant N availability have also altered the trends in the terrestrial C cycles. Variations in plant N availability occur mainly due to natural and anthropogenic N-deposition. Based on modeling studies, e.g., researchers (Townsend et al., 1996; Asner et al., 1997; and Holland et al., 1997) have demonstrated that N deposition is responsible for about 0.1-2.3 PgC yr^{-1} fixed by terrestrial vegetation which is almost half of the magnitude of C flux due to fossil fuel emission. Another factor that determines the nature of terrestrial C balance of an ecosystem is the age of the vegetation. Schimel et al. (1995) have demonstrated that forest re-growth can account for part of terrestrial C uptake as much as 0.5 ± 0.5 PgC yr^{-1}, especially in northern mid and high latitudes. This is because younger vegetation actively grows and hence sequesters more atmospheric CO_2 as opposed to mature forest stands. There are many other processes that directly and indirectly affect photosynthesis and thus, the C cycle. They include land use and land cover change (Caspersen et al., 2000; Houghton and Hackler, 2006; Easter et al., 2007), reforestation (House et al., 2002; Paul et al., 2002), agricultural and grazing activities (Cerri et al., 2005), insect attack (Chapman et al., 2003; Throop et al., 2004) and invasive species (Szlavecz et al., 2006). Respiration is a process by which C is added to the atmosphere from the biosphere. There are studies that indicate that total ecosystem respiration is a major determinant of terrestrial C balances (Valentini et al., 2000). Total ecosystem respiration includes respiration by aboveground plant parts (boles, branches, twigs, and leaves) and soil respiration, which is the sum of the heterotrophic respiration, and root respiration including respiration of symbiotic microorganisms. The temporal variability of respiratory metabolism is influenced mostly by temperature and humidity conditions (Davidson and Janssens, 2006). Although ecosystem respiration has received considerable attention in recent decades, much less is known about the relative contributions of its sub- components (Jassal et al., 2007), and our understanding of how they will respond to global warming is poor. Soil respiration (root + heterotrophic respiration) is a dominant component of C exchange in terrestrial ecosystems which accounts for more than half of the total ecosystem respiration (Black et al. 2005). This is because soils of terrestrial ecosystems contain more C than the atmosphere and live biomass together (Eswaran et al., 1993). Components of respiration can have different responses to temperature and soil water content (Boone et al., 1998; Lavigne et al., 2004), thus the effects of these environmental controls needs to be understood in order to fully comprehend the soil C cycling mechanism. There are many other mechanisms that can release terrestrial C to atmosphere. This includes both natural and anthropogenic reasons. Emission of large amounts of C to the atmosphere from vegetation can occur during forest fires (Amiro et al., 2002; Soja et al., 2004; Amiro et al., 2004) or biomass burning (Fernandez et al., 1999; Tanaka et al., 2001). These C emissions are of very high magnitudes although their duration is very short. Forest fires and biomass burning also affect the nutrient status of the soil which could have positive effects on the succeeding vegetation (Prietofernandez et al., 1993; Deluca and Sala, 2006). Another form of C flux in almost all terrestrial ecosystems is the import and export of dissolved organic carbon (DOC) (Neff and Asner 2001; Hornberger et al. 1994). DOC fluxes include C in the form of simple amino acids to large molecules that are transported through water flows. Fluxes of DOC into the ocean via runoff from terrestrial ecosystems are estimated to be 0.2 (Harrison et al., 2005) to 0.4 Pg C per year (IPCC, 2001). Since these fluxes are very small compared to the C fluxes due ecosystem is a net C sink due to the presence of soil C-pools having much longer residence times (Thompson et al., 1996; Chen et al., 2003; Canadell et al., 2007; Schulze, 2006). The strength of the terrestrial C sink was estimated to be 0.5-2.0 Pg C yr^{-1} (Schimel et al., 1995). By sequestering atmospheric C, the terrestrial ecosystems help decrease the rate of accumulation of anthropogenic CO_2 in the atmosphere, and its associated climate change (Cihlar, 2007). Terrestrial C sinks may be responsible for taking up about one-third of all the CO_2 that is released into the atmosphere (Canadell et al., 2007). The terrestrial C sink, inferred based on our current understanding, may not be permanent (Luo et al., 2003; Cox et al., 2000 ; Friedlingstein et al., 2003). Over the last few years there have been several studies suggesting that the size of this terrestrial C sink is vulnerable to global warming (Martin et al., 1998; Nemani et al., 2002; Canadell et al., 2007). The metabolism of terrestrial ecosystems is complex and highly dynamic because ecosystems consist of coupled, non-linear processes that possess many positive and negative feedbacks (Levin, 2002; Ma et al. 2007). How the C budget of major ecosystems will respond to changes in climate is not quantitatively well understood (Baldocchi & Meyers 1998, Goulden et al., 1998; Black et al., 2000; Baldocchi et al., 2001a; Baldocchi & Wilson, 2001; Law et al., 2002; Barr et al., 2004, 2007). A detailed understanding of the interactive relationships in atmosphere-biosphere exchange is relevant to ecosystem-scale analysis and is needed to improve our knowledge of the global C cycle (Falk,M et al., 2008). The metabolism of terrestrial ecosystems is complex and highly dynamic because ecosystems consist of coupled, non-linear processes that possess many positive and negative feedbacks (Levin et al., 2002; Ma et al., 2007). Complex features of ecosystem metabolism are relatively unknown and how C budget of major ecosystems will respond to changes in climate is not quantitatively well understood (Black et al., 2000; Baldocchi et al., 2001; Baldocchi et al.,2001; Barr et al., 2004; Law et al., 2002).

2.3 Terrestrial water cycling

Most of the Earth is covered by water, amounting to more than one billion km^3. The vast majority of that water, however, is in forms unavailable to land-based or freshwater ecosystems. Less than 3 percent is fresh enough to drink or to irrigate crops, and of that total, more than two-thirds is locked up in glaciers and ice caps. Freshwater lakes and rivers hold 100,000 km^3 globally, less than one ten-thousandth of all water on earth (Jackson et al, 2001).

Water vapor in the atmosphere exerts an important influence on climate and on the water cycle, even though only 15,000 km^3 of water is typically held in the atmosphere at any time. This tiny fraction, however, is vital for the biosphere. Water vapor is the most important of the so-called greenhouse gases (others include CO_2, CH_4 and N_2O) that warm the Earth by trapping heat in the atmosphere. Water vapor contributes approximately two-thirds of the total warming that greenhouse gases supply. Without these gases, the mean surface temperature of the earth would be well below freezing, and liquid water would be absent over much of the planet. Equally important for life, atmospheric water turns over every ten days or so as water vapor condenses and falls as rain to the Earth and the heat of the Sun evaporates new supplies of vapor from the liquid reservoirs on earth. Solar energy typically evaporates about 425,000 km^3 of ocean water each year. Most of this water returns back directly to the oceans as precipitation, but approximately 10% falls on land. If this were the only source of rainfall, average precipitation across the earth's land surfaces would be only 25 cm a year, a value typical for deserts or semi-arid regions. Instead, a second, larger source of water is recycled from plants and the soil through evapotranspiration. The water vapor from this source creates a direct feedback between the land surface and regional climate. This second source of recycled water contributes two-thirds of the 70 cm of precipitation that falls over land each year. Taken together, these two sources account for the 110,000 km^3 of renewable freshwater available each year for terrestrial, freshwater, and estuarine ecosystems. Because the amount of rain that falls on land is greater than the amount of water that evaporates from it, the extra 40,000 km^3 of water returns to the oceans, primarily via rivers and underground aquifers. A number of factors affect how much of this water is available for human use on its journey to the oceans. These factors include whether the precipitation falls as rain or snow, the timing of precipitation relative to patterns of seasonal temperature and sunlight, and the regional topography. For example, in many mountain regions, most precipitation falls as snow during winter, and spring snowmelt causes peak flows that flood major river systems. In other regions, excess precipitation percolates into the soil to recharge ground water or is stored in wetlands.

2.4 Coupling of the C and water cycles

The cycling of other materials such as C and N is strongly coupled to this water flux through the patterns of plant growth and microbial decomposition, and this coupling creates additional feedbacks between vegetation and climate. Thermodynamically, a terrestrial ecosystem is an open system. Therefore, hydrological and C cycles are closely coupled at various temporal and spatial scales (Betts, 2007; Ball et al., 1987; Levis et al., 1999; Rodriguez-Iturbe, 2000; Joos, 2001; Arain et al., 2006; Blanken and Black, 2004; Snyder 2004). C uptake for example, is closely coupled to water loss by ecosystems mainly through leaf stomatal pathway governed principally through leaf conductance (Jarvis, 1976; Harris et al., 2004; Rodriguez-Iturbe, 2001). Soil organic C decomposition is very sensitive to soil moisture content via microbial activity and other processes (Betts, 2007; Levis 1999; Snyder et al., 2004; Parton et al., 1993; D'odorico 2004). The flux of terrestrial organic C by river runoff to the ocean and wetland discharge is an important component of the global organic C cycle (Hedges, 1992; Wang et al., 2004). It is estimated that 0.25×10^{15} g dissolved organic carbon (DOC) is discharged to the ocean by the world rivers each year (Meybeck, 1982). The land surface hydrological processes (in particular the terrestrial river systems) play an important role in transport of dissolved and particulate organic C from terrestrial to marine ecosystems (Wang et al., 2004). However, the interactions between C and water cycles and the mechanisms how these interactions will shape future climatic and biosphere conditions are far from well understood.

3. The array of airborne and satellite sensors developed for monitoring of the coupled C and water cycles

3.1 Satellite monitoring

RS is the observation of a phenomenon from a distance, using devices that detect electromagnetic radiation. Satellite-borne remote sensing offers unique opportunities to parameterize land surface characteristics over large spatial extents at variable spatial and temporal resolutions. There has been a substantial increase in the number of satellite sensors for Earth observations that cover a large range of the electromagnetic radiation spectrum (Tables 1 and 2) since 1960s when the earlier Landsat satellites were launched into orbit, such as the Television Infrared Observation Satellite (TIROS-1) launched in 1960. None of these sensors have been designed exclusively for C, water or vegetation applications. For example, the TIROS-1 was focused on weather analysis and forecasting (Natl. Res. Counc., 2008). However, scientists were applying these observations to vegetation studies by the next decade (Rouse et al., 1974; Tucker et al., 1979). Tuker et al. (1986) exploited the properties of chlorophyll pigments to absorb wavelengths in the red spectral region and structural properties of leaves to reflect near-infrared spectra based on the imagery data obtained by the Advanced Very High Resolution Radiometer (AVHRR) sensor onboard TIROS. This pioneer study that synoptic view of the coupled atmosphere-biosphere as C sequestration by photosynthesis from the atmosphere in the Northern Hemisphere (Tuker et al., 1986) opened possibilities for global perspectives in ecology. The first Landsat satellite launched in 1972 carried the Multispectral Scanner System (MSS) sensors which were specifically designed to map land resources with finer spatial resolution (68 m × 82 m) than the AVHRR. The program was the first civil, non-weather satellite program and Landsat provided observations for any place on Earth once every 18 days, offering a wide range of studies on terrestrial vegetation and C and water cycles. The Landsat Thematic Mapper sensors carried onboard the Landsat series of satellites, acquire images at a 30-m spatial resolution with a 16-day interval. The acquired data have been the backbone for land-cover, vegetation and C cycle studies. NASA's Earth Observing System (EOS), launched in 1999 (Tilford S. 1984), brought new capabilities for monitoring vegetation productivity and other properties with near-daily and global coverage. The multispectral sensors---Moderate Resolution Imaging Spectroradiometer (MODIS), onboard the EOS platform, have built invaluable global observation dataset for C and water cycles research since the early 2000s. MODIS provides a global coverage every 1-2 days with 36 bands. The spatial resolution of MODIS (pixel size at nadir) is 250 m for channels 1 and 2 (0.6μm - 0.9μm), 500 m for channels 3 to 7 (0.4μm - 2.1μm) and 1000 m for channels 8 to 36 (0.4μm - 14.4μm), respectively. Data from the satellite-borne MODIS are currently used in the calculation of global weekly GPP and ET at 1-km spatial resolution (Running et al., 2004).

Sensors that have potential applications in C and hydrology studies fall into two groups---optical (Table 1) and microwave (Table 2). Optical sensors cannot penetrate vegetation or clouds. In contrast, microwave sensors are able to penetrate vegetation and can collect data independently of cloud cover and solar illumination. This is important because it is difficult to acquire cloud-free imagery using optical sensors. There are two types of microwave sensors: active sensors and passive sensors. The former send and receive their own energy; while the latter detect the microwaves emitted by the Earth's surface. The microwave bands, being useful for vegetation and carbon and water cycles, are K, X, C, and L, ranked in increasing wavelengths. K- and X-bands are useful for detecting surface temperature, snow density, and rainfall rates, whereas C- and L-bands are sensitive to soil moisture (Sass and Greed, 2011).

3.2 Other airborne measurements

Besides satellite monitoring, other airborne observation techniques (*e.g.* aircraft, airplane and land surface remote sensing) have been developed rapidly since the latest decade.

	Sensor	Bands (nm)				Spatial Resolution	Spatial Coverage	Temporal Resolution	Passive /Active	Operational years
		Vsible	NIR	SWIR	Thermal					
Very high resolution	LiDAR	-				0.15-1m	Global	No regular repeat cycle	Active	-
	GeoEye	450–510 510–580 655–690	780–920	-	-	Pan:0.41m Multi:1.65m	Global	Less than 3 days	Passive	2008-09-06 to now
	Worldview-2	400-450 585-625 705-745	860-1040	-	-	0. 5m	Global	1.1day	Passive	2009-10 to now
	Quikbird	450-520 520-660 630-690	760-900	-	-	Pan:0.61m Multi:2.44m	Global	1-3,5days depending on latitude	Passive	2001-10-18 to now
	Ikonos	450-530 520-610 640-720	770-880	-	-	Pan:1m Multi:4m	Global	1.5-2.9days	Passive	1999-09-24 to now
	Orbview-3	450-900 450-520 520-600 625-695	760-900	-	-	Pan:1m Multi:4m	Global	>3days	Passive	2003-06-26 to now
	KOMPSAT-2	500-900 450-520 520-600 630-690	760-900	-	-	Pan:1m Multi:4m	Global	3 days	Passive	2006-07-28 to now
	Resurs-DK1	580-800 500-600 600-700 700-800	-	-	-	Pan:0.9-1.7m Multi:1.5-2m	Global	5 days	Passive	2006-06-15 to now
	TopSat	500-700 450-500 500-600 600-700	-	-	-	Pan:2.5m Multi:5m	Global	4 days	Passive	2005-10-27 to now
	MTI	450-520 520-600 620-680	760-860 860-900 910-970	990-1040 1360-1390	1550-1750 3500-4100 4870-5070 8000-8400 8400-8850 10200-11500 20800-23500		Global	5-8 min	Passive	2000-03-12 to now
	RapidEye EOC	440-510 20-590 630-685	760-850	-	-	5m	Global	Daily	Passive	2008-8-29 to now
	Formosat-2	450-900 450-520 520-600 630-690	760-900	-	-	Pan:2m Multi:8m	Global	Daily	Passive	2004-04-21 to now

	Sensor	Bands (nm)				Spatial Resolution	Spatial Coverage	Temporal Resolution	Passive /Active	Operational years
		Vsible	NIR	SWIR	Thermal					
Fine	Spot4	510-730 500 - 590 610 - 680	780-890	1580-1750	-	Pan:10m Multi:20m	Global	2-3days	Passive	1998-03 to now
	Spot5	490-690 490-610 610-680	780- 890	1580-1750	-	Pan:2.5m,5m Multi:10m	Global	2-3days	Passive	2002-05-04 to now
	ALOS AVNIR-2	420-500 520-600 610-690	760-890	-	-	10m	Global	2days	Passive	2006-01-24 to now
	Terra ASTER	520-600 630-690	760-860	1600-1700	-	15m 30m 90m	Global	4-16days	Passive	1999-12-18 to now
	JERS-1 OPS	520-600 630-690	760-876	1600-1710 2010-2120 2030-2250 2270-2400	-	18m	Global	44days	Passive	1992-02-11 to now
	SPOT1-3IIRV	500-730 500-590 610-680	780-890	-	-	Pan:10m Multi:20m	Global	26days	Passive	1986-02 to now
	CBERS IRMSS	500-900	-	1550-1750 2080-2350 1040-1250	10400-12500	78m, 156m	Global	26days	Passive	1994-10-14 to now
	Deimos-1/ UK DMC-2	520-600 630-690	770-900	-	-	22m	Global	3 days	Passive	2009-07-29 to now
	IRS LISS3	20-590 620-680	770-860	1550-1700 2200-25000	-	23m	Global	24days	Passive	2003-10-17 to now
	Landsat7 ETM	520-900 450-520 520-600 620-690	760-960	1550-1750 2080-3350	1040-1250	Pan:15m Muti:30m,60m	Global	16days	Passive	1999-04 to now
	Landsat4-5TM	450-520 520-600 630-690	760-900	1550-1750 20800-23500	104000-125000	30m,120m	Global	16days	Passive	1984-03-16 to now
	Eo-1 ALI	520-900 450-520 520-600 620-690	760-960	1550-1750 2080-3350	1040-1250	Pan:10m Muti:30m,60m	Global	16days	Passive	2000-11-21 to now

	Sensor	Bands (nm)				Spatial Resolution	Spatial Coverage	Temporal Resolution	Passive /Active	Operational years
		Vsible	NIR	SWIR	Thermal					
	Eo-1 Hyperion	400-2500				30m	Global	16days	Passive	2000-11-21 to now
	Meteor 3M-1	450nm-1000 10500-12500 5700-7100				1.4km,3km,	Global	daily	Passive	2001-12-10 to now
	Mos-1,2 MESSR	510-590 610-690	720-800 800-1100 -	-		50m	global	17days	Passive	1987-02-19 to now
	Okean MSU-SK	530-590 610-690	700-800 900-1000	-	104000-126000nm	Visible:200m Ir:200m Thermal:600m	Global	17days	Passive	1983-7-10 to now
	landsat1-5 MSS	500-600 600-700	700-800 800-1100 -	-		80m	Global	1,2,3:18 days; 4,5:16days	Passive	1972 -07 to now
	CBERS IRMSS	500-1100	1550-1750 800-2350 -		104000-125000nm	Visible, Ir:78m Thermal :56m	Global	26days	Passive	1999-10-14 to now
Medium	SAC-C MMRS	480-500 540-560 630-690	795-835	1550 - 1700	-	175m	Global	16days	Passive	2000-06-08 to now
Medium	Terra MODIS	B1-B36:400-144000 B1-B2:250m B3-B7:500m B8-B36:1000m					Global	1/4day	Passive	1999-12-18 to now
Medium	Fengyun-3a MERSI	410-125000				B1-B5:250m B6-B20:1000m	Global	1/2 day	Passive	2008-05-27 to now
Medium	ENVISAT MERSI	410-125000				B1-B5:250m B6-B20:1000m	Global	35days	Passive	2002-03-1 to now
Medium	MOS-1,2 VTIR	520-590 620-680	-		115000-125000	Visibl:900m Thermal:2700m	Global	17days	Passive	1987-02-19 to now
COARSE	GEOS Imager	Shortwave:3800-4000 Vsible:550-750 Moisture:6500-7000 IR:10200-11200 IR:11500-12500				Shortwave:4km Vsible:1km Moisture:8km IR:4km	Global	-	Active	1985-04 to now
COARSE	Meteosat SEVIRI	12spectral channels in visible and near infrad red region				1.25km,5km	Global	daily	Passive	2002-12-28 to now
COARSE	NOAA AVHRR	550-680	725-1100 -	3550-3930 10500-11300	11500-12500	1.1km	Global	1/4 day	Passive	1978-10 to now
COARSE	GMS5	Visible:550-900 Water vapour:6500-7000 IR:10500-11500 IR: 11500-12500				Visible:1.25km Water vapour:5km IR:5km	Global	daily	Passive	1995-03-18~2003-05

Sensor	Bands (nm)				Spatial Resolution	Spatial Coverage	Temporal Resolution	Passive /Active	Operational years
	Vsible	NIR	SWIR	Thermal					
GMS1-4	Visible:550-900 Water vapour:6500-7000 IR:10500-11500 IR: 11500-12500				Visible:1.25km Water vapour:5km IR:5km	Global	daily	Passive	1997-7 to now
Fengyun-2CD	Visible:550-1050 Water vapour:6200-7600 IR:10500-12500				Visible:1.25km IR:5km	Global	1 hour	Passive	C:2004-10-19 D: 2006-12--08
fengYun-2AB	Visible:550-1050 Water vapour:6200-7600 IR:10500-12500				Visible:1.25km Water vapour:5km IR:5km	Global	1 hour	Passive	A:1997-06-10 B:1997-07-21
INSAT-2E	550-750	-	10500-12500	-	Visible:2km IR:8km	Global	daily	Passive	1992-04-02 to now
INSAT-2 VHRR	550-750	-	10500-12500	-	Visible:2km IR:8km	Global	daily	Passive	1995-12-06 to now
Meteosat MVIRI	Visible:450-1000 Water vapour:10500-12500 IR: 5700-7100				Visible:1.25km Water vapour:5km IR:5km	Global	daily	Passive	Meteosat–7:1993-11

Abbreviations: Ali, Advanced land Imager; ALOS AVNIR-2, Advanced Visible and Near Infrared Radiometer type 2; ALOS, Advanced land observing Satellite; ASTER, Japanese Earth Resources Satellite 1; AVHRR, The Advanced Very High Resolution Radiometer; CBERS, The *China–Brazil Earth Resources Satellite;* Deimos-1, Spanish Earth imaging satellite; DMC, Disaster Monitoring Constellation; Envisat, Environmental Satellite; EOS, Earth Observing System; Etm, Enhanced Thematic Mapper; Formosat-2, the first and only high-resolution satellite; GMS, Geosynchronous Meteorological Satellite; *HRV, High Resolution Visible* ; INSAT-2E, Indian geostationary communications and weather satellite; KOMPSAT, Korea Multi-Purpose Satellite; IRMSS, Infra-Red Multispectral Scanner; IRMSS, Infrared Multispectral Scanner Camera; IRS, Indian Remote Sensing; LISS-3, Linear Imaging Self-Scanning Sensor - 3. Satellites; Lidar, Light Detection And Ranging; MTI, moving target indication radar; MOS -1, Marine Observation Satellite 1; MESSR, Multi Spectral Electronic Self Scanning Radiometer; MSS, Multi-spectral Scanner; Meteor 3M-1, Meteorological Satellite; 3M, Monitoring of ocean and land surfaces, Meteorological observations, and Measurement of vertical profiles of aerosol, ozone and other constituents in the atmosphere; MMRS, Multispectral Medium Resolution Scanner; MODIS, The Moderate Resolution Imaging Spectroradiometer; MERSI, Medium Resolution Spectral Imager; *MVIRI, METEOSAT* Visible and Infrared Imager; NOAA, National Oceanic and Atmospheric Administration; Orbview, the satellite of Orbitally company; OPS, Optical System.
Okean MSU-SK: Multispectral Scanner - Conical Scanning; RapidEyeEOC, Electro-Optical Camera; SEVIRI, Spinning Enhanced Visible Infra-Red Imager; Spot, systeme probatoire d'observation de laterre, TM, Thematic Mapper; Topsat, Tactical Operational Satellite; UK-DMC 2, British Earth imaging satellite, operated by DMC International Imaging; VHRR, Very High Resolution Radiometer; VTIR, Visible and ThermalInfrared Radiometer.

Table 1. Optical Remote Sensing Systems

Spatial resolution	Sensor	Bands	Spatial Resolution	Spatial Coverage	Temporal Resolution	Passive/ Active	Operational years
Very high resolution	**Radarsat-2**	C	3m	Global	24days	Active	2007-12-14tonow
	Radarsat-1	C	10m,25m,30m,35m,50m, 100m	Global	24days	Active	1995-11-04to now
	COSMOS Skymed	X	1m	Global	several times a day	Passive	COSMO1:08.06.2007to now
							COSMO2:09.12.2007to now
							COSMO3:25.10.2008to now
							COSMO4:06.11.2010to now
	ALOS PALSAR	L	7-100m	Global	2days	Active	03.01.2002to now
FINE	**ERS-1,2**	C	30m,50km	Global	2days	Active	ERS-1:03.17.1991to now
							ERS-2:04.1995to now
	TerraSAR-X	X	1m,3m,16m	Global	11days	Active	06.15.2007to now
	7JERS-1	L	18mX18m	Global	44days		02.11.1992~10.12.1998
	ENVISAT ASAR	X	9mX6m;30X30m;150X150m;450mX450m;1800mX1800m	Global	35days	Active	03.01.2002to now
	SIR-C	X,C,L	50m 100m	Global	-	Active	04.091994~04.12.1999; 30.091994~11.10.1999;
	SRTM	X,C	30mX30m	60°N~56°S	16days	Active	02.11.2000~02.10.2000
medium	**Nimbus-7 SMMR**	K_a,K,K_u, X,C	30km,60km, 97.5km,156km	Global	5~6days	Passive	26.10.1978~08.21.1987
COARSE	**TRMM PR**	K_u,	4.3～5.0km	50°S~50°N 180°W~180°E	3h,1d, 3d,7d	Active	12.281997 to now
	TRMM TMI	K_a,K,C	6～50km	50°S~50°N 180°W~180°E	3h,1d, 3d,7d	Active	12.28.1997 to now
	FengYun 3a MWRI	K_a,K,K_u, X	250m	Global	5d	Passive	05.27.2008 to now
	Aqua AMSR-E	K_a,k,x	25km	Global	16day	Passive	05.04.2002 To now
	DMSP SSMI	K_a,k		Global	4h	Passive	01.19.1965to now
	DMSP SSMT	K_a k	174km	Global	4h	Passive	01.19.1965to mow
	MOS-1,2 MSR	K_a,K	32km,23km	Global	17days	Passive	02.19.1987~04.19.1996
	ADEOS-1 NSCAT	K_u,	25km	90%global sea	2days	Active	08.17.1996~06.30.1997 12.2002~10.24.2003
	SMOS MIRAS	L	50km,200km	Global	16days	Passive	02.11.2009 to now
	CRACE	K_a,K		Global	16days	Passive	03.17.2002to now

Abbreviations: ADEOS, Advanced Earth Observing Satellite; ALOS, Advanced land observing Satellite; *AMSR-E*, The Advanced Microwave Scanning Radiometer for EOS; ASAR, An Advanced Synthetic Aperture Radar; *COSMO*, COnstellation of small Satellites for the Mediterranean basin Observation; DMSP, Defense Meteorological Satellite Program; Envisat, Environmental Satellite; ERS, European Remote-Sensing Satellites; GRACE, Gravity Recovery And Climate Experiment; JERS-1, Japanese Earth

Resources Satellite 1; MIRAS, Instrument. Synthetic. Apperture Radiometer; MSR, MicrowaveScanning Radiometer; MWR, The Microwave Radiation Imager; **NSCAT, NASA SCATterometer;** PR, Precipitation Radar; SAR, Synthetic Aperture Rada; SIR-C, Space Imaging Radar; *SRTM*, Shuttle Radar Topography Mission; *SMMR*, Scanning Multichannel Microwave Radiometer; SMOS, Soil Moisture and Ocean Salinity; SSMI, Special Sensor Microwave Imager; SSMT, Special Sensor Microwave Temperature; TRMM, Tropical Rainfall Mearsuring Mission; *TMI,TRMM*'s Microwave Imager.

Table 2. Microwave Remote Sensing Systems

Microwave wavelengths penetrate greater depths into plant canopies than optical sensors (Kasischke et al. 1997). The potential for using RADAR (RAdio Detection And Ranging) for studying terrestrial carbon and water cycles, particularly for assessing standing woody biomass is promising. The sensitivity of RADAR to vegetation biomass strongly depends on wavelength: the longer wavelengths, the greater vegetation volumes and biomass levels. Single-band RADAR is able to detect aboveground biomass up to approximately 100 Mg per hectare (Dobson et al. 1992, Luckman et al. 1998). In addition, multiband RADAR enables to separate biomass into component fractions (e.g., stem and canopy) (Saatchi and Moghaddam 2000). Synthetic aperture RADAR (SAR) is also sensitive to vegetation structure and to the amount of biomass, including both photosynthetic (green) and nonphotosynthetic vegetation components (Turner et al., 2004). LiDAR (Light Detection and Ranging) is a remote sensing technology that determines distances to an object or surface using laser pulses, which is a relatively new technology compared to optical sensors, and has the added capability of characterizing the distribution of foliage with height in the canopy (Lefsky et al. 2002, Treuhaft et al. 2002, 2004; Turner et al., 2004). LiDAR data have proved to be highly effective for the determination of three dimensional forest attributes. The suitability of airborne LiDAR for the determination of forest stand attributes including LAI and the probability of canopy gaps within different layers of canopy has been widely acknowledged by various studies (Coops et al., 2004; Coops et al., 2007). The interpreted LiDAR data have been further used for landscape C modeling and scaling (Hilker et al. 2008; Chen et al., 2009). The number and types of sensors used for research on C and water cycles have multiplied many times over since the first sensor launched into orbit. Remote sensing provides consistency of coverage and repeat measurements through time are now indispensable in the C and hydrological scientist's toolbox.

3.3 Remote sensing of GPP

Satellite-based studies have used the light-use efficiency (ε) approach to estimate GPP (Prince & Goward, 1995; Running et al., 2000, 2004; Behrenfeld et al., 2001) or net primary production (NPP) (Field et al., 1995; Ruimy et al., 1999). Significant effort and progress have been made in developing the satellite-based GPP algorithms (Running et al., 2004; Xiao et al., 2004; 2005).The algorithm relies on ε approach relating GPP to the amount of absorbed photosynthetically active radiation (APAR) (Monteith, 1966, 1972), such that,

$$GPP = \varepsilon \times fPAR_{chl} \times PAR\,, \qquad (1)$$

where PAR is the photosynetically active radiation (in μmol photosynthetic photon flux density, PPFD), $fPAR_{chl}$ is the fraction of PAR absorbed by leaf chlorophyll in the canopy, and ε is the light use efficiency (μmol CO_2/μmol PPFD). Light use efficiency (ε) is affected by leaf phenology, temperature, and water:

$$\varepsilon = \varepsilon_0 \times P_m \times W_m \times T_m , \tag{2}$$

where ε_0 is the apparent quantum yield or maximum light use efficiency (μmol CO_2/μmol PPFD) for a given land cover type or vegetation function type, and P_m, W_m and T_m are the modifiers for the effects of leaf phenology, water and temperature on light use efficiency of vegetation, respectively.

Different parameters and inputs for the satellite-based algorithm were estimated in different ways: (i) the fraction of PAR absorbed by leaf chlorophyll in the canopy ($fPAR_{chl}$) and the modifiers (P_m, W_m); (ii) PAR and temperature modifier (T_m) were calculated using climate data (either from tower measurements or climate models); and (iii) the maximum light use efficiency (ε_0) was referred to the land-cover-related look-up table and then modified/optimized using EC tower C measurements and footprint climatology.

To accurately estimate $fPAR_{chl}$ in forests is a challenge to both radiative transfer modeling and field measurements. Significant efforts and progress have been made in developing advanced vegetation indices that are optimized for retrieval of *f*PAR from individual optical sensors (Gobron et al., 1999; Govaerts et al., 1999). The $fPAR_{chl}$ within the photosynthetically active period of vegetation was estimated as a linear function of the the Enhanced Vegetation Index (EVI),

$$fPAR = f(\text{EVI}) . \tag{3}$$

EVI is similar in design to NDVI but uses spectral information from the blue band (ρ_{blue}). Following Huete et al. (1997) it was computed,

$$EVI = G \times (\rho_{nir} - \rho_{red}) \,/\, (\rho_{nir} + C_1 \times \rho_{red} - C_2 \times \rho_{blue} + L), \tag{4}$$

where G = 2.5, C_1 = 6, C_2 = 7.5, and L =1. EVI is found to be significantly correlated with the fraction of the photosynthetically active radiation absorbed by leaf chlorophyll in the canopy providing a good surrogate of the spatial variability index for photosynthesis rate.

The parameter P_m was estimated using the Normalized Difference Vegetation Index (NDVI) and the Land Surface Water Index (LSWI) and was calculated at two different phases, depending upon life expectancy of leaves (deciduous versus evergreen):

$$P_m = \begin{cases} \frac{1+LSWI}{2} & \text{Duirng bud burst to leaf full expansion} \\ 1 & \text{After leaf full expansion} \end{cases} . \tag{5}$$

NDVI (Tucker 1979; Field et al., 1995) was calculated as,

$$NDVI = (\rho_{nir} - \rho_{red}) \,/\, (\rho_{nir} + \rho_{red}), \tag{6}$$

where ρ_{nir}, and ρ_{red} are the reflectance in the near infrared and red bands, respectively. NDVI is generally related to green vegetation cover or vegetation canopy density and has been shown to be well correlated with green LAI and biomass (e.g., Sellers, 1985; Myneni et al., 1995).

LSWI (Xiao et al. 2002) is a useful water index and was calculated as the normalized difference between the NIR (0.78-0.89 μm) and AWIR (1.58-1.75 μm) spectral bands:

$$LSWI = (\rho_{nir} - \rho_{swir}) \,/\, (\rho_{nir} + \rho_{swir}), \tag{7}$$

where ρ_{nir} and ρ_{swir} are the reflectance of near infrared bands, red bands and short infrared bands, respectively.

The timings of bud burst and leaf full expansion can be identified using NDVI. The effect of water on plant photosynthesis (W_m) has been estimated as a function of available soil content in plant root zone and water vapor pressure deficit (VPD) in a number of process-based ecosystem models (e.g. Chen et al., 2007) and remote-sensing based models (e.g. Running et al., 2000). Soil moisture represents water supply to the leaves and canopy, and VPD represents evaporative demand in the atmosphere. Leaf and canopy water content is largely determined by the dynamics of both soil moisture and VPD. As the first order of approximation, here following the alternative and simple approach that uses a satellite-derived water index (Xiao et al., 2004), the seasonal dynamics of W_m was estimated,

$$W_m = \alpha \times (1 + \text{LSWI}) / (1 + \text{LSWI}_{\max}), \tag{8}$$

where α is a magnifier (its default value equals 1.0) and LSWI_{max} is the maximum LSWI within the plant growing season for individual pixels. The temperature modifier T_m was estimated at each time step, using the equation developed for the terrestrial ecosystem model (Raich et al., 1991),

$$T_m = \frac{(\text{T} - \text{T}_{\text{min}})(\text{T} - \text{T}_{\text{max}})}{[(\text{T} - \text{T}_{\text{min}})(\text{T} - \text{T}_{\text{max}})] \quad (\text{T} - \text{T}_{\text{opt}})^2} \tag{9}$$

where T_{min}, T_{max} and T_{opt} are the minimum, maximum and optimal temperature for photosynthetic activities, respectively. Their default values are respectively set to be 0, 35 and 20 °C in this study. If air temperature falls below T_{min}, T_m is set to be zero.

The ε_0 values vary with vegetation types, and the information about ε_0 for individual vegetation types can be obtained from a survey of the literature (Ruimy et al., 1995) and optimized using EC tower measurements. According to the work (Zhang et al. 2006), the default ε_0 value was estimated to be 0.032 μmol CO_2/μmol PPFD in this study stand in 2004.

3.4 Remote sensing of ET

We follow a drop of water traveling through a watershed from input, storage, and finally output and assess how RS can be used to track water fluxes and reservoirs. Table 3 summarizes the potential application of RS to study of hydrology. ET, the largest component of water loss from ecosystems, plays an important role in affecting soil moisture, vegetation productivity, C cycle, and water budgets in terrestrial ecosystems (Dirmeyer, 1994; Hilker et al. 2008; Chen et al., 2009). In this section, I mainly discuss application of RS to ET.

Verstraeten (Verstraeten et al., 2008) provided a comprehensive review of remote sensing methods for assessing ET and soil moisture content across different scales and Kalma (Kalma et al., 2008) reviewed satellite-based algorithms for estimating ET and land surface temperatures at local, regional and continental scales, with particular emphasis on studies published since the early 1990s.

In general, water evapotranspired from ecosystems into the atmosphere will reduce the land surface temperature (T_a). Reduction in soil moisture will decrease plant transpiration and evaporation from soil and plant surfaces. Reduction in ET will increase T_a. T_a can be derived from remotely-sensed thermal-infrared (TIR) band (8-14 microns) from various operational satellites. Based on the relationship between T_a and ET, remotely sensed T_a has been used to estimate regional ET (Gillies et al,. 1997; Kite et al., 2000; Su et al., 2000., Coops et al., 2002). The existing thermal imaging sensors provide adequate coverage of thermal dynamics that are useful for operational monitoring applications of ET. For example, thermal images at 15 minutes intervals and at a spatial resolution of 5 kilometers can be obtained from the NOAA Geostationary Operational Environmental Satellites (GOES), and TIR data at a fine spatial resolution (60 m or 120 m) with a much longer time interval (16 days) have been provided by the Thematic Mapper (TM) and ETM+ instruments on Landsat 5 and Landsat 7.

4. Modeling of C and water dynamics in terrestrial ecosystems based on remote sensing

The land surface of the Earth represents significant sources, sinks, and reservoirs of C, heat and moisture to the atmosphere. C and energy fluxes and water cycles at soil-atmosphere and plant-atmosphere interfaces are therefore important land surface processes. Due to the complexity and non-linearity of C, N and water dynamics in terrestrial ecosystems, various modeling tools are needed for better understanding of these biogeochemical and hydrological processes and their feedback mechanisms with the land surface climate system (Rannik et al., 2006). The rapidly proliferating volume of spatial data generated by RS has created a significant challenge in terms of designing model algorithms. A spatially distributed process-based model uses spatial data for computing ecohydrological and biophysical processes. The model algorithms represent hypotheses that can be assessed and potentially revised after confrontation with RS and land surface-based observations. It is well known that realistic simulations of C and water dynamics in terrestrial ecosystems is of critical importance, not only for the surface microclimate, but also for the large-scale physics of the atmosphere (Cox et al., 1999; Gedney et al., 2006; Dickinson et al., 2002). Depending on the scientific objectives or applications, C and water cycle models have been designed with varying degrees of aggregation with respect to ecosystem processes, components, and RS data as model inputs. Such models can be flagged by land surface, ecosystem and hydrological models based on their objectives and emphases. The former focus on ecosystem processes and the interactions between ecosystems and the atmosphere; while the latter place emphasis on the land surface hydrology processes, including lateral flow resulting from catchment topography.

4.1 Land surface and ecosystem modeling

Global climate and the global carbon cycle are controlled by exchanges of water, carbon, and energy between the terrestrial biosphere and atmosphere. Thus land surface models (LSMs) are essential for the purpose of developing predictive capability for the Earth's climate on all time scales (Matthews et al., 1998). Most current LSMs can be associated with three broad types (Seth et al., 1994): soil-vegetation-atmosphere transfer schemes (SVATS), potential vegetation models (PVMs), and terrestrial biogeochemistry models (TBMs).

The first generation of SVATS evolved from simple bucket schemes focusing on soil water availability （Manabe et al., 1969）, through the schemes of Deardorff (Deardorff et al., 1978).

Marked improvements of the second generation (*e.g.*, BATS (Seth et al., 1994), SiB (Sellers et al., 1997; Sellers et al., 1986), and CLASS (Verseghy et al., 1999; Verseghy et al., 1993) from the first generation are the separation of vegetation from soil and the inclusion of multiple soil layers for dynamic heat and moisture-flow simulations (Chen et al., 2007). The second generation SVATS firstly modeled plant physiology in an explicit manner in GCMs (General Circulation Model or Global Climate Model) (Henderson et al., 1993). For most second-generation SVATS, land cover was fixed, with seasonally-varying prescriptions of parameters such as reflectance, leaf area index or rooting depth (Wang et al., 2002; Kickert et al., 1999; Kley et al., 1999; Schwalm et al., 2001). Some SVATS incorporated satellite data to characterize more realistically the seasonal dynamics in vegetation function (Kickert et al., 1999; Bonan et al., 1994). The latest (third generation) SVATS used more recent theories relating photosynthesis and plant water relations to provide a consistent description of energy exchange, ET, and C exchange by plants (Chen et al., 2007; Sellers et al., 1996). In our effort in understanding the impact of climate change on terrestrial ecosystems, energy, water, and C cycles need to be modelled simultaneously (Sellers et al., 1996; Williams et al., 2001). Recently, most of SVATS have thus been enhanced to include the CO_2 flux between the land surface and the atmosphere, such as SiB2 (Sellers, P.J et al, 1996), IBIS (Foley et al., 1996), NCAR-LSM (Bonan et al., 1995), BATS (Dickinson et al., 2002), CLASS-C (Wang et al., 2002) and EASS (Chen et al., 2007).

The earlier generation of PVMs comprised a suite of schemes that focus on modeling distributions of vegetation as a function of climate (Holdridge et al., 1947; Prentice et al., 1990) without influences of anthropogenic or natural disturbance. The second generation of PVMs included more sophisticated modules to account for factors controlling vegetation distributions, such as competition, varying combinations of plant functional types, and physiological and ecological constraints (Prentice et al., 1992).

TBMs developed from scaling up local ecological models, are process-based models that simulate dynamics of energy, water, and carbon and nitrogen exchange among biospheric pools and the atmosphere (Seth et al., 1994). Few of the existing TBMs incorporate PVMs. These models are not applicable to transient climate change experiments without coupling with PVMs.

In recent decades, the interactions among soil, vegetation and climate have been studied intensively and modeled successfully on the basis of water and energy transfer in the soil-vegetation-atmosphere system (Seth et al., 1994; Sellers et al., 1986; Verseghy et al., 1999; Verseghy et al., 1993; Zhang et al., 2003). Also the construction and refinement of LSMs have received increasing attention (Sellers et al., 1996; Viterbo et al., 1995; Christopher et al., 2004). Combination of these three different LSMs and utilization of remotely sensed land surface parameters are critical in the future LSM development, because of (1) the tight coupling of exchanges of water, energy and carbon between the land surface and the atmosphere; (2) the sophisticated impact/feedback mechanisms between climate change and terrestrial ecosystems; and (3) increasingly strong anthropogenic alterations to land cover. On-line coupling of a LSM with a GCM is needed for studying interannual to multi-decadal climate variations.

Several model intercomparisons have focused on evaluating SVATS and TBMs with particular objectives. For instance, the Project for Intercomparison of Land-surface Parameterization Schemes (PILPS) was initiated to evaluate an array of LSMs existing in GCMs (General Circulation Model or Global Climate Model) (Henderson et al., 1993); while the AMMA (African Monsoon Multidisciplinary Analysis) Land Surface Model Intercomparison Project (ALMIP) is being conducted to get a better understanding of the role of soil moisture in land surface processes in West Africa (de Rosnay et al., 2009). Coordinated land surface modeling activities have improved our understanding of land surface processes (de Rosnay et al., 2009).

4.2 Spatially-distributed hydrological processes modeling

Hydrology and ecosystem have, for the most part, been studied independently. Most LSMs and ecosystem models make an assumption of "flat Earth" with the absence of lateral redistribution of soil moisture. On the other hand, hydrological models have mostly been concerned with runoff production. Spatially-distributed models are needed, especially for hydrological simulation objective, because of heterogeneity of land surface and non-linearity of hydrological processes. Spatially-distributed hydrological models are not only able to account for spatial variability of hydrological processes, but enable computation of internal fluxes and state variables. Such kinds of models are increasingly applied to simulate spatial variability of forcing variables (*e.g.* precipitation), physiographic characteristics, detailed processes and internal fluxes within a catchment (Liang et al., 1994; Liang et al., 2004; Beldring et al., 2003; Brath et al., 2004; Christensen et al., 2007; Reed et al., 2004).

4.3 Modeling dynamics of stable C isotopic exchange between ecosystem and the atmosphere

It is recognized that the atmospheric measurements are still too sparse, relative to its spatial variability, to be used for inferring the surface flux at high spatial resolution (Ciais et al., 1995). The use of the isotope ratio as an additional constraint to identify various C sources and sinks can contribute to a significant reduction in the uncertainty. Though available isotopic datasets are being accumulated quickly (Griffis et al., 2005; Ponton et al., 2006; Lai et al., 2006; Lai et al., 2005) isotope measurements are still lacking considering land surface diversity and heterogeneity. This shortage of long-term measurements and of sampling frequency still limits C isotopic studies.

Mechanistic ecosystem models that couple micrometeorological and eco-physiological theories have the potential to shed light on how to extend efforts and applications of stable isotopes of CO_2 to global C budgeting, because biophysical models have the capacities of simulating isotope discrimination in response to environmental perturbations and can produce information on its diurnal, seasonal and interannual dynamics. Few biophysical models, however, have been developed to assess stable C discrimination between a plant canopy and the atmosphere (Suits et al., 2005; Oge'e et al., 2003; Baldocchi et al., 2003). Most existing biophysical models are based on individual leaf level discrimination equations given by Farquhar *et al.* (Farquhar et al., 1989; Farquhar et al., 1982) and only focus on the land surface layer (ignoring vertical and horizontal advection effects beyond 50~100 m above the ground (Baldocchi et al., 2003). However, in nature, the convective boundary layer (CBL) integrates the effects of photosynthesis, respiration, and turbulent transport of CO_2 over the landscape (Lloyd et al., 1996; Pataki et al., 2003). The influence of the CBL cannot be ignored when using isotope composition of CO_2 to investigate biological processes (Bowling et al., 1999), because the effect of atmospheric stability on turbulent mixing/diffusion has an important impact on scalar fluxes and concentration fields within and above canopies (Baldocchi et al., 1995; Leuning et al., 2000). Few such models considering the CBL effects on isotope fractionation have been developed to date (Lloyd et al.,1996; Lloyd et al., 2001; Chen et al,. 2006; Chen et al., 2006; Chen et al., 2007).

4.4 Modeling coupled C and water dynamics – An ecohydrological approach

C and N dynamics and hydrological processes are closely linked. The stomatal conductance (g_s) is the key linkage between C assimilation (photosynthesis) and transpiration. An empirical equation is used in the second-generation LSMs to calculate g_s, which is hypothesized to be controlled by the environmental conditions (Jarvis et al., 1976). While field and laboratory studies have documented that leaf photosynthesis also affects g_s. Therefore, Ball *et al.* (Ball et al., 1987) proposed a semi-empirical stomatal conductance formulation (Ball-Woodrow-Berry model), in which g_s is controlled by both photosynthesis and the environmental conditions. Most of third-generation LSMs (Ecological models, *e.g.* SiB2 (Sellers et al., 1997; Sellers et al., 1996); CN-CLASS (Arain et al., 2006); *Ecosys* (Grant et al., 2007; Grant et al., 1999; Chen et al., 2007) fully couple photosynthesis and transpiration processes by employing the Ball-Woodrow-Berry stamatal conductance formulation.

In addition to the coupling of hydrological condition and C assimilation through the linkage of g_s, C assimilation is also coupling with N dynamics through another biochemical parameter, $V_{c\max}^{25}$ --- maximum carboxylation rate at 25 °C. In the photosynthesis model proposed by Farquhar *et al.* (1980), the net photosynthetic rate A_{net} at leaf level is a function of two tightly-correlated parameters $V_{c\max}^{25}$ and $J_{c\max}^{25}$ (the maximum electron transport rate at 25 °C), and is calculated as,

$$A_{net} = \min(A_c, A_j) - R_d \tag{10}$$

where A_c and A_j are Rubiso-limited and light-limited gross photosynthesis rates, respectively, and R_d is the daytime leaf dark respiration and computed as R_d = 0.015 $V_{c\,max}$. A_c and A_j are expressed as,

$$A_c = V_{c\max} \frac{C_c - \Gamma^*}{C_c + K_c(1 + O_c / K_o)} \tag{11a}$$

and,

$$A_j = J_{\max} \frac{C_c - \Gamma^*}{4(C_c + 2\Gamma^*)} \tag{11b}$$

where C_c and O_c are the intercellular CO_2 and O_2 mole fractions (mol mol^{-1}), respectively; Γ^* is the CO_2 compensation point without dark respiration (mol mol^{-1}); K_c and K_o are Michaelis-Menten constants for CO_2 and O_2 (mol mol^{-1}), respectively. In the nutrient-limited stands, A_{net} is generally limited by A_c, while A_c is dominantly controlled by a parameter $V_{c\max}$ (see Eq. 11a). Many research results showed $V_{c\max}^{25}$ is very sensitive to leaf N status (more specifically leaf Rubisco-N) (Dickinson et al., 2002; Wilson et al., 2000; Wilson et al., 2001; Warren et al., 2001). As a result in some ecosystem models (*i.e.* C&N-CLASS (Arain et al., 2006)), $V_{c\max}^{25}$ is calculated as a nonlinear function of Rubisco-N following observations made by Warren and Adams (Dickinson et al., 2002):

$$V_{c\max}^{25}(N) = \alpha[1 - \exp(-1.8N_{r0}] \tag{12}$$

where α is the maximum value of $V_{c\max}^{25}$ and N_{r0} is the leaf Rubisco-N (g N m^{-2} leaf area) in the top canopy.

The coupled C, N and water processes have been carefully considered in most of the third-generation LSMs (*e.g.* SiB2 (Sellers et al., 1997; Sellers et al., 1986; Sellers et al., 1996); CN-CLASS (Arain et al., 2006) and *Ecosys* (Grant et al., 2007; Grant et al., 1999), the models' grids, however, are isolated from their neighboring grids mainly due to the availability of input data. Vertical soil hydrological processes are hard to be realistically simulated if the lateral flows are ignored by assuming that the Earth is "flat". However, Simulations of the topographically-driven lateral water flows are important components in most of spatially-distributed models, while the detailed ecophsiological processes are weakly represented (Govind et al., 2009). Much effort to bridge these two different models has been increasingly made (Rodriguez et al., 2001; D'odorico et al., 2004;Govind et al., 2009; Creed et al., 1998; Band et al., 2001; Porporato et al., 2002; Porporato et al., 2003; Daly et al., 2004; Chen et al., 2005). However, a model coupling approach --- a full combination of ecosystem model and hydrological model, *i.e.* ecohydrological modeling, is still lacking.

4.5 Applications of remotely-sensed data in ecohydrological modeling

Remote sensing techniques, which inherently have the ability to provide spatially comprehensive and temporally repeatable information of the land surface, may be the only feasible way to obtaining data needed for land surface and ecological modeling (Sellers et al., 1986; Gurney et al., 2003; Kite et al., 1996; Engman et al., 1996; Melesse et al., 2008). The most common rationale for interfacing remote sensing and land surface-ecosystem models is using remotely sensed data as model inputs (Plummer et al., 2000). These input data, corresponding to forcing functions or state variables in ecological modeling, include LC, LAI, normalized difference vegetation index (NDVI), and the fraction of photosynthetically active radiation (f_{PAR}) (Sellers et al., 1986; Running et al., 1998; Chiesi et al., 2002; Loiselle et al., 2001). Another effort is the direct estimation of GPP and net primary productivity (NPP) (Goetz et al., 1999; Seaquist et al., 2003) of biomass (Seaquist et al., 2003; Bergen et al., 1999) and of plant growth (Maas et al., 1988; Kurth et al., 1994), by making use of f_{PAR} and NDVI. It has been shown that the direct estimation has lower accuracy than the integration of remotely sensed data with process based models (Goetz et al., 1999).

Remote sensing data have also been used to parameterize hydrological models (Chen et al., 2005; Kite et al., 1996; Boegh et al., 2004). For instance, a hydrological model (TerrainLab) was further developed using remote sensing as inputs (Chen et al., 2005). TerrainLab is a spatially distributed, process-oriented hydrological model using the explicit routing scheme of Wigmosta et al. (Wigmosta et al., 2004). This model has been applied to flat areas such as boreal and wet land region, (Govind et al., 2009; Chen et al., 2005; Govind et al., 2009), but it has not yet been applied to mountainous areas.

Different from traditional hydrological models, which have coarse spatial resolutions, the grid-based-distributed ecohydrological models have a high demand for spatial data (Kite et al., 1996; Montzka et al., 2008). Some researchers highlight that the main obstacles in current distributed ecohydrological modeling is the lack of sufficient spatially distributed data for input and model validation (Stisen et al., 2008). Remote sensing can potentially fill in some of the gaps in data availability and produce means of spatial calibration and validation of distributed hydrological models. As a result the application of remote sensing techniques in hydrological studies and water resources management has progressed in the past decades (see review by (Kite et al., 1996)).

In general, the applications of remotely sensed data in ecohydrological modeling can be in the two ways (Kite et al., 1996; Chen et al., 2005; Boegh et al., 2004; Montzka et al., 2008; Stisen et al., 2008; Ritchie et al., 1996; Schultz et al., 1996; Melesse et al., 2007; Schmugge et al., 2002; Jain et al., 2004; Pietroniro et al., 2005; French et al., 2006): (i) multispectral remote sensing data are used to quantify surface parameters, such as vegetation types and density. Although the usefulness of remote sensing data is widely recognized, there remain few cases where remote sensing data have been actually used in ecohydrological simulations. Difficulties still exist in choosing the most suitable spectral data for studying hydrological processes as well as in interpreting such data to extract useful in formation (Chen et al., 2005; Kite et al., 1996; Engman et al., 1996); and (ii) processed remote sensing data are used to provide fields of hydrological parameters for calibration and validation of ecohydrological models, such as precipitation (Kite et al., 1996; Wang et al., 2001), and soil moisture (Jackson et al., 1993; Hollenbeck et al., 1996., 1996; Kim et al., 2002; Koster et al., 2006). Koster et al. (Koster et al., 2006) pointed out that remote sensing data take the form of emitted and reflected radiances and thus are not the type of data traditionally used to run and calibrate models. Hence, it is important to understand and develop relationships between the electromagnetic signals and hydrological parameters of interest (Chen et al., 2005). Kite and Pietroniro (Kite et al., 1996) stated that the use of remote sensing in hydrological modeling was limited. Even though a number of new sensors have been launched since then and research has documented that remote sensing data have promising perspectives, operational uses of satellite data in hydrological modeling still appear to be in its infancy (Stisen et al., 2008).

5. Research gaps in C and water flux estimates and scaling approaches

A variety of methods are being used in the C and water cycles studies. As shown in Figure 3, different approaches have different temporal and spatial scales. The most direct measurements of the terrestrial C flux are made either at the plot scale (10^{-2}-10^{1} m^2), *e.g.* using biometric methods and various forms of chamber, or at the ecosystem (patch) scale (10^4 - 10^6 m^2), using the EC technique. Ecohydrological / ecosystem modeling and remote sensing estimations are generally available across variable spatiotemporal scales. These estimates are normally available within a nested framework that permits a progressive comparison of measurements made by surface instrumentation (scale: 1 to 10 m), surface flux equipment (10 m to 1 km), airborne remote sensing equipment (100 m to several km), satellite remote sensing (30 m to global scale) and EC tower (1-3 km),

The atmosphere integrates surface fluxes over many temporal and spatial scales and links scalar sources and sinks with concentrations and fluxes. This principle has been successfully used to develop inverse models to estimate annual C budgets (Tans et al., 1990; Enting et al., 1995; Fan et al., 1998; Bousquet et al., 1999; Gurney et al., 2002; Gurney et al., 2003). However, due to model limitations and paucity of continental CO_2 observations these studies have yielded C fluxes only at coarse resolution, over large spatial regions (Gurney et al., 2004; Gurney et al., 2005; Gurney et al., 2008).

Progress in C balance studies has been achieved at both ends of the spatial scale spectrum, either large continents (larger than 10^6 km^2, e.g. global inverse modeling) or small vegetation stands (less than 1-3 km^2, e.g. EC-measurements). Methods to estimate CO_2 sources and sinks at the intermediate scale (i.e. landscape to regional scales) between continental and local scales are less well advanced. Moreover, the C cycle in different regions can vary markedly in response to changing climate (Friedlingstein et al., 2003). Reliable estimates of terrestrial C sources and sinks at landscape to regional spatial scales (finer than those used in global inversions and larger than local EC flux measurements and roughly defined as the range between 10^2 and 10^6 km^2) are required to quantitatively account for the large spatial variability in sources and sinks in the near-field of a measurement location (Gerbig et al., 2003), as well as fundamental to improving our understanding of the C cycle (Crevoisier et al., 2006).

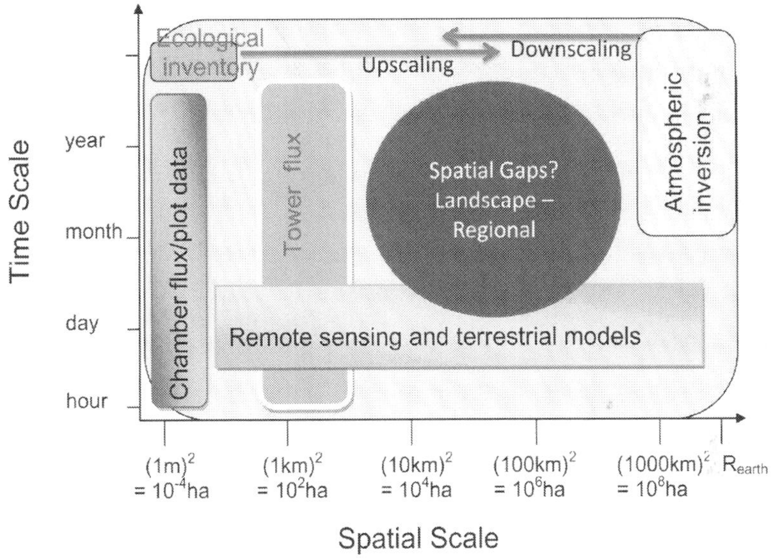

Fig. 3. Temporal and spatial scales of different approaches

It is generally considered unreliable to upscale stand-level fluxes (i.e. EC measurements) to a region by simple spatial extrapolation and interpolation because of the heterogeneity of the land surface and the nonlinearity inherent in ecophysiological processes (Levy et al., 1999). It is also challenging to apply atmospheric inversion technique to regional scales for quantifying annual C budgets because at such intermediate scales the atmosphere is often poorly constrained (Matross et al., 2006; Gloor et al., 1999). Moreover, aggregation errors and errors in atmospheric transport, both within the PBL and between the PBL and free troposphere, can also be obstacles to using these approaches to obtain quantitative estimates of regional C fluxes (Lin et al., 2004). Hence, there is a strong motivation to develop methods to quantify and validate estimates of the C balance at these intermediate scales (Lin et al., 2004; Chen et al., 2008; Bakwin et al., 2004; Matross et al., 2006). Observations of CO_2 over the continent within the PBL reflect exchange processes occurring at the surface at a regional scale (10^2 - 10^5 km^2). The flux information contained in CO_2 concentration data represents footprints of up to 10^5 km^2 (Gloor et al., 2001; Lin et al., 2004), which are several orders of magnitude larger than the direct EC-flux footprint. This information is therefore needed in our effort to upscale from site to region. Moreover, the number of CO_2 mixing ratio measurements above the land surface, made by either tower or aircraft, is steadily increasing. Previous efforts to interpret the signal of regional CO_2 exchange making use of tower concentration data have focused on simple one-dimensional PBL budgets that rely on gradients in CO_2 concentrations between the PBL and the free troposphere (Bakwin et al., 2004; Helliker et al., 2004). These methods are limited to monthly resolution because of the need to smooth and average over several synoptic events (Matross et al., 2006).

6. Future research directions

A synthetic research framework is needed to strength the less well researched areas as reviewed in Section 5: bottom-up and top-down approaches integrating scalable (footprint and ecosystem) models and a spatially nested hierarchy of observations which include multispectral RS, inventories, existing regional clusters of eddy-covariance flux towers and CO_2 mixing ratio towers and chambers.

The current research trends and the future directions in this field include: (i) A synthesis aggregation method --- integrating ecohydrological and isotopic models, remote sensing and component flux data, is becoming a pragmatic approach towards a better understanding of the coupled C, N and water dynamics at landscape/watershed scales; and (ii) The landscape- and regional-scale C fluxes are being estimated using an integrated approach involving direct land surface measurements, RS measurements, and ecosystem-, footprint- and inversion- modeling.

7. Summary

After comprehensive reviewing of a variety of approaches being used in research on the C/water cycles, the concluding remarks are summed the following:

Research gaps in this field are (i) The coupled terrestrial C and hydrological dynamics are far from well understood, especially at landscape (watershed) and regional scales; (2) Much progresses have been achieved at the extreme ends of the spatial-scale spectrum, either large regions/continents or small vegetation stands. Because of the heterogeneity of the land surface and the nonlinearity inherent in ecophysiological and ecohydrological processes in response to their driving forces, it is difficult to upscale stand level results to regions and the globe by extrapolation. Budgets of C and water at landscape intermediate regional scales (10^2–10^5 km^2) have large uncertainties.

A coupled spatially-explicit ecohydrological model is a powerful tool for quantitative and predictive understanding of the coupled C and water mechanism. This modeling framework can be used to infer aspects of the land surface system that are difficult to measure, and will be critical to improving the accuracy of forecasts of landscape change and C dynamics in the real world.

Combining and mutually constraining the bottom-up and top-down methods to reduce their uncertainties using data assimilation techniques is a practical and effective means to derive regional C and water fluxes with reasonably high accuracy. In the future upscaling framework, spatially nested hierarchy of observations, including multispectral RS, inventories, existing regional clusters of EC flux towers and CO_2 mixing ratio towers and chambers, are able to integrated using scalable (footprint and ecosystem and ecohydrological) models and data-model fusion techniques.

8. Acknowledgements

This research is financially supported by the National Science Foundation of China (Grant No. 41071059), "One hundred talents" program funded by Chinese Academy of Sciences and Alexander Graham Bell Canada Scholarship (CGS) funded by Natural Sciences and Engineering Research Council of Canada.

9. References

Amiro, B.D.; Barr, A.G.; Black, T.A. Carbon, energy and water fluxes at mature disturbed forest sites, Saskatchewan, Canada, *Agric. For. Meteor.* 2006, 136, 237-251.

Arain, M.A.; Yuan, F.M.; Black, T.A. Soil-Plant Nitrogen Cycling Modulated Carbon Exchanges in a Western Temperate Conifer Forest in Canada. *Agric. For. Meteor.* 2006, *140*, 171-192.

Bakwin, P.S.; Davis, K.J.; Yi, C.; Wofsy, S. C.; Munger, J. W.; Haszpra, L.; Barcza, Z. Regional carbon dioxide fluxes from mixing ratio data. *Tellus B* 2004, *56*, 301-311.

Baldocchi ,D.D.; Falge, E.; Gu, L. et al. FLUXNET: a new tool to study the temporal and spatial variability of ecosystem-scale carbon dioxide, water vapor, and energy flux densities. *Bull Amer. Meteorl. Soc.* 2001, *82*, 2415-2434.

Baldocchi, D.D. Breathing of the terrestrial biosphere: lessons learned from a global network of carbon dioxide flux measurement systems. *Aust. J. Bot.* 2008, *56*, 1-26.

Baldocchi, D.D., Bowling, D.R. Modeling the discrimination of ^{13}C above and within a temperate broad-leaved forest canopy on hourly to seasonal time scales. *Plant Cell Environ.* 2003, *26*, 231-244.

Baldocchi, D.D.; Harley, P.C. Scaling carbon dioxide and water vapour exchange from leaf to canopy in a deciduous forest: model testing and application. *Plant, Cell and Environment* 1995, *18*, 1157–1173.

Baldocchi, D.D.; Wilson, K.B. Modeling CO_2 fluxes and water vapor exchange of a temperate broadleaved forest across hourly to decadal time scales. *Ecol. Mod.* 2001, *142*, 155-184.

Ball, J.T.; Woodrow , *I.E.*; Berry J.A. A model predicting stomatal conductnance amd its contribution to the control of photosynthesis under different environmental conditions, In Progress in Photosynthesis Research, Martinus Nijhoff Publishers: *Dordrecht, Netherland,* 1987, pp 221-224.

Band, L.E.; Tague, C.L.; Groffman, P.; Belt, K. Forest Ecosystem Processes at the Watershed Scale: Hydrological and Ecological Controls of Nitrogen Export. *Hydro. Processes* 2001, *15*, 2013-2028.

Barford, C. C., Wofsy, S. C., and Goulden, M. L.: Factors controlling long- and short-term sequestration of atmospheric CO2 in a mid-latitude forest. *Science*, 2001, 294, 1688-1691.

Barr, A.G.; Black, T.A.; Hogg, E.H. *et al.* Inter-annual variability in the leaf area index of a boreal aspen-hazelnut forest in relation to net ecosystem production. *Agric. For. Meteor.* 2004, *126*, 237–255.

Behrenfeld, M. J., Randerson, J. T., McClain, C. R., Feldman, G. C., Los, S. O., and Tucker, C. J.: Biospheric primary production during an ENSO transition, *Science*, 2001, 291, 2594–2597.

Beldring, S.; Engeland, K.; Roald, L.A.; Sælthun, N.R.; Voksø, A. Estimation of parameters in a distributed precipitation runoff model for Norway. *Hydro. Earth System Sci.*, 2003, *7*, 304–316.

Bergen, K.M.; Dobson, M.C. Integration of remotely sensed radar imagery in modeling and mapping forest biomass and net primary production. *Ecol. Model.* 1999, *122*, 257–274.

Betts, A.K.; Desjardins, R.L.; Worth D. Impact of agriculture, forest and cloud feedback on the surface energy budget in BOREAS. *Agric. For. Meteor.* 2007a, *142*, 156–169.

Betts, R.A.; Boucher, O.; Collins, M.; Cox, P.M.; Falloon, P.D.; Gedney, N.; Hemming, D.L.; Huntingford, C.; Jones, C.D.; Sexton, D.M.H.; Webb, M.J. Projected Increase in Continental Runoff Due to Plant Responses to Increasing Carbon Dioxide, *Nature* 2007b, *448*, 1037-U5.

Betts, R.A.; Cox, P.M.; Woodward, F.I. Simulated responses of potential vegetation to doubled-CO_2 climate change and feedbacks on near-surface temperature, *Global Ecol. Biogeogr.* 2000, *9*, 171-180.

Black, T.A.; Chen, W.J.; Barr, A.G. *et al.* Increased carbon sequestration by a boreal deciduous forest in years with a warm spring. *Geophys. Res. Lett.* 2000, *27*, 1271-1274.

Blanken, P.D.; Black T.A. The canopy conductance of a boreal aspen forest, Prince Albert national park, Canada. *Hydro. Proces.* 2004, *18*, 1561–1578.

Boegh, E.; Thorsen, M.; Butts, M.B.; Hansen, S.; Christiansen, J.S.; Abrahamsen, P.; Hasager, C.B.; Jensen, N.O.; van der Keur, P.; Refsgaard, J.C.; Schelde, K.; Soegaard, H.; Thomsen, A. Incorporating remote sensing data in physically based distributed agro-hydrological modelling. *J. Hydro.* 2004, *287*, 279–299.

Bonan, G.B. Comparison of two land surface process models using prescribed forcings, *J. Geophys. Res.* 1994, 99, 25,803-25,818.

Bonan, G.B. Land-atmospheric interactions for climate system models: Coupling biophysical, biogeochemical and ecosystem dynamical processes. *Remote Sens. Environ.* 1995, *51*, 57-73.

Bousquet, P.; Ciais, P.; Peylin, P.; Ramonet, M.; Monfray, P. Inverse modeling of annual atmospheric CO_2 sources and sinks 1. method and control inversion. *J. Geophys. Res.* 1999, *104*, 26161-26178.

Bowling, D.R.; Baldocchi, D.D.; Monson, R.K. Dynamics of isotope exchange of carbon dioxide in a Tennessee deciduous forest. *Global Biogeochem. Cycles* 1999, *13*, 903–921.

Brath, A.; Montanari, A.; Toth, E. Analysis of the effects of different scenarios of historical data availability on the calibration of a spatially-distributed hydrological model. *J. Hydro.* 2004, *291*, 232–253.

Buermann, W.; Dong, J.; Zeng, X.; Myneni, R.B.; Dickinson, R.E. Evaluation of the utility of satellite-based vegetation leaf area index data for climate simulations, *J Clim*, 2001, 14, 3536-3550.

Canadell, J. G.; Que´ re´, C. Le; Raupacha, M. R.; Fielde, C. B.; Buitenhuisc, E.T.; Ciais, P.; Conwayg, T.J.; Gillettc, N. P.; Houghtonh, R. A.; Marlandi, G. Contributions to accelerating atmospheric CO_2 growth from economic activity, carbon intensity, and efficiency of natural sinks. *PNAS* 2007, *104*, 18866-18870.

Chen, B.; Black, A.; Coops, N.C.; Hilker, T.; Trofymow, T.; Nesic, Z.; Morgenstern, K. Assessing tower flux footprint climatology and scaling between remotely sensed and eddy covariance measurements. *Boundary-Layer Meteorology*, 2009a , 130, 137-167. DOI: 10.1007/s10546-008-9339-1.

Chen, B.; Black, A.; Coops, N.C.; Jassal, R.; Nesic Z. Seasonal controls on interannual variability in carbon dioxide exchange of a Pacific Northwest Douglas-fir forest, 1997 - 2006, *Global Change Biology*, 2009b , 15, 1962-1981, doi: 10.1111/j.1365-2486.2008.01832.x.

Chen, B.; Chen, J. M.; Mo, G.; Yuen, C-W.; Margolis, H.; Higuchi, K.; Chan D. Modeling and scaling coupled energy, water, and carbon fluxes based on remote sensing: An application to Canada's landmass, *J. Hydrometeorology*, 2007, 8, 123-143,.

Chen, B.; Chen, J.M.; Mo, G.; Black, T.A.; Worthy, D.E.J. Comparison of regional carbon flux estimates from CO2 concentration measurements and remote sensing based footprint integration. *Global Biogeochem Cycles*, 2008, 22, GB2012, doi:10.1029/ 2007GB003024.

Chen, B.; Chen, J.M. Diurnal, seasonal and inter-annual variability of carbon isotope discrimination at the canopy level in response to environmental factors in a boreal forest ecosystem. *Plant, Cell and Environment* 2007, *30*, 1223-1239, doi: 10.1111/j.1365-3040.2007.01707.

Chen, B.; Chen, J.M.; Huang, L.; Tans, P.P. Simulating dynamics of $\delta^{13}C$ of CO_2 in the planetary boundary layer over a boreal forest region: Covariation between surface fluxes and atmospheric mixing. *Tellus B* 2006, *58*,537-549.

Chen, B.; Chen, J.M.; Ju, W. Remote sensing based ecosystem-atmosphere simulation Scheme (EASS) --- model formulation and test with multiple-year data. *Ecol. Model.* 2007, *209*, 277-300.

Chen, B.; Chen, J.M.; Mo, G.; Black, T.A.; Worthy, D.E.J. Comparison of regional carbon flux estimates from CO_2 concentration measurements and remote sensing based footprint integration. *Global Biogeochem.Cycles*, 2008, 22, GB2012, doi:10.1029/ 2007GB003024

Chen, B.; Chen, J.M.; Tans, P.P.; Huang, L. Modeling dynamics of stable carbon isotopic exchange between a boreal ecosystem and the Atmosphere. *Global Change Biol.* 2006, *12*, 1842-1867.

Chen, J.M.; Chen, X.Y.; Ju, W.M.; Geng, X.Y Distributed Hydrological Model for Mapping Evapotranspiration Using Remote Sensing Inputs. *J. Hydrol.* 2005, *305*, 15-39.

Chiesi, M.; Maselli, F.; Bindi, M.; Fibbi, L.; Bonora, L.; Raschi, A.; Tognetti, R.; Cermak, J.; Nadezhdina, N. Calibration and application of FOREST-BGC in a Mediterranean area by the use of conventional and remote sensing data. *Ecol. Model.* 2002, *154*, 251-262.

Christensen, N.; Lettenmaier, D.P. A multimodel ensemble approach to assessment of climate change impacts on the hydrology and water resources of the Colorado River basin. *Hydro. Earth System Sci.* 2007, *11*, 1417–1434.

Christopher, S.R.; Ek, A.R., A process-based model of forest ecosystems driven by meteorology. *Ecol. Model.* 2004, *179*, 317-348.

Ciais, P; Tans P.P.; Trolier, M; White, J.W.C.; Francey, R.J. A large Northern Hemisphere terrestrial CO_2 sink indicated by the $^{13}C/^{12}C$ ratio of atmospheric CO_2. *Science* 1995, *269*, 1098-1102

Coops, N.; Wulder, M.; Culvenor, D.; St-Onge, B. Comparison of forest attributes extracted from fine spatial resolution multispectral and lidar data. *Can. J. Remot. Sens.* 2004, *30*, 855-866.

Coops, N.C.; Black, T.A.; Jassal, R.S.; Trofymow, J.A.; Morgenstern, K. Comparison of MODIS, eddy covariance determined and physiologically modelled gross primary production (GPP) in a Douglas-fir forest stand, *Remote Sens Environ*, 2007, 107, 385-401.

Coops, N.C.; Hilker, T; Wulder, M.A.; St-Onge, B.; Newnham G.; Siggins, A.; Trofymow, J.A. Estimating canopy structure of Douglas-fir forest stands from discrete-return lidar. *Trees* , 2007, *21*, 295-310.

Cosh, M. H.; Brutsaert, W. Microscale structural aspects of vegetation density variability, *J. Hydrol.*, 2003, 276,128-136.

Cox, P.M.; Betts, R.A.; Bunton, C.B.; Essery, R.L.H.; Rowntree, P.R.; Smith, J. The impact of new land surface physics on the GCM simulation of climate and climate sensitivity. *Climate Dyn.* 1999, *15*, 183-203.

Cox, P.M.; Betts, R.A.; Jones, C.D.; Spall, S.A.; Totterdell, I.J. Acceleration of Global Warming Due to Carbon-Cycle Feedbacks in a Coupled Climate Model. *Nature* 2000, *408*, 184.

Creed, I.F.; Band, L.E. Exploring Functional Similarity in the Export of Nitrate-N From Forested Catchments: a Mechanistic Modeling Approach. *Water Resour. Res.*, 1998, *34*, 3079- 3093.

Crevoisier, C.; Gloor, M.; Gloaguen, E.; Horowitz, L.W.; Sarmiento, J.; Sweeney, C.; Tans, P. A direct carbon budgeting approach to infer carbon sources and sinks: Design and synthetic application to complement the NACP observation network. *Tellus B*, 2006, *58*, 366-375.

Daly, E.; Porporato, A.; Rodriguez-Iturbe, I. Coupled Dynamics of Photosynthesis, Transpiration, and Soil Water Balance. Part I: Upscaling From Hourly to Daily Level. *J. Hydrometeo.* 2004, *5*, 546-558.

De Rosnay, P.; Drusch, M.; Boone, A.; Balsamo, G.; Decharme, B.; Harris, P.; Kerr, Y.; Pellarin, T.; Polcher, J.; Wigneron J.-P. AMMA Land Surface Model Intercomparison Experiment coupled to the Community Microwave Emission Model: ALMIP-MEM, *J. Geophys. Res.*, 2009, *114*, D05108, doi:10.1029/ 2008JD010724.

Deardorff, J.W. Efficient prediction of ground surface temperature and moisture, with inclusion of a layer of vegetation. *J. Geophys. Res.*, 1978, 83, 1889-1903.

Dickinson, R.E.; Berry, J.A.; Bonan, G.B. et al. Nitrogen controls on climate model evapotranspiration. *J. Climat.* 2002, *15*, 278–295.

Diego, S.; Peylin, P.; Viovy, N.; Ciais, P. Optimizing a process-based model with eddy-covariance flux measurements: A pine forest in southern France. *Global Biogeochemical cycles*, 2007, 21, GB2013, doi:1029/2006GB002834.

Dirmeyer, P.A. Vegetation stress as a feedback mechanism in mid-latitude drought. *J. Climate* , 1994, *7*, 1463-1483.

Dobson, M.C.; Ulaby, F.T.; Le Toan, T.; Beaudoin, A.; Kasischke, E.S.; Christensen, N.: Dependence of radar backscatter on coniferous forest biomass. IEEE Transactions on Geoscience and Remote Sensing, 1992, 30, 412–415.

D'odorico, P.; Porporato, A.; Laio, F.; Ridolfi, L.; Rodriguez-Iturbe, I. Probabilistic Modeling of Nitrogen and Carbon Dynamics in Water-Limited Ecosystems. *Ecol. Model.* 2004, *179*, 205-219.

Drolet, G.G.; Middleton, E.M.; Huemmrich, K.F.; Hall, F.G.; Amiro, B.D.; Barr, A.G. Regional mapping of gross light-use efficiency using MODIS spectral indices, *Remote., Sens. Environ.*, 2008, 112, 3064-3078.

Efron, B.; Tibshirani, R. J. An Introduction to the Bootstrap. *Chapman & Hall/CRC, Boca Raton*, 1993.

Engman, E.T. Remote sensing applications to hydrology: future impact. *Hydro. Sci. J.* 1996, *41*, 637-647.

Enting, I.G.; Trudinger, C.M.; Francey, R.J. A synthesis inversion of the concentration and $\delta^{13}C$ of atmospheric CO_2. *Tellus B* 1995, *47*, 35-52.

Falge, E.; Baldocchi, D.; Olson, R. et al. Gap filling strategies for defensible annual sums of net ecosystem exchange, *Agric. For. Meteorol.*, 2001, 107, 43–69.

Falk, M.; Wharton, S.; Schroeder, M. et al. Flux partitioning in an old-growth forest: seasonal and interannual dynamics. *Tree Physiol.* 2008, *28*, 509-520.

Fan, S.; Gloor, M.; Mahlman, J.; Pacala, S.; Sarmiento, J. et al. A large terrestrial carbon sink in North America implied by atmospheric and oceanic carbon dioxide data and models. *Science* 1998, *282*, 442-446.

Farquhar, G.D.; Ehleringer, J.R.; Hubick, K.T. Carbon isotope discrimination and photosynthesis. *Annual Review of Plant Physiology and Plant Molecular Biol.* 1989, *40*, 503–537.

Farquhar, G.D.; O'Leary, M.H.; Berry, J.A. On the relationship between carbon isotope discrimination and the intercellular carbon dioxide concentration in leaves. *Aust. J. Plant Physi.* 1982, *9*, 121–137.

Field, C. B.; Randerson, J. T.; Malmstrom, C. M. Global net primary production-combining ecology and remote sensing. *Remote Sensing of Environment*, 1995, 51, 74–88,.

Finnigan, J. The footprint concept in complex terrain, *Agric For Meteorol.*, 2004,127,117-129,.

Foken, T., Leclerc, M.Y.: Methods and limitations in validation of footprint models. *Agricultural and Forest Meteorology*, 2004, 127, 223-234.

Foley, J.A.; Prentice, I.C.; Ramankutty, N.; Levis, S.; Pollard, D.; Sitch, S.; Haxeltine, A.; An integrated biosphere model of land surface processes, terrestrial carbon balance, and vegetation dynamics. *Global Biogeochem. Cycles* 1996, *10*, 603-628.

French, R.H.; Miller, J.J.; Dettling, C.; Carr, J. Use of remotely sensed data to estimate the flow of water to a playa lake. *J. Hydro.*2006, *325*, 67–81.

Friedlingstein, P.; Dufresne, J.L.; Cox, P.M.; Rayner, P. How Positive Is the Feedback Between Climate Change and the Carbon Cycle? *Tellus B* 2003, *55*, 692-700.

Gao, F., Masek, J., Schwaller,M., and Hall, H: On the blending of the Landsat andMODIS surface reflectance: Predicting daily Landsat surface reflectance. *IEEE Transactions on Geosciences and Remote Sensing*, 2006, 44, 2207–2218,.

Gedney, N.; Cox, P.M.; Betts, R.A.; Boucher, O.; Huntingford, C.; Stott, P.A. Detection of a Direct Carbon Dioxide Effect in Continental River Runoff Records. *Nature,* 2006, *439*, 835-838.

Gerbig, C.; Lin, J. ; Wofsy, S.C.; Daube, B.C.; Andrews, A.E., *et al.* Towards constraining regional scale fluxes of CO_2 with atmospheric observations over a continent: 1. Observed spatial variability from airborne platforms. *J. Geophys. Res.* 2003, *108*, D4756,10.1029/2002JD003018.

Gillies, R.R.; Cui, J.; Carlson, T.N.; Kustas, W.P.; Humes, K.S. Verification of a method for obtaining surface soil water content and energy fluxes from remote measurements of NDVI and surface radiant temperature. *Inter. J. Remot. Sens.*, 1997, *18*, 3145–3166.

Gloor, M.; Bakwin, P.; Hurst, D.; Lock, L.; Draxler, R.; Tans, P. What is the concentration footprint of a tall tower? *J. Geophys. Res.* 2001, *106*, 17,831-17,840.

Gloor, M.; Fan, S.M.; Pacala, S.; Sarmiento, J.; Ramonet, M. A model-based evaluation of inversions of atmospheric transport, using annual mean mixing ratios, as a tool to monitor fluxes of nonreactive trace substances like CO_2 on a continental scale. *J. Geophys. Res.* 1999, *104*, 14245-14260.

Gobron, N.; Pinty, B.; Verstraete, M.; Govaerts, Y. The MERIS Global Vegetation Index (MGVI): Description and preliminary application, *International Journal of Remote Sensing*, 1999, 20, 1917- 1927.

Goetz, S.J.; Prince, S.D.; Goward, S.N.; Thawley, M.M.; Small, J. Satellite remote sensing of primary production: an improved production efficiency modelling approach. *Ecol. Model.* 1999, *122*, 239–255.

Gökede, M.; Rebmann, C.; Foken, T. A combination of quality assessment tools for eddy co-variance measurements with footprint modelling for the characterisation of complex sites, *Agric. For. Meteorol.*, 2004,127, 175–188.

Govaerts, Y. M.; Verstraete, M. M.; Pinty, B.; Gobron, N. Designing optimal spectral indices: A feasibility and proof of concept study, *International Journal of Remote Sensing*, 1999, 20, 1853- 1873.

Govind, A.; Chen, J. M.; Ju, W. Spatially explicit simulation of hydrologically controlled carbon and nitrogen cycles and associated feedback mechanisms in a boreal ecosystem, *J. Geophys. Res.*, 2009, *114*, G02006, doi:10.1029/2008JG000728.

Govind, A.; Chen, J.M.; Margolis, H.; Ju, W.; Sonnentag, O.; Giasson, M.-A. A spatially explicit hydro-ecological modeling framework (BEPS-TerrainLab V2.0): Model description and test in a boreal ecosystem in Eastern North America. *J. Hydro..* 2009, *367*, 200-216.

Grant, R.F.; Black, T.A.; den Hartog, G. et al. Diurnal and annual exchanges of mass and energy between an aspen-hazelnut forest and the atmosphere: testing the mathematical model Ecosys with data from the BOREAS experiment. *J. Geophys. Res.* 1999, *27*, 27699-27718.

Grant, R.F; Flanagan, L.B. Modeling stomatal and nonstomatal effects of water deficits on CO2 fixation in a semiarid grassland. *J. Geophys. Res.* 2007, *112*, G03011, doi:10.1029/2006JG000302.

Griffis, T. J.; Black, T. A.; Morgenstern , K. Ecophysiological controls on the carbon balances of three southern boreal forests. *Agricultural and Forest Meteorology*, 2003, 117, 53-71.

Griffis, T.J.; Baker, J.M.; Zhang, J. Seasonal dynamics and partitioning of isotopic CO_2 exchange in a C_3/C_4 managed ecosystem. *Agric. For. Meteor.* 2005, *132*, 1-19.

Gurney, K. R. et al. TransCom 3 CO_2 Inversion Intercomparison: 1. Annual mean control results and sensitivity to transport and prior flux information. *Tellus B* 2003, *55*, 555- 579.

Gurney, K.R.; Law, R.M.; Denning, A.S.; Rayner, P.J.; Baker, D. et al. Towards robust regional estimates of CO_2 sources and sinks using atmospheric transport models. *Nature* 2002, *415*, 626-630.

Hall, F. G.; Hilker, T.; Coops, N. C.; Lyapustin, A.; Huemmrich, F.; Middleton, E.; Margolis, H.; Drolet, G.; Black, T. Multi-angle remote sensing of forest light use efficiency by observing PRI variation with canopy shadow fraction. *Remote Sensing of Environment.*, 2008, 112:3201-3211.

Hall, F.; Sellers, P.; Strebel, D.; Kanemasu, E.; Kelly, R.; Blad, B.; Markham, B.; Wang, J.; Huemmrich, F. Satellite remote sensing of surface energy and mass balance results: Results from FIFE. *Remote Sensing of Environment*, 1991, 35, 187−199.

Harris, P. P.; Huntingford, C.; Cox, P. M.; Gash, J. H.C.; Malhi, Y. Effect of soil moisture on canopy conductance of Amazonian rainforest. *Agric. For. Meteor.* 2004, *122*, 215-227.

Hedges, J.I. Global biogeochemical cycles: progress and problems. *Mar. Chem.*, 1992. 39, 67−93.

Helliker, B.R.; Berry, J.A.; Betts, A.K.; Bakwin, P.S.; Davis, K.J.; Denning, A.S.; Ehleringer, J. R.; Miller, J.B.; Butler, M.P.; Ricciuto D.M. Estimates of net CO_2 flux by application of equilibrium boundary layer concepts to CO_2 and water vapor measurements from a tall tower. *J. Geophys. Res.* 2004, *109*, D20106, doi:10.1029/2004JD004532.

Henderson-Sellers, A.; Yang, Z.-L.; Dickinson, R.E. The Project for Intercomparison of Land-surface Parameterization Schemes, *Bull. Amer. Meteorol. Soc.*, 1993, 74, 1335.

Hilker, T.; Coops, N.C.; Hall, F.G.; Black, T.A.; Chen, B.; Krishnan, P.; Wulder, M.A.; Sellers, P.J.; Middleton, E.M.; Huemmrich, K.F. A modeling approach for upscaling gross ecosystem production to the landscape scale using remote sensing data. *J. Geophys Res.*,2008 113, G03006, doi:10.1029/2007JG000666.

Holdridge, L. Determination of world plant formations from simple climatic data, *Science* 1947,105, 367-368.

Hollenbeck, K.J.; Schmugge, T.J.; Homberger, G.M.; Wang, J.R. Identifying soil hydraulic heterogeneity by detection of relative change in passive microwave remote sensing observations. *Water Resour. Res.* 1996, *32*, 139–148.

Hollinger, D. Y.; Goltz, S.M.; Davidson, E. A. Seasonal patterns and environmental control of carbon dioxide and water vapour exchange in an ecotonal boreal forest. *Global Change Biology*, 1999, 5, 891-902.

Hollinger, D. Y.; Richardson, A. D. Uncertainty in eddy covariance measurements and its application to physiological models. *Tree Physiology*, 2005, 25, 873-885.

Horst, T.W.; Weil, J.C. Footprint estimation for scalar flux measurements in the atmospheric surface layer. *Boundary-Layer Meteorol.*, 1992, 59, 279-296.

Huang, M.; Ji, J.; LI, K.; Liu, Y.; Yang, F.; Tao, B. The ecosystem carbon accumulation after conversion of grasslands to pine plantations in subtropical red soil of South China, *Tellus*, 2007, 59B, 439-448.

Huete, A. R.; Liu, H. Q.; Batchily, K.; vanLeeuwen, W. A comparison of vegetation indices over a global set of TM images for EOS-MODIS, *Remote Sensing of Environment*, 1997, 59, 440−451.

Huete, A.; Didan, K.; Miura, T.; Rodriguez, E. P.; Gao, X.; Ferreira, L. G. Overview of the radiometric and biophysical performance of the MODIS vegetation indices, *Remote Sensing of Environment*, 2002, 83, 195- 213.

Jackson, T.J. Measuring surface soil moisture using passive microwave remote sensing. *Hydro. Processes* 1993, *7*, 139–152.

Jain, M.K.; Kothyari, U.C.; Raju K.G. A GIS based distributed rainfall-runoff model. *J. Hydro.*2004, *299*, 107–135.

Jarvis, P.G. Interpretation of variations in leaf water potential and stomatal conductance found in canopies in field. *Philos Trans R Soc Lond B Biol Sci*, 1976, *273*, 593-610.

Joos, F.; Prentice, I.C.; Sitch, S.; Meyer, R.; Hooss, G.; Plattner, G.K.; Gerber, S.; Hasselmann, K. Global Warming Feedbacks on Terrestrial Carbon Uptake under the

Intergovernmental Panel on Climate Change (Ipcc) Emission Scenarios, *Global Biogeochem. Cycles*, 2001, *15*, 891-907.

Ju, J.C.; Roy, D.P. The availability of cloud-free Landsat ETM+ data over the conterminous United States and globally. *Remote Sensing of the Environment*, 2007 112, 1196–1211.

Kalma, J. D.; McVicar, T. R.; McCabe, M. F. Estimating Land Surface Evaporation: A Review of Methods Using Remotely Sensed Surface Temperature Data, *Surv. Geophys.*, 2008, *29*, 421-469.

Kalvelage, T.; Willems, J. Supporting users through integrated retrieval, processing, and distribution systems at the Land Processes Distributed Active Archive Center. *Acta Astronautica*, 2005, 56, 681–687.

Kasischke, E.S.; Melack, J.M.; Dobson, M.C. The use of imaging radars for ecological applications—a review. *Remote Sensing of Environment*, 1997, 59, 141-156

Kickert, R.N.; Tonella, G.; Simonov, A.; Krupa, S. Predictive modeling of effects under global change. *Environ. Pollut.* 1999, *100*, 87-132.

Kim, G.; Barros, A.P. Space-time characterization of soil moisture from passive microwave remotely sensed imagery and ancillary data. *Remote Sens. Environ.* 2002, *81*, 393-403.

Kite, G.W.; Pietroniro, A. Remote sensing applications in hydrological modelling. *Hydro. Sci. J.* 1996, *41*, 563-591.

Kite, G.W.; Droogers, P. Comparing evapotranspiration estimates from satellites, hydrological models and field data. *J. Hydrol.* 2000, *229*, 3-18.

Kley, D.; Kleinmann, M.; Sandermann, H.; Krupa, S. Photochemical oxidants: state of the science. *Environ. Pollut.* 1999, *100*, 19-42.

Kljun, N.; Kormann, R.; Rotach, M.W. Comparison of the Langrangian footprint model LPDM-B with an analytical footprint model. *Boundary-Layer Meteorology* , 2003, 106, 349-355.

Koster, R.D.; Guo, Z.C.; Dirmeyer, P.A.; Bonan, G.;Chan, E.; Cox, P.; Davies, H.; Gordon, C.T.; Kanae, S.; Kowalczyk, E., *et al.* Glace: the Global Land-Atmosphere Coupling Experiment. Part I: Overview. *J. Hydrometeor.* 2006, *7*, 590-610.

Kurth, W. Morphological models of plant growth: possibilities and ecological relevance. *Ecol. Model.* 1994, *75*, 299-308

Lai C.T.; Ehleringer, J.; Schauer, A.; Tans , P.P.; Hollinger, D.; Paw, K.T.U.; Munger, J.; Wofsy, S. Canopy-scale $\delta^{13}C$ of photosynthetic and respiratory CO_2 fluxes: observations in forest biomes across the United States. *Global Change Biol.* 2005, *11*, 633-643, doi: 10.1111/j.1365-2486.2005.00931.x.

Lai, C.-T.; Schauer, A.J.; Owensby, C. et al. Regional CO_2 fluxes inferred from mixing ratio measurements: estimates from flask air samples in central Kansas, USA. *Tellus B* 2006, *58*, 523-536.

Law, B. E.; Waring, R. H.; Anthoni, P. M. Measurements of gross and net ecosystem, productivity and water vapour exchange of a Pinus ponderosa ecosystem, and evaluation of two generalized models, *Global Change Biol.*, 2000, 6, 155-168.

Law, B.E.; Falge, E; Gu, L. *et al.* Environmental controls over carbon dioxide and water vapor exchange of terrestrial vegetation. *Agric. For. Meteor.* 2002, *113*, 97-120.

Lee, X.; Fuentes, J. D.; Staebler, R. M. Long-term observation of the atmospheric exchange of CO2 with a temperature deciduous forest. *Journal of Geophysical Research*, 1999, 104, 15975–15984.

Lefsky, M.A.; Cohen, W.B.; Parker, G.G.; Harding, D.J. Lidar remote sensing for ecosystem studies. *BioScience,* 2002, 52, 19–30.

Leuning, R. Estimation of scalar source/sink distributions in plant canopies using Lagrangian dispersion analysis: Corrections for atmospheric stability and comparison with a multilayer canopy model. *Boundary-Layer Meteor.* 2000, *96,* 293-314.

Levin, S.A. Complex adaptive systems: exploring the known, unknown and unknowable. *Bull. Amer. Meteor. Soc.* 2002, *40,* 1-19.

Levis, S.; Foley, J.A.; Pollard D. Potential high-latitude vegetation feedbacks on CO_2-induced climate change, *Geophys. Res. Lett,* 1999, *26,* 747–750.

Levy, P. E.; Grelle, A.; Lindroth, A. et al. Regional-scale CO2 fluxes over central Sweden by a boundary layer budget method. *Agric. For. Meteor.* 1999, 98-99, 169-180.

Li, Z.; Yu, G.; Xiao, X. et al. Modeling gross primary production of alpine ecosystems in the Tibetan Plateau using MODIS images and climate data, *Remote Sensing of Environment,* 2007, 107, 510–519.

Liang, X.; Guo, J.; Leung, L.R. Assessment of the effects of spatial resolutions on daily water flux simulations. *J. Hydrol.* 2004, *298,* 287-310.

Liang, X.; Lettenmaier, D.P.; Wood, E.F.; Burges, S.J. A simple hydrologically based model of land surface water and energy fluxes for GSMs. *J. Geophys. Res.,* 1994, *99,* 14415-14428.

Lin, J.C.; Gerbig, C.; Wofsy, S.C.; Andrews, A.E.; Daube, B.C.; Grainger, C.A. Stephens, B.B., Bakwin, P.S.; Hollinger D.Y. Measuring fluxes of trace gases at regional scales by Lagrangian observations: Application to the CO_2 Budget and Rectification Airborne (COBRA) study, *J. Geophys. Res.* 2004, *109,* D15304, doi:10.1029/2004JD004754.

Liu, Y., Yu, G., Wen, X., Wang, Y., Song, X, Li, J., Sun, X., Yang, F., Chen, Y., Liu Q.: Seasonal dynamics of CO2 fluxes from subtropical plantation coniferous ecosystem. *Science in China Series D,* 2006, 49 Supp.II, 99-109.

Lloyd, J.; Francey, R.J.; Mollicone, D. et al. Vertical profiles, boundary layer budgets, and regional flux estimates for CO_2 and its $^{13}C/^{12}C$ ratio and for water vapor above a forest/bog mosaic in central Siberia. *Global Biogeochem. Cycles* 2001, *15,* 267-284.

Lloyd, J.; Kruijt, B.; Hollinger, D.Y., *et al.* Vegetation effects on the isotopic composition of atmospheric CO_2 at local and regional scales: theoretical aspects and a comparison between rain forest in Amazonia and a boreal forest in Siberia. *Austr. J. Plant Physi.* 1996, *23,* 371-399.

Loiselle, S.; Bracchini, L.; Bonechi, C.; Rossi, C. Modelling energy fluxes in remote wetland ecosystems with the help of remote sensing. *Ecol. Model.* 2001, *145,* 243–261.

Luckman, A.; Baker, J.; Honzak, M.; Lucas, R. Tropical forest biomass density estimation using JERS-1 SAR: Seasonal variation, confidence limits and application to image mosaics. *Remote Sensing of Environment,* 1998, 63, 126–139.

Ma, S.; Baldocchi, D.D.; Xu, L. *et al.* Interannual variability in carbon dioxide exchage of an oak/grass sacanna and open grassland in California. *Agric. For. Meteor.* 2007, *147,*157-171

Maas, S.J. Use of remotely sensed information in agricultural crop growth models. *Ecol. Model.* 1988, *41,* 247–268.

Marquis, M.; Tans, P. Carbon Crucible, *Science,* 2008, *320,* 460-470.

Matthews, E.; Hansen, J. Eds. Land Surface Modeling: A Mini-Workshop, May 1998: http://www.giss.nasa.gov/meetings/landsurface1998/section3.html

Melesse A.M.; Vijay N.; Jon H. Analysis of Energy Fluxes and Land Surface Parameters in Grassland Ecosystem: Remote Sensing Perspective. *Int. J. Remot. Sens.* 2008, *29*, 3325-3341.

Melesse, A.M.; Weng, Q.; Thenkabail, P.; Senay, G. Remote sensing sensors and applications in environmental resources mapping and modeling. *Sensors* 2007, *7*, 3209-3241

Meybeck, M. Carbon, nitrogen, and phosphorous transport by world rivers. *Am. J. Sci.* 1982, 282, 401–450.

Monteith, J. L. Solar radiation and productivity in tropical ecosystems, *J. of Applied Ecology*, 1972, 9, 747–766.

Monteith, J. L. The photosynthesis and transpiration of crops, *Experimental Agriculture*, 1966, 2, 1–14.

Montzka, C.; Canty, M.; Kunkel, R.; Menz, G.; Vereecken, H.; Wendland F. Modelling the water balance of a mesoscale catchment basin using remotely sensed land cover data. *J. Hydro.* 2008, *353*, 322-334.

Natl. Res. Counc. *Earth Observations from Space: The First 50 Years of Scientific Achievements*. 2008, Washington, DC: Natl. Acad. Press. 142 pp.

Oge´e, J.; Brunet, Y.; Loustau, D.; Berbigier, P.; Delzon, S. MuSICA, a CO_2, water and energy multi-layer, multi-leaf pine forest model: Evaluation from hourly to yearly time scales and sensitivity analysis. *Global Change Biol.* 2003, *9*, 697-717.

Papale, D.; Valentini, R. A new assesment of European forests carbon exchanges by eddy fluxes and artificial neural network spatialization. *Global Change Biology*, 2003, 9, 525-535.

Parton, W.J.; Ojima, D.S.; Cole, C.V.; Schimel, D.S. A general model for soil organic matter dynamics: sensitivity to litter chemistry, texture and management. *Soil Sci. Soc. Am. J.* 1993, *39*, 147 167.

Pataki, D.E.; Ehleringer, J.R.; Flanagan, L.B.; Yakir, D.; Bowling, D.R.; Still, C.J.; Buchmann, N.; Kaplan, J.O.; Berry, J.A. The application and interpretation of Keeling plots in terrestrial carbon cycle research. *Global Biogeochem.Cycles* 2003, *17*, 1022, doi:10.1029/2001GB001850.

Pielke, R.A. Sr. Influence of the spatial distribution of vegetation and soils on the prediction of cumulus convective rainfall. *Rev. Geophys.* 2001, *39*, 151-177.

Pietroniro, A.; Leconte, R. A review of Canadian remote sensing and hydrology, 1999–2003. *Hydro. Processes* 2005, *19*, 285–301.

Plummer, S.E. Perspectives on combining ecological process models and remotely sensed data. *Ecol. Model.* 2000, *129*, 169-186.

Ponton ,S; Flanagan, L.B.; Alstard, K; Johnson, B.G.; Morgenstern, K; Kljun, N; Black, T.A.; Barr, A.G. Comparison of ecosystem water-use efficiency among Douglas-fir forest, aspen forest and grassland using eddy covariance and carbon isotope techniques. *Global Change Biol.* 2006, *12*, 294-310.

Porporato, A.; D'odorico, P.; Laio, F.; Rodriguez-Iturbe, I. Hydrologic Controls on Soil Carbon and Nitrogen Cycles. I. Modeling Scheme. *Adv. Water Resour.* 2003, *26*, 45-58.

Prentice, I.C.; Cramer, W.; Harrison, S.P.; Leemans, R.; Monserud, R.A.; Solomon, A.M. A global biome model based on plant physiology and dominance, soil properties, and climate. *J. Biogeog.* 1992, 19, 117-134.

Prentice; I.C.; Farquhar, G.D.; Fasham, M.J.R. et al. In The carbon cycle and atmospheric carbon dioxide; Houghton, J.T., et al. Eds.; Cambridge University Press: *Cambridge, UK*, 2001, pp.183-238.

Prentice, K.; Fung, I.Y. The sensitivity of terrestrial carbon storage to climate change, *Nature* 1990, 346(6279), 48-51.

Prince, S. D.; Goward, S. N. Global primary production: A remote sensing approach, *J. Biogeography*, 1995, 22, 815–835.

Raich, J. W.; Rastetter, E. B.; Melillo, J. M. et al. Potential net primary productivity in South-America—application of a global-model, *Ecological Applications*, 1991, 1, 399–429.

Rannik, U.; Kolari, P.; Vesala, T.; Hari, P. Uncertainties in Measurement and Modelling of Net Ecosystem Exchange of a Forest. *Agric. For. Meteor.* 2006, *138*, 244-257.

Ranson, K. J.; Kovacs, K.; Sun, G.; Kharuk, V. I. Disturbance recognition in the boreal forest using radar and Landsat-7. *Canadian Journal of Remote Sensing*, 2003 ,29, 271−285.

Reed, S.; Koren, V.; Smith, M.; Zhang, Z.; Moreda, F.; Seo, D.J. Overall distributed model intercomparison project results. *J. Hydro.* 2004, *298*, 27–60.

Reichstein, M.; Falge, E.; Baldocchi, D. et al. On the separation of net ecosystem exchange into assimilation and ecosystem respiration: review and improved algorithm, *Glob. Change Biol.*, 2005, 11, 1424–1439.

Reichstein, M., Tenhunen, J. D., Roupsard, O., Ourcival, J. M., Rambal,S., Dore, S., and Valentini, R.: Ecosystem respiration in two Mediterranean veergreen Holm Oak forests: drought effects and decomposition dynamics, *Funct. Ecol.*, 2002, 16, 27–39.

Richardson, A. D.; Hollinger, D.Y.B.; George, G. et al. A multi-site analysis of random error in tower-based measurements of carbon and energy fluxes, *Agricultural Forest Meteorology*, 2006, 136, 1-18.

Ritchie, J.C.; Rango, A. Remote sensing applications to hydrology: introduction. *Hydro. Sci. J.* 1996, *41*, 429–431.

Rodriguez-Iturbe, I. Ecohydrology: A hydrologic perspective of climate-soil-vegetation dynamics. *Water Resour. Res.* 2000, *36*, 3–9.

Rodriguez-Iturbe, I.; Porporato, A.; Laio, F.; Ridolfi, L. Plants in Water-Controlled Ecosystems: Active Role in Hydrologic Processes and Response to Water Stress - I. Scope and General Outline. *Adv. Water Resour.* 2001, *24*, 695-705.

Rouse, J.W.; Haas, R.H.; Schell, J.A. et al. Monitoring the vernal advancement and retrogradation (greenwave effect) of natural vegetation. In *NASA/GSFC Type III Final Report*. Greenbelt, 1974, MD: NASA

Roy, P.; Junchang, J.; Lewis, P. et al. Multi-temporal MODIS-Landsat data fusion for relative radiometric normalization, gap filling, and prediction of Landsat data. *Remote Sensing of Environment*, 2008, 112, 3112−3130.

Ruimy, A.; Jarvis, P. G.; Baldocchi, D. D.; Saugier, B. CO2 fluxes over plant canopies and solar radiation: A review, *Advances in Ecological Research*, 1995, 1-68.

Ruimy, A.; Kergoat, L.; Bondeau, A. Comparing global models of terrestrial net primary productivity (NPP): Analysis of differences in light absorption and light-use efficiency, *Global Change Biology*, 1999, 5, 56−64.

Running, S. W.; Baldocchi, D. D.; Turner, D. P. et al. A global terrestrial monitoring network integrating tower fluxes, flask sampling, ecosystem modeling and EOS satellite data, *Remote Sensing of Environment*, 1999, 70, 108−127.

Running, S. W.; Nemani, R. R.; Heinsch, F. A. et al. A continuous satellite-derived measure of global terrestrial primary productivity: Future science and applications, *Bioscience*, 2004, 56, 547–560.

Running, S. W.; Thornton, P. E.; Nemani, R.; Glassy, J. M. Global terrestrial gross and net primary productivity from the Earth Observing System. In O. E. Sala, R. B. Jackson, & H. A. Mooney (Eds.), *Methods in Ecosystem Science*, (pp. 44–57). , 2000, New York: Springer.

Running, S.W.; Coughlan, J.C. A general model of forest ecosystem processes for regional applications I. Hydrological balance, canopy gas exchange and primary production processes. *Ecol. Model.* 1998, *42*, 125-154.

Running, S.W.; Nemani, R.R.; Heinsch, F.A.; Zhao, M.; Reeves, M.; Jolly, M. A continuous satellite-derived measure of global terrestrial primary productivity: Future science and applications. *Bioscience* 2004, *56*, 547-560.

Saatchi, S.S.; Moghaddam, M. Estimation of crown and stem water content and biomass of boreal forests using polarimetric SAR imagery. *IEEE Transactions on Geoscience and Remote Sensing*, 2000, 38, 697–709.

Sass, G.Z.; Creed, I.F.: Bird's eye view of forest hydrology: novel approaches using remote sensing techniques. In D. Levia, D. Carlyle-Moses, & T. Tanaka (eds), *Forest Hydrology and Biogeochemistry: Synthesis of Past Research and Future Directions*,(pp.45-68) 2011, New York: Springer.

Schmid, H. P. Experimental design for flux measurements: matching scales of observations and fluxes, Agric For Meteorol. , 1997, 87, 179-200.

Schmugge, J.S.; Kustas, W.P.; Ritchie ,J.C.; Jackson, T.J.; Rango, A. Remote sensing in hydrology. *Adv. Water Resour.* 2002, *25*, 1367–1385.

Schultz, G.A. Remote sensing applications to hydrology: runoff. *Hydro. Sci. J.* 1996, *41*, 453–475.

Schwalm, C.R.; Ek, A.R. Climate change and site: relevant mechanisms and modeling techniques. *For. Ecol. Manage.* 2001, *150*, 241-258.

Seaquist, J.W.; Olsson, L.; Ard¨ O, J. A remote sensing-based primary production model for grassland biomes. *Ecol. Model.* 2003, *169*, 131–155.

Sellers, P.J., *et al.* Modeling the exchange of energy, water, and carbon between continentals and the atmosphere. *Science* 1997, *275*, 502-509.

Sellers, P.J.; Mintz, Y.; Sud, Y.C.; Dalcher, A. A Simple Biosphere Model (SiB) for Use Within General-Circulation Models. *J. Atmos. Sci.* 1986, *43*, 505-531.

Sellers, P.J.; Randall, D.A.; Collatz, G.J. *et al.* A revised land surface parameterization (SiB2) for atmospheric GCMs. Part I: model formulation. *J. Climat.* 1996, *9*, 676-05.

Seneviratne, S.I.; Luthi, D.; Litschi, M.; Schar C. Land-atmosphere coupling and climate change in Europe. *Nature* 2006, *443*, 205–209.

Seth, A.; Giorgi, F.; Dickinson, R.E. Simulating fluxes from heterogeneous land surfaces: Explicit subgrid method employing the biosphere-atmosphere transfer scheme (BATS). *J. Geophys. Res.*, 1994, *99*(D9), 18,651–18,667.

Shen, S.; Leclerc, M. Y. How large must surface layer inhomogeneities be before they influence the convective boundary layer structure? A case study, *Quart J Roy Meteorol Soc.* , 1995, 121, 1209-1228.

Snyder, P.K.; Delire, C.L.; Foley, J.A. Evaluating the influence of different vegetation biomes on the global climate. *Clim. Dyn.* 2004, *23*, 279-302.

Sogachev, A.; Rannik, U.; Vesala, T. Flux footprints over complex terrain covered by heterogeneous forest, Agric. For. Meteorol. , 2004, 127, 142–158.

Stisen, S.; Jensen, K.H.; Sandholt, I.; Grimes D.I.F. A remote sensing driven distributed hydrological model of the Senegal River basin. *J. Hydro.* 2008, *354*, 131-148.

Stoy, P. C.; Katul, G.; Siqueira, G. Separating the effects of climate and vegetation on evapotranspiration along a successional chronosequence in the southeastern US, Global Change Biol. , 2006, 12, 2115-2135.

Su, Z. Remote sensing of land use and vegetation for mesoscale hydrological studies. *Inter. J. Remot. Sens.* 2000, *21*, 213–233.

Su, Z., The surface energy balance system (SEBS) for estimation of the turbulent heat fluxes. *Hydro. Earth Sci.* 2002, *6*, 85–99.

Suits, N.S.; Denning, A.S.; Berry, J.A.; Still, C.J.; Kaduk, J.; Miller, J.B.; Baker I.T. Simulation of carbon isotope discrimination of the terrestrial biosphere. *Global Biogeochem. Cycles* 2005, *19*, GB1017, doi:10.1029/2003GB002141.

Tans, P.P.; Fung, I.Y.; Takahashi, T. Observation Constraints on the global atmospheric CO_2 budget. *Science* 1990, *247*, 1431-1438.

Tarantola, A. Inverse Problem: Theory Methods for Data Fitting and Parameter Estimation. *Elsevier*, 1987, New York.

Tilford, S. Global habitability and Earth remote sensing. *Philos. Trans. R. Soc.* 1984 *A* 312(1519):115–18.

Treuhaft, R.N.; Asner, G.P.; Law, B.E.; Van Tuyl, S. Forest leaf area density profiles from the quantitative fusion of radar and hyperspectral data. *Journal of Geophysical Research*, 2002, 107, 4568–4580.

Treut, L.; Somerville, H. R.; Cubasch, U.; Ding, Y.; Mauritzen, C.; Mokssit, A.; Peterson T.; Prather, M. Historical Overview of Climate Change. In *Climate Change 2007: The Physical Science Basis. Contribution of Working Group I to the Fourth Assessment Report of the Intergovernmental Panel on Climate Change* Solomon, S., D. Qin, M. Manning, Z. Chen, M. Marquis, K.B. Averyt, M. Tignor and H.L. Miller Eds. Cambridge University Press, Cambridge, United Kingdom and New York, NY, USA, 2007; pp. 104-105.

Tucker, C.J.; Fung, I.Y.; Keeling, C.D.; Gammon, R. The relationship of global green leaf biomass to atmospheric CO2 concentrations. *Nature* , 1986 319:159–99.

Tucker, C.J. Red and photographic infrared linear combinations for monitoring vegetation. *Remote Sens. Environ.* 1979, 8(2):127–50.

Urbanski, S.; Barford, C.; Wofsy, S. Factors controlling CO2 exchange on timescales from hourly to decadal at Harvard Forest, *J. Geophys. Res.*, 2007, 112, G02020, doi:10.1029/2006JG000293.

Verseghy, D.L. Class-a Canadian Land Surface Scheme for Gcms.1. Soil Model. *Int. J. Climatol.* 1999, *11*, 111-133.

Verseghy, D.L.; McFarlane, N.A.; Lazare, M., CLASS: A Canadian land surface scheme for GCMs: II. Vegetation model and coupled runs. *Int. J. Climatol.* 1993, *13*, 347-370.

Verstraeten, W. W.; Veroustraete, F.; Feyen, J. Assessment of evapotranspiration and soil moisture content across different scales of observation, *Sensors*, 2008, *8*, 70–117.

Viterbo, P.; Beljaars, A.C.M.; An improved land surface parameterization scheme in the ECMWF model and its validation. *J. Climat.* 1995, *8*, 2716-2748.

Wang, J.; Price, K.P.; Rich, P.M. Spatial patterns of NDVI in response to precipitation and temperature in the central Great Plains. *Int. J. Remote Sens.* 2001, *22*, 3827–3844.

Wang, S.; Grant, R.F.; Verseghy, D.L.; Black, T.A. Modeling carbon-coupled energy and water dynamics of boreal aspen forest in a general circulation model land surface scheme. *Int. J. Climatol.* 2002, *22*, 1249-1265.

Wang, S.; Grant, R.F.; Verseghy, D.L.; Black, T.A. Modeling carbon dynamics of boreal forest ecosystems using the Canadian land surface scheme. *Clim. Chang.* 2002, *55*, 451-477.

Wang, X.-C.; Chen, R.F.; Gardner, G.B. Sources and transport of dissolved and particulate organic carbon in the Mississippi River estuary and adjacent coastal waters of the northern Gulf of Mexico, *Marine Chemistry*, 2004, 89, 241- 256.

Warren, C.R.; Adams, M.A. Distribution of N. Rubisco and photosynthesis in Pinus pinaster and acclimation to light. *Plant Cell Environ.* 2001, *24*, 597–609.

Wen, X.; Yu, G.; Sun, X.; Li, Q. et al. Soil moisture effect on the temperature dependence ofecosystem respiration in a subtropical Pinus plantation of southeastern China, *Agric. For. Meteor.*, 2006, 137, 166–175.

Williams, M.S.; Law, B.E.; Anthoni, P.M.;Unsworth, W.H. Use of a simulation model and ecosystems flux data to examine carbon-water interactions in ponderosa pine. *Tree Physiol.* 2001, *21*, 287-298.

Wilson, K.B.; Baldocchi, D.D.; Hanson, P.J. Leaf age affects the seasonal pattern photosynthetic capacity and net ecosystem exchange of carbon in a deciduous forest. *Plant Cell Environmen.* 2001, *24*, 571–583.

Wilson, K.B.; Baldocchi, D.D.; Hanson, P.J. Spatial and seasonal variability of photosynthetic parameters and their relationship to leaf nitrogen in a deciduous forest. *Tree Physi.* 2000, *20*, 565–578.

Wu, J. D.; Wang, D.; Bauer, M. E. Image-based atmospheric correction ofQuickBird imagery of Minnesota cropland. *Remote Sensing of Environment*, 2005,99, 315–325.

Xiao, X. M.; Zhang, Q. Y.; Hollinger, D.; Aber, J.; Moore, B. Modeling gross primary production of an evergreen needleleaf forest using MODIS and climate data, *Ecological Applications*, 2005, 15, 954–969.

Xiao, X.; Boles, S.; Liu, J. Y.; Zhuang, D. F.; Liu, M. L. Characterization of forest types in Northeastern China, using multi-temporal SPOT-4 VEGETATION sensor data, *Remote Sensing of Environment*, 2002, 82, 335–348.

Xiao, X.; Hollinger D.; Aber, J. D.; Goltz, M.; Davidson, E.; Zhang, Q. and Moore III B. Satellite-based Modeling of Gross Primary Production in an Evergreen Needle leaf Forest, Remote Sensing Envir., 2004, 89, 519 - 534.

Zhang, L. M.; Yu, G. R.; Sun, X. M. et al.; Liu, Y. F.; Xin, D.; Guan D. X.; Yan, J. H. Seasonal variations of ecosystem apparent quantum yield (α) and maximum photosynthesis rate (Pmax) of different forest ecosystems in China, *Agric. For. Meteor.*, 2006, 137, 176–187.

Zhang, Y.; Chen, W.; Cihlar, J. A process-based model for quantifying the impact of climate change on permafrost thermal regimes. *J. Geophys. Res.* 2003, *108(D22)*, 4695, doi:10.1029/2002JD003354.

Zhu, X.; Chen, J.; Gao, F.; Chen, X.; Masek J. An enhanced spatial and temporal adaptive reflectance fusion model for complex heterogeneous regions, *Remote Sensing of Environment*, 2010, doi:10.1016/j.rse.2010.05.032.

10

Oceanic Evaporation: Trends and Variability

Long S. Chiu[1], Si Gao[1] and Chung-Lin Shie[2]
[1]*Department of Atmospheric, Oceanic and Earth Sciences, George Mason University*
[2]*UMBC/JCET, NASA/GSFC*
USA

1. Introduction

The global water and energy cycles are strongly coupled as two essential components of earth system. They play important roles in altering the Earth's climate.

Oceanic evaporation, or sea surface latent heat flux (LHF) divided by latent heat of vaporization (L_v), is a key component of global water and energy cycle. In a bulk aerodynamic formulation, it is determined by the transfer coefficient of evaporation, C_E, and bulk parameters such as surface wind speed (U), surface saturated and near-surface air specific humidity (Q_s and Q_a),

$$LHF = \rho L_v C_E U(Q_s - Q_a) \tag{1}$$

where sea surface saturated humidity is determined by sea surface temperature (SST) and salinity, and ρ is density of moist air. The transfer coefficient is dependent on the stability of the atmosphere and the sea state (Liu et al., 1979; Zeng et al., 1988). Historically, marine surface observations have provided the basis for estimating these oceanic turbulent fluxes (e.g. Bunker, 1976; Cayan, 1992; da Silva et al., 1994; Esbensen & Kushnir, 1981; Hastenrath, 1980; Hsiung, 1985; Isemer & Hasse, 1985, 1987; Josey et al., 1998; Oberhuber, 1988; Renfrew et al., 2002; Weare et al., 1981). The advent of remote sensing techniques offers means to retrieve a number of surface bulk variables. Microwave radiation interacts directly with water molecules and hence is effective in providing water vapor information. The sea surface emissivity is affected by the sea state and foam conditions, which is related to surface wind. For instance, global microwave measurement of the Special Sensor Microwave Imager (SSM/I) on board a series of Defense Meteorological Satellite Program (DMSP) satellites has been used to retrieve near-surface air humidity and winds over the ocean.

At present there are several remote sensing products of global ocean surface latent heat flux. They include the NASA/Goddard Satellite-based Surface Turbulent Flux (GSSTF) dataset version 1 (Chou et al., 1997) and version 2 (GSSTF2, Chou et al., 2003), the Japanese Ocean Flux utilizing Remote Sensing Observations (J-OFURO) dataset (Kubota et al., 2002) and the Hamburg Ocean Atmosphere Parameters and Fluxes from Satellite (HOAPS) dataset (Grassl et al., 2000). Chiu et al. (2008) examined "trends" and variations in these global oceanic evaporation products for the period 1988-2000. They found a long-term increase in global average LHF that started around 1990 in GSSTF2. They argued that the dominant patterns may be related to an enhancement of Hadley circulation and El Niño-Southern Oscillation

(ENSO), respectively. An updated version of SSM/I version 6 (V6) data released by Remote Sensing Systems (RSS) in 2006 [as used by Wentz et al. (2007), see http://www.ssmi.com] that calibrates all SSM/I sensors is available in 2008. Shie et al. (2009) have reprocessed and forward processed GSSTF2 to version 2b (GSSTF2b, Shie et al. 2010; Shie 2010) using the SSM/I V6 data (including total precipitable water, brightness temperature, and wind speed retrieval), covering the period July 1987-December 2008. We provide an assessment of these data products and examine their "trends" and variability.

The data and methodology are described in Section 2. Section 3 presents the trends of these products, compares GSSTF2 and GSSTF2b for the pre 2000 periods, assesses the post 2000 performance, and examines the GSSTF2b Set1 and Set2 differences. Summary and discussion are presented in Section 4.

2. Data and methodology

Earlier version of these flux products have been described elsewhere (Chiu et al., 2008). The product versions described here represent the most updated versions as of the writing of this report.

2.1 HOAPS

Detail descriptions of the latest version of HOAPS, (version 3, or HOAPS-3) are given in Andersson et al. (2010). Bulk variables are derived from SSM/I data except for the SST which is derived from the Advanced Very High Resolution Radiometer (AVHRR) Oceans Pathfinder SST product. A neural network algorithm is used to derive U. The Q_a is obtained using the linear relationship of Bentamy et al. (2003). The Q_s is computed from the AVHRR SST using the Magnus formula (Murray, 1967) with a constant salinity correction factor of 0.98. The near-surface air temperature (T_a) is estimated from the SST using the assumptions of 80% constant relative humidity and a constant surface-air temperature difference of 1 K. Latent and sensible heat fluxes are calculated using the Coupled Ocean-Atmosphere Response Experiment (COARE) 2.6a bulk algorithm (Fairall et al., 1996, 2003).

The HOAPS-3 data sets cover the time period from July 1987 to December 2005. HOAPS-G pentad and monthly data sets with 0.5-degree resolution and HOAPS-C twice daily data set with 1-degree resolution are available at the website (http://www.hoaps.zmaw.de).

2.2 J-OFURO

The updated version of J-OFURO, (version 2, J-OFURO2) is described in Tomita et al. (2010). Bulk variables U, Q_a and SST (Q_s) are determined by multi-satellite and multiple satellite sensors. U is obtained from a combination of microwave radiometers (SSM/I, AMSR-E and TMI) and scatterometers (ERS-1, ERS-2 and QuikSCAT). Q_a is derived from SSM/I measurements. SST is taken from the Merged satellite and in-situ data Global Daily SST (MGDSST, Sakurai et al. 2005) analysis provided by Japanese Meteorological Agency (JMA). T_a is obtained from NCEP/DOE reanalysis. COARE 3.0 bulk algorithm (Fairall et al., 2003) is used to estimate LHF and SHF. The J-OFURO2 covers the time period from January 1988 to December 2006. Daily and monthly means with 1-degree resolution are available at the website (http://dtsv.scc.u-tokai.ac.jp/j-ofuro).

2.3 GSSTF

The GSSTF2 product has daily and monthly fields with a 1°x1° resolution for July 1987-December 2000 (Chou et al., 2001), based on the method of Chou et al. (1997) with some improvements (Chou et al., 2003). The temporal and spatial resolutions of GSSTF2b are the same as those of GSSTF2, except that GSSTF2b product covers a longer period (July 1987-December 2008).

GSSTF2b dataset is processed using improved input datasets, namely the recently released NCEP SST analysis, and a uniform (across satellites) surface wind and microwave brightness temperature (TB) V6 dataset from the SSM/I produced by RSS. Table 1 summarizes characteristics of input data and parameters for HOAPS3, J-OFURO2, GSSTF2 and GSSTF2b, in that order. As we focus on LHF, only detailed descriptions and discussions on input parameters of LHF are presented.

A major improvement in the input parameters of GSSTF2b is the use of the newly released SSM/I V6 product (see discussions in http://www.ssmi.com). The SSM/I V6 product removes the spurious wind speed trends found in the Wentz/RSS SSM/I V4 wind speed retrievals. To be consistent, the SSM/I V4 total precipitable water (*W*) and bottom-layer precipitable water (*WB*) used in GSSTF2 are replaced by the corresponding SSM/I V6 products in the production of GSSTF2b. Moreover, the weekly 1° spatial resolution Optimum Interpolation (OI) SST version 1 (V1) dataset (Reynolds & Smith, 1994) used in GSSTF2 is replaced by the improved OI SST version 2 (V2) dataset. The OI SST V2 has a lower satellite bias, a new sea ice algorithm, and an improved OI analysis (Reynolds et al., 2002) resulting in a modest reduction of the satellite bias and global residual biases of roughly −0.03°C. The major improvement in the V2 analysis shows up at high latitudes where local differences between the old and new analysis can exceed 1 °C due to the application of a new sea ice algorithm. There are two GSSTF2b sets, Set1 and Set2 (Shie et al., 2010; Shie 2010). Set1 is developed using all available DMSP SSM/I sensor data. In a preliminary analysis, it was noted that there are large trends associated with LHF which are mostly attributed to the DMSP F13 and F15 satellites. Set2 was produced by excluding satellite retrievals that are judged to caused relatively large artificial trends in LHF (mostly post 2000) from Set1. Consequently Set2 is identical to Set1 before 1997 and shows a smaller trend than Set1 for the whole period, while Set1 has better spatial coverage (less missing data). Hilburn & Shie (2011) further found a drift in the Earth incidence angle (EIA) associated with the SSM/I sensors on the DMSP satellites that introduces artificial trends in the SSM/I TB data. These artificial trends introduce large changes in the boundary water (*WB*), which affects the Q_a, and thus the LHF retrievals. An improved version, GSSTF2c, incorporating the corrected SSM/I brightness temperature, has been produced as of this writing (Shie et al., 2011). The retrieved *WB*, Q_a and LHF have genuinely improved, particularly in the trends post 2000 (Shie & Hilburn, 2011). An extensive study involving the GSSTF2c will be presented in a separate paper.

In this chapter, "trend" is used to indicate results from linear regression analysis and/or Empirical Mode Decomposition (EMD) for the period of study. Linear regression of the time series with time is used to detect linear trends and the significance can be estimated from the slope of the regression. EMD is based on local characteristic time scales of the data and is therefore applicable for analyzing nonlinear and non-stationary processes (Huang et al., 1998). It decomposes the time series into a finite and often small number of intrinsic mode

Datasets	HOAPS3	J-OFURO2	GSSTF2	GSSTF2b
C_E (transfer coefficient)	Fairall et al. (1996, 2003)	Fairall et al. (2003)	Chou (1993)	Chou (1993)
U (speed)	SSM/I V6 TB and neural network algorithm	SSM/I, AMSR-E, TMI, ERS-1, ERS-2 and QuikSCAT	Wentz V4 (1997)	Wentz V6 (2007)
U (vector)	N/A	N/A	Atlas et al. (1996)	CCMP Level-2.5 (SSM/I, TMI, and AMSR-E)
W/WB	N/A	N/A	Wentz V4 (1997)	Wentz V6 (2007)
SST	AVHRR	MGDSST Sakurai et al. (2005)	NCEP/NCAR Reanalysis (V1) Reynolds & Smith (1994)	NCEP/DOE Reanalysis (V2) Reynolds et al. (2002)
Q_a	Bentamy et al. (2003)	Schlussel et al. (1995)	Chou et al. (1995, 1997)	Chou et al. (1995, 1997)
T_a	Estimated from SST with assumptions of 80% humidity and 1 K surface-air temperature difference	NCEP/DOE Reanalysis (V2)	NCEP/NCAR Reanalysis (V1)	NCEP/DOE Reanalysis (V2)
Spatial resolution	1°x1°, 0.5°x0.5°	1°x1°	1°x1°	1°x1°
Spatial coverage	Global Oceans	Global Oceans	Global Oceans	Global Oceans
Temporal resolution	Twice daily, pentad and monthly	Daily and monthly	Daily and monthly	Daily and monthly
Temporal coverage	Jul. 1987 - Dec. 2005	Jan. 1988 - Dec. 2006	Jul. 1987 - Dec. 2000	Jul. 1987 - Dec. 2008

Table 1. Characteristics of input data and parameters for HOAPS3, J-OFURO2, GSSTF2 and GSSTF2b.

functions (IMFs) of increasing time scales. The existence of a trend is dependent on the length of the dataset. If the last IMF (one with longest time scale) is monotonically increasing or decreasing, a trend is indicated. The EMD is a more stringent test for significance of "trends."

Non-seasonal variability is examined using Empirical Orthogonal Function (EOF) analysis. Non-seasonal data are obtained by subtracting the monthly climatology of the study period from the monthly data. EOF analysis decomposes a spatio-temporal dataset into a series of orthogonal spatial EOF patterns and the associated time series (also called principal components). The test proposed by North et al. (1982) is used to judge the EOFs to see if they are significant and distinct. To examine the significance of each EOF, the logarithm of the variance explained by the EOF is plotted against its EOF number. The variance explained by the nth EOF is given by $\lambda_n / \sum_{i=1}^{N} \lambda_i$ where λ_n is the nth eigenvalue and N is the number of time samples. Linear regression between the logarithms of λ_n vs n is computed. EOFs above the regression line are judged to be significant.

3. Results

3.1 Global and zonal average

Figure 1 shows the area weighted global (60°N-60°S) average LHF of HOAPS3, J-OFURO2, GSSTF2, and Set1 and Set2 of GSSTF2b. The merged OAFLUX product is included for comparison. Visual inspection of the data products does not show large missing gaps in the spatial distribution of the products. All satellite products-HOAPS3, J-OFURO2 and all GSSTF datasets show increases while there is no obvious trend in OAFLUX. All products have similar global means in the early period 1988-1991. The means of GSSTF2b (both Set1 and Set2) are generally lower than that of GSSTF2 which is the highest among all products. All products show a dip in 1991 and an increase afterwards. The dip is clearly evident in HOAPS3. GSSTF2b Set1 and Set2 are identical up to 1998 after which they diverge, but tend to come close again after 2006.

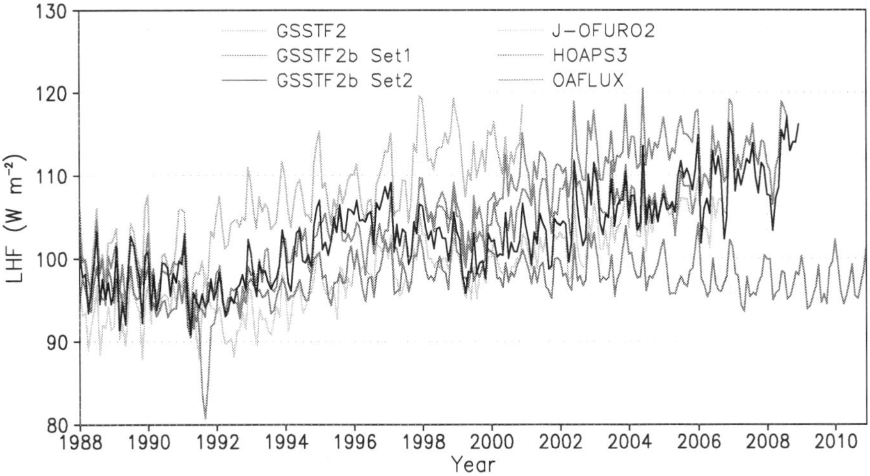

Fig. 1. Time series of global (60°N-60°S) oceanic average latent heat flux derived from HOAPS3, J-OFURO2, GSSTF2, GSSTF2b Set1 and Set2 and OAFLUX.

The zonal annual averages of LHF are depicted in Figure 2. The general features of the zonal means are quite similar among these products: they showed maxima in the subtropics, minima at the poles and relative minima at the equator. The subtropical maxima in the southern hemisphere are slighter higher than those in the northern hemisphere for the same product. While GSSTF2 is the highest among these estimates, the

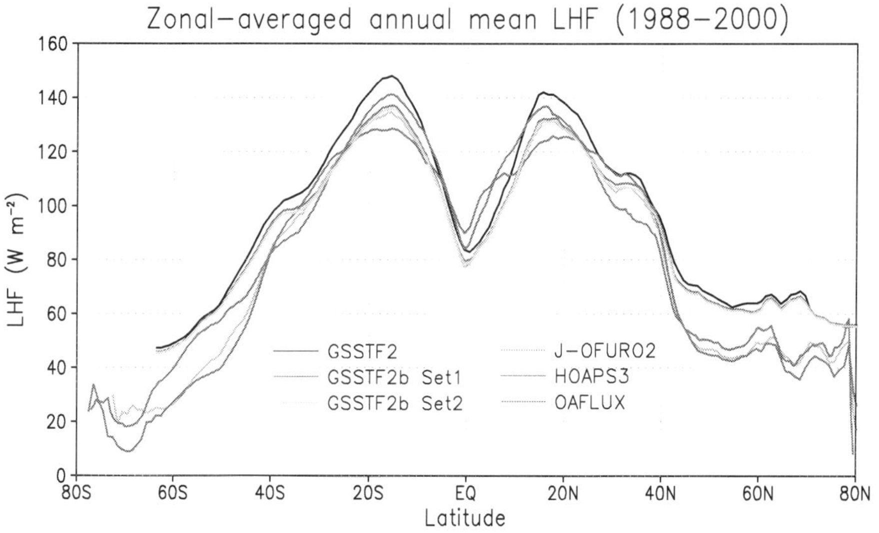

Fig. 2. Zonal annual mean of LHF for oceanic evaporation computed from HOAPS3, J-OFURO2, GSSTF2, GSSTF2b Set1 and Set2 and OAFLUX.

GSSTF2b Set2 is slightly lower than HOAPS3 but higher than J-OFURO2 and OAFLUX at their maxima. Poleward of 30°, the GSSTF zonal means are generally higher than the other products.

3.2 Trend analysis

Linear regression analyses of the time series with time were performed on the global mean time series. The significance of the slopes of the regression (trend) is tested using a t-test. The degree of freedom for the significance test takes into account the serial correlation of the time series (Angell, 1981; Chiu & Newell, 1983). GSSTF2 shows the largest trend. It is followed by GSSTF2b Set1 while the GSSTF2b Set2 trend is comparable to HOAPS3 and J-OFURO2 for the period of overlap (1988-2005). OAFLUX exhibits the smallest trend. Table 2 summarizes our results.

To map out the geographic differences, Figure 3 compares the spatial distribution of linear trends of GSSTF2b Set1 and Set2, HOAPS3, J-OFURO2 and OAFLUX. The linear trends are calculated for the common time period 1988-2005. While the magnitudes of the trends are different, the locations of maximum change are similar among HOAPS3, J-OFURO2 and GSSTF2b Set1 and Set2 - all show increasing trends in the storm tracks in the north Atlantic and north Pacific, the oceanic dry zones off the Inter-tropical Convergence Zone (ITCZ) in the western south Pacific and in latitude bands between 30−40°S off the coast of Australia and in the Indian Ocean. OAFLUX shows increasing trends in the storm tracks in both the North Atlantic and North Pacific, and in the eastern coastal regions off South America and Australia. There are large areas showing a decrease, notably in the south Indian Ocean, tropical eastern North Pacific and in North Atlantic.

Period	GSSTF2	GSSTF2b Set1	GSSTF2b Set2	HOAPS3	J-OFURO2	OAFLUX
1988-2000	10.44	9.88	5.98	7.75	6.51	3.71
1988-2005	N/A	11.69	6.34	7.35	7.62	2.41
1988-2008	N/A	10.45	7.08	N/A	N/A	1.50

Table 2. Linear trends of LHF products (in W m^{-2} decade^{-1}) for the different periods. All values are significant at 99% level.

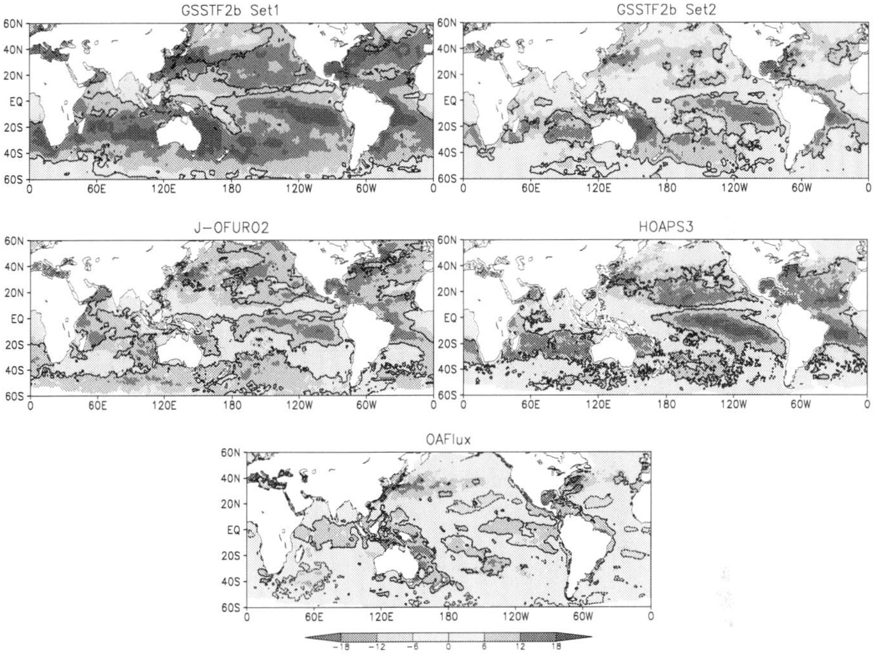

Fig. 3. Linear trends of LHF (W m^{-2} decade^{-1}) over the period 1988–2005. Contours give the trends above 95% confidence level.

3.3 EMD analysis of global average LHF

We performed EMD analyses for both GSSTF2 and GSSTF2b for the periods 1988-2000 and 1988–2008, respectively. Figure 4 shows the last IMFs from EMD analysis of GSSTF2b global average (60°N–60°S) LHF for the period 1988-2008, along with those of HOAPS3, J-OFURO2 and OAFLUX. EMD analysis of GSSTF2 global average LHF for the period 1988-2000 (Chiu et al., 2008; Xing, 2006) showed an increase starting around 1990 before the Mt Pinatubo eruption event in 1991. The last IMF of GSSTF2b Set2 and HOAPS3 show a dip around 1990 while J-OFURO2 and GSSTF2b Set1 show monotonic increases. From EMD analysis, a "trend" cannot be ascertained for the whole period for GSSTF2b Set2 and HOAPS3.

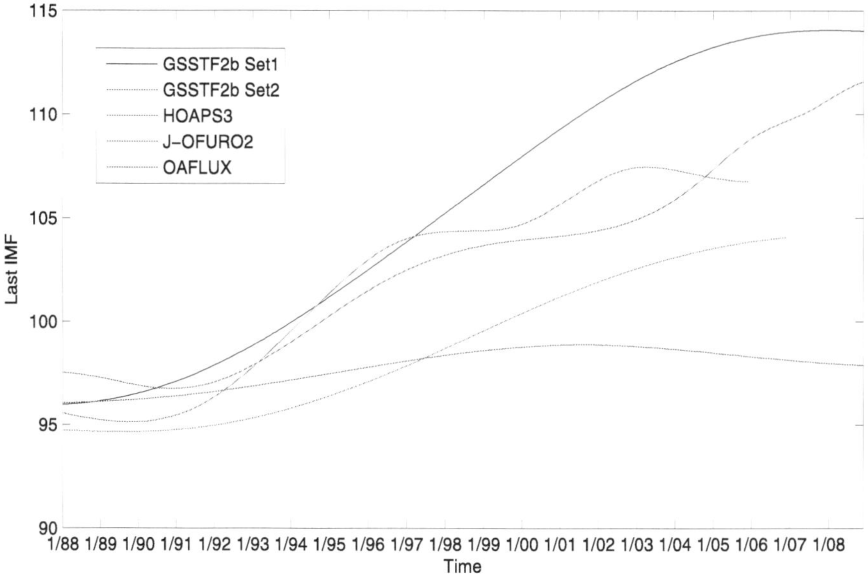

Fig. 4. The last IMFs (W m-2) from EMD analysis of global (60°N–60°S) average LHF for GSSTF2b Set1 (black), GSSTF2b Set2 (red), and OAFLUX (blue) for the period 1988–2008, HOAPS3 (purple)for the period 1988–2005 and J-OFURO2 (green) for the period 1988–2006.

3.4 GSSTF2b Set1 and Set2

In this subsection, we analyze the difference between GSSTF2b Set1 and Set2. The linear trends of GSSTF2b LHF for the entire period 1988-2008 are depicted in Figure 5. The magnitudes of the trends are reduced when compared to GSSTF2 (see Chiu et al., 2008) and Set1 shows larger trends than Set2, as anticipated.

Trends of the zonal averages are shown in Figure 6. Both Set1 and Set2 show similar trend patterns while the magnitude in Set2 is much reduced. Maximum increasing trends are located in the subtropics, with maxima at 30°S and 35°N, poleward of the latitude of maximum zonal evaporation, while the trend at the thermal equator (~5°N) shows a minimum.

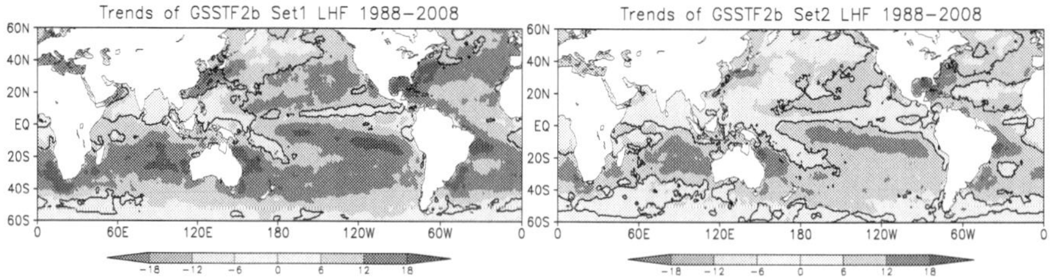

Fig. 5. Linear trends of GSSTF2b Set1 and Set2 LHF (1988–2008). Unit in W m^{-2} $decade^{-1}$. Contours give the trends above 95% confidence level.

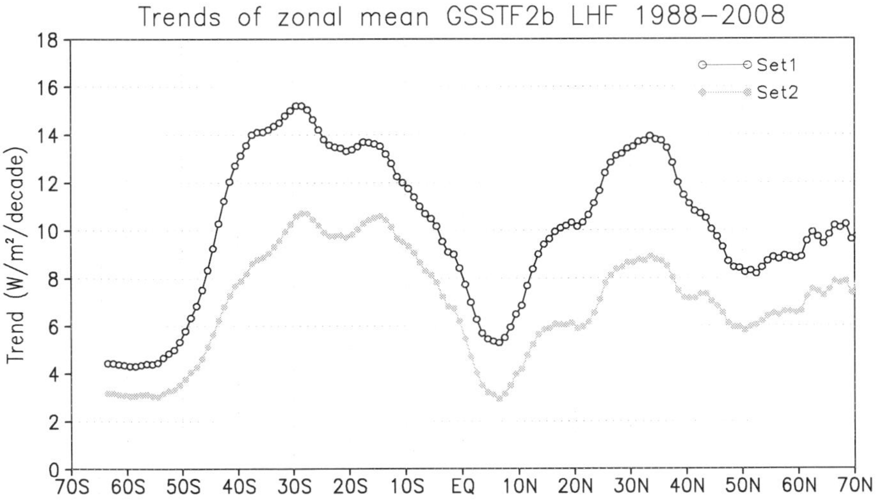

Fig. 6. Linear trends of zonal mean GSSTF2b Set1 and Set2 LHF.

Figure 7 shows the spatial distribution of trends in *U*, *DQ* and Q_a in Set 1 and Set2. It can be seen that the trend pattern in *U* is almost identical in both Set1 and Set2. Major difference is found in the *DQ* field, the magnitudes of the trends in Set1 are larger than that in Set2. In the equatorial eastern Pacific, in portion of the South China Sea, and the Bay of Bengal, Set2 actually shows a decreasing but non-significant trend. For the Q_a trend, Set1 shows large areas of decreases. While the patterns are similar for Set2, the magnitudes are much reduced, and in some areas, such as the eastern North Pacific and the bulk of the North Atlantic, the decreasing trends actually reverse to increasing trends.

LHF is a product of the surface wind (*U*) and the humidity difference ($DQ=Q_s-Q_a$) if we assume the variation in C_E is small (Equation 1). Judging from the change pattern of *U* and *DQ* they are essentially decoupled. Equation (1) can be integrated globally to get

$$\frac{\delta\overline{LHF}}{\overline{LHF}} \approx \frac{\delta\overline{U}}{\overline{U}} + \frac{\delta\overline{DQ}}{\overline{DQ}} \tag{2}$$

$$\delta\overline{DQ} = \delta\overline{Q_s} - \delta\overline{Q_a} \tag{3}$$

where δx represent the change in the quantity x, $\delta x/x$ represent fractional changes in x, and the over-bar $\overline{x}$ represents global average of x. For GSSTF2 (1988-2000), the terms in equation (2) is approximately 17%, 6%, and 11%, in that order (Xing, 2006). Most of the increase in $\overline{DQ}$ was attributed to increase in $\overline{Q_s}$ and decrease in $\overline{Q_a}$.

Figure 8 shows the time series of global average *U*, *DQ* and Q_a for GSSTF2b Set1 and Set2, respectively. It clearly indicates the divergence of Set1 and Set2 in late 1997. The large difference in LHF between Set1 and Set2 is mostly attributed to *DQ*, which is due to the higher Q_a in Set2. The difference in *U* between the datasets is small. There is a large decrease in Q_a at the end of 2008.

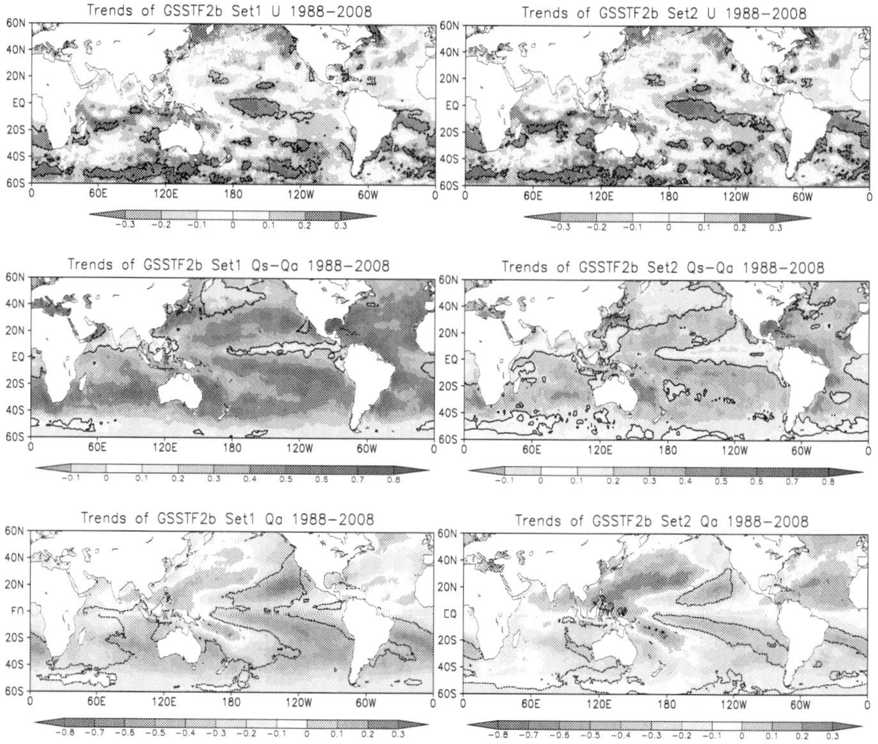

Fig. 7. Linear trends of surface wind speed (U, upper panel), surface humidity difference (DQ=Q_s-Q_a, middle panel), and surface air humidity (Q_a, lower panel) for GSSTF2b Set1 (left column) and Set2 (right column). Units are in m s^{-1} $decade^{-1}$ for U and in g kg^{-1} $decade^{-1}$ for DQ and Q_a. Contours give the trends above 95% confidence level.

Table 3 shows the changes in LHF in Set1 and Set2 of GSSTF2b (1988–2008) and the associated changes in U, DQ and the changes in Q_s and Q_a. The change in DQ contribute most to the change in LHF for both Set1 and Set2, while the changes in DQ is due both to an increase in Q_s and decrease in Q_a. The difference in the change of LHF between Set1 (23.1%) and Set2 (15.5%) is mostly attributed to DQ (20.0% vs. 12.3%) and changes in Q_a (-0.51 g kg^{-1} vs. -0.24 g kg^{-1}). It is clear that the impact of DMSP F13 is the introduction of a much lower Q_a, thus affecting DQ and ultimately LHF.

	$\frac{\delta\overline{LHF}}{\overline{LHF}} \approx \frac{\delta\overline{U}}{\overline{U}} + \frac{\delta\overline{DQ}}{\overline{DQ}}$	$\frac{\delta\overline{U}}{\overline{U}}$	$\frac{\delta\overline{DQ}}{\overline{DQ}}$	$\delta\overline{DQ}$ (g kg^{-1})	$\delta\overline{Qs}$ (g kg^{-1})	$\delta\overline{Qa}$ (g kg^{-1})
Set1	23.1%	3.1%	20.0%	0.73	0.22	-0.51
Set2	15.5%	3.1%	12.3%	0.46	0.22	-0.24

Table 3. Summary of changes in LHF, U and DQ and Q_s and Q_a for GSSTF2b Set1 and Set2.

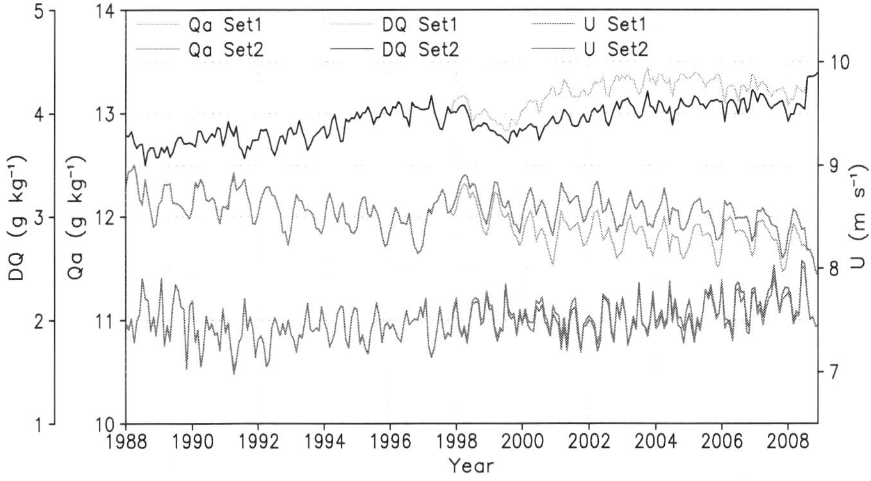

Fig. 8. Time series of global average surface humidity difference (*DQ*), surface air humidity (Q_a) and wind speed (*U*) for GSSTF2b Set1 and Set2.

3.5 EOF analyses and teleconnections

Empirical Orthogonal Function (EOF) analyses are performed on LHF of both GSSTF2b Set1 and Set2 for the period 1988-2008. Monthly means for the entire period are first removed from the data to form the non-seasonal dataset. The first, second and third EOF of Set1

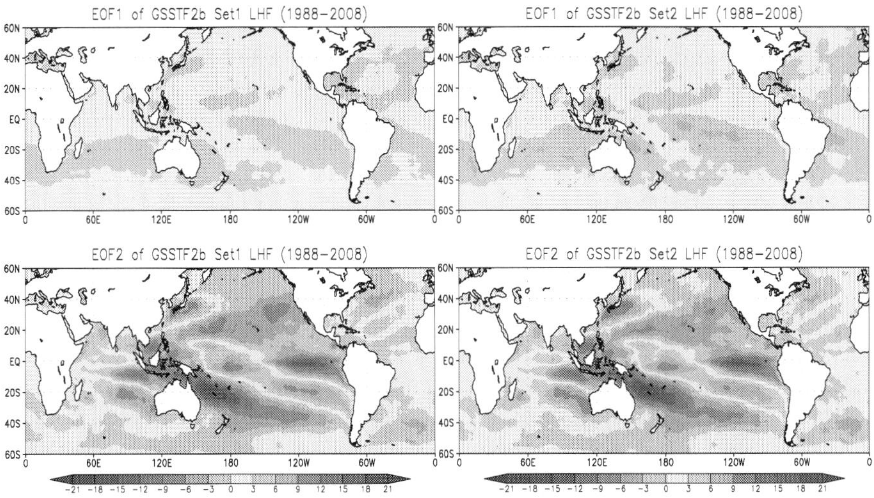

Fig. 9. Spatial patterns of the first (top) and second (bottom) EOF of GSSTF2b LHF for the period 1988-2008. The variances explained are 8.6% (11.5%) and 4.3% (4.3%) for EOF1 and EOF2 for Set2 (Set1), respectively.

explains 11.5%, 4.3%, and 3.4% of the total variance, and the corresponding variance explained for Set2 are 8.6%, 4.3% and 3.4%. Figure 9 shows the spatial patterns of the first (EOF1) and second (EOF2) EOF for Set1 and Set2, respectively. Their associated time series, accompanied by a rescaled Southern Oscillation Index (SOI), are presented in Figure 10. The general patterns of EOF1 of Set1 and Set2 are very similar, with large weights in subtropical Indian Ocean, the dry zone in the eastern tropical South Pacific and South Atlantic. They also bear striking resemblance to the first EOF pattern computed from GSSTF2 which shows

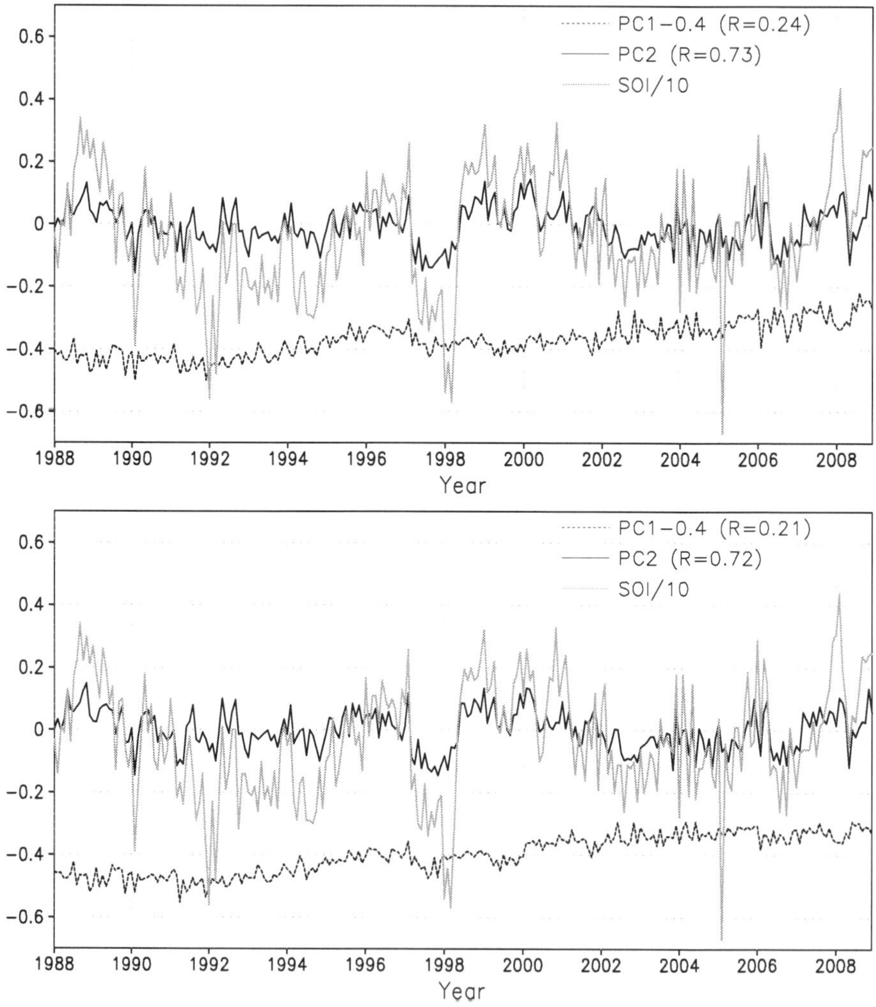

Fig. 10. Time series associated with EOF of non-seasonal GSSTF2b Set2 LHF for the period 1988–2008 and the SOI for the same period. The SOI is divided by 10 and the time series of the first EOF is shifted by -0.4 units for clarity. The lower panel shows the same for Set1.

weights of the same sign everywhere (Chiu et al., 2008). However, in GSSTF2b, there are regions of opposite sign in the South China Sea area in both Set1 and Set2, with higher weights in Set2 in the centers of highs. The EOF2 patterns for Set1 and Set2 are almost identical. Similarity is also noted for the EOF2 pattern derived from GSSTF2 (Chiu et al., 2008).

The time series of EOF1 also show a slight increasing trend, suggesting an increase of LHF over most of the global ocean. The associated time series of EOF2 have a significant correlation of 0.73 (0.72 for Set1) with SOI, reaffirming that EOF2 - ENSO events association.

The third EOF pattern (not shown) is characterized by a negative - positive - negative (- + -) zonal changes centering around 10–20°N, 20–30°N and north of 40°N in the north Pacific and a (+ - +) centering at 20°N, 30°N and 50°N in the North Atlantic. The pattern correlations for EOF1, EOF2 and EOF3 between Set1 and Set2 are 0.08, 0.49, and 0.39, and the corresponding temporal correlation for the associated time series are 0.92, 0.97, and 0.83, in that order.

EOF3 is reminiscent of the North Atlantic Oscillation (NAO) and the North Pacific Oscillation (NPO) patterns (Walker & Bliss, 1932; Wallace & Gultzer, 1981, Hurrell et al., 2003). These teleconnection patterns are further discussed in terms of an atmospheric annular mode, or Arctic Oscillation (AO), showing the opposition between subtropical highs and the polar lows (Wallace, 2000; Deser, 2000; Aubaum et al., 2001). We compute the correlations between an AO index with the time series of EOF3 and found no significant correlation with Set2, however a correlation of 0.32 is found for EOF3 Set1, significant at 95% level.

4. Summary and discussion

Four satellite based sea surface latent heat flux (LHF) products, HOSAPS3, J-OFURO2, GSSTF2, and GSSTF2b (Set1 and Set2) and a merged analysis OAFLUX are compared. Linear trend analysis of all satellite based products show large increasing trend, with GSSTF2 the largest, followed by GSSTF2b Set1, J-OFURO2, HOAPS3, and GSSTF2b Set2. OAFLUX exhibits the lowest linear increasing trend. Most of the satellite products used SSM/I as input. Small drifts in the SSM/I brightness temperature (TB) associated with changes in Earth incidence angle (EIA) was noted in most of the SSM/I data (Hilburn & Shie, 2011; Shie & Hilburn, 2011). Because of the sensitivity of the boundary layer water (*WB*) to the TB, these small drifts can introduce artificial trends in bulk quantities such as Q_a. A second data set, GSSTF2b Set2, which excludes satellite retrievals that were judged to introduce these biases, was introduced. The most-excluded satellite data are SSM/I onboard the DMSP F13 and F15 satellites. The new set, GSSTF2b Set2, was found to have a much reduced increasing trend, the magnitude of which is comparable to HOAPS3 and J-OFURO2 for the period of overlap. To account for the drift in the EIA, a new version of GSSTF, GSSTF2c, that takes account of the correction in EIA, has been completed as of this writing, and will be officially released to the public via NASA/GES DISC by the end of October 2011 (Shie et al., 2011). These trend issues will be revisited after its release.

Empirical Mode Decomposition (EMD) analyses, which are designed for examining non-stationary non-homogeneous time series, are performed on the global LHF. The last IMF of GSSTF2b Set1 shows a monotonic increase indicating the existence of a trend in this period.

The corresponding IMF of the other data products do not show monotonic increases, hence trends cannot be ascertained.

To examine the attribution of the increase in GSSTF2b Set1 and Set2, the linear trends in both the surface wind (*U*) and surface humidity difference (*DQ*) are computed. There is no significant difference between the trend patterns in the wind field. However, large difference in the *DQ* trend is noted. The *DQ* trend difference is attributed to a reduction in the negative trend in surface air humidity (Q_a) in Set2.

A major difference between GSSTF2 and GSSTF2b is the use of RSS SSM/I V4 for GSSTF2 and RSS SSM/I V6 for GSSTF2b. The changes in LHF, *U*, and *DQ* are 16%, 6%, and 11% for GSSTF2 (Xing, 2006). The corresponding changes are 23%, 3%, and 20% for GSSTF2b Set1 and 16%, 3%, and 12% for Set2 (Table 3). The use of RSS SSM/I V6 products reduces the wind trend from 6% to 3% but increases the *DQ* trend from 11% to 20% for Set1. The exclusion of F13 and F15 data reduces the LHF trend to 16%, mostly due to a reduction in the *DQ* trend to 12% (from 20%) with *U* changes remain at 3%.

Interannual variability is examined using EOF analyses. The first three significant non-seasonal EOF patterns are similar, and they explaining 10.5%, 4.3% and 3.4% for Set1 and 8.6%, 4.3% and 3.4% for Set2, respectively. The first EOF pattern of GSSTF2 for 1998-2000, with opposite changes between the equatorial eastern Pacific and the subtropics in the Pacific and Indian ocean, may be indicative of an enhance Hadley circulation (Chiu & Xing, 2004; Chiu et al., 2008). Observations also indicate large decadal variability in the Hadley Circulation (Wielicki et al., 2002; Cess & Udelhofen, 2003; Chen et al., 2002; Mitas and Clements, 2005).

This seesaw pattern is much reduced in the EOF1 pattern in both GSSTF2b Set1 and Set2 of 1998-2008, which may indicate a reduction, change of phase, or mixing of the signal with the trend in GSSTF2b. The contribution to the total variance is smaller for Set2, which excluded DMSP datasets that contains large long-term trends introduced by drifts in the Earth incidence angle in the SSM/I sensors. The difference in the fraction of variance explained in Set1 and Set2 is attributed to the artificial trend in F13 and F15 and the EIA drift effect.

Examination of the trends of the zonal means show that the latitude of maximum increase, situated in the subtropics, is found poleward of the LHF maximum in the tropic. This pattern is consistent with the expansion of the Hadley Circulation associated with global warming as predicted in climate models (Lu et al., 2007).

The EOF2 patterns of Set1 and Set2 are almost identical, both contributed to 4.3% of the variance of the dataset. The association with the El Nino/Southern Oscillation phenomena is corroborated by a high correlation between their time series and an index of the Southern Oscillation (SOI). The patterns for EOF3 and their associated time series are also similar, indicating that both GSSTF2b Set1 and Set2 are useful for examining interannual variability.

5. Acknowledgment

This study is supported by the MEaSUREs Program of NASA Science Mission Directorate Earth Science Division. The authors are especially grateful to their program manager M. Maiden and program scientist J. Entin for their valuable supports of this research.

6. References

Andersson, A.; Fennig, K.; Klepp, C.; Bakan, S.; Graßl, H. & Schulz, J. (2010). The Hamburg Ocean Atmosphere Parameters and Fluxes from Satellite Data—HOAPS-3. *Earth System Science Data*, Vol. 2, No. 1, pp. 215-234, ISSN 1866-3508

Angell, J. K. (1981). Comparison of Variations in Atmospheric Quantities with Sea Surface Temperature Variations in the Equatorial Eastern Pacific. *Monthly Weather Review*, Vol. 109, No. 4 , pp. 230-243, ISSN 0027-0644

Aubaum, M. H. P.; Hoskins, B. J. & Stephenson, D. B. (2001). Arctic Oscillation or North Atlantic Oscillation. *Journal of Climate*, Vol. 14, No. 16, pp. 3495-3507, ISSN 0894-8755

Atlas, R.; Hoffman, R. N.; Bloom, S. C.; Jusem, J. C. & Ardizzone, J. (1996). A multiyear global surface wind velocity dataset using SSM/I wind observations. *Bulletin of the American Meteorological Society*, Vol. 77, No. 5, pp. 869-882, ISSN 0003-0007

Bentamy, A.; Katsaros, K. B.; Mestas-Nuñez, A. M; Drennan, W. M.; Forder, E. B. & Roquet, H. (2003). Satellite estimates of wind speed and latent heat flux over the global oceans. *Journal of Climate*, Vol. 16, No. 4, pp. 637-656, ISSN 0894-8755

Bunker, A. F. (1976). Computations of surface energy flux and annual air-sea interaction cycle of the north Atlantic Ocean, *Monthly Weather Review*, Vol. 104, No. 9, pp. 1122-1140, ISSN 0027-0644

Cayan, D. R. (1992). Latent and sensible heat flux anomalies over the Northern Oceans: The connection to monthly atmospheric circulation. *Journal of Climate*, Vol. 5, No. 4, pp. 354-369, ISSN 0894-8755

Cess, R. D. & Udelhofen, P. M. (2003). Climate change during 1985–1999: Cloud interactions determined from satellite measurements. *Geophysical Research Letters*, Vol. 30, No. 1, 1019, ISSN 0094-8276

Chen, J., Carlson B. E., & Del Genio A. D. (2002). Evidence for strengthening of the tropical general circulation in the 1990s. *Science*, Vol. 295, No. 5556, pp. 838-841, ISSN 0036-8075

Chiu, L. S. & Xing, Y. (2004). Modes of interannual variability of oceanic evaporation observed from GSSTF2. *Gayana: International Journal of Biodiversity, Oceanology and Conservation*, Vol. 68, No. 2, pp. 115-120. ISSN 0717-652X

Chiu, L. S. & Newell, R. E. (1983). Variations of zonal mean sea surface temperature and large-scale air-sea interaction. *Quarterly Journal of the Royal Meteorological Society*, Vol. 109, No. 459, pp. 153-168, ISSN 0035-9009

Chiu, L. S.; Chokngamwong, R.; Xing, Y.; Yang, R. & Shie, C.-L. (2008). Trends and Variations of Global Oceanic Evaporation Datasets from Remote Sensing. *Acta Oceanologica Sinica*, Vol. 27, No. 3, pp. 1-12, ISSN 0253-505X

Chou, S.-H. (1993). A comparison of airborne eddy correlation and bulk aerodynamic methods for ocean-air turbulent fluxes during cold-air outbreaks. *Boundary-Layer Meteorology*, Vol. 64, No. 1-2, pp. 75-100, ISSN 0006-8314

Chou, S.-H.; Shie, C.-L.; Atlas R. M. & Ardizzone, J. (1997). Air-sea fluxes retrieved from special sensor microwave imager data. *Journal of Geophysical Research*, Vol. 102, No. C6, pp. 12706-12726, ISSN 0148-0227

Chou, S.-H., Atlas R. M., Shie C.-L. & Ardizzone J. (1995). Estimates of surface humidity and latent heat fluxes over oceans from SSM/I data. *Monthly Weather Review*, Vol. 123, No. 8, pp. 2405-2425, ISSN 0027-0644

Chou, S.-H.; Nelkin, E.; Ardizzone, J.; Atlas, R. & Shie, C.-L. (2001). The Goddard Satellite-Based Surface Turbulent Fluxes Dataset Version 2 (GSSTF2.0) [global (grid of 1° x 1°) daily air-sea surface fluxes from July 1987 to December 2000]. Available from <http://disc.gsfc.nasa.gov/precipitation/gsstf2.0.shtml>

Chou, S.-H.; Nelkin, E.; Ardizzone, J.; Atlas, R. M. & Shie, C.-L. (2003). Surface turbulent heat and momentum fluxes over global oceans based on the Goddard satellite retrieval, version 2 (GSSTF2). *Journal of Climate*, Vol. 16, No. 20, pp. 3256-3273, ISSN 0894-8755

da Silva, A. M.; Young, C. C. & Levitus, S.(1994). *Atlas of Surface Marine Data. Vol. 3: Anomalies of Heat and Momentum Fluxes*. NOAA/NESDIS, Washington, D.C., USA

Deser, C. (2000). On the teleconnectivity of the "Arctic oscillation". *Geophysical Research Letters*, Vol. 27, No. 6, pp. 779-782, ISSN 0094-8276

Esbensen, S. K. & Kushnir,V. (1981). *The heat budget of the global oceans: An atlas based on estimates from marine surface observations*. Climatic Research Institute, Report No. 29, Oregon State University, 27 pp

Fairall, C. W.; Bradley, E. F.; Rogers, D. P.; Edson, J. B. & Young, G. S. (1996). Bulk parameterization of air-sea fluxes for Tropical Ocean Global Atmosphere Coupled Ocean-Atmosphere Response Experiment. *Journal of Geophysical Research*, Vol. 101, No. C2, pp. 3747-3764, ISSN 0148-0227

Fairall, C. W.; Bradley, E. F.; Hare, J. E.; Grachev, A. A. & Edson, J. B. (2003). Bulk parameterization of air-sea fluxes: Updates and verification for the COARE algorithm. *Journal of Climate*, Vol. 16, No. 4, pp. 571-591 , ISSN 0894-8755

Grassl, H.; Jost, V.; Kumar, R.; Schulz, J.; Bauer P. & Schluessel, P. (2000). *The Hamburg Ocean-Atmosphere Parameters and Fluxes from Satellite Data (HOAPS): A climatological atlas of satellite-derived air-sea interaction parameters over oceans*. Max Planck Institute for Meteorology, Report No. 312, ISSN 0937-1060, Hamburg, Germany

Hastenrath, S. (1980). Heat budget of tropical ocean and atmosphere. *Journal of Physical Oceanography*, Vol. 10, No. 2, pp. 159-170, ISSN 0022-3670

Hilburn, K. A. & Shie, C.-L. (2011). Decadal trends and variability in Special Sensor Microwave Imager (SSM/I) brightness temperatures and Earth incidence angle. Report No. 092811, Remote Sensing Systems, 53 pp

Huang, N. E.; Shen, Z.; Long, S. R.; Wu, M. C.; Shih, H. H.; Zheng, Q.; Yen, N.-C.; Tung, C. C. & Liu, H. H. (1998). The empirical mode decomposition and the Hilbert spectrum for nonlinear and non-stationary time series analysis. *Proceedings of the Royal Society of London. Series A*, Vol. 454, No. 1971, pp. 903-995, ISSN 1364-5021

Hsiung, J. (1985). Estimates of global oceanic meridional heat transport. *Journal of Physical Oceanography*, Vol. 15, No. 11, pp. 1405-1413, ISSN 0022-3670

Hurrell, J. W.; Kushnir, Y.; Ottersen, G. & Visbeck, M. (Eds.). (2003). *The North Atlantic Oscillation-Climatic Significance and Environmental Impact*, Geophysical Monograph Vol. 134, American Geophysical Union, ISBN 0-875-90994-9, Washington D.C., USA

Isemer, H.-J. & Hasse, L. (1985). *The Bunker Climate Atlas of the North Atlantic Ocean: 1. Observations*. Springer-Verlag, ISBN 0-387-15568-6, New York, USA

Isemer, H.-J. & Hasse, L. (1987). *The Bunker climate atlas of the North Atlantic Ocean: 2. Air-sea interactions*. Springer-Verlag, ISBN 0-387-17594-6, New York, USA

Josey, S. A.; Kent,E. C. & Taylor,P. K. (1998). *The Southampton Oceanography Centre (SOC) Ocean-Atmosphere Heat, Momentum and Freshwater Flux Atlas*. Southampton Oceanography Centre, Report No.6, Southampton, United Kingdom

Kubota, M.; Ichikawa, K.; Iwasaka, N.; Kizu, S.; Konda, M. & Kutsuwada, K. (2002). Japanese Ocean Flux Data Sets with Use of Remote Sensing Observations (J-OFURO). *Journal of Oceanography*, Vol. 58, No. 1, pp. 213-225, ISSN 0916-8370

Liu, W. T.; Katsaros, K. B. & Businger,J. A. (1979). Bulk Parameterizations of Air-Sea Exchanges of Heat and Water Vapor Including Molecular Constraints at the Interface. *Journal of Atmospheric Science*, Vol. 36, No. 9, pp. 1722-1735, ISSN 1520-0469

Lu, J.; Vecchi, G. & Reichler, T. (2007). The expansion of the Hadley Cell under global warming. *Geophysical Research Letters*, Vol. 34, L06805, ISSN 0094-8276

Murray, F. W. (1967). On the computation of saturation vapor pressure. *Journal of Applied Meteorology*, Vol. 6, No. 1, pp. 203-204, ISSN 0021-8952

Mitas, C. M. & Clement, A. (2005). Has the Hadley cell been strengthening in recent decades?. *Geophysical Research Letters*, Vol. 32, L03809, ISSN 0094-8276

North, G. R.; Bell, T. L.; Cahalan, R. F. & Moeng, F. J. (1982). Sampling Errors in the Estimation of Empirical Orthogonal Functions. *Monthly Weather Review*, Vol. 110, No. 7, pp. 699-706, ISSN 0027-0644

Oberhuber, J. M. (1988). *An Atlas Based on the 'COADS' Data Set: The Budgets of Heat, Buoyancy and Turbulent Kinetic Energy at the Surface of the Global Ocean*. Max-Planck-Institut für Meteorologie, Report No. 15, Hamburg, Germany

Renfrew, I. A.; Moore, G. W. K.; Guest, P. S. & Bumke, K. (2002). A comparison of surface layer and surface turbulent flux observations over the Labrador Sea with ECMWF analyses and NCEP reanalyses. *Journal of Physical Oceanography*, Vol. 32, No. 2, pp. 383-400, ISSN 0022-3670

Reynolds, R. W. & Smith, T. S. (1994). Improved global sea surface temperature analyses using optimum interpolation. *Journal of Climate*, Vol. 7, No. 6, pp. 929-948, ISSN 0894-8755

Reynolds, W. R.; Rayner, N. A.; Smoth, T. M.; Stokes, D. C. & Wang, W. (2002). An improved in situ and satellite SST analysis for climate. *Journal of Climate*, Vol. 15, No. 13, pp. 1609-1625, ISSN 0894-8755

Sakurai, T.; Kurihara, Y. & Kuragano, T. (2005). Merged satellite and in-situ data global daily SST. *Procceeding of International Geoscience and Remote Sensing Symposium*, pp. 2606-2608, ISBN 0-7803-9050-4, Seoul, South Korea, November 14, 2005

Schlussel, P.; Schanz, L. & Englisch, G. (1995). Retrieval of latent-heat flux and longwave irradiance at the sea-surface from SSM/I and AVHRR measurements. *Advances in Space Research*, Vol. 16, No. 10, pp. 107-116, ISSN 0273-1177

Shie, C.-L.; Chiu, L. S.; Adler, R.; Nelkin, E.; Lin, I-I; Xie, P., Wang, F-C; Chokngamwong, R.; Olson, W. & Chu, A. D. (2009). A note on reviving the Goddard Satellite-based Surface Turbulent Fluxes (GSSTF) dataset. *Advances in Atmospheric Sciences*, Vol. 26, No. 6, pp. 1071-1080, ISSN 0256-1530

Shie, C.-L.; Chiu, L. S.; Adler, R.; Lin, I-I; Nelkin, E. & Ardizzone, J. (2010). The Goddard Satellite-Based Surface Turbulent Fluxes Dataset --- Version 2b (GSSTF2b), In: *NASA Goddard Earth Sciences (GES) Data and Information Services Center (DISC)*, October 2010, Available from
<ftp://aurapar1u.ecs.nasa.gov/data/s4pa/GSSTF/> or
<http://disc.sci.gsfc.nasa.gov/daac-bin/DataHoldingsMEASURES.pl?PROGRAM_List=ChungLinShie>

Shie, C.-L. (2010). Science background for the reprocessing and Goddard Satellite-based Surface Turbulent Fluxes (GSSTF2b) Data Set for Global Water and Energy Cycle Research, In: *NASA GES DISC*, 18 pp, October 12, 2010, Available from < http://disc.sci.gsfc.nasa.gov/measures/documentation/Science-of-the-data.pdf>

Shie, C.-L. & Hilburn, K. (2011). A satellite-based global ocean surface turbulent fluxes dataset and the impact of the associated SSM/I brightness temperature, *Proceedings of The 2011 EUMETSAT Meteorological Satellite Conference*, Oslo, Norway, September 5-9, 2011

Shie, C.-L.; Hilburn, K. A.; Chiu, L. S.; Adler, R.; Lin, I-I; Nelkin, E. & Ardizzone, J. (2011). The Goddard Satellite-Based Surface Turbulent Fluxes Dataset --- Version 2c (GSSTF2c). In: *NASA GES DISC*, by the end of October 2011.

Tomita, H.; Kubota, M.; Cronin, M. F.; Iwasaki, S.; Konda, M. & Ichikawa, H. (2010). An assessment of surface heat fluxes from J-OFURO2 at the KEO and JKEO sites. *Journal of Geophysical Reserach*, Vol. 115, C03018, ISSN 0148-0227

Walker, G. T. & Bliss, E. W. (1932). World weather V. *Memoirs of the Royal Meteorological Society*, Vol. 4, No. 36, pp. 53-84, LCCN 7021-1603

Wallace, J. M. (2000). North Atlantic Oscillation/annular mode: Two paradigms—one phenomenon. *Quarterly Journal of the Royal Meteorological Society*, Vol. 126, No. 564, pp. 791-805, ISSN 0035-9009

Wallace, M. & Gutzler, D. S. (1981). Teleconnections in the geopotential height field during the Northern Hemisphere winter. *Monthly Weather Review*, Vol. 109, No. 4, pp. 784-812, ISSN 0027-0644

Weare, B. C.; Strub, P. T. & Samuel, M. D. (1981). Annual Mean Surface Heat Fluxes in the Tropical Pacific Ocean. *Journal of Physical Oceanography*, Vol. 11, No. 5, pp. 705-717, ISSN 0022-3670

Wentz, F. J. (1997). A well calibrated ocean algorithm for Special Senor Microwave/Imager. *Journal of Geophysical Research*, Vol. 102, No. C4, pp. 8703-8718, ISSN 0148-0227

Wentz, F. J.; Ricciardulli, L.; Hilburn K. A. & Mears C. A. (2007). How Much More Rain Will Global Warming Bring? *Science*, Vol. 317, No. 5835, pp. 233-235, Supporting material, ISSN 0036-8075

Wielicki B.; Wong, T.; Allen, R. P.; Slingo, A.; Kiehl, J. T.; Soden, B. J.; Gordon, C. T.; Miller, A. J.; Yang, S.-K.; Randall, D. A.; Robertson, F.; Susskind, J. & Jacobowitz, H. (2002). Evidence for large decadal variability in the tropical mean radiative energy budget. *Science*, Vol. 295, No. 5556, pp. 841-844, ISSN 0036-8075

Xing, Y. (2006). *Recent changes in oceanic latent heat flux from remote sensing*. Ph.D. dissertation, George Mason University, 119 pp

Zeng, X.; Zhao, M. & Dickinson, R. E., (1998). Intercomparison of bulk aerodynamic algorithms for the computation of sea surface fluxes using TOGA COARE and TAO data. *Journal of Climate*, Vol. 11, No. 10, pp. 2628-2644, ISSN 0894-8755

11

Multi-Wavelength and Multi-Direction Remote Sensing of Atmospheric Aerosols and Clouds

Hiroaki Kuze
Centre for Environmental Remote Sensing (CEReS), Chiba University
Japan

1. Introduction

Aerosols are liquid and solid particles floating in the atmosphere. Aerosol particles are originated from both natural and anthropogenic origins (Seinfeld & Pandis, 1998). In regard to the radiation balance of the Earth's atmosphere, aerosols reflect solar radiation back to space (direct effect), thus reducing the influence of greenhouse gases, though some type of aerosol causes opposite effects due to absorption of radiation. At the same time, aerosol particles work as nuclei for cloud condensation (indirect effect). Knowledge on these radiative effects of aerosol and cloud, however, is still insufficient so that uncertainties remain in the prediction of future global warming trends (IPCC, 2007). In this respect, intensive efforts are needed to evaluate the optical/physical properties of aerosols and clouds by means of both ground- and satellite-based remote sensing observations.

In order to obtain better understanding of these particulate matters, what is obviously needed is the monitoring technique that enables the retrieval of their optical properties. In this chapter, we propose multi-wavelength and multi-directional remote sensing of atmospheric aerosols and clouds. The proposed method consists of the application of ground-based radiation measurement, lidar measurement, differential optical absorption spectroscopy (DOAS), and satellite observations using natural as well as artificial light sources. Such combinatory approach makes it possible to measure various aspects of radiation transfer through the atmosphere, especially the influence of tropospheric aerosols and clouds. Also, the data provided from the ground-based solar irradiance/sky radiance measurement and DOAS are valuable for precisely characterizing the optical property of aerosol particles near the ground level, including the information from both particulate scattering and gaseous absorption. Such ground data are also indispensable for the atmospheric correction of satellite remote sensing data in and around the visible range of the radiation spectrum. The multi-wavelength and multi-directional observation schemes treated in the present chapter are summarized in Table 1.

2. Differential optical absorption spectroscopy (DOAS)

The method of differential optical absorption spectroscopy (DOAS) provides a useful tool for monitoring atmospheric pollutants through the measurement of optical extinction (i.e., the sum of absorption and scattering) over a light path length of a few kilometres (Yoshii et al., 2003, Lee et al., 2009; Si et al., 2005; Kuriyama et al., 2011). The DOAS method in the

Scheme	Wavelength	Direction	Aerosol	Trace gas
DOAS	UV, VIS, and NIR, with the resolution of array detector	Nearly horizontal measurement	Measurable through the spectral intensity	Measurable (e.g. NO_2 around 450 nm)
solar/skylight	UV and VIS, with the resolution of array detector	Solar direction/ any direction including the zenith	Measurable through the spectral intensity	Measurable (e.g. H_2O around 720 nm)
Lidar	Fundamental and harmonics of Nd:YAG laser (1064, 532 and 355 nm)	Vertical and slant path observations	Profiling by solving the lidar equation	Not applicable unless tunable lasers are employed
Satellite	Spectral bands in UV, VIS, and NIR	Nadir or near-nadir directions	Evaluated and removed in the process of atmospheric correction	Spectral bands are usually too wide to retrieve trace gases

Table 1. Various schemes of atmospheric observation discussed in this chapter.

visible spectral region is quite suitable for urban air pollution studies, since both nitrogen dioxide (NO_2) and aerosol, the most important pollutants originated from human activities, can directly be measured using a near horizontal light path in the lower troposphere.

Although conventional approach in the DOAS measurement is to install a light source, our group at the Centre for Environmental Remote Sensing (CEReS), Chiba University, has established a unique DOAS approach based on aviation obstruction lights (white flashlights) equipped at tall constructions such as smokestacks (Yoshii et al., 2003, Si et al., 2005; Kuriyama et al., 2011). Since those xenon lamps produce flash pulses every 1.5 s during the daytime, they can easily be recognized with the coverage of the whole visible spectral range. Thus, a simple setup consisting of an astronomical telescope and a compact spectro-radiometer can be employed for the measurement of NO_2. Also, the stable intensity of the light source makes it possible to retrieve aerosol, or suspended particulate matter (SPM) concentration in the lower troposphere, since the intensity variation of the detected light is mostly ascribable to the aerosol extinction over the light path (Yoshii et al., 2003).

As shown in Figs. 1 and 2, the principle of DOAS analysis of NO_2 concentration is based on matching high-pass filtered spectral (wavelength) features between the DOAS-observed optical thickness ($\Delta\tau$) and laboratory-observed molecular absorption spectrum ($\Delta\sigma$). Because of the Lambert-Beer's law, the optical thickness, τ, is expressed as

$$\tau = -\ln(I / I_0), \quad (1)$$

where I and I_0 stand for the observed and reference spectrum, respectively. The reference spectrum can be obtained by either operating the DOAS spectrometer at a short distance

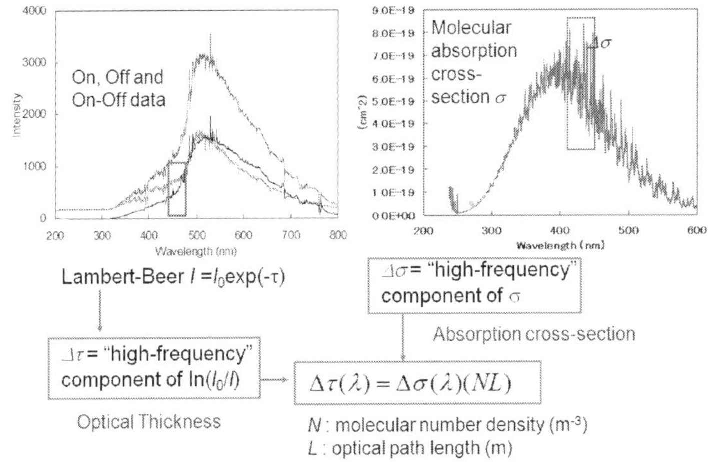

Fig. 1. Schematic flow of the DOAS analysis. The net radiation from the pulsed light source can be retrieved by subtracting the background due to sky radiation, and an appropriate portion of the spectrum is compared with the molecular cross-section spectrum obtained from laboratory measurement. Then, the "high-frequency" components of the observed optical thickness (Δτ) and cross-section data (Δσ) are compared to derive the molecular number density along the optical path length, *L*.

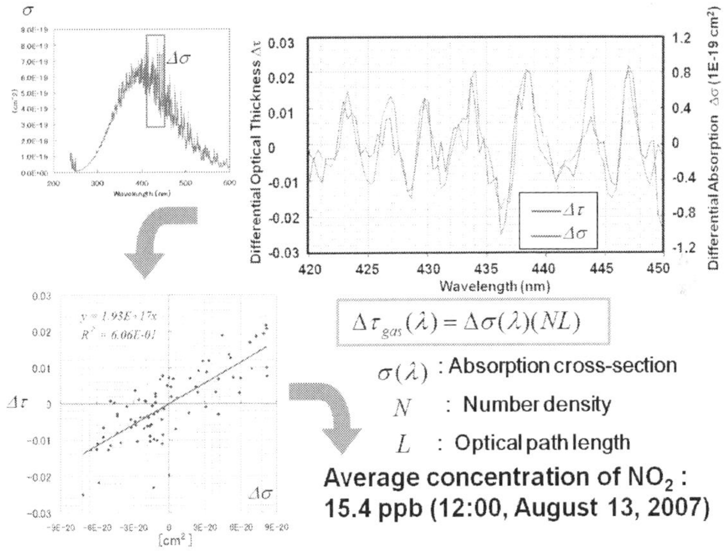

Fig. 2. An example of DOAS spectral matching, in which the correlation between the differential optical thickness from the DOAS data and differential absorption is examined to determine the average volume concentration of NO_2 molecules.

from the light source, or observing the spectrum under very clear atmospheric conditions with minimal aerosol loading. The optical thickness is generally proportional to the product of extinction coefficient, α, and the light path length, L, i.e., $\tau = \alpha L$. In the case of molecular absorption, the extinction coefficient is equal to the absorption coefficient, which can be given as the product of absorption cross-section, σ, and the molecular number density, N, i.e., $\alpha = N\sigma$. Although molecular scattering (Rayleigh scattering) and aerosol scattering (Mie scattering) also exist, their contribution can be eliminated by applying the high-pass filtering to both τ (λ) and σ (λ) (where λ is wavelength), since the absorption feature of NO_2 is a rapidly varying function with wavelength (see insets in Figs. 1 and 2), while the wavelength dependence of Rayleigh or Mie scattering is much more moderate. Thus, after the high-pass filtering, one obtains

$$\Delta\tau = (NL)\Delta\sigma. \quad (2)$$

This indicates that the correlation analysis between the rapidly varying components of the optical thickness and NO_2 cross-section in an appropriate wavelength range can lead to the determination of the molecular number density, hence the volume concentration ratio, of NO_2 along the DOAS observation light path. An example of the retrieval of NO_2 in the Chiba city area is shown in Fig. 3. In this case, the DOAS result shows the average concentration over a light path length of 5.5 km. From Fig. 3, it is seen that the DOAS data show good temporal correlation with the ground sampling data from nearby sampling stations. Note that the temporal resolution (5 min) of the DOAS observation is much better than that of the ground sampling (1 h). The observation of DOAS spectra is limited to daytime, since the white flashlight (Xe light) is replaced with blinking red lamps during night time.

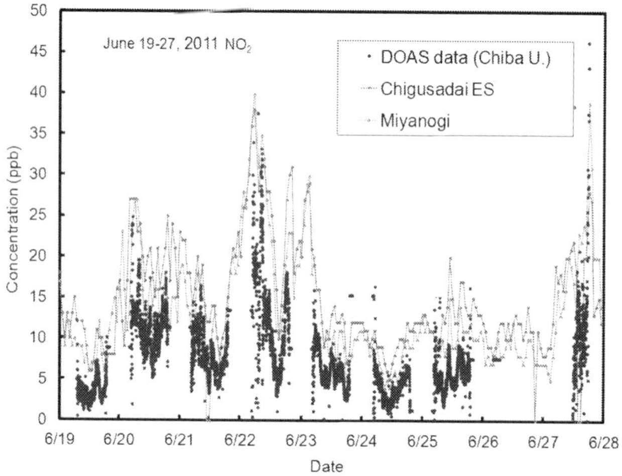

Fig. 3. Comparison of NO_2 volume concentration between the DOAS and conventional ground sampling measurements during June 19 - 27, 2011. The DOAS data are based on the measurement at CEReS, Chiba University, using an aviation obstruction flashlamp located around 5.5 km in the north direction. The ground sampling data are from two nearby sampling stations (Chigusadai Elementary School and Miyanogi stations) operated by the municipal government.

The analysis of light intensity detected with a DOAS spectrometer can yield information also on aerosol extinction along the light path. The wavelength dependence of each atmospheric component is exemplified in Fig. 4(a), where it is apparent that the contribution from aerosol extinction is much more significant than that from either NO_2 or molecular Rayleigh scattering. The optical thickness associated with aerosol extinction can generally be given as

$$\tau(\lambda) = B(\lambda / \lambda_0)^{-A}, \tag{3}$$

where $A=\alpha_{ang}$ is called the Angstrom exponent and B the turbidity constant. The value of A changes with the aerosol size distribution in such a way that a smaller value (~ 0.5) indicate the dominance of relatively coarse particles (such as sea salt or dust), while a large value (~ 2) that of relatively fine particles (such as ammonium sulfate or ammonium nitrate). The value of B, on the other hand, is equal to the aerosol optical thickness as wavelength λ_0,

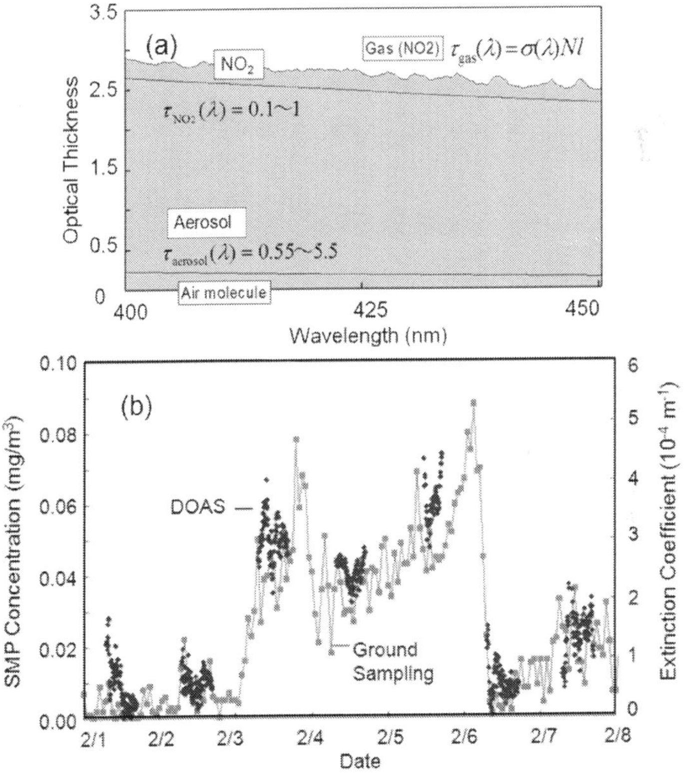

Fig. 4. Aerosol measurement from DOAS data: (a) comparison of contributions of gas (NO_2) absorption, aerosol extinction, and molecular extinction (Rayleigh scattering) to DOAS optical thickness, and (b) temporal change of SPM concentration from ground sampling and aerosol extinction coefficient from DOAS during February 1 to 7, 2011 observed in Chiba.

which is chosen to be 550 nm or some appropriate value within the observation wavelength range. Figure 4(b) shows the result of analysis based on eq. (4). As seen from Fig. 4(b) the temporal variation shows good agreement between the DOAS-derived aerosol optical thickness and the SPM mass concentration observed from the ground sampling.

3. Solar and skylight radiation measurement

For aerosols, network observation activities have been undertaken in terms of skyradiometer measurements (Takamura & Nakajima, 2004). Alternatively, the use of a compact, stand-alone spectroradiometer (EKO, MS-720) enables the spectral measurements of direct solar radiation (DSR), solar aureole (AUR) and scattered solar radiation (SSR) (Manago & Kuze, 2010; Manago et al., 2010). Since the instrument is powered by batteries with no PC requirement during measurements, it provides better portability compared to a skyradiometer. The wavelength coverage between 350 and 1050 nm with a resolution of 10 nm is useful for precise evaluation of the aerosol optical properties as well as that of the water vapor column amount. The wide dynamic-range measurement of both the direct and scattered solar radiation is attained by means of a thick diffuser and a stable photodiode array, in combination with the automatic exposure control equipped to the handy spectroradiometer (MS-720) (Manago & Kuze, 2011). In order to facilitate the radiation transfer calculation in the retrieval procedure, home-made baffle tubes are used to limit the field of view (FOV) of the observation to 20 deg (SSR) and 5 deg (DSR).

The radiation measurements were conducted at the CEReS site (35.62°N, 140.10°E) under clear-sky conditions, mostly around noon. The SSR measurements were made in 24 directions (north, east, south, and west directions, each with 6 elevation angles). The DSR and AUR components were measured before and after the SSR measurements. The total time required for a set of DSR, AUR, and SSR data was 30 - 40 minutes. Approximately 130 datasets were obtained during the observation period from August 2007 to March 2009.

Independent measurement of AOD was carried out with a sunphotometer (Prede, PSF-100). This instrument has four channels centred at 368, 500, 675, and 778 nm, each having the bandwidth of 5 nm. The wavelength dependence of AOT is analyzed with eq. (3) to obtain the Angstrom exponent. During the daytime the AOD is retrieved from the solar radiation intensity within a FOV of 1 deg at an interval of 10 s. From the sunphotometer measurement, $A=\alpha_{ang}$ and the coefficient B (turbidity constant at the reference wavelength λ_0= 550 nm in this case) can be retrieved.

In our ground observation with the battery-operated spectroradiometer, the direct solar irradiance, aureole radiance, and scattered solar radiance were measured in various directions as mentioned above. Even with these detailed measurements, however, usually it is not possible to determine the complete composition of aerosol particles. Thus, we rely on the three-component aerosol model (TCAM), in which three aerosol types of water soluble, oceanic, and soot components are considered as a basis set which is "quasi-complete" to describe the aerosol optical parameters, namely the wavelength dependence of extinction coefficient, single scattering albedo, asymmetry parameter, and scattering phase function. Figure 5 shows the wavelength dependence of the real and imaginary parts of the aerosol refractive index for the three aerosol components. It has been shown that most of the irradiance/radiance values are well reproduced by appropriately adjusting the total and relative contributions of these three basis components as well as the size distribution of each component (Manago et al, 2011). As seen from Fig. 5, the soot component shows remarkably high value of the imaginary part of the refractive index. This indicates that the absorption property is higher (single scattering albedo is lower) for aerosol with more contribution of soot particles. Figure 6 shows an example of the results of the irradiance and radiance observations. Figure 7 shows an example of aerosol optical parameters derived from the TCAM analysis of the data: Fig. 7(a) shows the wavelength dependence of the aerosol extinction coefficient (normalized to the value at 550 nm), (b) single scattering albedo, (c) asymmetry parameter, and (d) scattering phase function at wavelength 550 nm. In Sec. 5 below, we describe the application of these aerosol characteristics to the atmospheric correction of satellite remote sensing data.

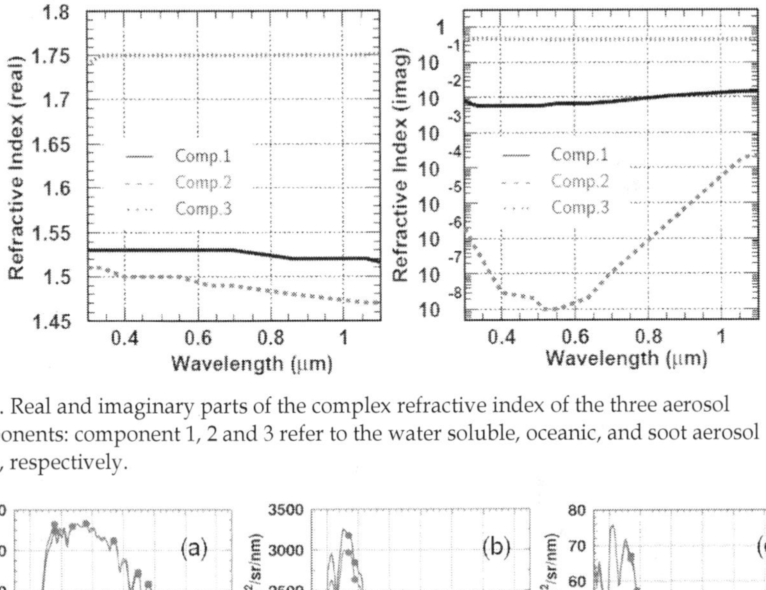

Fig. 5. Real and imaginary parts of the complex refractive index of the three aerosol components: component 1, 2 and 3 refer to the water soluble, oceanic, and soot aerosol types, respectively.

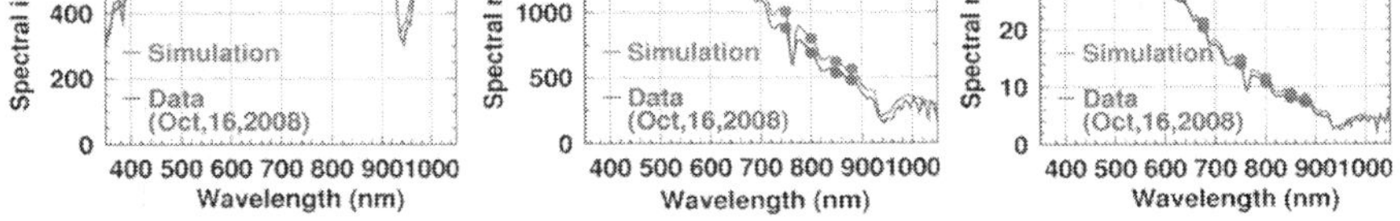

Fig. 6. Spectra observed around noon on October 16, 2008: (a) direct solar radiation (DSR), (b) aureole (AUR), and (c) scattered solar radiation (SSR). Acceptance angle of the instrument is 5 deg for DSR, 5-20 deg for AUR, and 20 deg for SSR. Simulation curves based on the TCAM best fitting are also shown with data points (circles) used for the fitting.

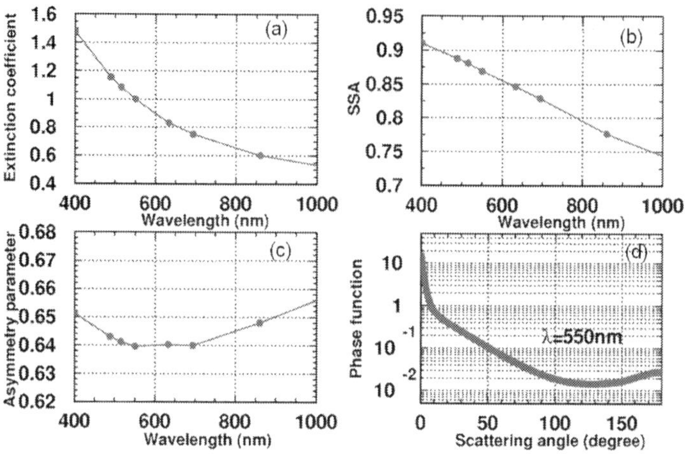

Fig. 7. Aerosol optical parameters derived from the TCAM analysis of the data shown in Fig. 6: (a) wavelength dependence of the aerosol extinction coefficient (normalized to the value at 550 nm), (b) single scattering albedo, (c) asymmetry parameter, and (d) scattering phase function at wavelength 550 nm.

4. Lidar measurement of aerosols and clouds

While the DOAS method and skylight/solar radiation measurement lead to the retrieval of atmospheric information integrated over optical paths, the lidar measurement makes it possible to measure aerosol and cloud distributions (profiles) along the optical path. Here we report the result of multi-wavelength lidar measurement conducted at CEReS. Conventionally lidar data have been analyzed by means of the Fernald method (Fernald, 1984), in which the lidar equation

$$P(R) = P_0 \frac{c\tau}{2} AK \frac{G(R)}{R^2} \beta(R) \exp\left[-2\int_0^R \alpha(R')\, dR'\right] \tag{4}$$

is solved by starting the integration from the far-end boundary. In eq. (4), $P(R)$ is the power detected by the lidar system corresponding to a distance R, P_0 is the power of the emitted laser radiation, c is the speed of light, τ is the time duration of the laser pulse, A is the area of the lidar telescope, $G(R)$ is the function describing the overlap between the laser beam and telescope field of view, $\beta(R)$ is the backscattering coefficient, and $\alpha(R)$ is the extinction coefficient. Since both air molecules and aerosol particles contribute to the scattering and extinction, one needs to separate these two components in solving the lidar equation. This can be achieved by introducing the ratio between the extinction coefficient and the backscattering coefficient. Thus, for aerosols,

$$S_1(R) = \alpha_1(R) / \beta_1(R) = \sigma_1(R) / \left(\frac{d\sigma_1}{d\Omega}\right)_{\theta=\pi} \tag{5}$$

is assumed, whereas for air molecules,

$$S_2 = \alpha_2(R) / \beta_2(R) = 8.52 \text{ (sr)} \tag{6}$$

is used as a constant. In eqs. (5) and (6), suffix 1 and 2 refer to aerosol and air molecule, respectively. In eq. (5), $\sigma_1(R)$ and $(d\sigma_1/d\Omega)_{\theta=\pi}$ indicate the total cross-section and backward differential cross-section of aerosol scattering, respectively. The parameter S_1 is often called the lidar ratio. In eq. (6), the range dependence of S_2 is omitted, since the composition of air molecules is stable througout the troposphere. Under these assumptions, the lidar equation can be analytically solved as

$$\alpha_1(R) = -\frac{S_1(R)}{S_2}\alpha_2(R) + \frac{S_1(R)\, X(R) \exp I(R)}{\dfrac{X(R_c)}{\dfrac{\alpha_1(R_c)}{S_1(R_c)} + \dfrac{\alpha_2(R_c)}{S_2}} + J(R)}. \tag{7}$$

Here

$$X(R) = R^2 P(R) \tag{8}$$

is the range-corrected signal, and functions $I(R)$ and $J(R)$ are defined as

$$I(R) = 2\int_R^{R_c} \left[\frac{S_1(R')}{S_2} - 1 \right] \alpha_2(R')\, dR' \tag{9}$$

and

$$J(R) = 2\int_R^{R_c} S_1(R')\, X(R') \exp I(R')\, dR'. \tag{10}$$

In eqs. (7), (9) and (10), R_c denotes the range of a far-end boundary, at which each integration is started. The reason that a far-end boundary value is assumed rather than a near-end boundary value is the stability of the numerical evaluation of eq. (7) (Fernald 1984).

Usually signals of a vertically looking lider are analyzed assuming that the aerosol property does not change with the altitude. Under this assumption, the range dependence in eq. (5) can be neglected. Even in this case, however, it is necessary to determine the value of lidar ratio as a function of wavelength [$S_1 = S_1(\lambda)$] in order to analyze multi- wavenength lidar data. One way to accomplish this is to use ancillary data from a sunphotometer (Kinjo et al., 1999), since the wavelength dependence of optical thickness provides a constraint to the intagration of $\alpha_1(R, \lambda)$ from R=0 to $R = R_c$. Another approach is to employ the aerosol properties measured at the ground level. In the case of Fig. 8, for example, the S_1 values of 54.7, 53.0, 46.0 and 43.2 sr are used for λ = 355, 532, 756 and 1064 nm, respectively, as derived from the chemical analysis of ground sampling data taken monthly at CEReS (Fukagawa et al., 2006). In Fig. 8, panel (a) shows the temporal variation of the aerosol extinction profile measured for 1064 nm and relative humidity (RH) at the ground level, while panel (b) depicts that of the profile of the Angstrom exponent, α_{ang}, as derived from the analysis of lidar data for the four wavelengths. The features in these panels indicate

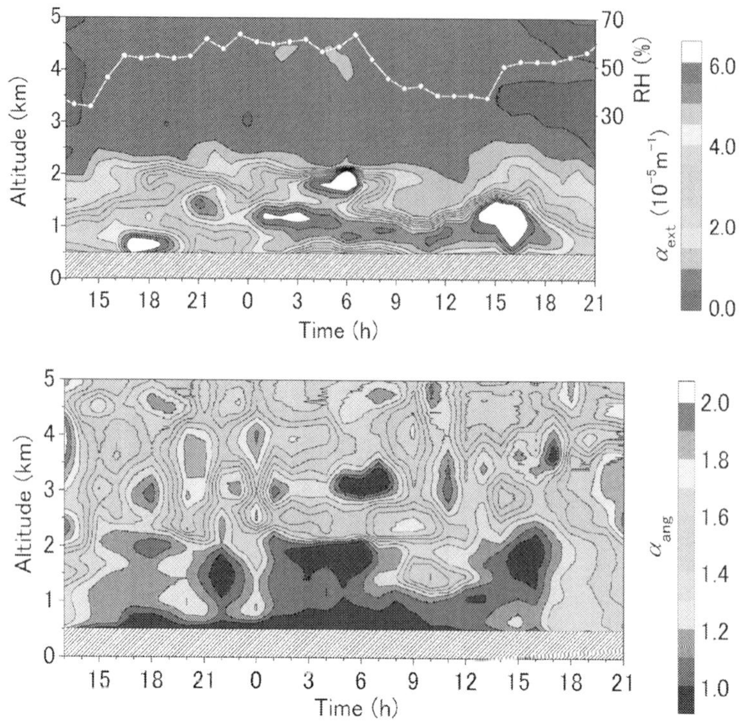

Fig. 8. Analysis of vertical looking multi-wavelength lidar data: (a) extinction profile for wavelength λ = 1064 nm observed at CEReS on 17-18 November 2005, and Angstrom exponent derived from extinction coefficients observed for λ = 355, 532, 756 and 1064 nm. The analysis is based on the Fernald method with lidar parameters S_1 = 54.7, 53.0, 46.0 and 43.2 sr for each lidar wavelength (based on sampling result at CEReS) and the reference altitude of R_c = 5.5 km.

that relatively higher extinction near the ground level is observed when RH increases, and at the same time, smaller values of α_{ang} are observed. It is likely that both of these observations are due to the aerosol growth associated with the increase of RH.

5. Atmospheric correction of satellite remote sensing data

Images taken from satellite sensors are affected by both the ground reflectance and atmospheric conditions, which include the influence of scattering and absorption of air molecules and aerosol particles. The process of atmospheric correction, in which such atmospheric effects are precisely evaluated and removed, is indispensable for extracting the intrinsic information of the ground reflectance from satellite imagery (Tang et al., 2005; Kaufman et al., 1997). Although it is rather straightforward to make corrections on the Rayleigh scattering of air molecules, aerosol particles are quite variable both temporally and spatially. This is due to the variable origins of aerosols, consisting of relatively coarse particles of natural origins (such as sea-salt and soil particles) and relatively fine particles of anthropogenic origins (such as sulphate and soot particles).

In standard atmospheric correction, it is customary to assume some representative aerosol models such as maritime, rural, continental, or urban aerosol, to implement the radiative transfer calculation of a satellite scene. This approach has an obvious disadvantage that if the assumed aerosol properties are different from those of real aerosols included in the satellite scene, the resulting information on the ground reflectance is inaccurate. To overcome this difficulty, here we use the aerosol information derived from the ground observation implemented nearly simultaneously with the satellite overpass. Such ancillary information ensures better separation of the ground and atmospheric effects from satellite imagery. Figure 9 shows the schematic drawing of radiation components considered in the radiative transfer calculation (Kotchenova et al., 2006). In this scheme, the radiance originated from the target area is denoted as L_{tar}, which consists of the ground direct (L_{gd}) and ground indirect (L_{gi} and L_{gi}') components. The environmental radiance, L_{env}, is the component associated with the surface reflection that takes place in adjacent pixels. The atmospheric radiance, L_{atm}, consists of two terms, namely, the path radiance due to single scattering (L_{ps}) and that due to multiple scattering (L_{pm}).

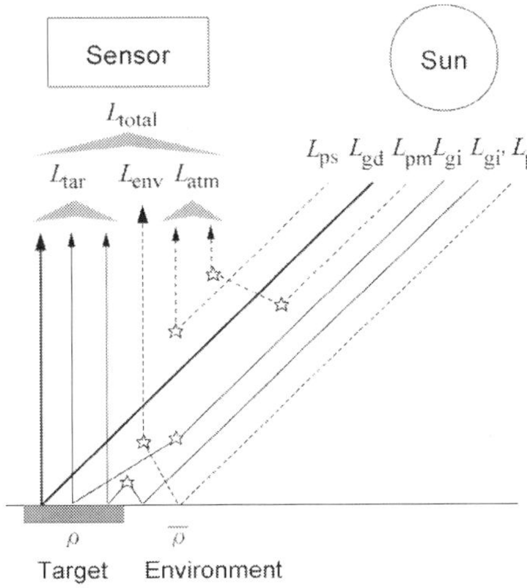

Fig. 9. Schematic drawing of radiation components considered in the radiative transfer calculation. See text for the explanation of radiance components shown in this figure.

In the present research, the ground measurement by means of a compact spectroradiometer was implemented in synchronous with the overpass of the satellite observation. The aerosol optical parameters were derived by analyzing both the direct solar radiation (DSR) and scattered solar radiation (SSR) through the Mie-scattering and radiative transfer calculations, as explained in Sec. 3 of this chapter. When the aerosol loading is relatively small (clear days), it is likely that the aerosol model resulting from this procedure can be applicable to the whole region of the Moderate Resolution Imaging Spectroradiometer (MODIS) image, and the atmospheric correction is applied to the image. Since this correction is based on the aerosol model from the simultaneous measurement, the resulting distribution of the surface reflectance is considered to be more reliable than the result that would be obtained by assuming usually available "standard" aerosol models such as urban, rural, or oceanic models. The surface reflectance map (ρ_{clear} map) on such a "clear" day, in turn, can be used as a standard for that particular season of the year, and the atmospheric correction of MODIS data taken on more turbid days can be implemented on the basis of these standard ρ_{clear} maps. This process leads to the derivation of the distribution of aerosol optical thickness (τ map).

For each of the visible bands of MODIS, a lookup table of the radiance at the top of the atmosphere, $L_{total}(\rho, \tau_{550})$, was constructed on the basis of the aerosol optical parameters and the geometric data describing the observational conditions of each image. Here, ρ is the diffuse reflectance of each pixel, and τ_{550} is the aerosol optical thickness (AOT) at wavelength 550 nm. The reflectance property of the surface was assumed to be Lambertian, and the radiative transfer calculation was carried out using the 6S code (Kotchenova et al., 2006).

The atmospheric correction was applied to channels 1 through 4 covering wavelength range between 0.450 and 0.876 μm of the Terra/MODIS and Aqua/MODIS images. The ground resolution of the MODIS sensor is 0.5 km×0.5 km/pixel. The region of 600×600 pixels around Chiba University was extracted from each of the MODIS images, which were taken from the satellite data archiving system of CEReS, Chiba University. The ground observations using the spectroradiometer were carried out at CEReS around noon on nearly cloud-free days from 2007 to 2009 (around 130 days). In order to take the time lag of around 2 h between the satellite overpass (10:00 local time) and the spectroradiometer observation (12:00) into account, the sunphotometer data taken at CEReS were employed to examine the temporal stability of atmospheric conditions. If the AOT derived from the spectroradiometer was close to the AOT value measured with the sunphotometer at the time of satellite overpass, the data were employed in the atmospheric correction. Otherwise, the data were not included in the clear-day analysis lest the instability in the atmospheric condition might also degrade the regional stability (i.e. homogeneity) of the aerosol distribution. Figure 10 shows the wavelength dependence of the surface reflectance (ρ_{clear} map) of the pixel including the location of Chiba University for various months in the year 2008.

Figure 11 shows the seasonal variation of the surface reflectance for the Chiba University pixel obtained from the analysis of MODIS band 4 centred at 550 nm. For the sake of comparison, our previous result obtained from the Landsat-5 analysis is also depicted (Todate et al., 2004). Note that the Landsat reflectance was obtained assuming a standard aerosol model (maritime), whereas the TCAM aerosol model is used in the present MODIS analysis. Pixels with vegetation and soil coverage are shown for the Landsat data, since the ground resolution associated with this sensor (30 m) is much better than the MODIS resolution of 500 m. From Fig. 11, it is seen that the surface reflectance decreases from November to December, due to the decrease in the vegetation coverage during winter. In winter the reflectance shows no critical dependence on the aerosol model assumed in the atmospheric correction because of the fact that the AOT tends to be small. In summer, on the contrary, the AOT generally increases so that the resulting value of surface reflectance varies in accordance with the aerosol model employed in the analysis.

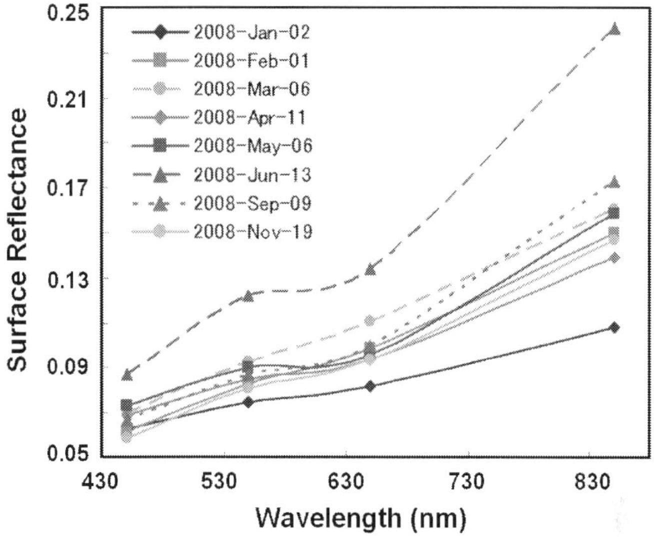

Fig. 10. Surface reflectance at Chiba University (2008)

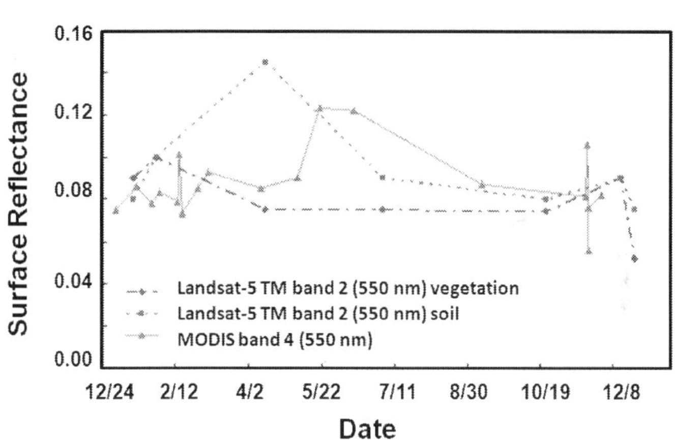

Fig. 11. Seasonal variation of surface reflectance at the MODIS pixel including the Chiba university campus (λ =550 nm).

From the present TCAM analysis of MODIS data, monthly reflectance image ($\rho_{monthly}$) is generated for each month as a composite of pixels that exhibit the lowest reflectance. This process ensures the removal of cloud pixels that might contaminate the resulting ρ map. These monthly ρ maps, in turn, are employed in the radiative transfer analysis to derive the aerosol distribution (τ map) from images taken on relatively turbid days. Examples of the reflectance and aerosol distribution images are shown in Fig. 12.

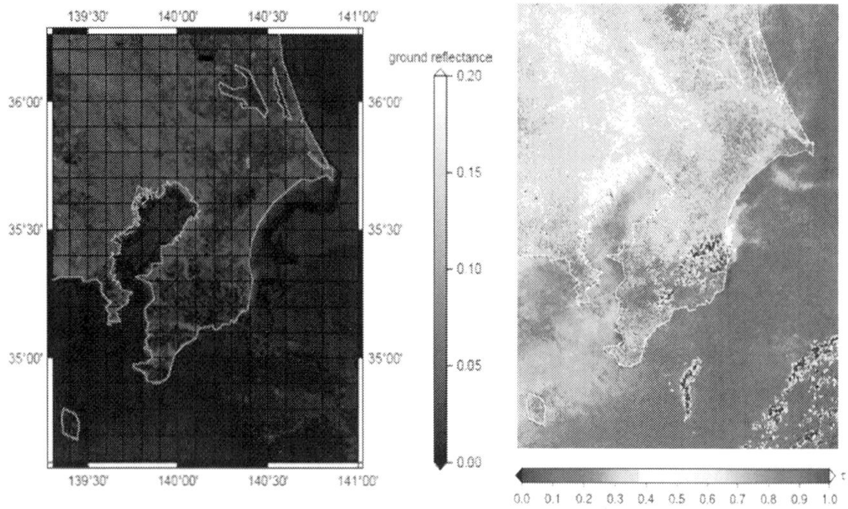

Fig. 12. Analysis of MODIS data in November 2007: (a) surface reflectance map (540 - 570 nm), and (b) aerosol optical thickness at 550 nm on 24 November 2007.

6. Conclusion

Optical properties of aerosols and clouds play an important role in the consideration of the Earth's radiation budget. In this chapter, we have described multi-wavelength and multi-directional remote sensing of the troposphere, putting emphasis on the visible part of the spectrum. The DOAS approach enables the direct observation of air pollutants by employing a nearly horizontal light path in the lowest part of troposphere, where the highest concentrations of pollutants such as NO_2 and aerosol (SPM) are found. The observation of direct solar radiation and scattered solar radiation (sky light), on the other hand, is useful for retrieving detailed aerosol optical properties under clear-sky conditions. Thus, the data can be quite useful for implementing precise atmospheric correction on satellite-observed imagery that includes the ground observation point. The multi-wavelength lidar observation provides an efficient tool to elucidate the vertical profiles of aerosol particles. By combining the lidar data with some appropriate ancillary data such as the ground-level characterization of aerosol properties, it becomes possible to derive useful information on temporal as well as spatial information on aerosol and cloud characteristics in the atmosphere.

7. Acknowledgment

We acknowledge the financial support of the Grant-in-Aid from the Ministry of Education, Culture, Sports, Science & Technology in Japan (#21510006). Also, contributions of a number of graduate students who participated in various researches presented in this chapter are gratefully acknowledged.

8. References

Fernald, F. G., Analysis of atmospheric lidar observation: some comments. Appl. Opt. Vol 23, (1984), pp. 652-653, ISSN 0003-6935

Fukagawa, S., Kuze, H., Bagtasa, G., Naito, S., Yabuki, M., Takamura, T., Takeuchi, N., Characterization of seasonal and long-term variation of tropospheric aerosols in Chiba, Japan. Atmospheric Environment, Vol. 40, No.12, (2006), pp. 2160-2169, ISSN 1352-2310

Intergovernmental Panel on Climate Change (IPCC), *Climate Change 2007: The Physical Science Basis*, Cambridge Univ. Press (2007) ISBN-13: 978-0521705967, Cambridge, U.K.

Kaufman, Y. J., Tanré, D., Gordon, H. R., Nakajima, T., Lenoble, J., Frouins, R., Grassl, H.,Herman, B. M., King, M. D., Teillet, P. M., Passive remote sensing of tropospheric aerosol and atmospheric correction for the aerosol effect. J. Geophys. Res., Vol.102, No.D14, (1997), pp. 16,815-16,830, ISSN 0148-0227

Kinjo, H., Kuze, H., Takeuchi, N., Calibration of the Lidar Measurement of Tropospheric Aerosol Extinction Coefficients. Jpn. J. Appl. Phys. Vol. 38, (1999), pp. 293-297, ISSN 0021-4922

Kotchenova, S.Y., Vermote, E.F., Matarrese, R., Klemm, Jr. F.J., Validation of a vector version of the 6S radiative transfer code for atmospheric correction of satellite data. Part I: Path radiance. Applied Optics Vol. 45, No. 26, No.10 (September 2006), pp.6762-6774, ISSN 0003-6935

Kuriyama, K., Kaba, Y., Yoshii, Y., Miyazawa, S., Manago, N., Harada, I., Kuze, H., Pulsed differential optical absorption spectroscopy applied to air pollution measurement in urban troposphere. J. Quantitative Spectroscopy & Radiative Transfer, Vol. 112, (2011), pp. 277-284, ISSN 0022-4073

Lee, H.; Kim, Y.J., Jung, J., Lee, C., Heue, K.P., Platt, U., Hu, M., 6 Zhu, T., Spatial and temporal variations in NO_2 distributions over Beijing, China measured by imaging differential optical absorption spectroscopy. J. Environ. Management, Vol. 90, No.5, (April 2009), pp. 1814-1823, ISSN 0301-4797

Manago, N., Kuze, H., Determination of tropospheric aerosol characteristics by spectral measurements of solar radiation using a compact, stand-alone spectroradiometer. Appl. Opt. Vol. 49, (2010), pp. 1446-1458, ISSN 1539-4522

Manago, N., Miyazawa, S., Bannu, Kuze, H., Seasonal variation of tropospheric aerosol properties by direct and scattered solar radiation spectroscopy. J. Quantitative Spectroscopy & Radiative Transfer, Vol. 112, Issue 2, (January 2011), pp. 285-291, ISSN 0022-4073

Miyazawa, S., Manago, N., Kuze, H., Precice atmospheric correction of MODIS data using aerosol optical parameters derived from simultaneous ground measurement, in Proceedings of the 49th Annual Meeting of the Remote Sensing Society of Japan, (2010), pp.245-246 (*in Japanese*)

Seinfeld, H. and Pandis, S.N. (1998). *Atmospheric Chemistry and Physics, from Air Pollution to Climate Change*, John Wiley, ISBN 0-471-17815-2, New York, U.S.A.

Si, F., Kuze, H., Yoshii, Y., Nemoto, M., Takeuchi, N., Kimura, T., Umekawa, T., Yoshida, T., Hioki, T., Tsutsui, T., Kawasaki, M., Measurement of regional distribution of atmospheric NO_2 and aerosol particles with flashlight long-path optical monitoring. Atmos. Environ. Vol. 39, (2005), pp. 4959-4968, ISSN 1352-2310

Takamura, T., Nakajima, T., Overview of SKYNET and its activities. Optica Pura y Aplicada Vol. 37, (2004), pp. 3303–3308, ISSN 0030-3917

Tang, J., Xue, Y., Yu, T., Guan,Y. Aerosol optical thickness determination by exploiting the synergy of TERRA and AQUA MODIS. Remote Sensing of Environment Vol. 94, (2005), pp. 327–334, ISSN 0034-4257

Todate, Y., Minomura, M., Kuze, H., Takeuchi, N., On the retrieval of aerosol information on the land surface from Landsat-5/TM data, in Proceedings of the 36th Annual Meeting of the Remote Sensing Society of Japan, (2004), pp.57 - 58. (*in Japanese*)

Yoshii, Y., Kuze, H., Takeuchi, N., Long-path measurement of atmospheric NO_2 with an obstruction flashlight and a charge-coupled-device spectrometer. Appl. Opt., Vol. 42, (2003), pp. 4362-4368, ISSN 0003-6935

Section 3

Oceans and Cryosphere

12

Remote Sensing of Submerged Aquatic Vegetation

Hyun Jung Cho[1], Deepak Mishra[2,3] and John Wood[4]
[1]*Department of Integrated Environmental Science, Bethune-Cookman University, Daytona Beach, FL*
[2]*Department of Geosciences, Mississippi State University*
[3]*Northern Gulf Institute and Geosystems Research Institute, Mississippi State University, MS State, MS*
[4]*Harte Research Institute for Gulf of Mexico Studies, Texas A&M University-Corpus Christi, Corpus Christi, TX*
USA

1. Introduction

Remote sensing has significantly advanced spatial analyses of terrestrial vegetation for various fields of science. The plant pigments, chlorophyll *a* and *b*, strongly absorb the energy in the blue (centered at 450 nm) and the red (centered at 670 nm) regions of the electromagnetic spectrum to utilize the light energy for photosynthesis. In addition, the internal spongy mesophyll structures of the healthy leaves highly reflect the energy in the near-infrared (NIR) (700- 1300) regions (Jensen, 2000; Lillesand et al., 2008). The distinctive spectral characteristics of the green plants, low reflectance in the visible light and high reflectance in NIR have have been used for mapping, monitoring and resource management of plants; and also have been used to develop spectral indices such as Simple Vegetation Index (SVI = NIR reflectance - red reflectance) and Normalized Difference Vegetation Index (NDVI = (NIR reflectance - red reflectance)/(NIR reflectance + red reflectance)) (Giri et al., 2007).

The simplicity and flexibility of vegetation indices allow comparison of data obtained under varying light conditions (Walters et al., 2008). NDVI was first suggested by Ruose et al. (1973) and is one of the earliest and most popular vegetation index used to date. It is usually applied in an attempt to decrease the atmospheric and surface Bidirectional Reflectance Distribution Function (BRDF) effects by normalizing the difference between the red and NIR reflectance by total radiation. Index values have been associated with various plant characteristics, including vegetation type (Geerken et al., 2005), vegetation cover (du Plessis, 1999), vegetation water content (Jackson et al., 2004), biomass and productivity (Fang et al., 2001), chlorophyll level (Wu et al., 2008), PAR absorbed by crop canopy (Goward & Huemmrich, 1992), and flooded biomass (Beget et al., 2007) at a broad span of scales from individual leaf areas to global vegetation dynamics.

2. Remote sensing of submerged aquatic vegetation (SAV)

2.1 Submerged aquatic vegetation (SAV)

Submerged aquatic vegetation (SAV) is a group of vascular plants that grow underwater which can grow to the surface of, but not emerge from shallow waters. SAV includes seagrass species that are a vital component of ecological processes, dynamics, and productivity of coastal ecosystems. Healthy beds of SAV provide nursery and foraging habitats for juvenile and adult fish and shellfish, protect them from predators, provide food for waterfowl and mammals, absorb wave energy and nutrients, produce oxygen and improve water clarity, and help settle suspended sediment in water by stabilizing bottom sediments (Jin, 2001; Findlay et al., 2006). Assessment of SAV distribution, composition, and abundance has been of a particular interest to coastal environmental managers, scientists, developers, and recreational users as the information serves as an excellent indicator of aquatic environmental quality.

2.2 Remote sensing of underwater habitats

Remote sensing is a valuable tool for monitoring benthic habitats such as SAV, benthic algae, and coral-reef ecosystems, and several researchers have tested airborne and space-borne sensor systems for such studies (e.g., Mishra et al., 2005). Spatial resolutions of these systems range from 30 m for the Landsat Thematic Mapper (TM) to 2.44 m for QuickBird multispectral data and 1 m or less for airborne hyperspectral data. Those evaluating the utility of TM have mapped subtidal coastal habitats (Khan et al., 1992), delineated sand bottoms (Michalek et al., 1993), classified coral reef zones (Mishra et al., 2005, 2006), evaluated the benthos (Matsunaga & Kayanne, 1997), and performed time series analyses (Dobson & Dustan, 2000). Similarly, researchers have used IKONOS (4 m) and QuickBird (2.44 m) imagery with radiative transfer models to map benthic habitats (Mishra et al., 2005, 2006) and apply a similar model to Airborne Imaging Spectroradiometer for Applications (AISA) hyperspectral data to identify benthic habitats (Mishra et al., 2007).

Mapping of SAV using satellite data has focused on supervised and unsupervised classifications based on signal variations in the multispectral bands, especially those in the short visible wavelengths with high water penetration (Ackleson & Klemas, 1987; Lyzenga, 1981; Marshall & Lee, 1994; Maeder et al., 2002; Ferguson & Korfmacher, 1997; Pasqualini et al., 2005). The NIR region is seldom used due to its high attenuation in water. When SAV beds are dense, the water is clear, and depth and sediment relatively constant, fine-scale spectral variation is often overlooked during classification. In other cases, the radiative transfer model is used to correct the solar angle, atmospheric perturbation, substrate type, and depth, but requires extensive *in situ* measurements (Zimmerman & Mobley, 1997).

Most of the currently available radiative transfer models or physics-based models have been applied to map benthic features in relatively clear aquatic environments (i.e. relatively deep, pristine coral reefs or seagrass meadows) and do not adequately correct for the strong NIR absorption by water (Mumby et al., 1998; Holden & LeDrew, 2001; Holden & LeDrew, 2002; Ciraolo, 2006; Brando et al., 2009). However, NIR reflectance serves as the primary cue for discriminating vegetation type and as the critical component for the widely used vegetation indices.

2.3 Dilemmas in remote sensing of shallow aquatic system and SAV

Remote sensing of benthic habitats is complicated because of several factors including (1) atmospheric interferences, (2) variability in water depth, (3) water column attenuation, and (4) variability in bottom albedo or bottom reflectance. In the case of aquatic remote sensing, the total signal received at satellite altitude is dominated by radiance contributed by atmospheric scattering, and only 8-10% of the signal corresponds to the water reflectance and reflectance from benthic features (Kirk, 1994, Mishra et al., 2005). Therefore, it is advisable to correct for atmospheric effects to retrieve any quantitative information for surface waters or benthic habitats from satellite images. Therefore, the lack of a rigorous absolute atmospheric correction procedure can introduce significant errors to a satellite derived benthic habitat map. There is also a tendency among benthic mapping researchers to use a relative atmospheric correction procedure such as a deep-water pixel correction, especially when local aerosol data and validation data are lacking. This often causes mediocre classification results.

Knowledge of the optical properties of the water column can help eliminate changes in reflectance attributable to variable depth and water column attenuation effects, which often lead to misclassification of the benthos (Mishra et al., 2005). Mishra et al (2005) proved that to derive accurate bottom albedo or bottom reflectance using a radiative transfer model, water depth and water column optical properties (absorption and backscattering) should be known for the study area. Knowledge of optical bottom albedo for shallow waters is necessary to model the underwater and above-water light field, to enhance underwater object detection or imaging, and ultimately to determine the distribution of benthic habitats (Gordon & Brown, 1974). Mishra et al (2005, 2006, 2007) also point out that the signals measured by a sensor from above the water surface of a shallow marine environment are highly affected by phytoplankton abundance (chlorophyll absorption), water column interactions (absorption by water and scattering by suspended sediments), and radiance reflected from the bottom. For the bottom contribution to be retrieved by a sensor the water column contributions have to be removed and the optical properties have to be known or at least be derivable. However, it is very challenging to measure these optical properties accurately because of logistical issues and instrumentation errors, which also leads to an inaccurate benthic mapping project.

Variability in bottom types and hence albedo gives rise to a mixed spectral response that often reduces the classification accuracy. Specific problems such as complex benthic combinations (e.g., sandy areas with variable amounts of algal cover; variation in color, texture, size), and error in depth estimation can also have a considerable impact on the classification results. Mishra et al (2005) proposed several solutions to increase the number of elements separable by a classification scheme and the classification accuracies including an extensive field campaign acquiring substantial samples to enable statistical evaluation for each class and deriving detailed ecological and biological information for each *in situ* data point. Close-range hyperspectral studies that may aid in discriminating between different types of benthic features can be used to develop baseline spectra to help minimize spectral confusion in satellite imagery.

Shallow littoral areas (generally the areas between the shoreline to a water depth 2 m) are one of the most productive habitats, yet the most sensitive landscape to human-induced environmental alteration and global climate changes. Modeling of optical water properties for the littoral zone is more complicated due to rapidly changing water depth and/or substrate and higher amounts of Colored Dissolved Organic Matter (CDOM) and/or suspended particles (phytoplankton, seston, and inorganic particles) compared to the deeper portions of oceans. In addition, bottom backscattering in the shallow areas is more significant, which makes the NIR signals more important, especially in areas that contain substantial amount of seagrasses, benthic algae, or phytoplankton (Kutser et al., 2009) and that the conventional Beer-Lambert's exponential light attenuation with depth is not applicable (Holden & LeDrew, 2002).

Upwelling signals from water bodies contain several components including reflectance from water surface, water column (suspended matter), and bottom backscattering (SAV and substrate) (Spitzer & Dirks, 1987). The aforementioned conventional vegetation indices also are not effectively used for plants that grow underwater or that are temporarily flooded (Beget & Di Bella, 2007; Cho et al. 2008) because the water overlying the vegetation canopies reduces the vegetation effects of 'red absorption' and the 'NIR reflectance' (Han & Rundquist, 2003; Cho, 2007; Cho et al., 2008; Fig. 1). Differentiation of the SAV spectral signature from bare substrate or algae is further limited in shallow coastal waters that are more turbid than open ocean waters (Bukata, 1995) due to higher levels of phytoplankton, suspended sediment, and dissolved color. According to our on-going study using hyperspectral data obtained over both experimental tanks and field seagrass habitat, the SAV signal rapidly decreases as water depth increases, and almost completely disappears within a depth of 0.5 meter in even mildly turbid waters (turbidities > 12 NTU).

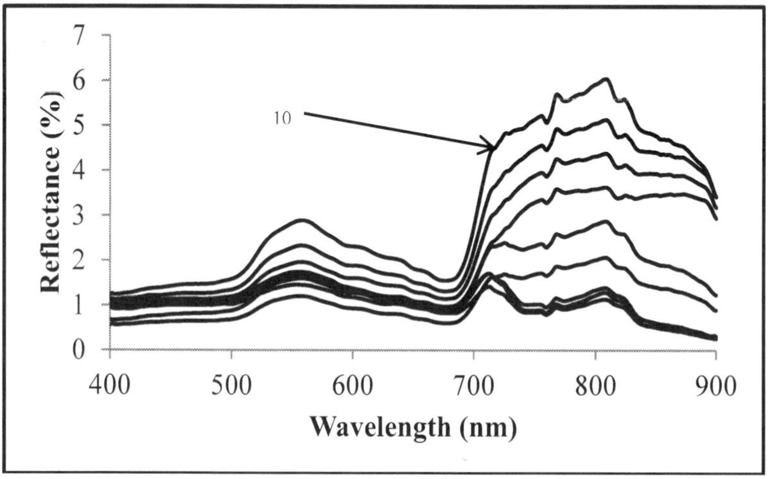

Fig. 1. Depth-induced reflectance variation of submerged aquatic vegetation (SAV) in clear water between 10 cm and 50 cm above the SAV canopy. The line for the highest reflectance is at 10 cm and the reflectance continuously decreases with water depth increases.

2.4 SAV mapping using hyperspectral data

Two decades ago, only spectral remote sensing experts had access to hyperspectral images or the software tools necessary to take advantage of such images. Over the past decade, hyperspectral image analysis has matured into one of the most powerful and fastest growing technologies in the field of remote sensing (Phinn et al., 2008). While multispectral remote sensing systems detect radiation in a small number of broad regions of the electromagnetic spectrum, hyperspectral sensors acquire numerous very narrow, contiguous spectral bands throughout the visible, near-infrared, mid-infrared, and thermal infrared portions of the electromagnetic spectrum for every pixel in the image, yielding much more detailed spectral data (Govender et al., 2009).

Collection and processing of hyperspectral imagery can be quite costly, depending on the size of the area to be studied. In order for the imagery to be usable for sub-aquatic analysis, the following guidelines are suggested by Finkbeiner et al. (2001):

- The best time of year for collecting hyperspectral imagery may occur in early summer, during the season of maximum biomass, and when there is less epiphytic coverage.
- The imagery should be collected when turbidity is low; this is often during times of low or no winds. High turbidity may also be caused by heavy rains, winds on previous days, and localized dredging. Often, boat traffic may cause a localized but far-spreading plume of turbidity, as sediments are re-suspended.
- Winds can also cause problems other than turbidity, such as wrack lines, debris lines, whitecaps, and areas with unacceptable amounts of glint. As a general rule, winds less than 8 kph are acceptable, winds between 8-15 kph may be acceptable depending on the locality, and winds higher than 15 kph are generally unacceptable.
- Tidal stage can play an important role in the success of imagery collection. Consult local and/or NOAA tide gauges to plan for acquisition within 2 hours of the lowest tide for the collection area, unless the estuary drains an area of highly turbid or tannic water, in which case, a rising tide may be desirable.
- Collection times should be planned to adjust for sun angle, to avoid both sun glint and shadows. As a general rule, sun angles between 30° and 45° are recommended; different sensors may allow or require more or less angle.
- Clouds and haze create areas of shadows and distortion as well as white or gray streaks in the imagery, and should be avoided as much as possible.
- Field work should occur simultaneously with the sensor flight. Since it is virtually impossible to collect all the field data needed for signature development and accuracy assessment in the same time frame as the flights occur, every effort should be taken to gather field measurements as close to the actual flight as possible, and under similar conditions.
- Field data should include measurements of reflectivity, turbidity, empirical or anecdotal data on epiphytic coverage, bottom type and reflectance, classification of the field point, and precise location. Locate these field measurements within a large enough patch that there will be no ambiguity, and consider the spatial sphere of uncertainty. For instance, if the imagery will have a radiometric accuracy of approximately 2 m, the location should be consistent out to a four meter radius.

The unique spectral signatures of vegetation are often used as training data for hyperspectral imagery classification. Chlorophyll and other pigments are found in SAV as in other photosynthesizing vegetation, however, the ratio of these to each other will differ by species, as well as with changes in conditions and stressors (Govender et al., 2009).

While these minor differences can be detected above the surface in spite of epiphytic coverage (Fyfe, 2003), detection of these differences below the surface may be hampered or dampened by the effects of the water column. Depth, water clarity or turbidity, organic and inorganic materials within the water column, the surface of the water itself, and physical properties such as the absorption of energy in the NIR and beyond can all affect the ability to discriminate the relatively small differences in ratios of accessory pigments and chlorophyll (Kutser 2004).

3. Case study in SAV mapping using hyperspectral data

3.1 Hyperspectral algorithm to correct overlying water effects

A new water-depth correction algorithm was developed to improve detection of underwater vegetation spectra signals. The algorithm was developed conceptually, calibrated, and validated using experimental and field data. The conceptual model was based on the idea that the upwelling signals measured from a water surface is the sum of the energy reflected from the water surface, the water column, and also from the water bottom surface. The energy reflected from the water surface and the water column (the volumetric reflectance) was combined as a single term because the surface reflectance is a constant and does not change with water depth changes (Lu & Cho, 2011).

The effects of the overlying water column on upwelling hyperspectral signals were modeled by empirically separating the energy absorbed and scattered by the water using data collected through a series of controlled experiments using hypothetical bottom surfaces that either 100% absorbs or 100% reflects (Cho & Lu, 2010). Later, the white surface (the 100% reflecting surface) was replaced with a gray surface with a known reflectance to reduce problems associated with enhanced multi-path scattering (Lu & Cho, 2011). The experimental setting allowed the calculation of water absorption and scattering values for up to a water depth of 60 cm, and the light remaining at water depths that were beyond the experimentally measured points were estimated by establishing the mathematical relationships between water depth and the vertical attenuation coefficient (K_d) derived from the experimental data (Washington et al., 2011). The depth- and wavelength-dependent water absorption and volumetric scattering factors (0-100 cm; 400 - 900 nm) were calculated and applied to independently-measured underwater vegetation signals and airborne hyperspectral data taken over shallow seagrass beds, to remove the effects of the overlying water.

3.2 Successful water correction in the infrared region

The empirically driven correction algorithm significantly restored the vegetation signals, especially in the NIR region, when applied to independently measured reflectance of underwater plants taken over indoor and outdoor tanks (Cho & Lu, 2010; Washington et al. 2011). The algorithm was also successful in restoring the NIR signals originating from seagrass-dominated sea floors when applied to airborne hyperspectral data of Mississippi and Texas coastal waters (Cho et al., 2009; Lu & Cho, 2011; Fig. 2). As stated earlier, NIR reflectance serves as the primary cue for discriminating benthic vegetation from other substrates. Due to the restored NIR reflectance, the correction algorithm increased the NDVI values for the seagrass pixels (Lu & Cho 2011).

3.3 Seagrass classification using water corrected image

3.3.1 Ground truth data collection

Several hundred ground data points were collected over seagrass beds in Redfish Bay, Texas, in the summer of 2008 (June - July). Seagrass species makeup, water depth, vegetation percent coverage, and bottom substrate type were recorded at each site. Site

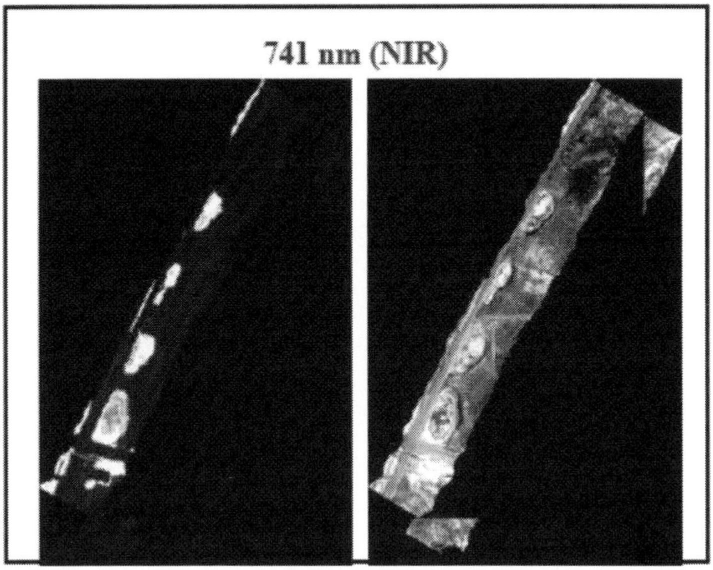

Fig. 2. The original (left) and water-corrected (right) airborne AISA Eagle hyperspectral image at 741 nm obtained over seagrass beds in Redfish Bay, TX in 2008.

location was recorded to accuracies within 1 m using a Real Time Kinetic (RTK) GPS. When necessary, the preselected random sites were shifted to avoid dry or unreachable locations.

The field collected data were entered into a spatial database along with descriptive attributes to help determine which class each sampling site would be assigned to. Data points were then randomly divided into training or accuracy assessment points.

3.3.2 Image processing and vector classification

Image data were obtained in 63 bands of the AISA Eagle Hyperspectral sensor over the seagrass beds in October 2008 and corrected for atmospheric effects. Since selection of the proper bands for analysis helps reduce noise introduction and processing burden (Borges et al. 2007), several selection techniques were used within this project, including Principal Component Analysis and regression analysis. Ultimately, seven bands recommended by Fyfe (2003) were used. To reduce noise, these were again reduced to 5 bands, as recommended by Cho et al. (2009).

Image segmentation is performed prior to image classification. Segmentation groups like pixels into homogenous areas. Initially, an unsupervised classification using ISODATA (Iterative Self Organized Data Analysis Technique A) (De Alwis et al., 2007) was performed. After the initial image classification, each segmented vector in the output was assigned to a seagrass/substrate class (i.e. *Thalassia testudinum, Halodule wrightii, Ruppia maritima,* Mixed Beds, Bare, or Unclassified). Those which contain only one type of point ('*Halodule*') were considered to be finally classified. Those classified as mixed or unknown were removed from the classified vector set, a mask created of their spatial footprint, and the entire classification procedure re-run on the image for that footprint area only. When it became impossible to further classify the image by this method, a supervised classification was performed, using a selection of training points as training data, which produced monospecific vectors. The same mixed-method classification procedure was performed to the image data after the water correction algorithm was applied.

3.3.3 Improved accuracy assessment

The water correction algorithm improved the classification accuracy results in an image subset from an overall accuracy of 28% to approximately 36%. Identification of the species *Halodule wrightii* improved from 33% user's accuracy to almost 78%, and *Thalassia testudimun* from 0% users accuarcy to almost 17%. Although these numbers appear to be somewhat low, several factors must be considered: this analysis only used a subset of the imagery, which allowed the area analysed to have less variation, but there was also less training and accuracy assessment data to work with.

Using the full set of imagery and the combination of supervised and unsupervised classification, results have improved considerably; we have achieved overall accuracies of over 60%. In addition, we have calculated seagrass presence/absence using the complete corrected imagery, achieving overall accuracies of over 95%. With the future addition of *in situ* turbidity and bathymetric data, these accuracies should continue to improve.

4. Current efforts in developing water correction module and graphical user interface

We have continued improving the algorithm by including turbidity (measured in NTU, Nephelometric Turbidity Unit) as an additional function. In addition, the water correction algorithm is currently being implemented as a module that can be called from the ENVI (ENvironment for Visualizing Images, ITT Visual Information Solutions)'s programming environment using a Graphical User Interface (GUI) (Gaye et al. 2011). Under the current module and the GUI, users are able to select water depth (0 - 2.0 m) and turbidity (0-20 NTU) using slide bars, for which (a) given hyperspectral band(s) can be corrected. The corrected reflectance can be generated and compared to the original one at either a given pixel, within a small subset; and the original and corrected images can be displayed.

5. Conclusion

Remote sensing has significantly advanced spatial analyses on terrestrial vegetation for various fields of science. However, mapping of benthic vegetation or submerged aquatic vegetation (SAV), using remotely sensed data is complicated due to several factors including atmospheric interferences, variability in water depth and bottom albedo, and water column attenuation by scattering and absorption. Hence, correction for the atmospheric and the overlying water column effects is necessary to retrieve any quantitative information for SAV from satellite and airborne images, especially when using hyperspectral data. Significant misclassification of the SAV often occurs due to the lack of information on *in situ* water depths and water column optical properties. Most of the currently available radiative transfer models only work well when applied to mapping of benthic features in relatively clear aquatic environments, but they do not correct for strong water absorption of the near infrared energy. The fluctuating water depths, high amounts of suspended particles and colored dissolved organic matter in shallow littoral zones make it even more challenging to map benthic vegetation using remotely sensed data. A new water-depth correction algorithm was developed conceptually, and calibrated and validated using experimental and field data. The effects of the overlying water column on upwelling hyperspectral signals were modeled by empirically separating the energy absorbed and scattered by the water using data collected through a series of controlled experiments. The empirically driven algorithm significantly restored the vegetation signals, especially in the NIR region. Due to the restored NIR reflectance, which serves as the primary cue for discriminating SAV from other substrates, use of the water corrected airborne data increased the NDVI values for the SAV pixels and also improved the seagrass classification accuracy. Our continuing efforts to incorporate turbidity and CDOM into the algorithm, in developing a graphical user interface, and in implementing the algorithm into a module that can be called from commercially available image processing software promise a user-friendly application and wide use of the algorithm in the near future.

6. Acknowledgment

The work was supported by grants from the National Geospatial-Intelligence Agency (Grant No. HM1582-07-1-2005 and HM1582-08-1-0049) and National Oceanic and Atmospheric Administration-Environmental Cooperative Sciences Center (Grant No.NA17AE1626).

7. References

Ackleson, S. G., & Klemas, V. (1987). Remote Sensing of Submerged Aquatic Vegetation in Lower Chesapeake Bay: A Comparison of Landsat MSS to TM imagery. *Remote Sensing of Environment*, Vol.22, No.2, pp. 235−248, ISSN 0034-4257

Beget, M. & Di Bella, C. (2007). Flooding: The Effects of Water Depth on Spectral Response of Grass Canopies. *Journal of Hydrology*, Vol.335, No.3-4, pp. 285-294. ISSN 0022-1694

Borges, J.; Marçal, A. & Dias, J. (2007). Evaluation of feature extraction and reduction methods for hyperspectral images. In: *New Developments and Challenges in Remote Sensing*, Bochenek, Z., pp. 265-284, Millpress, ISBN 978-90-5966-053-3, Rotterdam

Brando, V.E.; Anstee, J.M.; Wettle, M.; Dekker, A.G.; Phinn, S.R. & Roelfsema, C. (2009). A Physics Based Retrieval and Quality Assessment of Bathymetry from Suboptimal Hyperspectral Data. *Remote Sensing of Environment*, Vol.113, No.4, pp. 755-770, ISSN 0034-4257

Bukata, R.; Jerome, J.; Kondratyev, K. & Pozndyakov, D. (1995). Optical Properties and Remote Sensing of Inland Coastal Waters. New York: CRC Press.ISBN 978-0849347542 Cho, H.J. & Lu, D. (2010). A Water-depth Correction Algorithm for Submerged Vegetation Spectra. *Remote Sensing Letters*, Vol.1, No.1, pp. 29-35, ISSN 2150-7058

Cho, H.J., & Lu, D. (2010). A Water-depth Correction Algorithm for Submerged Vegetation Spectra. Remote Sensing Letters, Vol.1, No.1, pp. 29-35

Cho, H.J.; Lu, D. & Washington, M. (2009). Water Correction Algorithm Application for Underwater Vegetation Signal Improvement. *Proceedings of the 2009 MS Water Resource Conference*, pp. 154-157

Cho, H.J.; Kirui, P. & Natarajan, H. (2008). Test of Multi-spectral Vegetation Index for Floating and Canopy-forming Submerged Plants, *International Journal of Envirnmental research and Public Health*, Vol.5, No.5, pp. 447-483, ISSN 1661-7827

Cho, H. (2007). Depth-variant Spectral Characteristics of Submerged Aquatic Vegetation (SAV) detected by Landsat 7 ETM+. *International Journal of Remote Sensing*, Vo.28, No.7, pp. 1455-1467, ISSN 1366-5901

Ciraolo, G.; Cox, E.; La Loggia, G. & Maltese, A. (2006). The Classification of Submerged Vegetation Using Hyperspectral MIVIS Data. *Annals of Geophysics*, Vol. 49, No.1, pp. 287-294, ISSN 2037-416X

De Alwis, D.; Easton, Z., Dahlke, H., Philpot, W., & Steenhuis, T. (2007). Unsupervised Classification of Saturated Areas using a Time Series of Remotely Sensed Images. *Hydrology and Earth System Sciences* Vol 11, No. 511, pp. 1609-1620, ISSN 1027-5606

Dobson, E. & Dustan, P. (2000). The Use of Satellite Imagery for Detection of Shifts in Coral Reef Communities. *Proceedings, American Society for Photogrammetry and Remote Sensing*, Washington, DC, USA, May 22-26, 2002, (CD-ROM)

du Plessis, W.P. (1999). Linear Regression Relationships between NDVI, Vegetation and Rainfall in Etosha National Park, Namibia. *Journal of Arid Environments*, Vol.42, No.4, pp. 235-260, ISSN 0140-1963

Fang J.Y.; Piao S.L.; Tang Z.Y.; & Peng J.W. (2001). Interannual Variability in Net Primary Production and Precipitation. *Science*, Vol.29, No.5526, pp. 293:1723, ISSN 0036-8075

Ferguson, R.L.; & Korfmacher, K (1997). Remote Sensing and GIS Analysis of Seagrass Meadows in North Carolina, USA. *Aquatic Botany*, Vol.58, No. 3-4, pp. 241-258, ISSN 0304-3770

Findlay, S.E.G.; Nieder, W.C.; Blair, E.A. & Fischer, D.T. (2006). Multi-scale Controls on Water Quality Effects of Submerged Aquatic Vegetation in the Tidal Freshwater Hudson River. *Ecosystems*, Vol.9, No.1, pp. 84-96, ISSN 1432-9840

Finkbeiner, M.; Stevenson, B., & Seaman, R. (March 2001). *Guidance for Benthic Habitat Mapping: An Aerial Photographic Approach*. U.S. NOAA Coastal Services Center. http://www.csc.noaa.gov/benthic/mapping/pdf/bhmguide.pdf

Fyfe, S. (2003). Spatial and Temporal Variation in Spectral Reflectance: Are Seagrass Species Spectrally Distinct? *Limnology and Oceanography* Vol. 48, No. 1, pp. 464-479, ISSN 1939-5590

Gaye, G.; Kim, H.; Cho, H.J. (2011). Utilizing the Significant Spectral Bands for Image Outputs for SAV from the Water Effect Correction Module. Proceeding of the 2011 WORLDCOMP (CD-ROM)

Geerken, R.; Zaitchik, B.; & Evans, J.P. (2005). Classifying Rangeland Vegetation Type and Fractional Cover of Semi-arid and Arid Vegetation Covers from NDVI Time-series. *International Journal of Remote Sensing*, Vol.26 No.24, pp. 5535–5554, ISSN 1366-5901

Giri, C.; Pengra, B.; Zhu, Z.; Singh, A. & Tieszen, L. (2007). Monitoring Mangrove Forest Dynamics of the Sundarbans in Bangladesh and India Using Multitemporal Satellite Data from 1973-2000. *Estuarine, Coastal and Shelf Science*, Vol. 73, No.1-2, pp. 91-100. ISSN 0272-7714

Gordon, H.R. & Brown, O.B. (1974). Influence of Bottom Albedo on the Diffuse Reflectance of a Flat Homogeneous Ocean. *Applied Optics*, Vol.13, No.9, pp. 2153-2159, ISSN 1559-128X

Govender, M.; Chetty, K. & Bulcock, H. (2009). A Review of Hyperspectral Remote Sensing and Its Application in Vegetation and Water Resource Studies. *Water SA* Vol. 33, pp. 145-152, ISSN 1816-7950

Goward, S.N.; & Huemmrich, K.F. (1992). Vegetation Canopy PAR Absorptance and the Normalized Difference Vegetation Index: An Asssessment Using the SAIL model. *Remote Sensing of Environment*, Vol.39, No.2, pp. 119-140, ISSN 0034-4257

Han, L. & Rundquist, D.C. (2003). The Spectral Responses of Ceratophyllum demersum at Varying Depths in an Experimental Tank. International Journal of Remote Sensing, Vol. 24, pp. 859-864, ISSN 1366-5901

Holden, H. & LeDrew, E. (2001). Effects of the Water Column on Hyperspectral Reflectance of Submerged Coral Reef Features. *Bulletin of Marine Science,* Vol.69, No.2, pp. 685-699, ISSN 0007-4977

Holden, H. & LeDrew, E. (2002). Measuring and Modeling Water Column Effects on Hyperspectral Reflectance in a Coral Reef Environment. *Remote Sensing of Environment,* Vol. 81, No. 2-3, pp. 300-308, ISSN 0034-4257

Jackson, T.J.; Chen, D.; Cosh, M.; Li, F.; Anderson, M.; Walthall, C.; Doriaswamy, P. & Hunt, E.R. (2004). Vegetation Water Content Mapping Using Landsat Data Derived Normalized Difference Water Index for Corn and Soybeans. *Remote Sensing Of Environment,* Vol.92, No..3 pp. 475–482, ISSN 0034-4257

Jensen, J.R. (2000). *Introductory Digital Image Processing: A Remote Sensing Perspective.* Prentice Hall, Inc. Upper Saddle River, NJ 07458, ISBN-13: 978-0131453610, Upper Saddle River, NJ

Jin, X. (2001). *Technologies of Lake Eutrophication Control management.* Beijing: Chemistry Industry Press.

Khan, M.A.; Fadlallah, Y.H. & AL-Hinai, K.G. (1992). Thematic Mapping of Subtidal Coastal Habitats in the Western Arabian Gulf Using Landsat TM Data - Abu Ali Bay, Saudi Arabia. *International Journal of Remote Sensing.* Vol.13, No.4, pp. 605 - 614, ISSN 1366-5901

Kirk, J.T.O. (1994). *Light and Photosynthesis in Aquatic Ecosystems.* Cambridge University Press, ISBN 978-0-521-15175-7, Cambridge, UK

Kutser, T. (2004). Quantitative Detection of Chlorophyll in Cyanobacterial Blooms by Satellite Remote Sensing. *Limnology and Oceanography* Vol.49, No. 6, pp. 2179-2189, ISSN 1939-5590

Kutser, T.; Vahtmae, E. & Praks, J. (2009). A Sun Glint Correction Method for Hyperspectral Imagery Containing Areas with Non-negligible Water Leaving NIR Signal. *Remote Sensing of Environment,* Vol.113, No.110, pp. 2267-2274, ISSN 0034-4257

Lillesand, T.M., Kiefer, R.W., and Chipman, J.W. (2008). *Remote Sensing and Image Interpretation.* John Wiley and Sons, Inc., 111 River Street, Hoboken: NJ , ISBN-13 978-0470052457

Lu, D. & Cho, H.J. (2011). An Improved Water-depth Correction Algorithm for Seagrass Mapping Using Hyperspectral Data. *Remote Sensing Letters,* Vol.2, No.2, pp. 91-97, ISSN 2150-7058

Lyzenga, D. (1981). Remote Sensing of Bottom Reflectance and Water Attenuation Parameters in Shallow Water Using Aircraft and Landsat data. *International Journal of Remote Sensing,* Vol.2, No.1, pp. 71-82, ISSN 1366-5901

Maeder, J.; Narumalani, S.; Rundquist, D.C.; Perk, R.L.; Schalles, J.; Hutchins, K. & Keck. J. (2002). Classifying and Mapping General Coral-reef Structure Using Ikonos Data. *Photogrammetric Engineering & Remote Sensing,* Vol.68, No.12, pp. 1297-1305, ISSN 0099-1112

Marshall, T., & Lee, P. 1994. An Inexpensive and Lightweight Sampler for the Rapid Collection of Aquatic Macrophytes. *Journal of Aquatic Plant Management,* Vol.32, pp. 77-79, ISSN 0146-6623

Matsunaga, T. & Kayanne, H. (1997). Observation of Coral Reefs on Ishigaki Island, Japan, Using Landsat TM Images and Aerial Photographs, *Proceedings of the Fourth*

International Conference on Remote Sensing for Marine and Coastal Environments, Vol. I, pp. 657-666, ISSN 1066-3711, Orlando, Florida, USA, March 17-19, 1997

Michalek, J.; Wagner, T.; Luczkovich, J. & Stoffle, R. (1993). Multispectral Change Vector Analysis for Monitoring Coastal Marine Environments. *Photogrammetric Engineering and Remote Sensing,* Vol. 59, No.3, pp. 381-384, ISNN 0099-1112

Mishra, D., Narumalani, S.; Rundquist, D. & Lawson, M. (2005). High Resolution Ocean Color Remote Sensing of Benthic Habitats: A Case Study at the Roatan Island, Honduras. *IEEE Transactions in Geosciences and Remote Sensing,* Vol. 43, No.7, pp. 1592-1604, ISSN 0196-2892

Mishra, D., Narumalani, S.; Rundquist, D. & Lawson, M. (2006). Benthic Habitat Mapping in Tropical Marine Environments Using QuickBird Imagery. *Photogrammetric Engineering and Remote Sensing,* Vol.72, No.9, pp. 1037-1048, ISSN 0099-1112

Mishra, D., Narumalani, S.; Rundquist, D.; Lawson, M. & Perk, R. (2007). Enhancing the Detection and Classification of Coral Reef and Associated Benthic Habitats: A Hyperspectral Remote Sensing Approach. *Journal of Geophysical Research,* Vol.112, C08014, doi:10.1029/2006JC003892

Mumby, P.; Clark, C.; Green, E. & Edwards, A. (1998). Benefits of Water Column Correction and Contextual Editing for Mapping Coral Reefs. *International Journal of Remote Sensing,* Vol.19, No.1, pp. 203-210, ISSN 1366-6901

Pasqualini, V.; Pergent-Martini, C.; Pergent, G.; Agreil, M.; Skoufas, G.; Sourbes, L. & Tsirika, A. (2005). Use of SPOT 5 for Mapping Seagrasses: An Application to *Posidonia oceanica. Remote Sensing of Environment,* Vol.94, No.1, pp. 39-45, ISSN 0034-4257

Phinn, S.; Roelfsema, C.; Dekker, A.; Brando, V. & Anstee, J. (2008). Mapping Seagrass Species, Cover and Biomass in Shallow Waters: An Assessment of Satellite Multi-spectral and Airborne Hyper-spectral Imaging Systems in Moreton Bay (Australia). *Remote Sensing of Environment 2008,* Vol.112, No.8, pp. 3413-3425, ISSN ISSN 0034-4257

Rouse, J.; Haas, R.; Schell, J. & Deering, D. (1973). Monitoring Vegetation Systems in the Great Plains with ERTS. *3rd. ERTS Symposium,* NASA SP-351 I, 309-317

Spitzer, D. & Dirks, R. (1987). Solar Radiance Distribution in Deep Nataural Waters Including Fluorescence Effects. *Applied Optics,* Vol. 26, Issue 12, pp. 2427-2430, ISSN 1559-128X

Walters, B.; Ronnback, P., Kovacs, J., Crona, B., Hussain, S., Badola, R., Primavera, J.H., Barbier, E. & Dahdouh-Guebas, F. (2008). Ethnobiology, Socio-economics and Adaptive Management of Mangroves: A Review. *Aquatic Botany,* Vol. 89, No.2, pp. 220–236, ISSN 0304-3770

Washington, M.; Kirui, P.; Cho, H. & Wafo-Soh, C. (2011). Data-driven Correction for Light Attenuation in Shallow Waters. *Remote Sensing Letters,* Vol.3, No.4, pp. 335-342, ISSN 2150-7058

Wu, C.; Niu, Z.; Tang, Q. & Huang, W. (2008). Estimating Chlorophyll Content from Hyperspectral Vegetation Indices: Modeling and Validation. *Agricultural and Forest Meteorology,* Vol. 48, No. 8-9, pp. 1230-1241, ISSN 0168-1923

Zimmerman, R. & Mobley, C. (1997). Radiative Transfer within Seagrass Canopies: Impact on Carbon Budgets and Light Requirements. *Proceedings of SPIE Ocean optics XIII,* 2963: 331-336

13

Remote Sensing and Environmental Sensitivity for Oil Spill in the Amazon, Brazil

Milena Andrade[1] and Claudio Szlafsztein[2]
[1]*Federal University of Pará, Amazon Advance Studies (NAEA)*
[2]*Federal University of Pará, Center of Environment (NUMA)*
Brazil

1. Introduction

The use of remote sensing has become a fundamental tool for the identification and analysis of different types of risks in coastal zones. The numerous and, in some cases, recent incidents of oil spills have encouraged companies and government agencies to improve methods, both anticipatory and corrective, to minimize damages. The term 'risk' denotes the possibility that adverse effects may occur as a result of natural events or human activities (Kates et al., 1985). Risk is defined as an association between the hazard´s characteristics (e.g. frequency, magnitude and location) and the vulnerability of affected human populations, environment and infrastructure (Wisner et al., 2004). Risk can be classified by their origin, such as natural, social, or technological (Renn, 2008). Oil spills are an example of the last category, and the coastal areas are one of the most impacted. Environmental sensitivity to oil impacts can be defined through the coastal Environmental Sensitivity Index (ESI), which considers: (i) the geomorphologic aspects such as type and slope of coastline and the degree of exposure to the energy of waves and tides; (ii) oil sensitive biological resources; and (iii) the socio-economic activities that can be affected by oil spills (Gundlach & Hayes, 1978; Dutrieux et al., 2000).

In Brazil, environmental sensitivity mapping has been carried out under the law 9966/2000, which gave the Ministry of the Environment (Climate Change and Environmental Quality Secretary) responsibility to identify, locate and define the boundaries of ecologically sensitive areas with respect to the spill of oil and other dangerous substances in waters within national jurisdiction. This way, based on PETROBRAS (2002) and NOAA (2002), the specifications and technical standards for preparing environmental sensitivity maps for oil spills in coastal and marine zones was elaborated upon (MMA, 2002). Such environmental sensitivity maps provide information in an easy format being useful to determine priorities to impact protection and mitigation. Identification and mapping is developed at three levels: (i) Strategic (1:500,000 for the entire area of a hydrographical basin); (ii) Tactical (1:150,000 for the entire coastline mapped); and (iii) Operational (up to 1:50,000 for a high-risk/sensitivity areas). Each of these mapping scales uses specific tools for remote sensing and GIS tools.

The Amazonian coastal zone extends along ~2250 km, not including the several inlets, islands and small estuaries, which punctuate the coastline (Souza Filho et al., 2005a). This coastal zone is placed in the context of the tropical humid regions, in a low-lying area with active processes of erosion, sedimentation and neotectonics. Also, it is marked by a great hydrologic influence; in a meso- to macrotidal area (Souza Filho, 2005). It is a high-density drainage network, in which the Amazon River discharges a volume of water of 6.3 trillion m^3/year and of sediment estimated at 1.2 billion tons/year (Meade et al., 1985).

Such environmental characteristics are responsible for the development of an extensive mud plain and mangrove area which is located in three States (Amapá, Pará and Maranhão), is approximately 8,386 km^2 wide, and contains 80% of all mangroves in Brazil (Herz, 1991). Where macrotides are present, the area of a flooded mangrove may extend for up to 30 km inland, and the estuaries themselves as much as 80 km (Souza Filho, 2005) (Figure 1). These extensive mud and mangroves plains are considered to be one of the most sensitive areas to oil spills. Also, these mangroves are along national and international ships routes. Transportation and storage are mainly responsible for oil spills in Amazonian coastal zone, since there is no expressive exploration. In 2001, in the state of Pará, approximately 1900 tons of oil sank near the Port of Vila do Conde (Berredo et al., 2001).

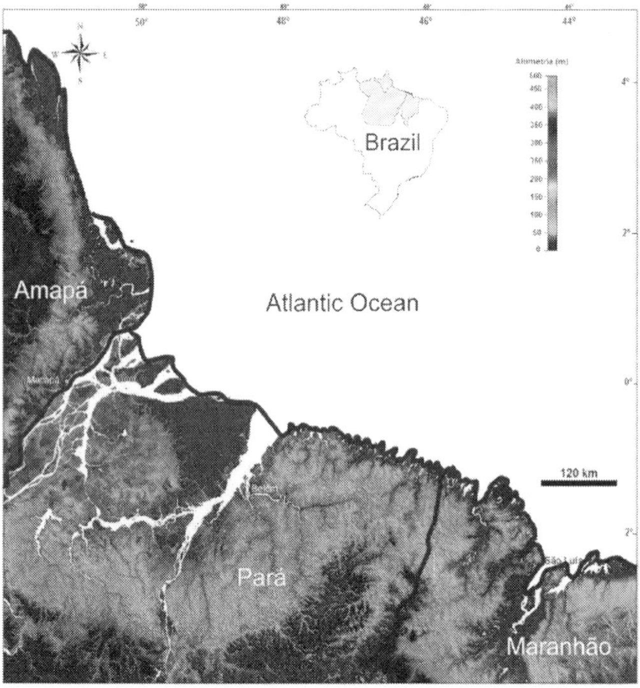

Fig. 1. Amazonian coastal zone in radar SRTM representation (source: modified from Souza Filho et al., 2005a)

In this sense, researches from Federal University of Pará have been working on several projects since 2001 aiming to study the Amazonian coastline and the impact of oil spills on the environment. Therefore, from 2004 to 2010, a large group of scientists were grouped in PIATAM-Mar project "Potential Environmental Impacts and Risks of the Oil and Gas Industry", financially supported by PETROBRAS[1], to map and analyse the vulnerability of the Amazonian coastline oil related disasters. Since 2012, the project "Elaboration of Environmental Sensitivity Maps (SAO maps) for oil spills in Pará-Maranhão and Barreirinhas Basins", founded by the National Research Council of Brazil (CNPQ) has been developed with similar objectives.

Remote sensing and GIS are principal tools aimed to enhance basic socio and environmental knowledge about Amazonian coast. Maps were prepared in strategic and tactical scales through the use of digital elevation models derived from the SRTM (Shuttle Radar Topographic Mission) and optical sensors data (Cohen & Lara, 2003; Souza Filho & El Robrini, 2000; Souza Filho, 2005; Szlafsztein & Sterr, 2007; Silva et al., 2009), synthetic aperture radar (SAR) data (Souza Filho & Paradella, 2002 and 2005; Costa 2004; Souza Filho et al., 2005b; Silva et al., 2009), and the combination of some of them (Souza Filho & Paradella, 2005; Gonçalves et al., 2009; Rodrigues & Souza Filho, 2011).

Oil spill environmental sensitivity maps, adapted to the peculiarities of the Amazonian region (Souza Filho et al., 2004) were drawn at tactical scales through the use of Radarsat and Landsat sensors (Gonçalves et al., 2009; Teixeira & Souza Filho 2009; Boulhosa & Souza Filho, 2009), and operational scale through the use of High Resolution remote sensing (Andrade et al., 2010; Rodrigues & Szlafsztein 2010; Andrade et al., 2009). Over the past decade were reached advances in identification and assessment of sensitivity through spatial maps, the impacts to oil spill analyses, and oil spill risk in Amazonian coastal. The goal of this book chapter is to present a review of the oil spills environmental sensitivity mapping activities using remote sensing and GIS tools in the Amazonian coastal zone of Brazil.

2. Remote sensing and coastal environmental sensitivity for oil spill

2.1 Remote sensing

Remote sensing tools are essential for the construction of maps. These tools help in the precise delimitation of coastlines and specific landforms. The selection of appropriate remote sensing data and applicable digital image processing techniques involves a compromise between costs and mapping capabilities, including coverage area, and spatial resolution (Green, 2000).

For risk maps, remote sensing are fundamental. Risk appears in a broader context in humans transform of the natural into a cultural environment, with the aim of improving living conditions and serving human wants and needs (Turner et al., 1990). There are several sources of hazards to the environment and to society, some of them originated in human activities (Smith & Petley, 2008).

Oil spills are an example of this technological risk. Information and detection about oil spills can be collected through remote sensing tools for prevention planning, as well as river/ocean pollution monitoring and restoration. Some reviews of the use of remote sensing and oil spills including Brekke & Solberg (2005) and Fingas & Brown (2000).

[1] PETROBRAS is the large oil company in Brazil

For a coastal environment, remote sensors can provide information about the physical characteristics of the shoreline, coastal ecosystems dynamics, water quality, and land use/occupation. This information could be mapped at different scales generating cartographic products using all types of sensors and specific digital image processing (Jensen, 1996). Sensors can provide timely and valuable information about oil spills, including the location and extent, thickness distribution, and oil type in order to estimate environmental damage, take appropriate response activities, and to assist in clean-up operations for oil spill contingency planning (Grüner, 1991).

The most common sensors utilized to detect oil spills and to map coastal environments are: optical (visible, infrared sensors and ultraviolet sensors) or radar. Both types of sensor may be acquired at terrestrial, sub-orbital or orbital levels. At terrestrial level, both still and video cameras are commonly used. At the sub-orbital level (or airborne remote sensing), airplanes is the most commonly utilized platform. At the orbital level, satellites are usually used as a platform for sensors. Satellite differs from airborne remote sensing due to timing and frequency of the data collection, the demand of good climate conditions and the time required for processing the dataset (Jha et al., 2008). Aiming to compare sensors, a brief description is given in Table 1.

2.1.1 Optical

Optical sensors can be composed by three bands in the electromagnetic spectrum. These sensors are usually composed by multispectral bands in visible and infrared intervals from the electromagnetic spectrum. In the visible region (350 to 750 nm), oil has a higher surface reflectance than water, but also shows limited nonspecific absorption tendencies (Jha et al., 2008). Instruments such as cameras, films and spectrometers are optical techniques for remote sensing with the benefit of low cost. Normally, visible sensors cannot operate at night as they depend on the reflectance of sunlight, but, in the case of oil spills they can be used to create environmental and logistic maps of the coast to subsidize field trips and first risk management decisions. The infrared sensors are at the 0,7-14 μm intervals in the electromagnetic spectrum. Solar radiation is partially absorbed and emitted as thermal energy by oil. This is thermal energy concentrated in the thermal infrared region with a distinct spectral signature; water has a higher emissivity (Salisbury et al., 1993). Infrared sensors can provide information about the relative thickness of oil slicks, but these sensors are unable to detect emulsions of oil in water when oil is diluted to 70% water (Fingas & Brown, 1997). Infrared is reasonably inexpensive, but has limitations related to false positive results generated by weeds and shorelines (Fingas & Brown, 2000).

Ultraviolet sensor scanners capture ultraviolet radiation (0,003 - 0,38 μm) reflected by the sea surface for detecting oil spills. Oil is more reflective than water in the ultraviolet region. Limitations of this sensor are related to undetected information greater than 10 microns and false images produced by such hindrances as wind slicks, sun glints, and biogenic material (Grüner, 1991).

2.1.2 Radar

Radar is an active sensor (not dependent on electromagnetic radiation from the sun) and operates in a radio wave region (1m - 10^4m). Radar sensors can have two principal

Sensors		Platform		Spatial resolution (m)	Over-pass Frequency (days)	Imagery area	Application
RADAR	SAR	Spaceborne	ERS-2	30	3, 35 and 176 days	100km	Identify large offshore spills and coastal environments - Strategic planning and monitoring
	SAR		RADARSAT-1	8-100	24 days	45-500km	
	SLAR	Airbone	Airplane	10-50	As required	60-80km	Detect and identify the polluter, the extent and type of oil spill and the cleaning necessity; Environmental mapping - Strategic and tactical planning
	SAR		Airplane	1-10	As required	-	
OPTICAL	MSS,TM, ETM, ETM +	Spaceborne	Landsat 5 Landsat 6 Landsat 7	15-120	16 days	183-185km	Detect oil spill if the weather conditions are good; can discriminate false positives; identify and mapping environments - Strategic and tactical planning
	HRV		Spot-2 Spot-3	10-20	26 days	60x60km /100km	
	CCD		Cbers-1; Cbers-2	20	26 days	113 km	Detect oil spill if the weather conditions are good; identify and mapping environments - Strategic and tactical planning
	IRMSS		Cbers-1; Cbers-2	80-160	26 days	120 km	Detect oil spill if the weather conditions are good; capable to detect thermal surface differentiations - Strategic and tactical planning
	WFI		Cbers-1; Cbers-2	260	5 days	890 km	Detect oil spill if the weather conditions are good; monitoring; identify and mapping environments - Strategic planning
	Video camera	Airbone	Airplane	Altitude Dependent	As required	-	Oil spill and coastal environmental documentation. The infrared sensor for measure the thickness of oil slicks - Operational planning
	Still camera		Airplane	Altitude Dependent	As required	-	

Table 1. Characteristics of some existing sensors for oil spill management applications.

instruments: Side-Looking Airborne Radar (SLAR) and Synthetic Aperture Radar (SAR). Radar is a very powerful and useful sensor for searching large areas, observing oceans at night, and capturing images during cloudy weather conditions. The presence of an oil spill can be detected without thickness estimation or oil type recognition. In the radar image, the leak appears as a dark area in contrast to the bright image of the ocean because radar waves are reflected by capillary waves on the ocean (Brown et al., 2003). For a coastal environment, mapping SAR is already considered to be a powerful tool for geomorphologic mapping, providing relevant information about the emergence and submergence of the coast (Souza Filho et al., 2009a).

SLAR is an old technology predominantly used for airborne remote sensing (Fingas & Brown, 2000). Airborne surveillance is limited by high costs and is less efficient for wide area observation due to its limited coverage. SAR has greater spatial range and resolution than the SLAR because it uses the forward motion of the aircraft to synthesize a very long antenna, thereby achieving very good spatial resolution, at the expense of sophisticated electronic processing (Mastin et al., 1994). SAR can be used to provide an initial warning because aircrafts are more suitable to identify the polluter, the extent, and the type of spill.

For large scale oil spill detection, satellite platforms, including ERS-1 and -2, Radarsat, and JERS-1, are commonly used for large scales oil spills (Fingas & Brown, 2005). Radar satellites, including ERS-1 and -2, Radarsat, and JERS-1, have been useful for mapping known large offshore spills (Biegert et al., 1997). On the other hand, optical satellite imagery does not offer much potential for oil spill detection (Fingas & Brown, 2000). However, to map coastal environments, geomorphology and its sensitivity, multi-sensor data fusion such as optical and radar has proved to be a successful tool (Souza Filho et al., 2009b).

2.2 Coastal environmental sensitivity to oil spills

The inter-relationships involving natural resources and human societies have led to a concentration of human activities, services and survival strategies in the coastal environment (Viles & Spencer, 1995; Muehe & Neves, 1995; Pernetta & Elder, 1992). The unique natural geodynamics, the highly productive and extremely diverse biological systems extending from coastal lands to deep water regions (Malthus & Mumby, 2003), the growing land use changes and the pressure on natural resources (MEA, 2005) transform the coastal zone into a conflict area. Oil exploration, transportation and storage have increased the technological risk in this zone.

Areas neighboring major ports (environmental and human populated) may be affected by oil transportation, tank cleaning and oil storage procedures in a port area (Noernberg & Lana, 2002). One of the initial concerns about oil spills result in a necessity for the construction of maps that indicate which type of environment and human resources will be affected. In the mid 1970s, scientists from the National Oceanic and Atmospheric Administration (NOAA) and the American Coast Guard of the United States began to study and classify the sensitivity of coastal environments to oil spill.

This classification was based, initially, on the vulnerability index to oil spills proposed by Gundlach & Hayes (1978). Coastal area is segmented considering environmental and geomorphologic characteristics and then classified using the Vulnerability Index, scaled from 1 (low) to 10 (high). This Vulnerability Index became the standard for coastal management, planning and research about the effects of oil spills on different types of coastline. Over time, this index evolved and was modified, leading eventually to the development of the Environmental Sensitivity Index (ESI).

The ESI should be represented cartographically as maps in different scales for different goals. The first ESI map was produced in 1979, in response to the advance toward the coast of oil resulting from the blowout of the IXTOC 1 oil-well in the Gulf of Mexico. In the 1980s, ARPEL produced an innovative ESI atlas for the whole coast of the United States, including Alaska and the Great Lakes, to be used for the planning of contingency measures in response to oil spills (NOAA, 2002). From this moment on, ESI maps have been an integral component of response and contingency planning for oil spills, looking for the protection of life, the reduction of environmental impacts, and facilitation of the response efforts. These atlases were integrated by color printed maps on a two dimensional representation of a three-dimensional world and high production costs.

After the 1990's, NOAA (2002) standardized output formats and symbols for ESI maps construction. The basic necessary information is 1) shoreline classification; 2) biological resources; 3) human-use resources. The shoreline classification scheme is based on an understanding of the physical and biological characteristics of the shoreline environment. Relationships among physical processes, exposure to wave and tidal energy, slope, substrate type (i.e. grain size, mobility, penetration and/or burial, and mobility), and associated biota produce specific geomorphic/ecologic shoreline types. Shoreline classification helps to identify oil spill origin and impacts and the best cleanup methods for a specific shoreline type. The sensitivity ranking was developed for the estuarine settings and is slightly modified for lakes and rivers. The human use resources relate to specific, valuable specific areas because of their use, such as beaches, parks and protected marine areas, water intakes, fisheries, tourism, economic sectors, and archaeological sites. The biological resources include the study and maps of oil-sensitive biological and ecological resources.

3. Remote sensing and coastal environmental sensitivity in Brazil

Brazil has an expansive coastline through the equatorial region to the subtropical latitude of the south hemisphere. The length is approximately 8.500 km with 17 of the 26 states of the country lying on the coast of the Atlantic Ocean. The Brazilian coast is defined by the National Plan for Coastal Management (law 7661/1988), as the geographic space where there are air, sea and land interacts, which includes renewable and non-renewable resources along a maritime and terrestrial border.

A diversity of coastal environments and population densities are found along the Brazilian coast. Population is higher in state capitals than in the other coastal municipalities. Environments vary from very productive, such as mangroves, to rocky and artificial man-made structures. Man-made structures, such as ports are established along the entire coastline of Brazil (Figure 2).

Ports are high-risk areas, and oil spill monitoring is clearly important there. In 2000, two large oil spills occurred at Baía de Guanabara (Rio de Janeiro) and Paraná, both resulting from pipeline ruptures. After these accidents, fundamental changes have been made to

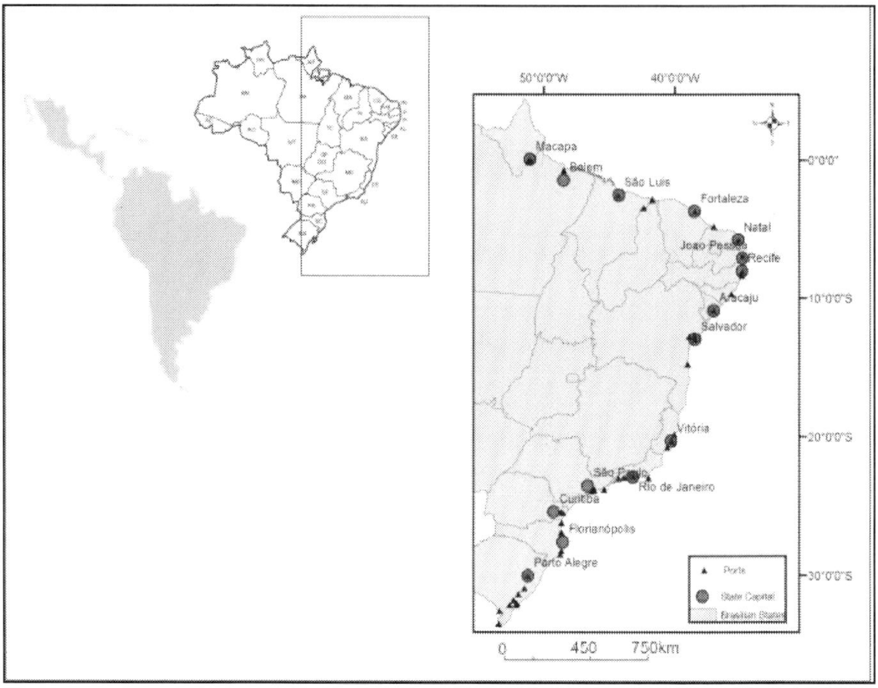

Fig. 2. Main ports of the Brazilian coast (IBAMA, 2011).

the environmental policies of Brazil and PETROBRAS (Souza Filho et al., 2009b), in order to give priority to prevention and mitigation activities.

After these accidents, the prevention and mitigation oil spill impacts became a priority. The law 9966/2000 was established to regulate the activities of prevention, control and supervision of pollution caused by oil and other dangerous substances in Brazilian waters. The Ministry of the Environment has the responsibility of identifying, locating and defining the boundaries of the ecologically sensitive areas to oil spills. Ecologically sensitive areas are defined as regions where special attention is needed in order to protect and preserve the environment from pollution by regulatory and preventive measures (MMA, 2002). In this context, Araújo et al. (2002) published the document "Basic Manual for the Elaboration of Maps of Environmental Sensitivity to Oil Spills in the Petrobras System: Coastal and Estuarine Environments" edited by the Ministry of the Environment (Climate Change and Quality Environmental Secretary) and based on PETROBRAS (2002) and NOAA (2002).

This change in attitude was reflected in a large production of oil spill environmental sensitivity maps for the Brazilian coast (Araújo et al., 2006) (Table 2). These maps are produced in order to support environmental management and the elaboration and implementation of contingency plans. The ESI preparation was intensified through Decrees 4136/2002 and 4871/2003.

Author	Spatial Resolution	Sensor	Method	Scale	Case study (State)
Romero, et al. (2010)	High	Aerial photography	Literature review, Field data collection	Operational	Cananéia-Iguape Estuary (São Paulo)
Silva et al. (2010)	Low	SRTM, Radarsat	Highly precision field data collection with DGPS, Previous database research, visual classification	Tactical, Operational	Potiguar Sedimentary Basin (Rio Grande do Norte)
	Moderate	Landsat 7 ETM+, Cbers-2, Cbers-2B			
Cantagalo et al. (2008)	High	Aerial photography	Visual classification, field data collection	Tactical, Operational	Santos Estuary (São Paulo)
Carvalho & Gherardi (2008)	Moderate	Landsat 7 ETM	Automatic classification, visual interpretation, Field data collection	Tactical, Operational	Potiguar Sedimentary Basin (Ceará and Rio Grande do Norte)
Bellotto & Sarolli (2008)	Moderate	Landsat 7 ETM+	Visual interpretation, Previous database research, Field data collection	Operational	Municipality of Imbatuba (Santa Catarina)
Poletto & Batista (2008)	Moderate	Cbers	Visual interpretation, Previous database research, Field data collection	Tactical, Operational	Municipality of Ubatuba (São Paulo)
	High	Aerial photography			
Rocha-Oliveira et al. (2008)	Moderate	Landsat 7 TM+	Visual interpretation, Field data collection	Operational	Southeast and south area (Santa Catarina)
Silva et al. (2008)	Moderate	Landsat 7 TM+	Visual interpretation, field works, literature review	Operational	Santa Catarina Island and surrounding areas (Santa Catarina)
Araújo et al. (2007)	Moderate	Landsat 7 ETM+	Visual interpretation, Field data collection	Operational	Municipalities of Itapoa, Barra Vellha, Piçarras, Itajaí, Balneário Camboriú (Santa Catarina)
	High	Aerial photography			
Chacaltana (2007)	High	Ikonos	Visual interpretation, Field data collection	Operational	Vitória Bay (Espírito Santo)
Lima et al. (2008)	High	Aerial photography	Visual interpretation, Field data collection	Operational	São Sebastião Island (São Paulo)
Wieczorek et al. (2007)	High	Aerial photography	Visual interpretation, Field data collection	Operational	Cananéia-Iguape Estuary (São Paulo)

Author	Spatial Resolution	Sensor	Method	Scale	Case study (State)
Castro et al. (2006)	Moderate	Landsat 5 TM, Landsat 7 ETM+	Database development, geomorphology; hydrodynamic, waves energy, currents direction; slope and grain size of profile beach	Operational	São Bento, Galinho Municipalities (Rio Grande do Norte)
	High	Aerial photography			
Souto et al. (2006)	Moderate	Landsat 5 TM, Landsat 7 ETM+	Normalized Difference Vegetation Index, Automatic classification, Visual interpretation, Field data collection	Operational	Ponta Macau (Rio Grande do Norte)
	High	Aerial photography			
Souza, et al. (2005)	Moderate	Landsat 5 TM, Landsat 7 ETM+, SPOT, Cbers-2	Database utilization; visual interpretation, Field data collection	Tactical	Northern coast (Rio Grande do Norte)
	High	Ikonos			
Noernberg & Lana (2002)	Moderate	Landsat TM	Database access, digital processing	Operational	Paranaguá Estuary (Santa Catarina)

Table 2. Principal studies of oil spill coastal sensitivity using remote sensing techniques in order to generate ESI maps in Brazil organized by date (Amazon Region are not included).

In Brazil, the ESI maps were also developed in a cartographic plan at strategic, tactical and operational scale for the role country. As an initial step, the tools of remote sensing and GIS are necessary to ESI maps construction and to comprehend differential spread of the technological risk for the country's coasts. Mostly moderate and high resolution images were used to produce these maps.

Moderate resolution images (e.g. RADARSAT-1 and Landsat TM/ETM/ETM+) and SRTM derived digital elevation models have been used to map the Brazilian coastal zone at strategic and tactical scales. Studying an oil spill emergency due to a pipeline rupture in Guanabara Bay (Rio de Janeiro), Bentz and Miranda (2001) found that RADARSAT-1 provided suitable temporal coverage. Once cloud cover, haze and the eight-day revisit schedule (using both Landsat-5 and -7) prevented Landsat from being used systematically for oil spill monitoring. In the same case Thematic Mapper (TM) sensor was used to capture images after the oil spill emergency where a pipeline ruptured (Bentz & Miranda, 2001).

Carvalho & Gherardi (2008) used Landsat 7 ETM+ images to generate land use and land cover maps, as well as ESI maps in Northeast Brazil, aiming for oil spill contingency planning and emergency responses. A fusion of multispectral and panchromatic ETM images via IHS (Intensity-Hue-Saturation) transformation was used. Then socioeconomic information was inserted using automated and visual image interpretation.

High resolution images have mostly been used for operational ESI maps production in the states of São Paulo and Rio Grande do Norte using aerial photography and Ikonos. Visual interpretation, together with field data collection, has been the principal methodological procedure. Most areas have mangroves, conservation units and are surrounding by intensive technological activities.

The methodology, standards and technical specifications for determining coastal sensitivity follow Araújo et al. (2002). The principal steps for shoreline identification are: 1) Analysis of the available literature, aerial photographs, maps of the entire area; 2) Aerial reconnaissance of the entire area and selection of detailed study areas; 3) Mapping of major features in representative areas; 4) Collection of sediment from the intertidal zone and biologic floral and faunal groups samples; construction of beach topographic profile; 6) Analysis of the sediment sample sizes; 7) Data compilation and classification; and 8) Construction of detailed sensitivity maps.

Colors are used indicate the ESI and symbols to the human and biological resources. Each number is represented by a color index. Two environments may occur at the same coastal segment; in that case, both colors of the separated lines should be displayed, one inside and the other outside. In the case of intertidal zones, for example, the intertidal plain should display colors according to the differences of sediment sizes to the high tidal line and the low tidal line. Table 3 compares ESI specification defined by the Ministry of the Environment (MMA, 2002) with the original defined by NOAA (2002).

ESI number	Color	NOAA (2002)	MMA (2002)
1		Exposed rocky shores and man-made structures; rocky cliffs with boulder talus base	Exposed rocky shores; exposed rocky sedimentary cliffs; exposed solid man-made structures
2		Exposed wave-cut platforms in bedrock, scarps and steep slopes in clay	Exposed medium to high declivity rocky shores; exposed sandy substrate with medium declivity
3		Fine to medium grained sand beaches; Scarps and steep slopes in sand; Tundra cliffs	Fine to medium grained sand in dissipative beaches; continuous and multiple beach strings; Scarps and steep slopes in sand; exposed dune field
4		Coarse-grained sand beaches	Coarse-grained sand beaches; exposed; exposed fine to medium grained sand intermediary beaches; sheltered fine- to medium- grained sand beaches
5		Mixed sand and gravel beaches	Mixed sand and gravel beaches, coral reefs fragments; vegetated abrasion platform; sandy reefs

ESI number	Color	NOAA (2002)	MMA (2002)
6		Gravel beaches; Riprap gravel beaches (cobbles and boulders)	Gravel beaches; dendritic limestone coast; platform with lateritic concretion
7		Exposed tidal flats	Exposed sandy tidal flats; low tide platform
8		Sheltered: scarps in bedrock, mud or clay, rocky shores (impermeable/permeable), solid man-made structures, riprap, rocky rubble shores; Peat shoreline	Sheltered scarps in bedrock (permeable and non permeable); Scarps and steep slopes in sand; permeable sheltered man-made structures (riprap)
9		Sheltered tidal flats; Vegetated low banks; hypersaline tidal flats	Sand tidal flats; sheltered mud tidal flats; coral reefs
10		Salt and brackish water marshes; Freshwater marshes; Swamps; Scrub-shrub wetlands: mangroves; Inundated low-lying tundra	Delta and vegetated sand bars; sheltered wetlands; salt marshes saline wetlands; mangroves

Table 3. ESI comparison between NOAA (2002) and Ministry of the Environment (MMA, 2002) classification.

4. Remote sensing and coastal environmental sensitivity in Amazon

The coastal zone of the Brazilian Amazon is composed by tree states: Amapá, Pará and Maranhão. According to the IBGE (2011), the total population of Amapá State is 669,526 distributed in 16 municipalities; the state capital, Macapá, concentrates 59% of this population. The state of Pará has a total population of 7,581,051 distributed in 143 municipalities; Belém comprises 18%. Maranhão state has a total population of 6,574,789 distributed in 217 municipalities; São Luis comprises 15% of this population. The population density in capital cities is over 100 hab/km^2, while in other coastal municipalities vary from 10 to 50 hab/km^2.

Until 21st century most of the coastal zone of the north of Brazil had sectors virtually devoid of information, or where data was available, it was non-systematized and both temporally and spatially non-continuous. The most important environmental dataset is related to the large continuous and well-developed mangroves - *Rhizophora mangle, Avicennia germinans* and *Laguncularia racemosa* (Szlafsztein, 2000). The mangroves have ecological and socioeconomic importance due to communities' livelihoods, and they are considered a protected ecosystem. The main activities are fishing, collecting shrimp and crabs (Andrade et al., 2010; Andrade et al., 2009).

However, port complexes and industries have been established alongside residential, protected areas and fishing grounds. For example, in Piatam Mar context the principal ports chosen to develop oil mapping were "Santana" (State of Amapá); "Itaqui" (State of Maranhão); "Outeiro", "Miramar" and "Vila do Conde" (State of Pará). The biological information was intensively identified in "Lago Piratuba biological reserve" (Amapá); "Soure extractive reserve" (Pará) and "Ilha dos Caranguejos Environmental Protection Area" (Maranhão). According to Souza Filho et al., (2009a) these conservation units work as control areas, given both their well-preserved conditions and their proximity to transportation routes due to proximity to protected areas along the ports mentioned above (Figure 3).

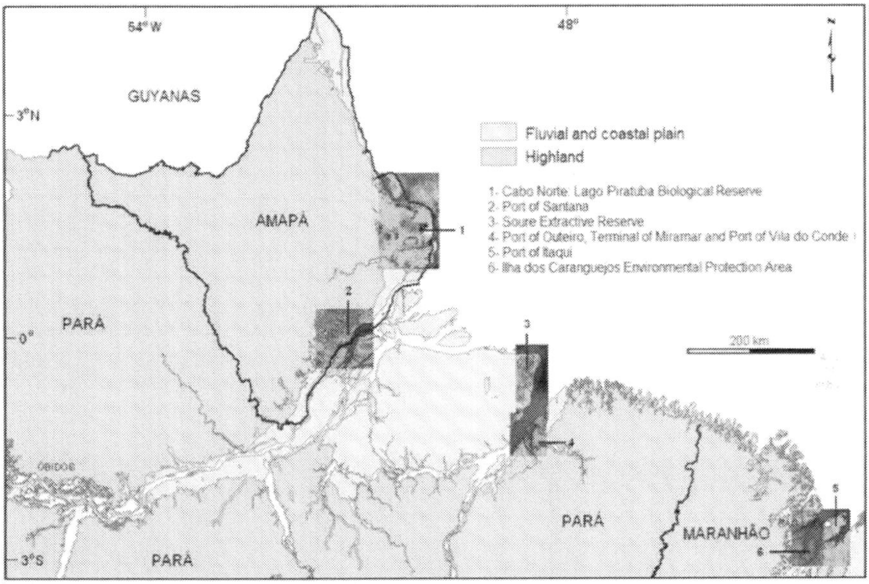

Fig. 3. Principal ports and environmental protected areas in the Amazon coast (source: Souza Filho et al., 2009a).

Oil spills are a potential risk around these port areas which can affect the environment, human population infrastructure and livelihood, resulting from the transportation process, as well as tank cleaning and oil storage procedures within the area of the port (Noernberg & Lana, 2002). To comprehend the oil impact, it is necessary to analyze the coastal Amazonian environment as a whole. PETROBRAS established and financed nine projects to deal with this subject, among them, the "Environmental Sensitivity Map to Oil Spill in Guajará Bay (PA)" (2001 - 2003), the "JERS-1, RADARSAT-1 and ALOS/PALSAR application in monitoring and mapping Amazon coastal environments: an approach for multi-temporal environmental sensitivity maps to oil spill" (2004 - 2006), PIATAM MAR (2004 - 2010) and currently "Elaboration of ESI maps for Pará-Maranhão and Barreirinhas basin" (2012 until 2014).

The PIATAM MAR project was implemented in Northern Brazil and was led by the Federal University of Pará, the Federal University of Rio de Janeiro and PETROBRAS. The general aims proposed are: the consolidation of a multidisciplinary researcher's network that are active in the Amazonian coastal zone; the development of technological tools and infrastructure to support local monitoring and environmental management; and ESI maps construction (Souza Filho et al., 2009a).

Initially, the researchers of PIATAM MAR project compiled environmental data and other information available on the Amazon coastal zone. The results are integrated in the book, "Bibliography of the Amazon Coastal Zone: Brazil" (Souza Filho et al., 2005a). Meanwhile, a computational database system using the MYSQL language was developed and used as a basis for the development of a geographic information system called SIGmar.

All of these initial steps support the subsequent aim of PIATAM MAR: the construction of ESI maps. From 2006 to 2010 socio-economic and environmental data were integrated in ESI maps. These maps have been developed through the SIGmar and the extensive use of remote sensing. ESI maps create an operational alternative for the monitoring and mapping of the Amazonian coastal zone and provide guidelines for the use of the InfoPAE (Computerized Emergency Action Plan Support) System on the Amazonian coast (Souza Filho et al., 2009a).

Two considerations should be taken into account when mapping and monitoring oil spills in the Amazonian coastal environments. First, the unique complex environmental dynamics of the Amazon basin have demanded an adaptation of ESI classification with values from 1 (low) to 10 (high) sensitivity (Souza Filho et al., 2004) (Table 4). Second, the Amazonian coast is situated in the intertropical convergence zone (ITCZ) that is located near the equator and has a broad area of low atmospheric pressure. Therefore, there is a huge cloud cover between December and May which limits the use of some kinds of sensors.

Coastal ESI mapping for the Amazon uses remote sensing as an indispensable and very powerful tool. Oil spill and environmental sensitivity to oil spills in the Amazon were mostly mapped during the PIATAM MAR project. Table 5 shows the most important scientific results in this study area. The perspectives of the ESI adaptation proposed by Souza Filho et al. (2004) were extensively used.

The whole Amazon coastal zone was mapped with spatial resolution of 90 m based on the processing and images mosaics of SRTM images and 30 m of RADARSAT-1 Wide 1 images and mosaics of JERS-1 SAR. This sensor was chosen given the six months of unfavorable climatic conditions; radar sensors (Synthetic Aperture Radar - SAR) are used for strategic scale.

On a tactical scale, multi-sensor data fusion between microwaves sensors and optical sensors are considered to be the most important source of spatial data for geomorphologic recognition and basic coastline characteristics. The commons sensors fusion are made in general with low resolution data from RADARSAT-1 Wide 1 and JERS-1 SAR mosaic, together with moderate spatial resolution data (10–30 km) from Landsat series (MSS, TM and ETM+) and Cbers-2 images (20m). The multi-fusion of optical (Landsat 5 TM) and radar (RADARSAT-1) sensors had a particular emphasis on the evaluation of the new hybrid sensor product combining PCA (Principal Component Analysis) and IHS components. In areas with little or no data, this fusion method from multi sensors to orbital images, together with field data are economically efficient and provide a good environment sensitivity characterization (Rodrigues & Souza Filho, 2011).

Hydrological dynamics with flood area delimitation could be differentiated by the use of JERS-1, L band (Santos et al., 2009), which is important in a region dominated by different tidal regimes that can amplify the area affected by oil spills. Methods include visual and automatic classification leading to good results in identifying widespread occurrence of

ESI	Amazon Coastal Environment
1A	Exposed rocky shores
1B	Exposed, solid man-made structures
1C	Exposed rocky cliffs with boulder talus base
2	Exposed scarps and steep slopes in clay
3A	Fine to medium grained sand beaches
3B	Scarps and steep slopes in sand
4	Coarse-grained sand beaches
5	Mixed sand and gravel banks and beaches
6	Riprap
7	Exposed tidal flats
8A	Sheltered scarps in bedrock, mud, or clay
8B	Sheltered, solid man-made structures
8C	Sheltered riprap
8D	Peat shorelines
9A	Sheltered tidal flats
9B	Vegetated low banks
9C	Hypersaline tidal flats
10A	Salt, and brackish-water marshes
10B	Freshwater marches, aquatic vegetation
10C	Intertidal mangrove
10D	Supratidal mangrove

Table 4. ESI shoreline classification for the Amazon Coast, modified by Souza Filho et al. (2004) based on the proposals of NOAA (2002) and Araújo et al. (2002).

flooded mangrove forests. This environment is considered to be the most oil-sensitive habitat described in Table 4 - ESI Ranking specification = 10c and 10d (Souza Filho et al., 2005a).

High resolution images, such as Ikonos, were used for operational scale mapping. The resolution of 1 m provides a detailed geomorphic map, and it's also possible to map the potentially hazardous industrial structures stratified by type of hazard. However, the use of Ikonos images is limited when cloud cover is higher than 25%. As a result the images are mostly inadequate between March and June (Andrade et al., 2010). On the other hand, biological and socioeconomic resources, risk areas and oil spill hazard zones of storage and platform transportation can be better identified and delimitated (Rodrigues & Szlafsztein, 2010; Andrade et al., 2009). This location contributes to planning and management strategies and cleaning efforts.

Author	Map Type	Spatial Resolution	Sensor	Method	Scale	Study case
Andrade et al. (2010)	Oil spill Vulnerability	High	Ikonos	Visual interpretation, Field data collection	Operational	Municipality of São Luis (Maranhão)
Rodrigues & Szlafsztein (2010)	Oil spill risk	High	Ikonos	Visual interpretation, Field data collection	Operational	Municipality Barcarena (Pará)
Andrade et al. (2009)	Oil spill hazard representation and susceptible socioeconomic resources	High	Ikonos	Visual interpretation, Field data collection	Operational	Municipality of São Luis (Maranhão)
Boulhosa & Mendes (2009)	ESI	Moderate	Spot-5	Visual interpretation, Field data collection	Operational	Municipality of Barcarena (Pará)
Boulhosa & Souza Filho (2009)	ESI	Moderate	Landsat 7 ETM+, SRTM High, aerial photography	Automatic classification, multi-fusion sensors, Field data collection	Tactical	Municipalities of Maracanã, Santarém Novo, Salinópolis, Cuiarana, São João de PIrabas, Santa Luzia and Primavera (Pará)
Gonçalves et al. (2009)	ESI	Moderate	Landsat 7 ETM, Radarsat 1	Automatic classification, multi-sensor fusion, field data collection	Tactical	Municipality of Belém (Pará)
Souza Filho et al. (2009)	ESI	Moderate	Landsat 5 TM, Radarsat 1	Visual classification	Tactical	Municipality of Bragança (Pará)
Novaes et al. (2007)	ESI	Moderate	Landsat 5 TM	Visual interpretation, Field data collection	Operational	Municipality of São Luis, (Maranhão)

Table 5. Results published in a scientific paper related to remote sensing use and sensitivity environment to oil spill in Amazon coast.

5. Conclusions

Remote sensing techniques are used for risk identification, assessment and analysis. The technological risk of oil spills needs continuous planning and monitoring actions. The availability of airborne and satellite remote sensing provides a diversity of resolution and sensors required to construct environmental sensitivity maps, using basic information about socioeconomic and biological resources and geomorphic characteristics.

Remote sensing and ground confirmation provide accurate information about this basic information. In particular, the coastline is usually mapped in detail with both optical and radar sensors. The multi-sensor data fusion of an optical moderate sensor with radar has been extensively used in the Amazon region to provide basic information about coastal environments. Radar is a very powerful tool once it can operate in difficult weather conditions. It provides detailed information about shoreline irregularities and geomorphic units if the texture and the altitude of this type of images are precise. Optical sensors are used for environmental differentiation once land cover and land use have different spectral responses.

Studies in Brazil regarding oil spills have increased after 2000, and ESI maps have been generated at different scales for different areas along the coast. Remote sensing tools were essential to achieve initial and advanced cartographic information in a context of the diversity of the environment, information and cartographic background. Particularly in the Amazon, little background information about the coastline existed before the PIATAM MAR project. In the context of this project, the Amazon coast was previewed at strategic scale with the use of a moderate sensor. ESI maps were produced at the tactical and operational scales and it was possible to map the coastal environment and organize information about socioeconomic and biological resources. A large, extensive mangrove system coexists with industrial port areas on the Amazon coast with a high sensitivity to oil spills, which should to be continuously monitored with remote sensing techniques.

6. References

Andrade, M.; Souza Filho, P. & Szlafsztein, C. (2009). High Resolution Images for Recognition of the Susceptibility of Social Economic Resources to Oil Spill in the Itaqui-Bacanga Port Complex, Maranhão, Brazil. *Journal of Integrated Coastal Zone Management*, vol. 9, pp. 127-133.

Andrade, M; Szlafsztein, C., Souza Filho, P., Araújo, A. & Gomes, M. (2010). A socioeconomic and natural vulnerability index for oil spills in an Amazonian harbor: A case study using GIS and remote sensing. *Journal of Environmental Management*, vol 91, pp. 1972-1980.

Araújo, S., Silva G. & Muehe, D. (2002). Manual básico para elaboração de mapas de sensibilidade ambiental a derrames de óleo no sistema Petrobras: Ambientes Costeiros e Estuarinos. CENPES/Petrobras, Rio de Janeiro, 134 pp.

Araújo, S.; Silva, G. & Muehe, D. (2006). *Mapas de sensibilidade ambiental a derrames de óleo - Ambiente costeiros, estuarinos e fluviais.* Rio de Janeiro Rio de Janeiro. CENPES/Petrobras, pp.168 Ed. 2ª, ISBN: 8599891014

Araújo, R.; Petermann, R.; Klein, A.; Menezes, J.; Sperb, R. & Gherardi, D. (2007). Determinação do ìnndice de sensibilidade do litoral (ISL) ao derramamento de óleo para regiões norte e centro-norte da costa de Santa Catarina (SC), *Gravel*, N°5, pp. 45-73.

Belloto, V. & Sarolli, V. (2008). Mapeamento da sensibilidade ambiental ao derramamento de óleo e ações de resposta para a região costeira e área portuária de Imbituba, SC, Brasil. *Brazilian Journal for Aquatic Sciences and Technology*, vol. 12, N°2, pp. 115-125.

Bentz, C. & Miranda, F. Application of remote sensing data for oil spill monitoring in the Guanabara Bay, Rio de Janeiro, Brazil. In: *Proceedings of the International Geosciences and Remote Sensing Symposium (IGARSS)*, Sydney, 2001.

Berredo J.; Mendes A.; Sales M. & Sarmento J. (2001). Nível de Contaminação por óleo nos sedimentos de Fundo e na água do Rio Pará, decorrente do acidente com a balsa Miss Rondônia. In: Prost M. & Mendes A. (Org.). *Ecossistemas Costeiros: Impactos e Gestão Ambiental*. Museu Paraense Emílio Goeldi, Belém, pp. 153 - 165.

Biegert, E., Baker, R., Berry, J., Mott, S. & Scantland, S. (1997). Gulf Offshore Satellite Applications Project Detects Oil Slicks Using Radarsat, In: *Proceedings of International Symposium: Geomatics in the Era of Radarsat*, Ottawa, Canada.

Boulhosa, M. & Mendes, A. (2009). Mapeamento dos Índices de Sensibilidade Ambiental ao derramamento de óleo através de Imagens SPOT 5, na região portuária de vila do conde - Barcarena - PA. *XIV Simpósio Brasileiro de Sensoriamento Remoto*, Natal, Brasil, pp. 3597-3603.

Boulhosa, M. & Souza Filho, P. (2009). Reconhecimento e mapeamento de ambientes costeiros para geração de mapas de índice de sensibilidade ambiental ao derramamento de óleo, Amazônia Oriental. *Revista Brasileira de Geofísica*, vol. 27, pp. 23-37.

Brekke, C. & Solberg, A. (2005). Oil spill detection by satellite remote sensing. *Remote Sensing of Environment*, vol. 95, pp. 1–13.

Brown, C. & Fingas, M. (2003). Development of airborne oil thickness measurements. *Marine Pollution Bulletin*, vol. 47, pp. 485 - 492.

Cantagallo, C.; Milanelli, J. & Dias-Brito, D. (2007). Limpeza de Ambientes Costeiros Brasileiros Contaminados pó Petróleo: uma revisão. *Pan-American Journal of Aquatic Sciences*, vol 2, pp. 1-12.

Castro, A.; Grigio, A.; Souto, M.; Amaro, V. & Vital, H. (2006). Modeling and development of a geographic database: application to the elaboration of oil-spill environmental sensitivity maps in coastal areas of the Rio Grande do Norte state. *Journal of Coastal Research*, vol. 39, pp. 1436-1440.

Carvalho, M. & Gherardi, D. (2008). Mapping the environmental sensitivity to oil spill and land use/land cover using spectrally transformed Landsat 7 ETM data. *Brazilian Journal for Aquatic Sciences and Technology*, vol. 12, N° 2, pp. 1-9.

Chacaltana, T. (2007). Mapeamento de áreas de sensibilidade ambiental ao derrame de óleo na Baía de Vitória, ES. *Master dissertation*. Federal University of Espírito Santo. p.140.

Cohen, M. & Lara R. (2003). Temporal changes of mangrove vegetation boundaries in Amazonia: Application of GIS and remote sensing techniques. *Wetlands Ecology and Management*, vol 11, pp. 223 -231.

Costa, M. (2004). Use of SAR satellite for mapping zonation of vegetation communities in the Amazon floodplain (10): 1817-1835. *International Journal of Remote Sensing*, vol 25, pp. 1817-1835.

Dutrieux, E.; Canovas, S.; Denis, J.; Hénocque, Y.; Quod, J.; Bigot, L. (2000). Guide méthodologique pour l'élaboration de cartes de vulnérabilité dês zones côitères de Océan Indien Réalisé par Créocéan, I fremer et Arvan pour le compte de l'UNESCO/IOC et lê PRE-COI/UE. *IOC Manuals and Guides*, No 38.

Fingas, M., & Brown, C., 1997, Remote sensing of oil spill, *Sea Technology*, vol. 38, pp. 37-46.

Fingas, M. & Brown, C. (2000), "Review of Oil Spill Remote Sensing", In *Proceedings of the Fifth International Conference on Remote Sensing for Marine and Coastal Environments*, Environmental Research Institute of Michigan, Ann Arbor, Michigan, pp. I211-218.

Fingas, M. & Brown, C. (2005). An update on oil spill remote sensors. In: *Proc. 28th Arctic and Marine Oil Spill Program (AMOP) Tech. Seminar*, Calgary, Canada, pp. 825-860.

Gonçalves F.; Souza Filho, P.; Paradella, W. & Miranda, F. (2009). Fusão de dados multisensor para a identificação e o mapeamento de ambientes flúvio-estuarinos da Amazônia. *Revista Brasileira de Geofísica*, vol. 27, pp. 57-67.

Gundlach, E. & Hayes, M. (1978). Classification of coastal environments in terms of potential vulnerability to oil spill impact. *Marine Technology Society Journal*, vol. 12, pp. 18-27.

Green, E. 2000. Satellite and airborne sensors useful in coastal applications. In: Edwards A.J. (Ed.) *Remote Sensing Handbook for Tropical Coastal Management*, Costal Management Sourcebooks 3, France: UNESCO Publishing, 41-56pp. ISBN: 9231037366

Grüner, K.; Reuter, R. & Smid, H. (1991). A new sensor system for airborne measurements of maritime pollution and of hydrographic parameters. *Geojournal*, vol. 24, n° 1, pp. 103-117.

Herz, R. (1991). *Manguezais do Brasil*. IOUSP, São Paulo, São Paulo. p.227.

IBAMA. 2011. Informações vetorial sobre porto no Brasil. Avaiable in: siscom.ibama.gov.br

IBGE. 2011. *Sinopse do censo demográfico*. Ministério do Planejamento, Orçamento e Gestão, Rio de Janeiro, pp. 261.

Jha, M.N.; Levy, J. & Gao, Y. (2008). Advances in remote sensing for oil spill disaster management: state-of-the-art sensors technology for oil spill surveillance, Sensors, vol. 8, pp. 236-255.

Jensen, J. (1996). *Introductory digital image processing*, 2 ed., Prentice Hall, London, 318p. ISBN: 0131453610.

Kates, R., Hohenemser, C. & Kasperson, J. (1985) *Perilous Progress: Managing the Hazards of Technology*, Westview Press, Boulder, CO, ISBN: 0813370256

Lima, M.; Dias-Brito, D. & Milanelli, J. (2008). Environmental sensitivity mapping for oil spills in Ilhabela, São Paulo, *Revista Brasileira de Cartografia*, N° 60, pp. 145-154.

Malthus, T. & Mumby, P. (2003). Remote sensing of the coastal zone: an overview and priorities for future research, *International Journal of Remote Sensing*, vol. 24, N°13, pp. 2805–2815.

Mastin, G.; Mason, J.;. Bradley, J.; Axline, R. & Hover, G.(1994). Comparative Evaluation of SAR and SLAR, In: *Proceedings of the Second Thematic Conference on Remote Sensing for Marine and Coastal Environments: Needs, Solutions and Applications*, ERIM Conferences, Ann Arbor, Michigan, pp. I-7-17.

MEA, Millenium Ecossystem Assessment. (2005). Ecossystem and human well-being. Island Press, Washington, DC, ISBN 1-59726-040-1.

Meade, R.; Dunne, T.; Richey, J.; Santos, U. & Salati, E. (1985). Storage and remobilization of suspended sediment in the lower Amazon River of Brazil. *Science*, 228, pp. 488-490.

MMA (2002) Especificações e normas técnicas para elaboração de cartas de sensibilidade ambiental para derramamentos de óleo. MMA, Brasília.

Muehe, D. & Neves, C. (1995). The implications of sea level rise in the Brazilian coast: A preliminary assessment. *Journal of Coastal Research*, vol. 14, pp.54-78.

NOAA. (2002). National Oceanic and Atmospheric Administration. *Environmental Sensitivity Index Guidelines*. Office of Response and Restoration Technical Memorandum No 11, Seattle, USA.

Novaes, R.; Tarouco, J.; Rangel, M. & Dias, L.(2007). *XIV Simpósio Brasileiro de Sensoriamento Remoto*, Florianópolis, Brasil, pp. 4089-4096.

Noernberg, M. & Lana, P. (2002). A sensibilidade de manguezais e marismas a impacto por óleo: fato ou mito? Uma ferramenta para avaliação da vulnerabilidade de sistemas costeiros a derrames de óleo. *Geografares*, vol.3, pp. 109-122.

Pernetta, J. & Elder, D. (1992). Climate, sea level rise and the coastal zone: Management and planning for global changes. *Ocean & Coastal Management*, vol. 18, N°1, pp. 113-160.

PETROBRAS. (2002). *Manual Básico para Elaboração de Mapas de Sensibilidade Ambiental a Derrames de Óleo no Sistema Petrobras: Ambientes Costeiros e estuarinos*. PETROBRAS, Rio de Janeiro.

Poletto, C. & Batista, G. (2008). Sensibilidade ambiental das ilhas costeiras de Ubatuba, SP, Brasil. *Revista Ambiente & Água - An interdisciplinary Journal of Applied Science*, vol. 3, N° 2, pp. 106-121.

Renn, O. (2008). *Risk governance: coping with uncertainty in a complex world*. 1°ed, Earthscan, ISBN 978-1-84407-292-7, USA.

Rocha-Oliveira, T.; Klein, A.; Petermann, R.; Menezes, J. & Sperb, R. (2008). Determinação do índice de sensibilidade do litoral (ISL) ao derramamento de óleo, para região sudeste e sul do estado de Santa Catarina. *Brazilian Journal for Aquatic Sciences and Technology*, vol. 12, N°2, pp. 91-114.

Rodrigues, S. & Souza Filho, P. (2011). Use of Multi-Sensor Data to Identify and Map Tropical Coastal Wetlands in the Amazon of Northern Brazil, *Wetlands*, vol. 31, 11-23pp.

Rodrigues, J. & Szlafsztein, C. (2010). Análise de vulnerabilidade do núcleo urbano da Vila do Conde frente a vazamento de óleo. *Olam: Ciência & Tecnologia*, vol 10, pp. 125-142.

Romero, A.; Riedel, P. & Milanelli, J. (2010). Environmental Oil Spills Sensitivity Mapping of the Cananéia-Iguape Estuarine System, Southern Coast of São Paulo, *Revista Brasileira de Cartografia*, N° 62, special edition 01, 299-238.

Salisbury, J.; D'aria, D. & Sabins, F. (1993). Thermal Infrared Remote Sensing of Crude Oil Slicks. *Remote Sensing of Environment*, vol. 45, pp. 225-231.

Santos, V.; Polidori, L.; Silveira, O. & Figueiredo Jr., A. (2009). Aplicação de dados multisensor (SAR e ETM+) no reconhecimento de padrões de uso e ocupação do solo em costas tropicais - costa Amazônica, Amapá, Brasil. *Brazilian Journal of Geophysics*, vol. 27, supl. 1, pp. 39-55.

Silva, D.; Amaro, V.; Souto, M.; Nascimento, M. & Pereira, B. (2010). Geomorfology of a High Sensitive Area on Potiguar Basin (NE Brazil). *Journal of Integrated Coastal Zone Management*, vol. 10, N°4, pp. 545-566.

Silva, A.; Klein, A.; Petermann, R.; Menezes, J.; Sperb, R. & Gherardi, D. (2008). Índice de sensibilidade do litoral (ISL) ao derramamento de óleo, para a ilha de Santa

Catarina e áreas do entorno. *Brazilian Journal for Aquatic Sciences and Technology*, vol. 12, N°2, pp. 73-89.

Smith, K. & Petley, D. (2008). *Environmental Hazards Assessing risk and Reducing disaster*. Routledge: Londres. ISBN: 0-203-88480-9.

Souto, M.; Castro, A.; Grigio, A.; Amaro; V. & Vital, H. (2006). Multitemporal analysis of geoenvironmental elements of the coastal dynamics of the region of the Ponta do Tubarão, city of Macau/RN, on the basis of remote sensing products and integration in GIS. *Journal of Coastal Research*, vol. 39, pp. 1618-1621.

Souza, C.; Castro, A. ; Amaro, V. & Vital, H. (2005). Sistema de Informações Geográficas para o monitoramento de derrames de óleo no litoral norte do estado do Rio Grande do Norte. In: *XII Simpósio Brasileiro de Sensoriamento Remoto*, Goiânia, Brasil, vol. 1, pp. 2383-2388.

Souza Filho, P. & Elrobrini, M. (2000). M. Coastal Zone Geomorphology of the Bragança Area, Northeast of Amazon Region, Brazil. *Revista Brasileira de Geociências*, vol. 30, No 3, pp. 518-522.

Souza Filho, P. & Paradella, W. (2002). Recognition of main geobotanical features in the in the Bragança mangrove coast (Brazilian Amazon Region) from Thematic Mapper and RADARSAT-1 Data. *Wetlands Ecology and Management*, vol 10, pp.123-132.

Souza Filho, P.; Miranda, F.; Beisl, C.; Almeida, E. & Gonçalves, F. (2004). Environmental sensitivity mapping for oil spill in the Amazon coast using remote sensing and GIS technology. In: *International Geoscience and Remote Sensing Symposium*, IEEE Geocience and Remote Sensing, Ancorage,Alaska.

Souza Filho P. & Paradella, W. (2005). Use of RADARSAT-1 fine mode and Landsat 5 TM selective principal component analysis for geomorphological mapping in a macrotidal mangrove coast in the Amazon Region. *Canadian Journal Remote Sensing*, vol 31, pp. 214-224.

Souza Filho, P. (2005). Costa de Manguezais de Macromaré da Amazônia: Cenários Morfológicos, Mapeamento e Quantificação a partir de Dados de Sensores Remotos. *Revista Brasileira de Geofísica*, Rio de Janeiro, vol. 23, No 4, pp. 427-435.

Souza Filho P.; Gonçalves F.D.; Beisl C.; Miranda F.; Almeida E. & Cunha E. (2005a). Sistema de Observação Costeira e o Papel dos Sensores Remotos no Monitoramento da Costa Norte Brasileira, Amazônia. *Revista Brasileira de Cartografia*, 57, pp. 79 - 86.

Souza Filho, P.; Paradella, W. & Silveira, O. (2005b). Synthetic Aperture Radar for Recognition of Coastal Features in the Wet Tropics: Applications in the Brazilian Amazon Coast. *Boletim do Museu Paraense Emilio Goeldi*. Zoologia, Belém, vol. 1, No 1, pp. 149-154.

Souza Filho, P.; Prost, M.; Miranda, F.; Sales, M.; Borges, H.; Costa, F.; Almeida, E.; & Nascimento Jr.; W.(2009a). Environmental sensitivity index (esi) mapping of oil spill in the amazon coastal zone: the piatam mar project. *Brazilian Journal of Geophysics*, vol. 27, supl. 1, pp. 7-22.

Souza Filho P.; Gonçalves, S.; Rodrigues, F. Costa, R. & Miranda, F. (2009b). Multi-sensor data fusion for geomorphological and environmental sensitivity index mapping in the Amazonian mangrove coast, Brazil. *Journal of Coastal Research*, vol. 56, pp. 1592-1596.

Silva, C.; Souza Filho, P. & Rodrigues, S. (2009). Morphology and modern sedimentary deposits of the macrotidal Marapanim Estuary (Amazon, Brazil). *Continental Shelf Research*, vol. 29, pp. 619-631.

Szlafsztein, C. (2000). Coastal management on the state of Pará and the MADAM project research. In: *German-Brazilian Workshop on Neotropical Ecosystems - Achievements and Prospects of Cooperative Research*, Hamburg, Germany.

Szlafsztein, C. & Sterr H. (2007). A GIS-based vulnerability assessment of coastal natural hazards, state of Pará, Brazil. *Journal of Coast Conservation*. vol. 11, pp. 53-66.

Teixeira, S. & Souza Filho, P. (2009). Mapeamento de ambientes costeiros tropicais (Golfão Maranhense, Brasil) utilizando imagens de sensores remotos orbitais. *Revista Brasileira de Geofísica*, vol. 27, pp. 69-82.

Turner, B., Clark, W., Kates, R., Richards, J., Mathews, J. & Meyer, W. (1990) The Earth as Transformed by Human Action, Cambridge University Press, Cambridge, MA. ISBN: 978521446303

Viles, H. & Spencer, T. (1995). *Coastal Problems Geomorphology, Ecology and Society at the Coast*. Edward Arnold, London. ISBN: 0340625406

Wieczorek, A.; Dias-Brito, D. & Milanelli, J. (2007). Mapping oil spill environmental sensitivity in Cardos Island State Park and surroundings areas, São Paulo, Brazil. *Ocean & Coastal Management*. Vol.50, 872-886 pp.

Wisner, B.; Blaikie, P.; Cannon, T. & Davis, I. (2004). *At risk: natural hazards, people´s vulnerability and disasters*, 2° ed., Routledge, London, ISBN 0415252164.

14

Satellite Remote Sensing of Coral Reef Habitats Mapping in Shallow Waters at Banco Chinchorro Reefs, México: A Classification Approach

Ameris Ixchel Contreras-Silva[1], Alejandra A. López-Caloca[1], F. Omar Tapia-Silva[1,2] and Sergio Cerdeira-Estrada[3]
[1]*Centro de Investigación en Geografía y Geomática "Jorge L. Tamayo" A.C., CentroGeo*
[2]*Universidad Autónoma Metropolitana, Unidad Iztapalapa*
[3]*Comisión Nacional para el Conocimiento y Uso de la Biodiversidad, CONABIO*
Mexico

1. Introduction

Interest in protecting nature has arisen in contemporary society as awareness has developed of the serious environmental crisis confronting us. One of the ecosystems most impacted is the coral reefs, which while offering a great wealth of habitats, diversity of species and limitless environmental services, have also been terribly damaged by anthropogenic causes. One example of this is the oil spill from petroleum platforms (in the recent case of the Gulf of Mexico). The effects of global warming—such as the increase in the incidence and intensity of hurricanes and drastic changes in ocean temperature—have caused dramatic damage, such as the bleaching and decrease of coral colonies. In light of this devastating situation, scientific studies are needed of coral reef communities and the negative effects they are undergoing.

The case study presented in this work takes place in the Chinchorro Bank coral reefs in Mexico. These are part of the great reef belt of the western Atlantic, with a biological richness that inherently provides environmental, economic and cultural services at the local scale as well as worldwide. Nevertheless, these services have been weakened for decades due to overexploitation, inducing imbalances and problems in the zone. Over recent decades, numerous biological communities that house constellations of species—whose natural evolutionary process dates back million of years (Primack et al., 1998)—have been alarmingly degraded. If this trend continues, the entire evolution that is sustained by the life of these communities will disappear in a relatively short period of time.

This study clearly demonstrates the application of state-of-art Remote Sensing (RS) in coral ecosystems. It includes an analysis based on the application of Iterative Self Organizing Data Analysis (ISODATA) as a classifier for generating classes of benthic ecosystems present in a coral reef system, using satellite images (Landsat 7-ETM+).

2. Use of remote sensing in coral reef ecosystems

The observation of the earth using remote sensors is a most complete method for monitoring the most significant natural risks (Xin et al., 2007). In general, RS has proven to be a powerful tool in the overall understanding of natural and anthropogenic phenomena. It is particularly appreciated as a non-invasive, non-destructive technique with global coverage. Thus, satellite, airborne and *in-situ* radiometry have become useful tools for tasks such as characterization, monitoring and the continuous prospecting of natural resources.

Research using RS has been strengthened in recent decades as a result of the growing concern worldwide for the preservation of coral reef systems as natural reservoirs. This has been observed to be an excellent method for analysis, which aids in the holistic study of this complex ecosystem. In order to develop an approach that helps to safeguard these ecosystems, it is necessary to understand the physical, chemical, biological and geological dynamics that occur therein (Brock et al., 2006). Andréfouët & Riegl (2004) refer to RS as a technology that is now virtually mandatory for research where spatial and temporal precision is required. RS has gone from being a tool with no application to coral reef systems to one that is *per se* indispensable. Andréfouët & Riegl (2004) discuss four reasons why this change has occurred:

- The proliferation of new sensors for acquiring direct and indirect data for monitoring coral reefs,
- The proliferation and improvement of analytical, statistical and empirical approaches,
- Recognition of global climate change due to anthropogenic human impacts that are lethal to coral reefs and
- Improved integration of technology for the conceptual design of coral reef research.

RS techniques offer an option for marine habitat mapping to determine not only the location and amount of different benthic habitats (Kirk, 1994) but also how these habitats are distributed and the degree of connectivity among them (Rivera et al., 2006). Nevertheless, the study of coral reefs using RS presents several important limitations. For example, intense cloud cover in optical images, optical similarities among spectral signatures of benthic communities, attenuation of the deep component (specific to each coral reef ecosystem) as well as the spatial and spectral resolution of remote sensors. In spite of these limitations, satellite sensors are highly useful for mapping the benthic bottom (Mumby et al., 1997), monitoring changes in its ecology (Krupa, 1999) and defining management strategies (Green et al., 1996).

2.1 Determination of ecological characteristics of coral reefs using remote sensors

Some of the characteristics of coral reefs that can be calculated using RS are temperature, wave height, sea level, turbidity, amount of chlorophyll and concentration of dissolved organic matter. In the case of atmospheric variables, it is possible to determine cloud cover, amount of seasonal rainfall, presence of contaminants and incidental solar energy (Andréfouët et al., 2003). All these factors directly and indirectly influence coral reefs and determine their health status (Andréfouët & Riegl 2004). In addition, it is possible to determine the different benthic ecosystems present in the coral reefs, such as seagrass, type of bottom, algae communities and different types of coral. If the reef is near a tourist or vacation area, anthropogenic impacts can be determined by calculating the growth of the urban stain, vegetation coverage, the structure of the hydrographic basins, etc. Intrinsic conditions of coral reefs can be described, which are largely defined by the inflows and outflow and their transport of sediments and export of dissolved organic matter. This enables us to understand the patterns involved in coral whitening, among other events (Brock et al., 2006).

The coral reefs—located in relatively clear water—allow us to use passive optic sensors (Benfield et al., 2007). The more common satellite sensors that have been used to study this are SPOT, Landsat TM and ETM+ (Andréfouët & Riegl 2004; Benfield et al., 2007; Mumby 2006; Mumby et al., 2004; Mumby and Harborne 1998). Studies previously conducted (Green, 2000; Mumby et al., 1999) have observed that Landsat and SPOT images are suitable for mapping corals, sands, and seagrass, depending on their resolution. Nevertheless, it is important to note that various types of habitats can be represented in one Landsat image pixel (or others with less spatial resolution), which may limit classification abilities (Benfield et al., 2007). Previous studies conducted (Green, 2000; Mumby et al., 1999) have observed that according to the resolution of Landsat images, they are suitable for mapping sea corals, sands and seagrass. Based on this assumption, the data obtained from Landsat and SPOT are adequate for simple complexity mapping (3-6 classes, such as seagrass, sand, dead corals and some species of corals) but for more complex targets (7-13 classes) they are limited by their spatial and spectral resolution. (Mumby, 1997; Andréfouët et al., 2003; Capolsini et al., 2003). To a lesser extent, SeaWiFS (sea-viewing wide field of view sensors) have also been used, as well as IKONOS with higher spatial resolution, LIDAR and SONAR, among others (Andréfouët & Riegl 2004; Andréfouët et al., 2003; Brock et al., 2006; Elvidge et al., 2004; Liceaga-Correa & Euan-Avila, 2002; Hsu et al., 2008; Lesser and Mobley, 2007). It is important to note that analytical methods as well as spatial modeling, statistics and empirical methods at different scales and for different applications have been used in direct relation to ecological processes of reefs (Andréfouët & Riegl 2004). The use of airborne remote sensors, such as CASI (Compact Airborne Spectrographic Imager) with a high spectral or hyperspectral resolution, has gradually been increasing in this type of studies, to the extent that the specialists mention that mapping reefs using air or satellite sensors have proven to be more effective than fieldwork (Mumby, 1999). Nevertheless, field measurements cannot be discarded, since they provide us with the basis for corroborating the information obtained from satellite images. In addition, images from satellite sensors provide the opportunity to conduct multi-temporal monitoring (Helge et al., 2005) in order to identify the status of an ecosystem and predict possible future changes.

According to the above, it can be stated that studies applying RS in coastal ecosystems and, specifically, in coral reef ecosystems provide information and knowledge that can successfully be applied to define management strategies for these important ecosystems, as well as to design viable alternatives for their conservation.

3. Spectral reflectance of coral

To make observations, we move vertically and gradually from the coral surface to the water surface, measuring the changes in the quantity of light in the water column that falls directly on the coral. The quantity of light present obviously affects the amount that is reflected by the coral, and is therefore a crucially important parameter for mapping it.

Spectral reflectance (ρ) is a key parameter for conducting studies of coral reefs using RS (Hochberg et al., 2004). Two factors clearly and concisely explain this. First, ρ represents the boundary of radiative transference in the water surface optics. Therefore, taking into account ρ can resolve the problem of inverse radiative transference presented by passive remote sensors when applied in this field. Second, ρ is the function that denotes the object, the composition of the material and its structure. Therefore, it serves as a bridge between the optics of the object and the shape of the sea bottom (Hochberg et al., 2004).

In the process of classifying images and generating thematic maps, large differences have been noted in spectral reflectance among the coral reefs' benthic communities (Brock et al., 2006). Variability in the vertical relief, or rugosity, is a significant aspect of the complexity of a habitat, a factor that both reflects and governs the spatial distribution and density of many reef organisms (McCormick 1994). These factors, which respond to these evaluations, vary according to the differences among sediments, the presence of different algae species and the coverage of atypical algae in surface water in some reef zones. Thus, Hochberg et al. (2004) mention the importance of creating a specific approach using RS to study the surface water mass presented by atypical algae, since it has been shown that the mere presence of these organisms indicates classes that are spectrally distinct from other reef communities, even when they represent the same species.

Differences among the spectral signatures of corals provide a high likelihood of satisfactorily delineating and defining their different features in a satellite image. The problem with the above process is that the ρ of the corals is a function of pigmentation, structure, the orientation of their branches and their internal characteristics (Newman et al., 2006). In addition, though the interactions between light and the atmosphere are well-studied, the challenge is to establish controls for the effects of the water column in which the coral is found that influence these factors. Taking into account the curvatures in order to correct the acquired data provides more valuable information about the conditions and health of the living communities sheltered by the coral. Newman et al. (2006) indicate that two categories have been defined by recent studies which were conducted to measure in situ the spectral signatures of the coral environment:

i. The spectral signatures are examined according to the variation in the pigment density, which characterizes the sensorial color of the different coral species (Newman et al., 2006). Some studies have analyzed the contribution of color to the measurement of radiance (R), in particular, by comparisons with unpigmented coral structures. These observations resulted in the spectrum of coral whitening and structures saturated with zooxanthellae (Newman et al., 2006), which provide a measure of the health status of the complex reef system. Color has been used as a comparison measurement among three coral species, five algae species and three benthic communities (Hochberg and Atkinson, 2000), and as a means to differentiate between dead coral in different stages and algae colonization (Clark et al., 2000).
ii. Spectral signatures were examined according to morphological characteristics (Newman et al., 2006).
   Corals exhibit distinct and complex structural morphologies, partially due to environmental conditions such as light availability, water motion and suspended sediment (Joyce & Phinn, 2002). Reflectance values measured over varying angles and azimuths were examined to determine the bidirectional reflectance distribution

function of coral species and the inter-species variation between rounded and branching types (Joyce and Phinn, 2002; Newman et al., 2006).

4. Mapping coral reefs using remote sensors

The worldwide importance of coral reefs in light of current threats has generated interest in developing methods to study this type of ecosystems at global scales (Kuhn 2006). The use of remote sensing to map underwater habitats is increasing substantially. This enables using the derived information to determine the status of these natural resources as a basis for planning, management, monitoring, conservation and evaluating their potential.

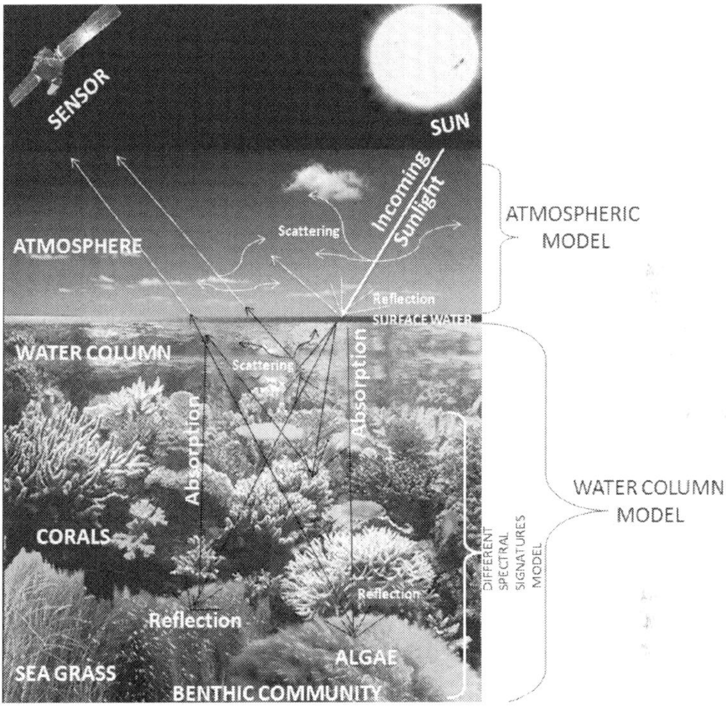

Fig. 1. Components of Remote Sensing in mapping coral reefs.

As was mentioned previously, high resolution spectral sensors exist that have elements that enable specific analysis with an excellent capacity for modelling environmental and structural variables in the coral reefs (Holden and LeDrew, 1998). The data produced by this type of sensors provide products that can be combined with models to photosynthetically calculate the radiation available through the photic zone and the surface of benthic substrates. Established models for calculating incident solar radiation are developed and evaluated based on routine satellite and meteorological observations (Brock et al., 2006). The spectral differences among corals, seagrass and algae are nearly imperceptible and not easy to detect with the three bands (blue, green and red) of the sensors that can penetrate the water column (Holden and LeDrew 1998; Hedley and Mumby 2002; Karpouzli et al., 2004). This is why RS studies applied to the mapping of submerged benthic ecosystems requires the generation of new processing methodologies. In addition, coral habitats present a heterogeneity that is inherent of their complexity, and therefore the task of discerning among the different spectral signatures is more complicated. That is, the pre-processing of images applied to this type of environments should not only incorporate the elimination of noise in the atmospheric and batimetric portions, but should also take into account the components of the water column, as shown in Figure 1.

5. Pre-processing of satellite images

All satellite images must undergo an initial processing of crude data to correct radiometric and geometric distortions of the image and eliminate noise. It must be taken into account that the energy captured by the sensor goes through a series of interactions with the atmosphere before reaching the sensor. As a result, the radiance registered by the sensor is not an exact representation of the actual radiance emitted by the covering. This means that the image acquired in a numerical form presents a series of anomalies with respect to the real scene being detected. These anomalies are located in the pixels and digital levels of the pixels that make up the data matrix. The purpose of correction operations is to minimize these alterations. The corrections are made during pre-processing operations, since they are carried out before performing the procedures to extract quantitative information. The product obtained is a corrected image that is as close as possible, geometrically and radiometrically, to the true radiant energy and spatial characteristics of the study area at the time the data are collected. Atmospheric correction is a process used to reduce or eliminate the effects of the atmosphere and allow for more precisely seeing the reflectance values of the surface being studied or analyzed.

Nevertheless, when attempting to map or derive quantitative information from subaquatic habitats, the depth of the water significantly affects the measurements taken by remote sensors, making it possible to generate confusion about spectral signatures. Therefore, atmospheric and geometric corrections are not sufficient when the objective is to extract features of the covering of the bottom of the water. That could be considered a characteristic and, in some cases, a limitation of passive sensors in remote sensor applications in marine environments. Thus, in this type of studies, a water column correction is performed to improve reliability when analyzing the results of the image and to eliminate the noise resulting from the variation in the ground's reflectance (Holden 2002; Holden and LeDrew, 1998; Mumby, 1998).

5.1 Correction of remotely sensed imagery

5.1.1 Radiometric correction

The radiance from the sensor (L) is calculated as:

$$L = c0 + c1 * ND \tag{1}$$

Where c0 and c1 are the offset and gain, respectively, of the radiometric calibration and ND is the digital number recorded in a particular spectral band. The process of obtaining L is called radiometric correction.

The total signal captured by the sensor consists of three parts: atmospheric scattering of radiation, radiation reflected by the pixel and radiation reflected by the vicinity of the pixel and scattered in different (adjacent) directions.

5.1.2 Atmospheric correction

The atmospheric conditions (water vapor, aerosols and visibility) in a scene can be calculated using algorithms that are performed using a database based on atmospheric functions. The surface spectral reflectance of an interaction target in a scene can thereby be seen as a function of the atmospheric parameters. ¶(6pt)

5.1.3 Geometric correction

The geometric correction consists of distinguishing the other types of radiation and only considering that which is reflected by the pixel. The objective is to remove geometric distortion; that is, to locate each pixel in its corresponding planimetric position. This enables associating the information obtained from a satellite image with thematic information from other sources.

5.2 Water column correction

The coral reefs generally develop in transparent or clear water, which facilitates study and analysis with passive optic, multispectral or hyperspectral sensors (Mumby et al., 1999). When light penetrates the water column, its intensity exponentially decreases as the depth increases. This process is known as attenuation, and it has an important effect on data obtained by remote sensors in aquatic environments (Green, 2000). The attenuation process is shown in Figure 2.

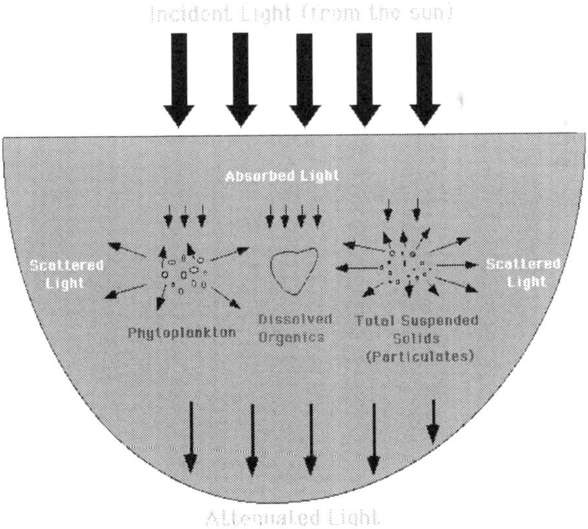

Fig. 2. Processes of light attenuation in the water column (SERC, 2011).

There are two reasons for this phenomenon:

- Absorption: light energy is converted into another type of energy, generally heat or chemical energy. This absorption is produced by the algae, which utilize the light as a source of energy, by suspended organic and inorganic particulate matter (OPM and IPM), dissolved inorganic compounds and the water itself.
- Scattering: This phenomenon results from the collision of light rays and suspended particles, causing multiple reflections. The more turbid the water (more suspended particles) the greater the scattering effect, making it difficult for light to penetrate.

The attenuation varies according to the wavelength of the electromagnetic radiation (EMR). For example, in the region of visible light, the red portion of the spectrum attenuates more quickly than the short wavelength, such as blue.

Figure 3 shows, for 4 spectral bands (blue, green, red and infrared), how the spectrum in a particular habitat (seagrass or macroalgae) can change as the depth increases. The spectral radiance registered by a sensor is dependent on the reflectance of the substrates and the depth. As the depth increases, the possibility to discriminate spectrums or spectral signatures of the habitats decreases. In practice, the spectrum of sand at a depth of 2 meters is very different than that at 20 meters. According to Mumby and Edwards (2000), the spectral signature of sand at 20 meters could be similar to that of seagrass at 3 meters. All these factors influence the signal and can create a good deal of confusion when using visual inspection or spectral classification to classify these habitats. Therefore, the influence of the variability in depth must be eliminated, which is known as water column correction or depth correction (Mumby and Edwards 2000).

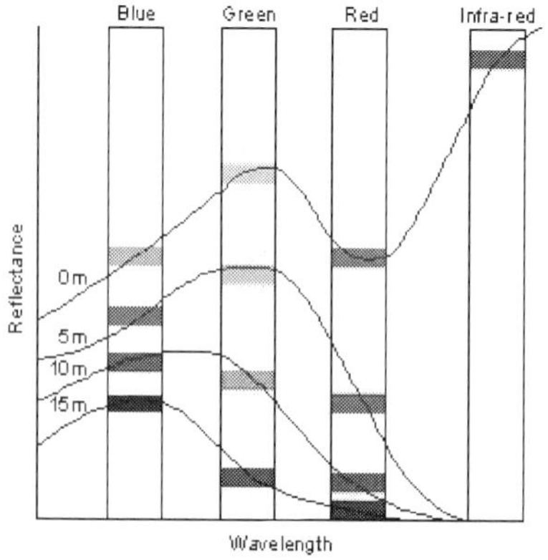

Fig. 3. Spectral differences for a habitat (seagrass or macroalgae) at different depths (Mumby and Edwards, 2000).

A variety of models exist that can be used to compensate for the effect of the water column. Nevertheless, many require optical measurements of the optical properties of the water itself, as well as information about the depth of water per pixel (Gordon, 1978; Philpot, 1989; Mobley et al., 1993; Lee et al., 1999; Maritorena et al., 1994; Maritorena 1996; Lee et al., 1999). Thus, the method proposed by Lyzenga (1981) is applied, which has been used and described by other authors, such as Mumby et al., 1997, 1998, Mumby and Edwards 2002, Andréfouët et al., 2003, etc. This approach has the advantage of taking into account the majority of the spectral information and not requiring data for the components of the water surrounding the reef. Instead of deriving the spectra of the different types of sea bottoms and water properties, this method transforms the spectral values into "depth-invariant indices." The primary limitation of this method, among others, is that it must be applied to clear water (i.e. type 1 or type 2); the study area meets this requirement.

To eliminate the influence of depth on sea bottom reflectance, the following need to be taken into account: the identification of the characteristics of attenuation of the water column and having digital models of the depth; although these are not very common, particularly for coral reef systems (Clark et al., 2000). This work used a bathymetric model provided by SEMAR (2008) that makes possible a good deal of reliability and precision to the measurements.

The procedure is divided into various steps:

1. Elimination of the atmospheric scattering and the external reflection from the water surface (atmospheric correction). This can be carried out using a variety of methods, such as dark pixel subtraction (Maritorena, 1996) and ATCOR (Richter, 1996, 1998).
2. Selection of pixel samples with the same substrate and different depths.
3. Selection of a spectral band pair, with good penetration of the water column (that is, bands found in the visible light spectrum—Landsat TM and ETM+ 1/2, 2/3 and 1/3).
4. Linearization of the relationship between depth and radiance, Xi = ln (Li), where Xi is the transformed radiance of the pixel in band i (band 1) and Li is the radiance of the pixel in band j (band 2). When the intensity of the light (radiance) is transformed using the natural logarithm (ln), this relationship becomes linear with the depth. Therefore, the transformed radiance values will decrease linearly as depth increases:

$$X_i = Ln(L_i) \tag{2}$$

5. Determination of the attenuation coefficient (quotient) using a biplot of the transformed radiance of the 2 bands (Li and Lj). The biplot contains data for one type of uniform bottom (sand) and variable depth. It is created using the following equations:

$$K_i/K_j = a + \sqrt{(a^2+1)} \tag{3}$$

$$a = \frac{\sigma_{jj} - \sigma_{ii}}{2\sigma_{ij}} \text{ and } \sigma_{ij} = \overline{X_i X_j} - \overline{X_i}\,\overline{X_j} \tag{4}$$

where σ_{ii} is the variance in band i and a is the covariance between bands i and j.

6. Lastly, the depth-invariant index is generated using the equation by Lyzenga (1981):

$$IIP_{ij} = \ln(L_i) - \left[\left(\frac{k_i}{k_j}\right)\ln\right](L_j) \tag{5}$$

The result of this operation generates a new band—the image with water column correction for a band pair (depth-invariant index). Since the values of this band are whole numbers with decimals and can be negative, in order to visualize them they need to be converted into an 8-bit format, that is, gray values between 0 and 255. To this end, minimum and maximum values for the resulting image must be found and linearly distributed between the values 1 and 255 (0 is not included because it is assigned to the masked surface area). The depth-invariant index is essential when the objective of the study is to extract spectral data for submerged aquatic environments.

6. Review of classification methodologies

The classification of a satellite image consists of assigning a group of pixels to specific thematic classes based on their spectral properties. The spatial classification of underwater coastal ecosystems is one of the most complex processes in thematic cartography using satellite images. As previously mentioned, this can be attributed primarily to the influence of the atmosphere and the ocean water column, through which electromagnetic radiation passes. In addition, it is worth mentioning that these ecosystems undergo constant variation, especially after significant events such as strong hurricanes. Nevertheless, different authors (Mumby et al., 1997; Andréfouët & Payri 2000; Mumby and Edwards 2002; Andréfouët et al., 2003; Pahlevan et al., 2006; Call et al., 2003, etc.) have been using remote sensing to develop different classification methods for these ecosystems and, in particular, for coral reefs.

The maximum likelihood classifier is the most common method, and has been used by authors such as Mumby et al. (1997), Andréfouët et al. (2000), Mumby and Edwards (2002), Andréfouët et al. (2003), Pahlevan et al. (2006), and Benfield et al. (2007). Its primary advantage is that it offers a greater margin for accounting for the variations in classes through the use of statistical analysis of data, such as the mean, variance and covariance. The results of the method can be improved with the incorporation of additional spatial information during the post-classification process, since this helps to spectrally separate the classes that had been mixed.

Another method also used by Mumby et al. (1997) is agglomerative hierarchical classification with group-average sorting. An alternative proposal is object-oriented classification, which consists of two steps, segmentation and classification. Segmentation creates image-objects and is used to build blocks for further classifications based on fuzzy logic. Another method that has been used is ISODATA (iterative self-organizing data analysis), which uses a combination of Euclidian squared distance and the reclassification of the centroid (Call et al., 2003). In this study, ISODATA was used to perform the classification.

6.1 ISODATA (Iterative Self Organizing Data Analysis)

ISODATA is an unsupervised classification method as well as a way to group pixels, and uses the minimum spectral distance formula. It begins with groups that have arbitrary means and each time the pixels in each of the iterations are regrouped and the means of the groups change. The new means are then used for the next iterations.

The algorithm for obtaining the classification is based on the following parameters:

a. The user decides on the number N of clusters to be used. For the first calculation, it is recommended to use a high number, which is then reduced by interpreting the image.
b. A set of N clusters in the space between the bands is selected. The initial location is in the zones with the highest reflectance.
c. The pixels are assigned to the closest cluster.
d. The clusters are associated, dispersed or eliminated depending on the maximum distance of the class or the minimum number of pixels in a class.
e. The grouping of pixels in the image is repeated until the maximum number of iterations has been reached, or a maximum percentage of pixels are left unchanged after two iterations. Both parameters can be specified

7. Case study

The Chinchorro Biosphere Reserve (Fig. 4) is located in the open Caribbean Sea, 30.8 km east of the coastal city of Mahahual, which is the closest continental point. The coral reef of Chinchorro Bank, Mexico, is part of the great reef belt in the western Atlantic, the second largest in the world, and is the biggest oceanic reef in Mexico. With a reef lagoon area of 864 km2, it is considered a pseudo-atoll or reef platform (Camarena, 2003). Chinchorro Bank is a reef complex that contains an extensive coral formation with a vast wealth and diversity of species and high ecological, social and cultural value. It inherently provides certain services, including the protection of the coast from battering by storms and hurricanes. The area has been exploited by fishing and tourist-related scuba diving over the past decades. The Chinchorro Bank supports pristine reefs, coral patches, extensive areas of seagrass, microalgae beds and sand beds. The reserve's ecosystems are marked by mangroves and reef zones. The composition of the taxocenosis of coral is known to contain hexacorals,

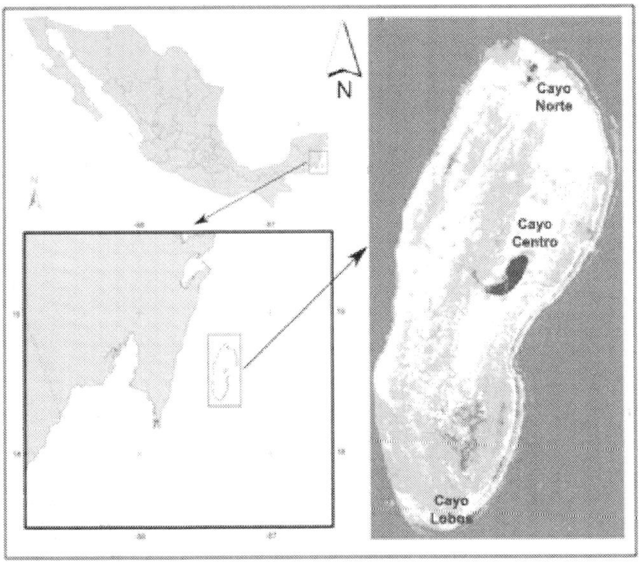

Fig. 4. Study Area: Chinchorro Bank, Mexico.

octocorals and hydrozoas and a reported 95 different species (Camarena, 2003). The diversity of the fauna in the Chinchorro Bank is very high and includes several phyla, families, genres and species, with at least 145 macro invertebrate and 211 vertebrate species, in addition to the corals (Bezaury et al., 1997).

The biogeographic region of Chinchorro Bank is delimited on the north by the Caribbean Province which extends along Central and South America. This province begins in Cabo Rojo, in southern Tampico, and extends into eastern Venezuela and the northern Orinoco delta. The land biota is greatly similar to that of the continent and is therefore considered to be part of the Yucateca Province. It is located in the Mexican Caribbean, across from the southeastern coast of the state of Quintana Roo, between the 18°47′-18°23′ N and 87°14′-87°27′ W parallels. It is 30.8 km from the continent and separated from it by a wide canal 1000 m deep. The shape of Chinchorro Bank is elliptic, with a reef lagoon that includes a sandy bank 46 m long (north-south) and 18 km wide (east-west) at its broadest part. The total area is 144360 ha. The periphery of the bank is bordered by active coral growth on the eastern (windward) margin, which forms a coral reef, or breaker, while along the western margin (leeward) the breaker disappears and the coral growth is semicontinuous and diffuse (Camarena, 2003). There are four emerged zones within the bank—known as "Cayo Norte" (two islands), "Cayo Centro" and "Cayo Lobos"—whose ecological value is very high because of their diverse species of land and water flora and fauna (Camarena, 2003).

8. Information resources

The geospatial database used in this study includes a Landsat 7-ETM+ image (Table 1), bathymetric information and in situ data for sand (Figure 5). The digital data were projected to UTM (Universal Transverse Mercator) zone 16 north with WGS-84 datum. ERDAS, GEOMATIC 10.2 and ArcMap 9.3 were used to process the data.

The importance of choosing the type of image with which to work is well-known, particularly because the users will need to make sure to use images that are suitable to the purpose of the study. The nature of a platform-sensor system determines the characteristics of the image's data (Green, 2000). The Landsat 7-ETM+ (Table 1) image obtained had no cloud cover. It is worth noting that this type of images provides adequate coverage of the area for regional and temporal monitoring studies.

Date	2000-03-29
Scan time	16:03:05
Path/Row	18/47
Spatial resolution (m)	30
Spectral bands used	3
Spectral range (µm)	0.5-0.69
Azimuth	116.29
Solar angle	59.43

Table 1. Characteristics of the Landsat 7-ETM+ image used

It is also very important to note that bathymetry is one of the most relevant factors in the dynamic ecology of coral reefs. Numerous reef studies show that coral species diversity tends to increase as a function of depth, reaching its maximum between 20–30 m and diminishing with greater depth (Huston, 1985). This depth effect results in a marked zonation of the reef community (Aguilar-Perera and Aguilar-Dávila, 1993). While the upper depth limits of corals are controlled by various physical and biological factors, their maximum depth depends largely upon light availability (García-Ureña, 2004). The bathymetric soundings for Chinchorro Bank used by this study were done in 2008 by the Mexican Navy (SEMAR, 2008). The depth of the interior of the bank varies. The northern portion is shallower, between 1 and 2 m, the depth of the central portion ranges between 3 and 7 m, and the southern portion is deepest, varying between 8 and 15 m (SEMAR, 2008). There are 4 emerged zones within the bank, known as keys, which have high ecological value because of their diverse species of flora and aquatic and land fauna (Camarena, 2003). Figure 5b shows bathymetry data for the Chinchorro Bank, where the depths of the zone can be seen.

In situ sampling data were provided by SEMAR. Data from Carricart-Ganivet et al. (2002) were also used. Based on these data, 4 of the most representative classes were determined: 1) coral mass, 2) coral patches, 3) seagrass and algae and 4) sand. The ocean and keys, or emerged areas, are not part of the classification criteria, though they are also represented. Unfortunately, the databases for the in situ sampling have disadvantages—such as mixing classes in the same point and lack of definition of the benthic bottom, among others—that prevent their being used for validation purposes. Only data for sand provided by SEMAR do not present these disadvantages and could be used for water column correction, as explained further below.

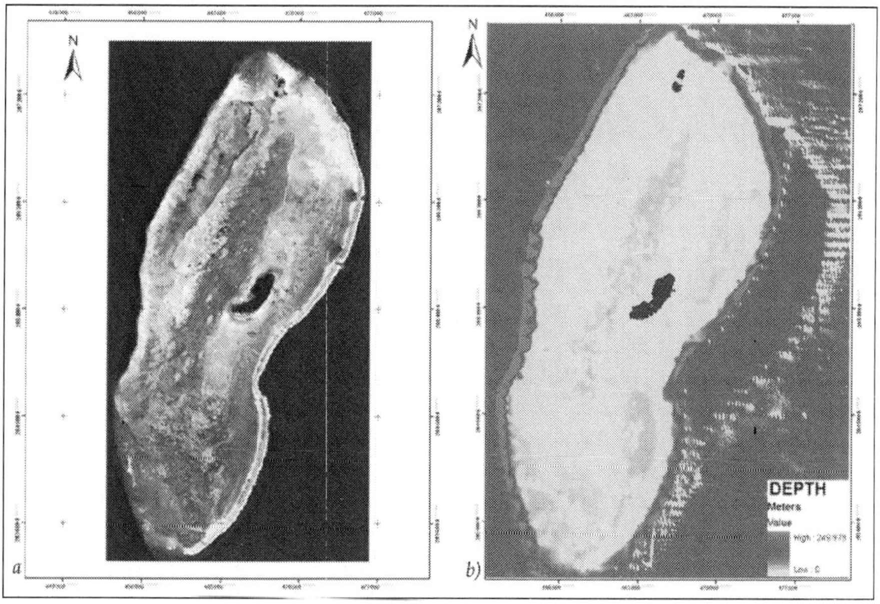

Fig. 5. Information resources. a) Landsat 7-ETM+ image and b) depth of the Chinchorro Bank

9. Results and discussion

9.1 Image processing

A Landsat 7-ETM+ image from March 29, 2000 was processed. Before conducting the quantitative analysis of the data, a post-calibration was performed of the constant gain and offset to convert the image ND to spectral radiance. The spectral radiance was also corrected for atmospheric effects to obtain the surface reflectance values. A geometric correction was not performed because the level of the processing of the Landsat images includes this correction. Only 3 of the 8 bands contained by Landsat were used (blue, green and red). The depth correction was developed with the Lyzenga (1981) method, which has been used and described by other authors (Mumby et al., 1997, 1998; Mumby and Edwards 2002; Andréfouët et al., 2003).

9.2 Water column correction

Lyzenga (1981) shows that when drawing a scatterplot of 2 of the logarithmically transformed bands in the visible spectrum (one on each axis), the pixels for the same type of bottom (i.e. sand at different depths) follow a linear trend. Repeating this process for different types of bottoms produces a series of parallel lines and the intersection of those lines generate a unique depth-invariant index which is independent of the type of bottom; all the pixels for a particular bottom have the same value as the index regardless of the depth at which they are found (Andréfouët et al., 2003). A group of pixels representative of the depth of the water column was selected for this study, therefore pixels very close to the surface (< 1m) were eliminated. Sand was the only substrate used since it is the most homogenous bottom in coral environments, and is the one most used by various authors (Mumby and Edwards 2002; Lyzenga, 1981) and the most easily recognizable for interpretation purposes. For the specific case of the Chinchorro Bank, 100 points of sand between 1 and 10m of depth were used to determine the attenuation coefficient (quotient) for the band pair ½, 99 points were used for bands 1/3 and 96 for bands 2/3. The data for point radiance to a type of bottom were extracted from the image and transferred to a spreadsheet. Figure 6a shows the graphic spectral radiance of bands 1 and 2 (atmospherically corrected) with respect to the depth for one specific type of bottom (sand) and variable depth.

Figure 6b shows the linearization of the exponential attenuation of the radiance for bands 1 and 2 using natural logarithms, since in practice it is virtually impossible for the points to adhere to a perfect line given the natural heterogeneity of the different types of bottoms, variations in the water quality, surface roughness of the water, etc. Figure 6c shows the biplot of bands 1 and 2 for a single substrate (sand) at different depths. To this end, the variance of band 1 and the covariance of bands 1 and 2 are evaluated (Table 2 and 3). Table 3 shows the different values for obtaining the attenuation coefficient, according to spectral band.

	Band 1	Band 2	Band 3
Variance (σ_{ii})	0.2628	0.6334	0.2761

Table 2. Variance of the radiance of each band

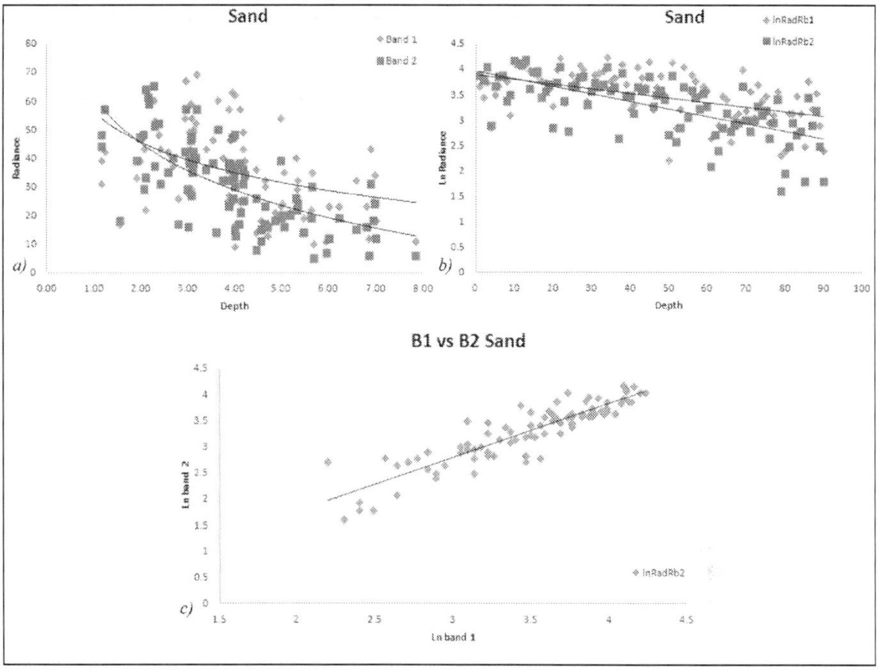

Fig. 6. Steps for water column correction: (a) spectral radiance of bands 1 and 2 (atmospherically corrected), (b) exponential decay of the radiance for bands 1 and 2 using natural logarithms and (c) biplot of bands 1 and 2 for a single bottom (sand) at different depths.

	Ratio 1/2	Ratio 1/3	Ratio 2/3
Covariance (σ_{ij})	0.3200	0.1178	0.2327
aij	-0.0593	-0.0031	0.0184
ki/kj	0.94	0.99	1.00

Table 3. Calculation of ratio of attenuation coefficients

Figure 6c shows the biplot of the logarithmically transformed bands 1 and 2, representing the attenuation coefficient (ki/kj) for bands 1 and 2. It is important to mention that if different types of bottoms are represented in a biplot, they would theoretically represent a line with a similar behavior, varying in position only due to differences in spectral reflectance. The gradient of the line would be identical since ki/kj does not depend on the type of bottom. The intersection of the line with the y-axis represents the depth-invariant index, since each type of bottom has a unique y-intersect regardless of depth. Each pixel is assigned an index depending on the type of bottom, which is obtained using the natural logarithm transformation for each band and the connection of the coordinate to the origin of the y-axis through gradient line ki/kj. The pixels are thus classified for different types of bottoms.

As mentioned before, the depth-invariant index is generated according to band pairs—1/2, 1/3 and 2/3, correponding to bands 1 (blue), 2 (green) and 3 (red) (Figure 7). The image

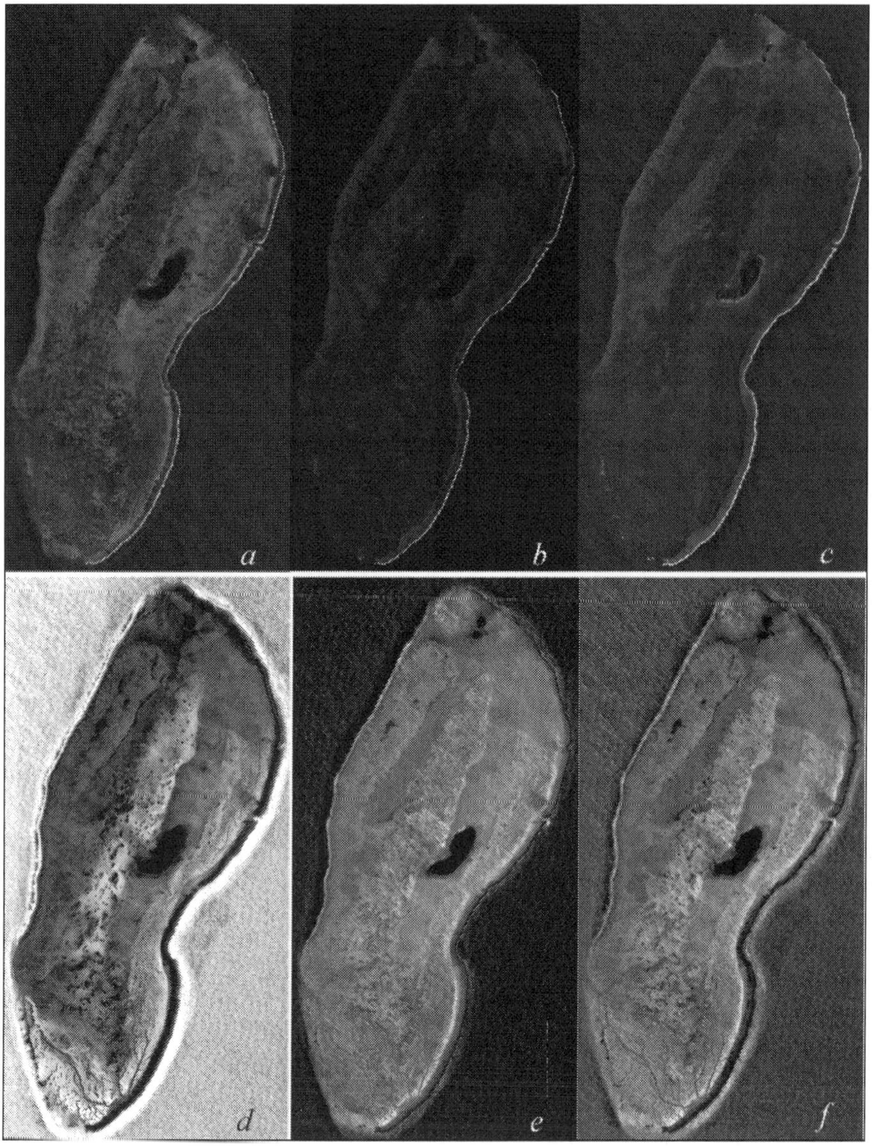

Fig. 7. Visualization of the Landsat 7-ETM+ image before and after water column correction. a) image of band 1 (blue, 450-520 nm), b) band 2 (green, 530-610 nm), c) band 3 (red, 630-655 nm), d) depth-invariant index combination of bands 1/2, e) 2/3 and f) 1/3.

resulting from the depth-invariant index was significantly different than the image without correction, since it reveals more details of the structures of the benthic bottom, especially in zones with greater depths.

9.3 ISODATA classification

As an initial approach to the classification of submerged benthic ecosystems in the Chinchorro Bank, ISODATA was used as a classification method, since not much needs to be known about that data beforehand. A little user effort is required to identify spectral clusters in data. The results of the benthic classification in the Chinchorro Bank were visually evaluated according to the quality of the segmentation using the classification by Aguilar-Perera & Aguilar-Dávila (1993), and with bathymetric data that greatly determine the ecology of the corals, as explained next.

Figure 8a shows the Landsat image with atmospheric correction for the RGB (1,2,3) combination and Figure 8b shows the image resulting from the depth-invariant index by bottom type. At the bottom of the figure, two images classified using ISODATA are included, both with the same type and number of classes. Figure 8c presents the classification performed without water column correction; that is, using the image from 8a as input. Figure 8d includes the classification performed based on the depth-invariant index (shown in 8b); that is, taking into account water column correction. To identify the categories resulting from the ISODATA process, benthic bottoms in the Chinchorro Bank as defined by Aguilar-Perera & Aguilar Dávila (1993) were used as a basis. It can be seen (8c) that the classification without water column correction produced a substantial mix of classes throughout the image, unlike the classification obtained by applying water column correction (8d). According to authors such as Aguilar-Perera & Aguilar Dávila (1993), Chávez and Hidalgo (1984) and Jordán (1979), the periphery of the Chinchorro Bank is surrounded by abundant coral growth on the eastern margin. A barrier reef is thereby formed that disappears along the western margin where the coral growth is semi-continuous and diffuse. This spatial distribution of the corals can be clearly seen in the results of the classification with water column correction (Figure 8d), unlike classification without correction (Figure 8c).

One known ecological characteristic of reef systems is that the zonation of the reef bottom and its ecological dynamics are strongly influenced by the depth (Huston, 1985; Loya, 1972; Gonzáles et al., 2003). The seagrasses constitute a type of benthic bottom normally present in shallower zones. These observations and the use of bathymetry enable corroboration of the validity of the spatial distribution of seagrasses obtained by classification with water column correction. The shallower zones are located in the northern (1-2m) and central (3 and 4 m) portions; these two zones best correspond to the zone with seagrass generated in the image shown in 8d, as opposed to the image in 8c where it can be seen that the seagrass class is distributed throughout the bank. In addition, 8c shows a mix between seagrass and corals, a result that is not justifiable since the corals normally develop at depths between 5 and 30m. Using the depth criterion again in order to define the zonation, it is possible to state that the classification with water column correction produces good results for identifying coral patches, since they are found at depths between 7 and 12 m, as can be seen in Figure 8d. As a general observation, we can state that the results of the classification with water column correction generate data that are consistent with the theory regarding the

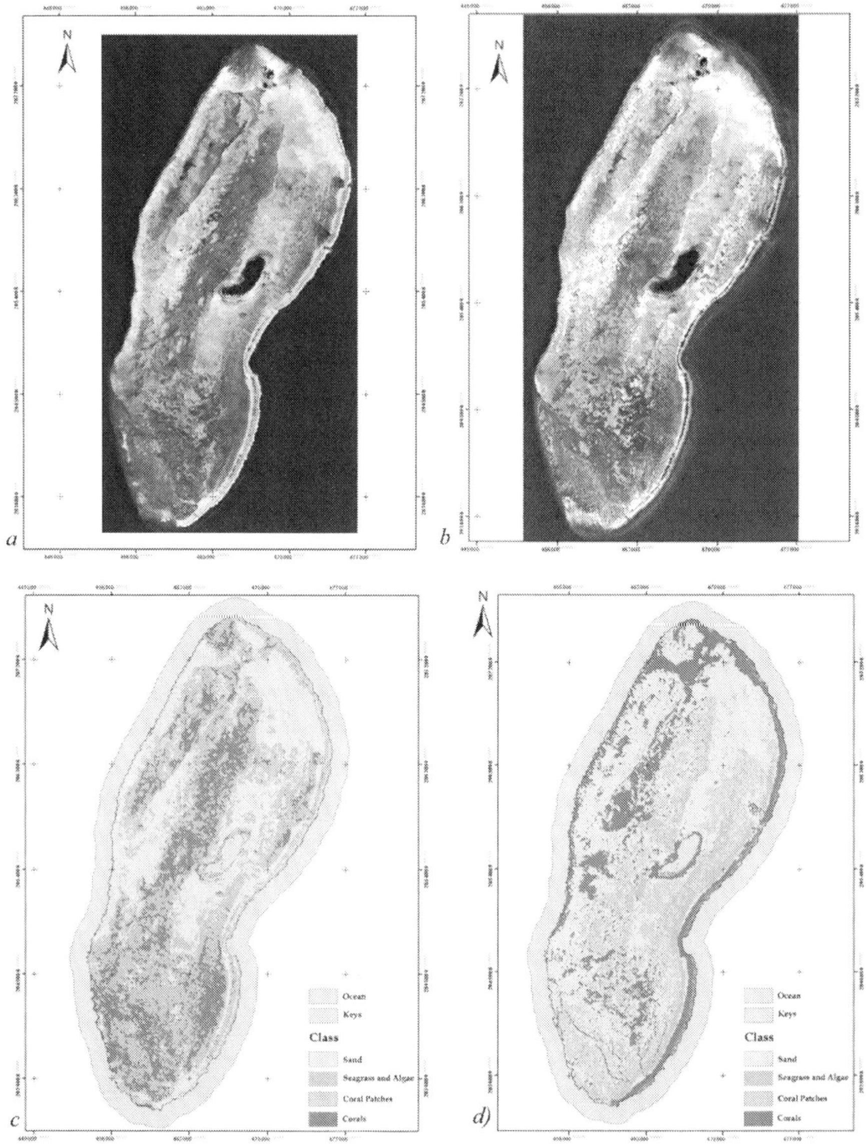

Fig. 8. a) Landsat 7-ETM+ image, RGB (1, 2, 3), b) image resulting from the depth-invariant index by bottom type using bands 1 and 2, and classification of the benthic bottom in the Chinchorro Bank using ISODATA, c)without water column correction and d) with water column correction.

influence of depth in defining the zonation of benthic bottoms, as well as observation of other authors regarding the spatial distribution of sea-bottoms.

Figure 9 shows a close-up to facilitate the visual analysis of the differences between the classes obtained using ISODATA, implemented with and without water column correction.

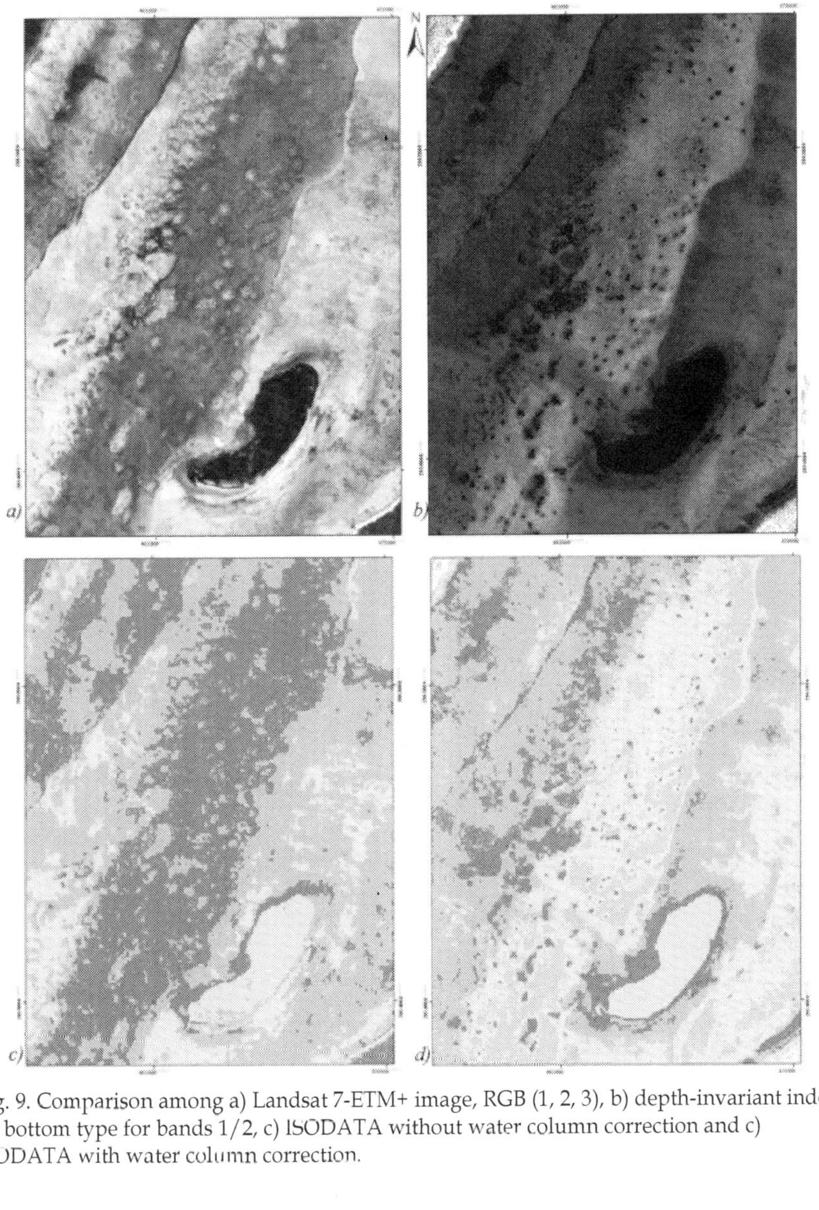

Fig. 9. Comparison among a) Landsat 7-ETM+ image, RGB (1, 2, 3), b) depth-invariant index by bottom type for bands 1/2, c) ISODATA without water column correction and c) ISODATA with water column correction.

In this figure, it can be seen that thanks to the water column correction, the classes are better defined, with mixing among them—caused by interference by the depth of the water column—avoided to whatever extent possible. The ISODATA algorithm more accurately selects and groups clusters, eliminating this problem. This visualization again confirms the advantage of performing water column corrections to obtain better results for the processes to classify benthic bottoms.

10. Conclusions

The study shows that the application of new remote sensing methods is crucial to the pre-processing of images in order to identify submerged aquatic ecosystems. This is because when quantitative information is mapped or derived from satellite images of aquatic environments, the depth of the water causes spectral confusion and therefore significantly affects the measurements of submerged habitats. Water column correction minimizes this effect, which enables distinguishing the classes of benthic ecosystems present in the Chinchorro Bank and demonstrates improvement especially in zones representing more variation in depth. Thus, water column correction is an indispensible pre-processing method in the cartography of submerged aquatic ecosystems.

The water column correction method used in this study uses the majority of the spectral information while disregarding the characteristics of the water surrounding the reef, such that the spectral values are transformed from a band pair into a depth-invariant index. This should be applied in relatively clear water (type 1 or type 2), as is the case of the Chinchorro Bank. Using this process, the attenuation effect of the water column was minimized, which is one of the primary problems with the segmentation of images of submerged ecosystems.

Traditional, unsupervised classification methods, such as ISODATA, have difficulty detecting subclasses, that is, this type of classifier makes it complicated to detect pixels between very close classes with distributions that share an overlapping zone. When classifying benthonic habitats in the Chinchorro Bank, it was possible to observe that the classes with less concentration of pixels were masked by those with greater amounts. This may be because standard methods, such as ISODATA, use moving mass center techniques to locate the classes and, thus, what are called subclasses become undetectable.

In general, the data from remote sensors are used for mapping reef habitats. Although the classification presented here was quite general—only 4 classes were determined—the results show that the Landsat 7-ETM+ images are able to identify different classes in submerged benthonic environments. Although the classification resulted in visually optimal results, the need to incorporate statistical validation of the data is important, so as to determine the accuracy of the classification performed in comparison to the reality; this was not possible for this study because an adequate database of in situ sampling was not available. Nevertheless, because of the visual comparison with classes identified by studies such as those by Aguilar-Perera & Aguilar Dávila (1993), Chávez and Hidalgo (1984) and Jordán (1979) and the consistency with the theory of the zonation of benthic bottoms based on depth, it can be concluded that the classifications obtained by ISODATA successfully determined the majority of the benthonic cases defined in this study of the Chinchorro Bank.

Coral reefs are being threatened worldwide by a combination of natural and anthropogenic impacts. Although the natural impacts are intense, there are intermediate time lapses that can contribute to maintaining biodiversity. On the other hand, the human impacts—which may seem to be less intense because they are not as perceivable to the eye—are chronic and can unleash a chain of negative effects. This sequence of negative effects normally does not give ecosystems the opportunity to recover and maintain their characteristic function and structure.

The search for new methodologies to process satellite images is indispensable to identifying the current trend in the degradation of marine habitats; methodologies that generate new and improved classifications that are highly reliable and with a level of detail that is adequate for mapping these ecosystems. Through this type of study, it is possible to organize, relate and manage information from satellite images in order to propose agreed-upon strategies to conserve natural resources, as part of comprehensive environmental policies to properly solve the problems. Thus, these data can be used as a basis to plan the monitoring of reefs in order to create scientific methods to generate knowledge and environmental awareness in the society and to contribute to the mitigation of the loss of reefs due to impacts from current global warming and other anthropogenic and global changes.

11. Acknowledgments

The authors would like to thank the Mexican Navy (SEMAR), Deputy Department of Oceanography, Hydrography and Meteorology (Dirección General Adjunta de Oceanografía, Hidrografía y Meteorología) for the information provided regarding bathymetry and the field sampling of sand data. We also thank Dr. Juan Pablo Carricart Ganivet and Janneth Padilla Saldívar for the information and geographic basis from the Comprehensive Management of the Chinchorro Bank: Geographic survey and geomorphologic characterization of the reef.

12. References

Aguilar-Perera A. & Aguilar-Dávila W., (1993). Banco Chinchorro: Arrecife Coralino en el Caribe. Pp. 807-816. In: *Biodiversidad Marina y Costera de México,* Salazar-Vallejo, S.I. & González, N.E. pp. 1-35 Com. Nal. Biodiversidad y CIQRO, México

Andréfouët, S. y Riegl, B. (2004). Remote sensing: a key tool for interdisciplinary assessment of coral reef precesses. *Coral reef,* Vol. 23, No. 1, (April 2004), pp 1-4, ISSN: 0722-4028

Andréfouët, S. ; Kramer, P. ; Torres-Pulliza, D. ; Joyce, K. ; Hochberg, E. ; Garza-Perez, R. ; Mumby, P. ; Riegl, B. ; Yamano, H. ; White, W. ; Zubia, M. ; Brock, J. ; Phinn, S. ; Naseer, A. ; Hatcher, B. & Muller-Karger, F. (2003). Multi-site evaluation of IKONOS data for classification of tropical coral reef environments. *Remote sensing of environment,* Vol. 88, No. 1-2. (November 2003), pp. 128-143, ISSN 0034-4257

Benfield, S., H. Guzman, J. Mair & A. Young (2007). Mapping the distribution of coral reefs and associated sublittoral habitats in Pacific Panama: a comparison of optical satellite sensors and classification methodologies. *Internatinal joural of remote sensing,* Vol. 28, No. 22, (November), pp. 5047-5070, ISSN 5047 5070

Brock, J., K. Yates & R. Halley (2006). Integration of coral reef ecosystem process studies and remote sensing, In: *Remote sensing of aquatic coastal ecosystem processes,* Richardson, L.L & LeDrew, E. F., 324 pp. Springer, ISBN-13 978-1-4020-3968-3, Netherlands

Call, A. K., T. J. Hardy & D. O. Wallin (2003). Coral reef habitat discrimination using multivariate spectral analysis and satellite remote sensing. *International Journal of remote sensing,* Vol. 24, No. 13, pp. 2627-2639

Camarena-Luhrs, T., (2003). Ficha Informativa de los Humedales de Ramsar: Reserva de la Biosfera Banco Chinchorro. Cancún, México.

Capolsini P, Andréfouët S, Cedric R, Payri C (2003) A comparison of Landsat ETM+, SPOT HRV, Ikonos, ASTER, and air- borne MASTER data for coral reef habitat mapping in South Pacific islands, *Canadian Journal of Remote Sensing,* Vol. 29, No. 2, pp. 187-200

Capolsini, P., B. Stoll, S. Andréfouët (2003). A comparison of classification algorithms for coral reef habitat mapping in South Pacific islands, IGARSS

Carricart-Ganivet J.P., E. Arias-González, G. García-Gil, G. Acosta-González, A. Beltrán-Torres, J. Castro-Pérez, N. Membrillo-Venegas y J. Padilla-Saldívar, (2002). Manejo integral de Banco Chinchorro: levantamiento geográfico y caracterización geomorfológica del arrecife, con especial énfasis en las comunidades coralina e íctica. SISIERRA, Mexico

Chávez, E. & Hidalgo, E. (1984). Spatial structure of benthic communities of Banco Chinchorro, México, In: *Advances in Reef Science, A Joint meeting of the Atlantic Reef Committee and the International Society for Reef Studies,* Miami

Clark, C. D., P.J. Mumby, J. R. M. Chisholm, J. Jaubert (2000). Spectral discrimination of coral mortality states following a severe bleaching. *International Journal of Remote Sensing,* Vol. 21, No. 11, (January 2000), pp. 2321-2327, ISSN 0143-1161

Elvidge, C.; Dietz J.; Berkelmans, R.; Andréfouët, S.; Skirving, William.; Strong, Alan. & Tuttle, B. (2004). Satellite observation of Keppel Islands (Great Barrier Reef) 2002 coral bleaching using IKONOS data. *Coral Reefs,* Vol. 23, No.1, pp 123-132, ISSN 07224028

García-Ureña, R. P., (2004). *Dinámica de crecimiento de tres especies de coral en relación a las propiedades ópticas del agua,* 24 September 2011, Available from: <http://grad.uprm.edu/tesis/garciaurena.pdf>

González, A.; Torruco, D.; Liceaga, A. & Ordaz, J. (2003). The shallow and deep bathymetry of the banco chinchorro reef in the mexican caribbean. *Bulletin of Marine Science,* Vol. 73, No. 1, pp. 15-22, ISSN 00074977

Gordon, H. (1978). Removal of atmospheric effects from satellite imagery of the oceans. *Applied Optics,* Vol. 17, No. 10, (May 1978) pp. 1631-1636

Green, E. P. ; Mumby, P. J. ; Edwards, A. J. & C. D., Clark, (1996). A review of remote sensing for the assessment and management of tropical coastal resources. *Coastal Management,* Vol. 24, No. 1, (January 1996), pp 1-40

Green, E., (2000). Satellite and airbone sensors useful in coastal applications. In: *Remote sensing handbook for tropical coastal management,* Green, E., P., Mumby, A. Edwards & C. Clark, Pp 41-56, UNESCO. ISBN 92-3-103736-6, France

Hedley, J.D. & Mumby, P.J. (2002). Biological and remote sensing perspectives on pigmentation in coral reefs. *Advances in Marine Biology,* Vol. 43, pp. 277-317,

Helge, J. ; Lindberg, B. ; Christensen, O. ; Lundälv, T. ; Svellingen, I. ; Mortensen, P. B. & Alvsvåg, J. (2005), Mapping of Lophelia reefs in Norway: experiences and survey methods, In: *Cold-water Corals and Ecosystems*, Freiwald, A. & Roberts J.M., pp 359-391, Springer-Verlag, ISBN 978-3-540-27673-9, Berlin Heidelberg

Hochberg, E. ; Atkinson, M. ; Apprill, A. & Andréfouët, S. (2004). Spectral reflectance of coral. *Coral Reefs*, Vol. 23. No. 1, (Abril 2004), Pp. 84-95, ISSN 0722-4028

Holden, H. (2002). Characterization of Optical Water Quality in Bunaken National Marine Park Indonesia. *Singapore Journal of Tropical Geography*, Vol. 23, No. 1, pp. 23-36, ISSN: 01297619

Holden, H. & LeDrew, E. (1998). Hyperespectral identification of coral reef features. *International Journal of Remote Sensing*, Vol. 20, No. 13, pp. 2545-2563

Hsu, M. K.; Liu, A. K.; Zhao, Y. & Hotta, K. (2008). Satellite remote sensing of Spratly Islands using SAR, *International Journal of Remote Sensing*, Vol. 29, No. 21, pp. 6427-6436, ISSN 01431161

Huston, M. (1985). Variation in coral growth rates with depth at Discovery Bay, Jamaica. *Coral Reefs*, Vol. 4, No. 1, (December 1985), pp. 19-25, ISSN 07224028

Jordán, E. (1979). Estructura y composición de los arrecifes coralinos en la región noreste de la península de Yucatán, México. *An. Centro de ciencias del mar y limnología, UNAM*, Vol. 16 No. 1, pp. 69-86

Karpouzli, E. ; Malthus, T. J. ; & Place, C. J. (2004). Hyperspectral discrimination of coral reef benthic communities in the western Caribbean, *Coral Reefs*, Vol. 23, (February 2004), pp. 141–151, ISBN 00338-003-0363-9

Kirk, J.T.O. (1994). *Light & photosynthesis in aquatic ecosystems* (2nd edition), Cambridge University Press, ISBN 0521 45966 4, New York

Krupa, S. (1999). *Polución, población y plantas*, Colegio de Postgraduados, México.

Kuhn, T., (2006). *La estructura de las revoluciones científicas*, Fondo de cultura económica, México

Lee, Z., K. L. Carder, C. D. Mobley, R. G. Steward & J. S. Patch (1999). Hyperspectral remote sensing for shallow waters: 2.Deriving bottom depths and water properties by optimization. *Applied Optics*, Vol. 38, No. 18, (June 1999), pp. 3831-3843

Lesser, M. P. & Mobley, C. D. (2007). Bathymetry, water optical properties, and benthic classification of coral reefs using hyperspectral remote sensing imagery. *Coral Reefs*, Vol. 26, No. 4, (January 2007) pp. 819-829, ISSN 07224028

Liceaga-Correa, M. A. & Euan-Avila, J. I. (2002). Assessment of coral reef bathymetric mapping using visible Landsat Thematic Mapper data. *International Journal of Remote Sensing*, Vol. 23, No. 1 pp 3-14, ISSN: 01431161

Loya, Y. (1972). Community structure and species diversity of hermatypic corals at Eilat, Red Sea. *Marine Biology*, Vol. 13, No. 2 pp. 100-123, ISSN 00253162

Lyzenga, D.R., (1981). Remote sensing of bottom reflectance and water attenuation parameters in shallow water using aircraft and Landsat data. *International Journal of Remote Sensing*, Vol. 2, No. 1, (Enero 1981), pp. 71-82. ISSN 0143-1161

Maritorena, S., A. Morel & B. Gentili (1994). Diffuse Reflectance of Oceanic Shallow Waters: Influence of Water Depth and Bottom Albedo. *Limnology and Oceanography*, Vol. 39, No. 7 (Nov., 1994), pp. 1689-1703

Mobley, C.D., B. Gentili, H. R. Gordon, Z. Jin, G. W. Kattawar, A. Morel, P. Reinersman, K. Stamnes & R. H. Stavn (1993). Comparison of numerical models for computing underwater light fields. *Applied Optics*, Vol. 32, No. 36, (December 1993) pp. 7484-7504

Mumby, P.J, E.P. Green, A.J. Edwards & C.D. Clark (1997). Coral reef habitat mapping: how much detail can remote sensing provide? *Marine Biology*, Vol. 130, No.4 (December 1997), pp. 193-202. ISSN: 0025-3162

Mumby, P.J. & Harborne, A.R. (1999). Development of a systematic classification scheme of marine habitats to facilitate regional management of Caribbean coral reefs. *Biological Conservation*, Vol. 88, No. 2, (May 1999), pp155-163, ISSN 0006-3207

Mumby, P.J., W. Skirving, A.E. Strong, J.T. Hardy, E.F. LeDrew, E.J. Hochberg, R.P. Stumpf & L.T. David. (2004). Remote sensing of coral reefs and their physical environment. *Marine Pollution Bulletin*, Vol.48, No. 3-4, (February 2004), pp. 219-228, ISSN 0025-326

Mumby, P.J., A.R. Harborne, J.D.Hedley, K. Zychualuk & P. Blackwell (2006). The spatial ecology of Caribbean coral reefs: revisiting the catastrophic die-off of the urchin Diadema antillarum. *Ecological Modelling*, Vol. 196, pp. 131-148

Newman, C., E. LeDrew & A. Lim. (2006). Mapping of coral reef for management of marine protected areas in developing nations using remote sensing. In: *Remote sensing of aquatic coastal ecosystem processes*, Richardson, L. y LeDrew, E., pp. 251-278, Springer, ISBN 978-1-4020-3968-3, Netherlands

Pahlevan, N.; Valadanzouj, M. J. & Alimohamadi, A. (2006). A quantitative comparison to water column correction techniques for benthic mapping using high spatial resolution data. *ISPRS Commission VII Mid-term Symposium "Remote Sensing: From Pixels to processes"*, Enschede, Netherlands, 8-11 May 2006

Philpot, W. D. (1989). Bathymetric mapping with passive multispectral imagery. *Applied Optics*, Vol. 28, No. 8 (April 1989), pp. 1569-1578

Primack, R., R. Rozzi, P. Feinsiger, R. Dirzo y F. Massardo. (1998). *Fundamentos de conservación biológica*. Fondo de cultura económica. México.

Richter, R. (1996). Atmospheric correction of satellite data with haze removal including a haze/clear transition region. *Computers & Geosciences*, Vol. 22, No. 6, pp. 675-681, ISSN 00983004

Richter, R. (1998). Correction of satellite imagery over mountainous terrain. *Applied Optics*, Vol. 37, No. 16, (June 1998), pp. 4004-4015

Rivera, J. A. ; Prada, M. C. ; Arsenault, J.L. ; Moody, G. & Benoit, N. (2006). Detecting fish aggregations from reef habitats mapped with high resolution side scan sonar imager, In: *Emerging technologies for reef fisheries research and management*, NOAA, pp. 88-104. NOAA Professional Paper NMFS, 5 Seattle, WA.

Secretaría de Marina, Dirección General adjunta de Oceanografía, Hidrografía y Meteorología (SEMAR) (2008). *Carta náutica SM-932 Majahual to Banco Chinchorro*, Scale 1: 100000

SERC (September 2011). Phytoplankton lab, 10.09.2011 Available from http://www.serc.si.edu/labs/phytoplankton/primer/hydrops.aspx

Xin, Y., J. Li y Q. Cheng (2007). Automatic Generation of Remote Sensing Image Mosaics for Mapping Large Natural Hazards Areas, In: *Geomatics solutions for disaster management*, Li, J., Zlatanova, S. y Fabbri, A. 451 pp. Springer, ISBN 13 978-3-540-72106-2, Springer Berlin Heidelberg New York.

15

Predictability of Water Sources Using Snow Maps Extracted from the Modis Imagery in Central Alborz, Iran

Seyed Kazem Alavipanah*, Somayeh Talebi and Farshad Amiraslani
Faculty of Geography, Department of Cartography, University of Tehran, Tehran Iran

1. Introduction

1.1 Snow reserves and remote sensing

Snow reserves in mountainous basins are important and reliable water resources in Iran. Identification of their quality is necessary because of an increasing value of freshwater and utilization of water recourses. About 60 percent of surface water and 57 percent of ground water sources in Iran flows in snowy regions (Rayegani, 2005). The water produced from snowmelt process provides soil water, ground water reserves and water in lakes and rivers. Since snow cover is one of the most important sources of provided water, an accurate prediction and timing of snow runoff is necessary for the efficient management and decision- making in water supply.

The science of snow hydrology, compared to other branches of hydrology science, has a relatively shorter history due to difficulties accompanied with snow measurement. The correct analysis of snow issues needs a set of observations and statistics in snow-gauging. Currently, however, there are no regular and comprehensive snow measurement procedures in most parts of Iran. Measurements are only limited to those snowy basins recharging important dams; even these measurements are carried out in scattered points rather than an entire dam catchment area.

The measurement range of these stations is limited to 2000-3000 m asl heights. Thus, in mountainous Iran, current distribution of stations would not seem to be adequate. In such conditions, study of snow reserves and identification of snow melting trend in most basins would be accompanied with limitations. Consequently, measuring snow cover using ground methods will be difficult and costly. Remote sensing technology has many applications in various environmental and earth resources studies including ice and snow research. These applications have been increased recently as a result of unique technical advantages such as multi-temporal imagery acquired in various wavelengths, extent of spatial coverage, and improvement of computer hardwares for interpretation and extraction of information. Regarding snow research, remote sensing technology can provide

* Corresponding Author

continuous information layers with higher accuracy and lower cost compared to the ground survey, so it can fill the information gaps in snow hydrological statistics. However, using ground data can increase the efficiency of remotely-sensed measurement of snow-gauging. Satellites are appropriate tools for gauging snow coverage, because of high reflection of snow that creates proper contrast to most of natural surfaces (with the exception of clouds). Therefore, using satellite imagery and GIS modeling one can produce snow-cover maps, assess the changes in snow cover area with various time series, discriminate snow from other features, and model it in a catchment area. These simplify decision-making process for engineers and hydrology managers.

One of the important issues in remotely-sensed snow-gauging is the selection of sensor. Some of optical sensors that have ever used in snow-measuring include sensors mounted to satellites namely TIROS-1 (1960), ESSA_3, NOAA (1996), LANDSAT (MSS and ETM), and MODIS (2000). Since each sensor has unique properties, a sensor with appropriate spectral, temporal and spatial resolution for snow-gauging must be selected. Since snow is a phenomenon with noticeable surface changes over time, it is necessary to select a sensor that produces proper multi-temporal series. Snow-gauging is done in vast areas, and snow surface is generally even; therefore, MODIS is an appropriate imagery for this purpose. From the view point of spectral resolution, MODIS is one of the best optical sensors for studying snow and discrimination of snow from phenomena such as cloud which has similar spectral reflectance.

One of the purposes of designing of MODIS is a global identification of various types of clouds; hence, several bands have been considered for it to identify various types of cloud cover, optical thickness, effective radius and thermal phase (King et al., 2004).

NASA (National Aeronautics and space administration) launched TERRA satellite to space on December 18th 1999, and MODIS as one of the five sensors mounted on TERRA transferred the first information to Earth on February 24th 2000. MODIS has 36 various bands in visible, infra-red and thermal parts of electromagnetic spectrum including 2 visible bands with 250 m resolution, 5 infra-red bands with 500 m resolution, and 29 thermal bands with 1000 m resolution (Hall et al., 2000).

1.2 A review on remotely-sensed snow measurement

Various methods have been used to estimate snow surface such as classification methods, threshold limit, decision-based methods, etc. One of the most applicable algorithms used to estimate snow surface is MODIS snow map algorithm. It was introduced in 1998 as a decision-based algorithm which uses group tests of threshold limit for detection of snow. This algorithm has very small volume from the calculation viewpoint and simple from the conceptual viewpoint, thus user can track how product has been created. In addition, this algorithm has an appropriate efficiency with global application (Hall et al., 1998).

Totally the properties of this algorithm include:

1. The precision of this method for various types of snow-covered surfaces for identification of snow surface is higher than other methods such as supervised classification, unsupervised classification and sub-pixel methods provided that atmospheric correction is considered (Dadashi Khaneghah, 2008).

2. This method is a completely automatic algorithm.
3. This algorithm is applicable for all regions in the globe.
4. This method is simple, accurate and easy to understand.

Snow Map algorithm uses normalized subtractive index (NDSI). Lee et al. (2001) compared MODIS snow maps created with NDSI index with maps prepared by National Operational Hydrologic Remote Sensing Center (NOHRSC, prepared automatically by GOES and NOAA images) in upper region of Rio Grande reservoir. In NOHRSC, the teta algorithm is used. In this algorithm, two classified images are subtracted to identify snow surface. In teta algorithm, two separate threshold limit is introduced for each image. Lee et al. (2001) concluded that both images are affected by cloudy condition and the main error is cloud coverage. They also mentioned that maps produced from MODIS were more accurate than the above-mentioned maps.

Ault et al. (2006) concluded that MOD10-L2 snow surface product, MODIS sensor, in clear sky condition had the highest accuracy. They showed that the highest error was associated with those conditions that snow depth was lower than 1 cm; thus the higher was the snow depth, the higher was the accuracy. Hall et al. (2000) also showed a low accuracy of low-mass and patchy snows in New England. Klein and Barnett (2003) carried out a snow cover study using MODIS in Rio Grande reservoir during the 2000-2001 period and compared their results to the ground-measuring methods such as snowpack telemetry (SNOTEL) and NOHRSC models. They ultimately concluded that the highest error associated with maps prepared by MODIS was related to the beginning and end of snowfall period. They showed that when the surface was completely covered by snow with no mixed cloud, ground survey or SNOTEL had the highest accuracy.

It can be mentioned that MODIS sensor and NDSI index are appropriate in snow map preparation, although cloud coverage and classification are regarded as constraints (Klein and Barnett, 2003; Zhou et al., 2005). In fact, in spite of various advantages, Snow Map algorithm has some limitations due to inseparability of snow cover from cloud and similarity of cloud behaviours to snow cover. This algorithm cannot completely distinguish clouds from snow (of course, this problem is relatively removable by using Cloud Mask algorithm). Also, this algorithm cannot detect coastal terrains which are similar to snow from viewpoint of whiteness and brightness. However, temperature can act as factor to discriminate snow from these terrains using MODIS bands 31 and 32. Since Cloud Mask and thermal Mask are used before applying algorithm, some error sources in snow map algorithm will be removed (Taghvakish, 2005; Adhami, 2005).

In snow-gauging using satellite imagery, the existence of cloud is problematic due to the following reasons (Riggs and Hall, 2002): first, clouds conceal Earth information; second, clouds create shades on area and change reflectance. Indeed, if clouds cannot be detected well, they will reduce the accuracy of snow map.

Clouds and snow have generally similar spectral reflectional properties in range of visible and infra-red spectra, so thermal properties is not enough for distinguishing them from each other as clouds may be cooler or hotter than snow (Singh and Singh, 2001). In order to detect clouds, a procedure called Cloud Mask algorithm is being used. Akerman et al. (1998) introduced MOD35 Cloud Mask algorithm. MOD35 algorithm is based on obstruction of Earth surface affected by cloud or dust particles that identifies water body, land and atmosphere (Strabala,

2003). In this process, based on land type, geographical position and available data, Cloud Mask algorithm uses 14 bands amongst 36 bands of MODIS to test 18 spectral and spatial features (Hall and Riggs, 2002). However, this procedure was modified by Hall and Riggs (2002) who presented a new version of Cloud Mask algorithm (Liberal). This algorithm can analyse the pixels located under thin and transparent clouds (Zhou et al., 2005; Ault et. al., 2006). This procedure identifies the darkness and if it faces to such darkness, it means that sun angle is higher than 85°. This algorithm is called Liberal Cloud Mask algorithm. In fact, Liberal Cloud Mask algorithm functions as subset of spectral tests of old Cloud Mask algorithm (MOD35) and uses 7 bands of MODIS and set 4 criteria (Hall and Riggs, 2002, 2004).

2. Material and methods

2.1 Study area

The Central Alborz mountainous range extends from 49° 5´- 53° 5´ longitude to 35° 5´- 37° 2´ latitude. Its area is about 40,000 km^2 and covers 64 sub-basins. The lowest, highest and average altitude of the basin is 48 m, 5671 m and 1870 m, respectively. The minimum, maximum and average slope is 5%, 25.4% and 23.56%, respectively. The main slope aspect of this region is directing towards north and south. Climatically, the region is classified to three classes as temperate in north, cold in center, and semi-arid in south. Geological structures mostly consist of mild outcrops that are inconsistent with general trend of east-west. In western part of Alborz, the structures have northwest-southeast trend, but in eastern part, the structures have northeast-southwest trend. These two inconsistent trends cross each other in Central Alborz.

2.2 Data

2.2.1 MODIS Data, TERRA satellite

MODIS encompasses noticeable number of spectral and thermal bands with narrow width, high radiometric resolution, proper width and collecting time, powerful and accurate calibration, and diverse land resolution (MODIS Home page) (http://modis.gsfc.nasa.gov).

In many cases, MODIS provides satellite snow-gauging requirements and therefore these data were used. In this research, images were provided from website (https://wist.echo.nasa.gov/wist) according to Table 1.

Year	2006	2006	2006	2006	2006	2006
Month	February	February	march	march	march	march
Day	21	25	4	8	13	15

Table 1. Temporal table applied images from MODIS sensor data

The imagery used in this research include MOD02 and MOD09. The MOD02 imagery include 36 bands while MOD09 include 6 bands. In MOD09GA imagery, atmospheric corrections have been done based on 6sv model, as one of the best models in atmospheric corrections with minimum error, suitable for measuring snow surface and detecting cloud from snow (Vermote and Kotchenova, 2008). In MOD09 imagery, corrections have been implemented in way that atmospheric diffusion and reflection were minimum. Since data with higher wavelength are being less influenced by aerosols, suspended particles and non-selective diffusion phenomenon, thermal bands of MOD02 imagery were used.

2.2.2 Digital Elevation Model

Digital Elevation Model (DEM) obtained from SRTM Shuttle was used. These data that was in format of GeoTiff had Lambert image system. For transformation of this data to UTM coordinates system, PCI Geomatica software was used.

2.2.3 Ground stations data

Snow-gauge station data were obtained from Water Organization- Department of Surface Water. Snow-gauge stations of Central Alborz are located in five basins namely Lattian, Lar, Taleghan, Karaj and Golpayegan basins. In some cases, ground data survey time was not consistent with the time of image acquisition. For solving this problem, in those dates that no ground statistics were available, previous and next day's information were interpolated. Of course interpolation was carried out for those stations where sampling time was close to image acquisition date, and the station had snow cover during the period (February and March). Among sampled stations, 18 stations with above-mentioned conditions were selected.

It is necessary to mention that snow depth data has been used to examine the presence or absence of snow cover. Those gauging stations located far from human interfering features (e.g. buildings) were selected and their snow depth measured. The snow surface was defined an area where surface is regular and even, with the minimum wind blowing effect to increase measurement accuracy (Pfister and Schneebli, 1999). The output spatial resolution of snow map algorithm and Cloud Mask Algorithm is one kilometre. Around each station, up to two kilometres was regarded to include 9 to 13 pixels covering snow. The most repeated pixel shows snowy or not snowy condition. Finally, snow surface obtained from Snow Map Algorithm with and without Liberal Cloud Mask was compared to ground data. Figure 1 illustrates distribution pattern of snow-gauge stations within five basins in Central Alborz.

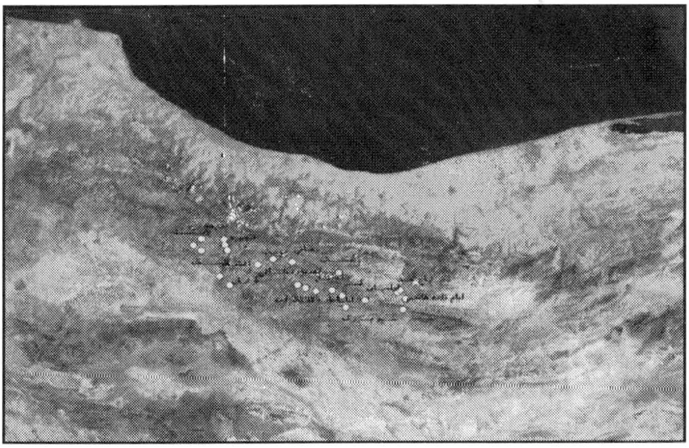

Fig. 1. The position Of snow gauge stations in Alborz-e-Markazi

2.3 Research method

Figure 2 illustrates overall flowchart of this research methodology.

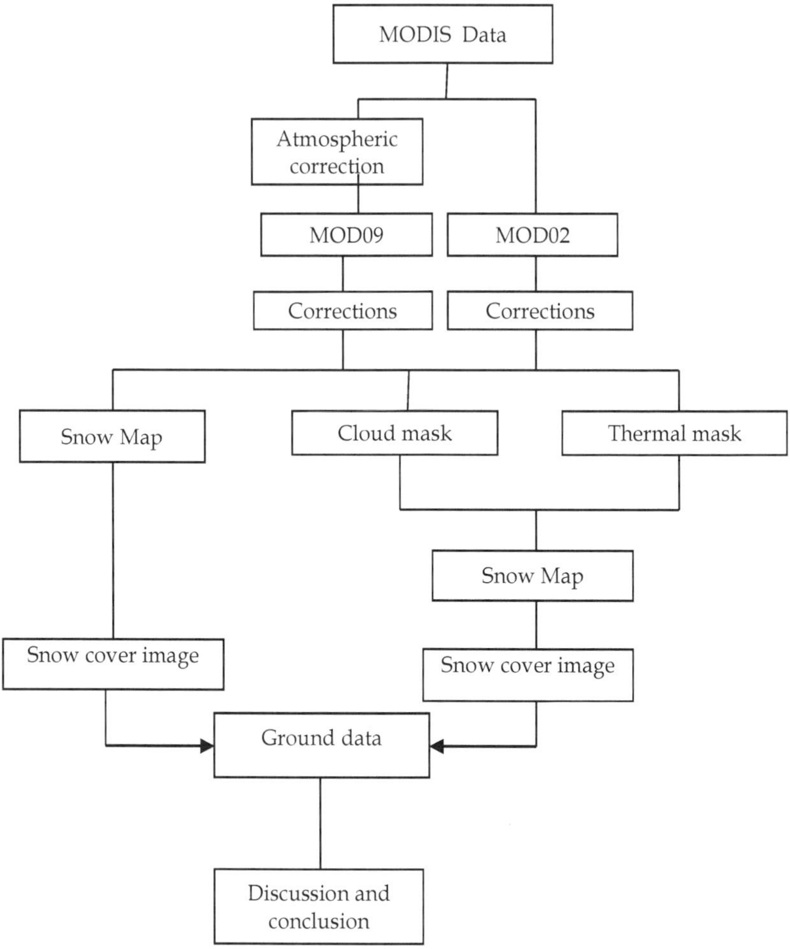

Fig. 2. The flowchart of this research methodology

2.3.1 Snow map algorithm

Snow map algorithm benefits from Normalized Difference Snow Index (NDSI). Because of low reflection of snow in infrared bands and high reflection in visible bands, NDSI can be useful for discrimination of snow from other phenomena. NDSI is calculated by equation below (Hall et al., 1998):

NDSI = (band4-band6) / (band4+ band6)

Snow map algorithm includes following thresholds:

If NDSI ≥ 0.4; and MODIS band2 > 11%; and MODIS band 4 ≥ 0.4.

NDSI index is used to recognize snow and ice and also to differentiate between cumulus and ice or snow. In fact this index represents relative differential reflectance value of visible and short wave-infrared channels emitted from snow. Pure snow has a high NDSI value but other materials such as soil, smoke, etc cause a NDSI reduction. The mentioned threshold for band 4 of MODIS is a key tool to prevent identifying pixels with low reflection, for example dark Cypress, instead of snow. Water and cloud are separable using mentioned threshold for band 2 of MODIS and finally, NDSI has the key role in investigating snow (Hall et al., 1998).

Necessary eligibilities of pixels for applying snow map algorithm are as follows (Riggs et al., 2003):

- Pixels should have level1B reflection (geo-reference process and radiometric correction should be done),
- They should belong to terrestrial region or water bodies surrounded by lands,
- Imagery should be taken in day light,
- Imagery should not be covered by cloud (applying cloud mask),
- Their approximate temperature should be less than 283^0 K (applying temperature mask)

2.3.2 Cloud isolation

Liberal cloud masking just uses 7 out of 36 MODIS bands as well as 4 out of 18 old cloud masking algorithm criteria. Before performing snow map as one of the preprocessing steps, liberal cloud masking was applied. With regards to spectral resemblances between snow and cloud, applying mask on image is inevitable. In liberal cloud masking, a pixel will be considered as cloud provided that it covers one of the following criteria:

1. High cloud index introduces it as cloud
2. Heat difference index consider it as cloud
3. Visual bands reflection index proves the existence of cloud when reflection of band (1/625, 1/628, μm) is more than 20 percent and visual band threshold is applied.
4. NDSI ≥ 0.4 and reflection of band 6 is more than 20 percent (Riggs and Hall, 2002).

In this research, since the study area is located on terrestrial area and consequently discrimination of snow from cloud is very important and also all MODIS imagery were taken in day time, thresholds related to terrestrial region in day time were applied. Furthermore, water bodies were eliminated before image processing.

2.3.3 Heat masking

Heat masking is the final step before using snow map algorithm. This method was introduced on 3rd October 2001 and resulted in eliminating many of incorrect land cover classified as snow. In MODIS version 3, a threshold of 277^0 K was used whereas in version 4 this value increased to 283^0 K. Every individual pixel of band 31 with a temperature more than threshold of version 4 is not classified as snow (Kamanpoon, 2004). Heat masking is used to remove ambiguity between snow and other phenomena such as water bodies, sand

and cloud (Zhou et al., 2005). In this part using calculated apparent temperature for band 31 and applying 283^{0}K threshold, heat masking is performed after new cloud masking algorithm and before snow map algorithm.

3. Results

Images resulted from snow map algorithm before and after applying Liberal cloud mask related to February and March are illustrated in Figure 3. Right column shows images before liberal cloud masking and middle column show them after masking. Left column shows false color images which are made by combining visual and infrared channels of MODIS according to method introduced by Miller et al. (2004). Lands without snow cover, with snow cover, low height clouds and higher clouds appear as green, white, yellow and violet tones, respectively. False color image help to recognize cloudy regions on image as well as cloud height.

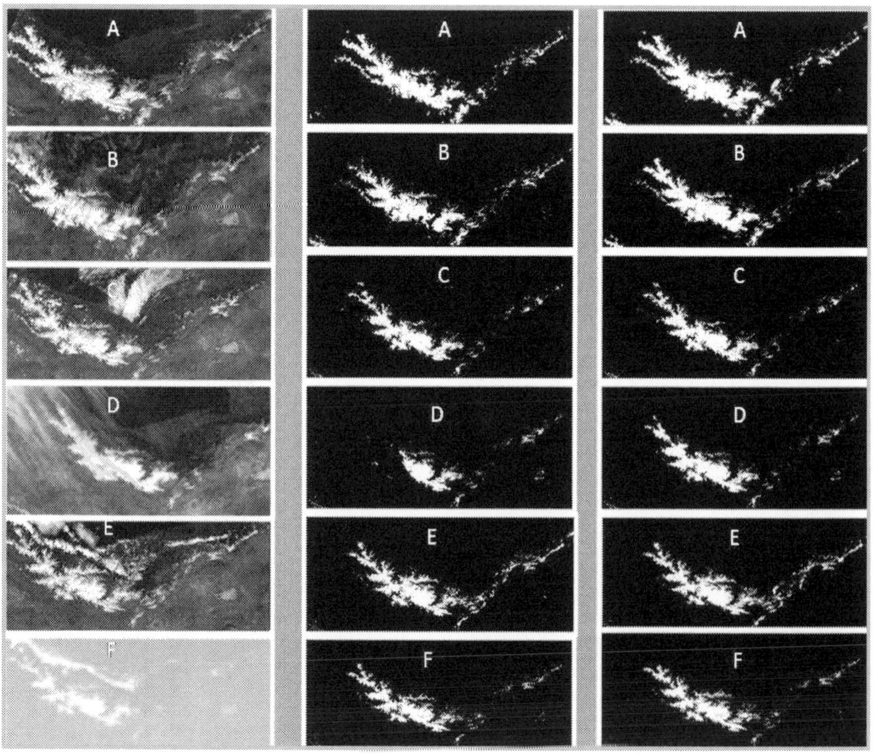

Fig. 3. Snow area before applying the liberal cloud mask (right column) and after applying the Liberal cloud mask (middle column) and false color composite (left column).

4. Discussion

In this part, using NDSI, topographic data and data gathered from snow measurement stations, snow map algorithm alone and together with Liberal cloud masking were separately interpreted. Ground-based snow measurement data and their corresponding points on images resulted from snow map algorithm as well as images resulted from snow map algorithm together with Liberal cloud masking in different dates are illustrated in table 2. In fact error matrix is drawn for each image and results have been surveyed.

Date of acquisition	Snow map	Snow	No snow	Total	Accuracy %	Snow map (masking)	Snow	No snow	Total	Accuracy %
13/3/2006	Snow	12	2	14	77	Snow	12	0	12	100
	No snow	0	4	4	100	No snow	0	6	6	100
	Total	12	6	18		Total	12	6	18	
	Accuracy %	100	66		88	Accuracy %	100	100		100
15/3/2006	Snow	12	2	14	85	Snow	12	0	12	100
	No snow	0	4	4	100	No snow	0	6	6	100
	Total	12	6	18		Total	12	6	18	
	Accuracy %	100	66		88	Accuracy %	100	100		100

Table 2. Evaluation of accuracy of snow gauge obtained from snow map algorithm and snow map algorithm with attending the Liberal cloud mask using earth data in 6th of February, 13th and 15th of March

Results demonstrate that in both images of 13th and 15th March in which snow map was applied, the number of points classified as snow is more than the time when applying snow map algorithm; adding cloud masking to snow map algorithm reduces this number. It means that regions which are incorrectly classified as snow by snow map algorithm can be categorized as cloud after adding cloud masking. Furthermore, no snow regions identified as snow in snow-gauging station and snow map algorithm with Liberal cloud masking are more than those no snow regions that are not classified as snow without applying Liberal cloud masking. So it can be concluded that snow map algorithm shows some regions as snow despite the fact that they are clouds. However, cloud masking can detect them and classify as cloud. Error matrix demonstrates that accuracy of snow map algorithm increases by applying cloud mask (Riggs and Hall, 2002; Ault et al., 2006; Hall and Riggs, 2007). Overall, results from snow map algorithm together with Liberal cloud masking are more compatible with data gathered from ground-based stations.

One of the factors affecting accuracy of snow detection is clouds which cover snow surface. These clouds are distinguishable by Liberal cloud masking provided that they are transparent and thin (Riggs and Hall, 2002; Ault et al., 2006). In images related to 21st February and 8th March, the observed cloud is thick and far from the Earth. False color images show that clouds are far from the Earth surface in both mentioned images so they can be detected and classified correctly by Liberal cloud masking. However, there is snow under these clouds and should be considered as snow. Data from ground-based snow measurement and their corresponding points on images related to 21st February and 8th March resulted from snow map algorithm before and after applying Liberal cloud masking is shown in Table 3. As is shown in Table 3, field survey data are different from results obtained as a result of snow map algorithm together with Liberal cloud masking. In this situation, considering neighborhood effect, topographic factors and false color images, clouds over snow can be distinguishable and classify them as snow. Of course, neighborhood and topographic factors can be helpful when the cloud is smaller that total area of snow.

Date of acquisition	Snow map	Snow	No snow	Total	Accuracy %	Snow map (masking)	Snow	No snow	Total	Accuracy %
21/2/2006	Snow	15	2	17	77	Snow	12	0	12	100
	No snow	0	1	1	100	No snow	3	3	6	50
	Total	15	3	18		Total	15	3	18	
	Accuracy %	100	66		88	Accuracy %	83	16		83
8/3/2006	Snow	12	2	14	85	Snow	8	0	8	100
	No snow	0	4	4	100	No snow	4	6	10	55
	Total	12	6	18		Total	1	6	18	
	Accuracy %	100	66		88	Accuracy %	66	100		77

Table 3. Evaluation of accuracy of snow gauge obtained from snow map algorithm

There is a negligible difference before and after applying Liberal masking images covered by low height clouds (e.g. image of 4th March) (Table 4). It means that in this situation snow map algorithm with and without liberal cloud masking has the same result. So it can be concluded that snow map algorithm is able to detect low height clouds because the spectral diagram of low height clouds are different from that of snow in visual and infrared spectrum range. Data gathered from ground-based snow measurement and its corresponding points on images resulted from snow map algorithm as well as images resulted from snow map algorithm together with liberal cloud masking is shown in Table 4.

Date of acquisition	Snow map	Snow	No snow	Total	Accuracy %	Snow map (masking)	Snow	No snow	Total	Accuracy %
4/3/2006	Snow	11	0	11	100	Snow	11	0	11	100
	No snow	1	6	7	85	No snow	1	6	7	85
	Total	12	6	18		Total	12	6	18	
	Accuracy %	91	100		94	Accuracy %	91	100		94

Table 4. Evaluation of accuracy of snow gauge obtained from snow map algorithm and snow map algorithm with attending the Liberal cloud mask using earth data in 4th of March

In order to show NDSI ability in isolation of cloud from snow, variation range of NDSI in regions which are identified as cloud using new cloud masking is compared with variation in regions where classified as snow by use of snow map algorithm. As it can be found in diagram NDSI variation in regions where identified as cloud and regions where classified as snow has some overlaps so NDSI cannot distinguish between snow and cloud in these regions (Figure 4).

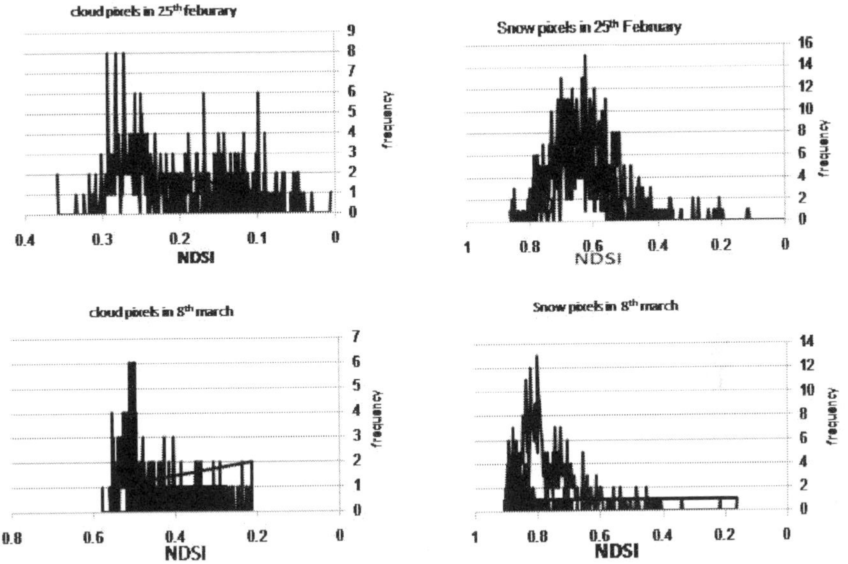

Fig. 4. The variation range of DSI in cloudy and snowy area on images of 25th February and 8th March

Figure 5 shows comparison of NDSI variation for image taken in 4th March which has only low height clouds in regions which are identified as cloud using new cloud masking with variation in regions where classified as snow using snow map algorithm. As it can be seen in diagram NDSI variation in regions where identified as cloud and regions where classified as snow are absolutely separable so NDSI can distinguish between snow and cloud in these regions and act similar to new cloud mask (Figure 5).

As a general rule, the amount of snow increases in higher elevations so if classification of snow and cloud is done perfectly, percentage of pixels related to snow should increase in higher elevation. This rule can be used to evaluate the accuracy of outputs resulted from snow map algorithm alone and together with Liberal mask algorithm. Figure 6 shows the relative frequency of snow pixels in each altitudinal zone. As it can be seen in this Figure, ascending trend occur whenever new cloud mask is applied together with snow map algorithm. In fact, cloud masking leads to the better identification of cloud pixels and prevents these pixels to be classified as snow. However ascending trend will not happen in mentioned diagram when snow map algorithm is used alone because some cloud pixels are categorized as snow incorrectly.

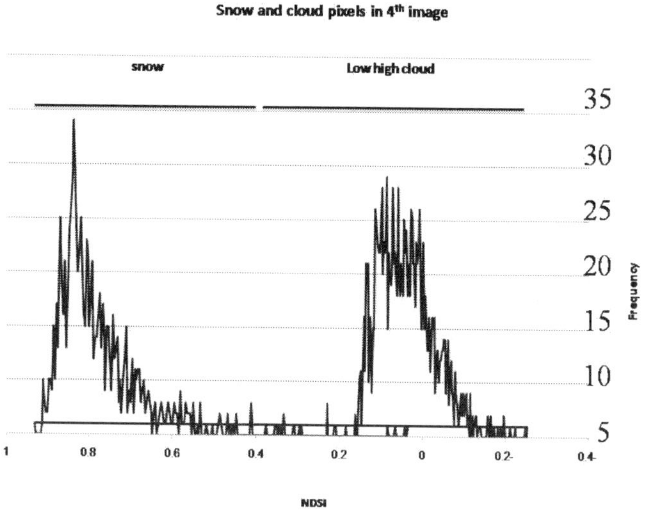

Fig. 5. The variation range of NDSI in cloudy and snowy area on images of 4th March

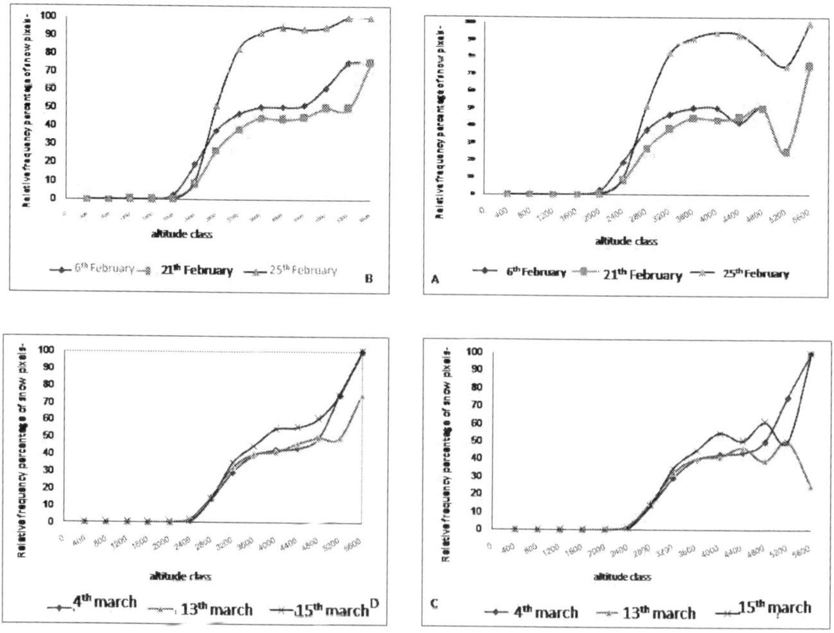

Fig. 6. Relative frequency percentage of snow pixels in each altitude class; right: the obtained images of snow map algorithm; left: the obtained images of snow map algorithm accompanying the Liberal clod mask.

5. Conclusion

Reviewing data resulted from ground-based snow measurements in addition to results from snow map algorithm and Liberal cloud mask, it can be concluded that snow map algorithm cannot detect some types of cloud and classify them as snow (Zhou et al., 2005; Riggs and Hall, 2002; Ault et al., 2006; Hall and Riggs, 2007), reducing the accuracy of maps produced for snow detection. Clouds which are not detected by snow map algorithm are those include ice particles in high elevations (Taghvakish, 2005). Using Liberal cloud masking can largely solve this problem and prevent some types of clouds to be categorized as snow. The accuracy of maps is increased approximately 10% in comparison with other methods. In images including only low elevation clouds, cloud masking cannot make better results; therefore it can be concluded that these kinds of clouds can be detected by snow map algorithm alone. Also, results from applying NDSI shows that some types of clouds are categorized in the same class as snow, so NDSI cannot distinguish between snow and cloud. However, those clouds in low elevation can be detected from snow.

Altitudinal parameter is another tool in order to evaluate the accuracy of snow map algorithm and Liberal cloud masking. An ascending trend in frequency of snow pixels is evident whenever cloud masking is used in addition to the snow map algorithm.

In summary, it can be said that although low height clouds are separable by snow map algorithm, some types of clouds cannot be detected by snow map algorithm alone and thus, application of cloud masking is inevitable. These are clouds which are in high elevation and include ice particles (Taghvakish, 2005). Finally, in some cases even cloud masking cannot distinguish between snow and ice particles (Ault et al., 2006, Riggs and Hall, 2002 , Taghvakish, 2005).

6. References

Ackerman S A, Strabala K I, Menzel P W P, Frey R A, Moeller C C and Gumley L E, 1998: Discriminating clear sky from clouds with MODIS, *Journal of Geophysical Research*, 103(D24):32,141-32,157.

Adhami S, 2005. Application of remote sensing and geographic information system in snow cover (Agichay). Unpublished MSc Thesis, University of Tabriz (In Persian).

Ault T W, Czajkowski K P, Benko T, et al., 2006. Validation of the MODIS snow product and cloud mask using student and NWS cooperative station observations in the Lower Great Lakes Region. *Remote sensing of Environment* 105: 341-353.

Dadashi Khanegha S, 2008. Appointment of snow cover using image processing techniques. Unpublished MSc Thesis. University of Shahid Beheshti (In Persian).

Hall D K, Tait A B, Riggs G A, Salomonson V V, Chien J, Andrew Y L, and Klein G. 1998. Algorithm Theoretical Basis Document (ATBD) for the MODIS Snow-, Lake Ice- and Sea Ice-Mapping Algorithms, MODIS Algorithm Theoretical Basis Document Number ATBD-MOD-10, NASA Goddard Space Flight Center, 1998.

Hall D K and Riggs G, 2007. Accuracy assessment of the MODIS snow products. *Hydrological Processes* 21:1534-154.

Hall D K, Tait A B, Foster J L, Change A T C, and Allen M, 2000. Intercomparison of satellite-derived snow-cover maps. *Annals of Glaciology* 31:396-376.

Kamanpoon S, 2004. Hydrological modelling using MODIS data for snow covered area in the Northern Boreal Forest of Manitoba, MSc Thesis.

King M D, Platnick S, Yang P, Arnold G T, Gray M A, Riedi J C, Ackerman S A, Liou K, 2004. Remote sensing of liquid water and ice cloud optical thickness and effective radius in the Arctic: Application of airborne multispectral MAS Data. *Journal of Atmospheric and Oceanic Technology* 21: 857-875.

Klein A G and Barnett A C, 2003. Validation of daily MODIS snow cover maps of the Upper Rio Grande River Basin for the 2000-2001 snow year. *Remote Sensing of Environment* 86:162-176.

Lee S, Klein A G and Over T M, 2001. A Comparison of MODIS and NOHRSC snow cover products for simulating streamflow using the Snowmelt RunOff Model in: http://www.modissnowice.gsfc.nasa.gov

Miller S D and Lee T F, 2004. Satellite-Based imagery techniques for daytime cloud/snow delineation from MODIS. *Journal of applied meteorology*. volume 44.

Pfister R and Schneebeli M, 1999. Snow accumulation on boards of different sizes and shapes. *Hydrological Processes* 13:2345-2355.

Hall D k, Foster J L, Robinson D A, Riggs G A, 2004. Merging the MODIS and RUCL monthly snow cover records. In Proceeding of IGARSS 04, Anchorage, AK, September 2004.

Rayegani B, 2005. Investigation on snow cover changes and estimation snowmelt runoff using MODIS data. Unpublished MSc Thesis. Isfahan University of Technology (In Persian).

Riggs G and Hall D K, 2002. Reduction of cloud obscuration in the MODIS snow data product, 59th Eastern Snow Conference, Stowe, Vermonte, USA.

Riggs G A , Haii D K and Salomonson V V, 2003. MODIS snow products users guide. Available at: http://www.modis-snow-ice.gsfc.nasa.gov/sug.pdf.

Strabala K, 1999. MODIS Cloud Mask User's Guide', *User's Guide*.

Singh P and Singh V P, 2001. *Snow and Glacier Hydrology*, Kluwer Academic Publishers, p742.

Taghvakish S, 2005. Supervision of snow cover using satellite images. Unpublished MSc Thesis, University Of Tehran (In Persian).

Vermote E F and Kotchenova S, 2008. Atmospheric correction for the monitoring of land surfaces. . *Journal of Geophysical Research -ATMOSPHERES*, 113(D23), D23S90.

Zhou X, Xie H and Hendrick J M H, 2005. Statistical evaluation of remotely sensed snow cover products with constraints from streamflow and SNOTEL measurements, *Remote Sensing of Environment* 94: 214–231.

16

Remote Sensing of Cryosphere

Shrinidhi Ambinakudige and Kabindra Joshi
Mississippi State University
USA

1. Introduction

The cryosphere is the frozen water part of the Earth's system. The word is derived from the Greek "kryos," meaning cold. Snow and ice are the main ingredients of the cryosphere and may be found in many forms, including snow cover, sea ice, freshwater ice, permafrost, and continental ice masses such as glaciers and ice sheets. Snow is precipitation made up of ice particles formed mainly by sublimation (NSIDC, 2011). Ice is the key element in glaciers, ice sheets, ice shelves and frozen ground. Sea ice forms when the ocean water temperature falls below freezing. Permafrost occurs when the ground is frozen for a long period of time, at least two years below 0^0 C, and varies in thickness from several meters to thousands of meters (NSIDC, 2011). Glaciers are thick masses of ice on land that are caused by many seasons of snowfall. Glaciers move under their own weight, the external effect of gravity, and physical and chemical changes. The cryosphere lowers the earth's surface temperature by reflecting a large amount of sunlight, stores fresh water for millions of people, and provides habitat for many plants and animals.

Apart from the Arctic and Antarctic regions, the cryosphere is mainly a high altitude phenomenon. It is found on Mount Kilimanjaro in Africa, the Himalayan mountain range, high mountains of United States, and in Canada, Russia, Japan, and China. Researchers in the cryosphere are often hindered by the lack of accessibility due to the rugged terrain. In such cases, remote sensing technologies play an important role in cryosphere research. These techniques are imperative for researchers studying glacial retreat and mass balance change in relation to global climate change.

The cryosphere has a significant influence on global climate and human livelihoods. Change in spatial and temporal distribution of the cryosphere influences the water flow in the world's major rivers. Among the various parts of the cryosphere, glaciers play the most important role in climate change studies since glacier recessions are indicators of global climate change (Oerlemans et al., 1998; Wessels et al., 2002; Ambinakudige, 2010). Retreating glaciers can pose significant hazards to people (Kaab et al., 2002). Glacier retreat often lead to the formation of glacial lakes at high altitudes, the expansion of existing lakes, and the potential for glacial lake outburst floods (GLOFs) (Fujita et al., 2001; Bajracharya et al., 2007). A GLOF is the sudden discharge of a huge volume of water stored in a glacial lake due to huge ice falls, earthquakes, avalanches, rock fall or failure of a moraine dam (Grabs & Hanisch, 1993). There are more than 15,000 glaciers and 9,000 glacial lakes in the Himalayan mountain ranges of Bhutan, Nepal, Pakistan, China and India (Bajracharya et al., 2007). All these countries within the Himalayan region have at some time or another suffered a flood from a glacial lake outburst causing loss of property and lives, and these floods can be disastrous for the downstream riparian area (Richardson & Reynolds, 2000; Bajracharya et al., 2007). The significance of the glaciers as fresh water resources for millions of people is another reason to justify the continuous monitoring of these glaciers (Shiyin et al., 2003). Therefore, monitoring glaciers has significant importance both in understanding global climate change and in sustaining the livelihoods of the people downstream of the glaciers.

This chapter explores the use of remote sensing technologies in studies of the cryosphere and particularly in glaciers. First, we will discuss remote sensing sensors that are effective in monitoring glaciers. Then we will discuss the global effort to create glacier data, using remote sensing tools to delineate glacier areas, estimate volume and mass balance.

2. Remote sensors for monitoring glaciers

Remote sensing methods are more convenient than field methods to measure changes in glaciers. Studies have used Landsat (Ambinakudige, 2010), SPOT (Berthier et al., 2007), Terra ASTER (Rivera & Casassa, 1999; Kaab, 2007), IRS (Kulkarni et al. 2011), and ALOS (Narama et al., 2007) sensors successfully to measure glacial parameters. High-resolution satellite data such as IKONOS (Huggel et al., 2004) and Quickbird (Schmidt, 2009) have also been used in studies on glaciers. ASTER, SPOT5, IRS-1C, Resourcesat1 and 2, CORONA KH-4, KH 4A and KH 4B satellites have the capability to acquire stereoscopic images from which elevation data can be extracted for monitoring glacial surfaces in three dimensions (Racoviteanu, et al. 2008). Digital elevation models developed from these stereoscopic images can be used in measuring volume and mass balance change in glaciers.

2.1 Glacial mapping

Glacial mapping using remote sensing initially involved manual digitization of glacier boundaries on a false color composite (FCC) of Landsat MSS and TM images in Iceland (Williams, 1987) and Austria (Hall et al., 1992). Figure 1, is a picture taken by the astronauts in the international space station shows the snow and ice in Colonia glacier and its surroundings in Chile (International Space Station, 2000).

Snow has high reflectivity in the visible wavelength region and relatively less reflectivity in middle and shortwave infrared regions (Pellika and Rees, 2010). Freshly fallen snow has the highest reflectance in the visible and near-infrared wavelengths. Firn (partially compacted snow) has 25-30 % less reflectance than snow. Ice in glaciers has high reflectance in blue (400- 500 nm) and green (500 - 600 nm) wavelengths. However, in red (600-700 nm) the reflectance of ice is near zero. Debris in glaciers will significantly reduce the reflectance (Pellika and Rees, 2010). The spectral reflectance properties of snow also depend on time and season of the year. Sharp changes in the reflectance of snow can be seen when the melted snow recrystallizes to form firn and the density of snow changes. Similarly, albedo, which is the ratio of radiation reflected from a surface to the radiation incident on that surface, also varies among different cryospheric surfaces. Snow has high albedo (0.8 - 0.97) while dirty ice has low albedo (0.15-0.25). A high albedo value helps to reflect a huge amount of sunlight, which otherwise would have heated the earth's surface. Snow and ice reflectance is the main characteristic that is measured using remote sensing techniques. This

Fig. 1. Astronaut photograph of Colonia Glacier, Chile. Photographer: International Space Station (2000).

characteristic assists in delineating glacier boundaries and classifying various cryospheric surface types (Pellika and Rees, 2010).

In figure 2, the Landsat TM bands 1 to 6, acquired on 25 April 2010, are shown to compare spectral characteristics of glaciers. This figure presents the area around the Imja glacier in the Sagarmatha National Park in the Himalayas of Nepal. TM1 (0.45 - 0.52 μm) is useful to distinguish snow/ice in cast shadow, and also in mapping glacier lakes. Snow and firn areas get saturated in TM1. TM2 (0.52 - 0.60 μm) and TM3 (0.63 - 0.69 μm) have very similar spectral reflectance. TM2 is also useful in distinguishing snow and ice in cast shadow. TM3 is used in band ratio such as the normalized difference vegetation index (NDVI), which helps in classification of ice and snow in areas of dense vegetation. TM4 (0.76-0.90 μm) in the near infrared wavelength region has less reflectance from snow than TM2 and TM3. The clean ice region looks darker in near infrared band, indicating lower reflectance due to the presence of water at the surface (Hall et al., 1988). In TM5, the snow-covered area absorbs nearly all radiation and appears almost dark. This band is also useful in identifying clouds. The thermal band TM6 (2.08-2.35 μm) registers thermal emissions from the surface. Debris has a higher temperature and thus brighter pixels. Thick debris on ice can be easily distinguished as it will have a higher temperature (Pellika and Rees, 2010).

The high reflectance of the snow compared to the ice makes it easy to separate snow and ice. Snow and clouds are often difficult to distinguish when single imagery is used. Clouds and snow have similar reflectance at wavelengths below 1 μm in the near infrared region. The distinction between snow and ice is clearer near 1.55 and 1.75 μm. Therefore a ratio of two

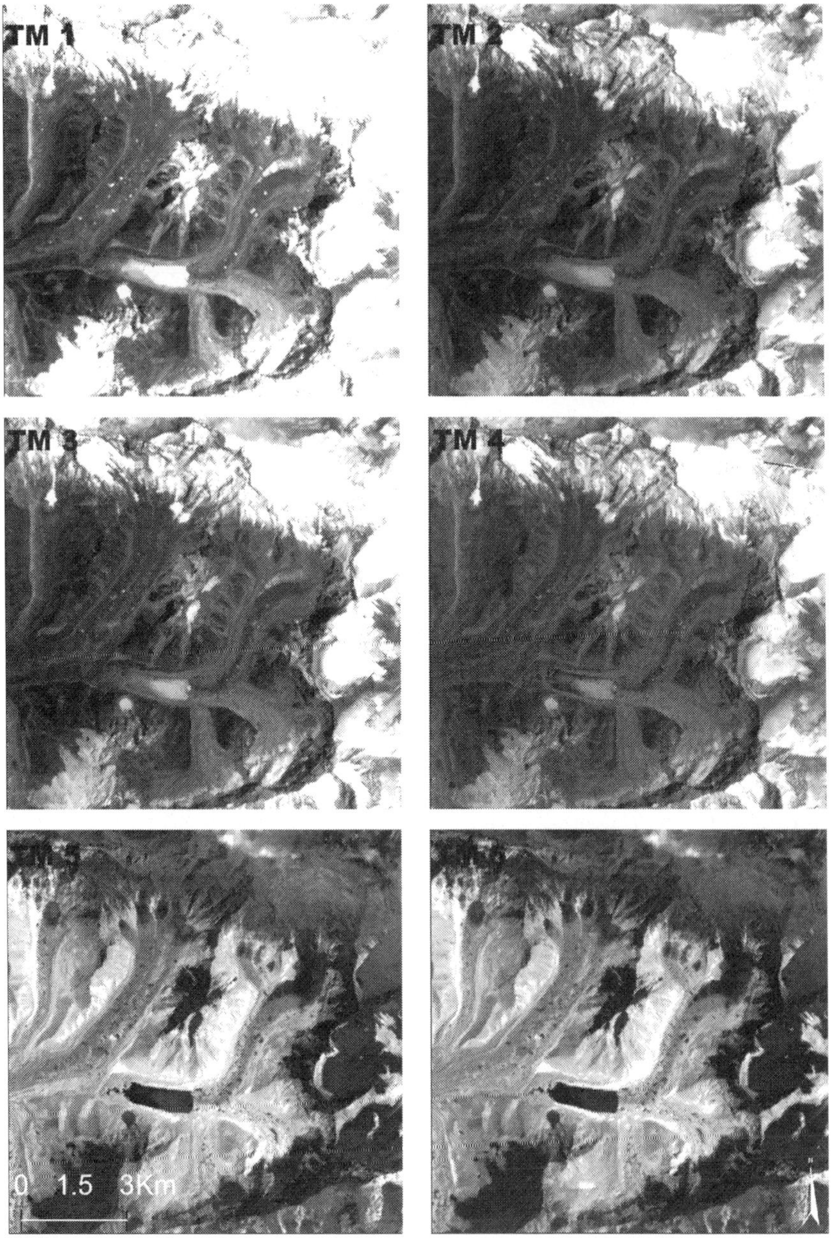

Fig. 2. Six Landsat TM bands showing Imja and surrounding glaciers in the Himalayas.

spectral bands in the visible (VIS) and shortwave near infrared (SWIR) regions are commonly used in automated mapping of snow and ice. A Normalized Difference Snow Index (NDSI) calculated as (VIS-SWIR)/(VIS+ SWIR) helps to separate snow and ice from darker areas such as rocks and soils. Whereas the visual spectrum band covers a wavelength of 0.57μm, the shortwave infrared band covers 1.65μm. If the NDSI value exceeds 0.4, it is assumed there is snow cover on the ground (Dozier 1984). However, seasonal variation in NDSI value for snow has also been observed. A threshold of 0.48 in July and 0.6 in September was observed as an optimal threshold during the field investigation in Abisko, Sweden (Vogel 2002). The Normalized Difference Water Index (NDWI) calculated as (NIR-VIS)/(NIR+VIS) is useful to differentiate water from snow, ice and other physical features. NDWI is very useful in detecting formation of new, as well expansion of existing, glacial lakes (Huggel et al., 2002).

Band ratios such as Landsat TM3/TM5 or TM4/TM5 are also helpful in mapping glacial areas. The TM4/TM5 ratio is more appropriate for clean-ice glacier mapping (Paul and Kaab, 2001), whereas TM3/TM5 works better in areas of dark shadow and thin debris cover (Andreassen et al., 2002). Both NDSI and ratio methods have similar robust outcomes in glacier mapping and are recommended. NDSI and band ratio methods often misclassify debris-covered glacier ice because of the similarity in spectral signature to the surrounding debris Band ratio VIS/NIR also often misclassifies proglacial lakes. The band ratio NIR/SWIR is good for classifying only clean glacier ice (Bhambri and Bolch, 2009). Therefore, manual corrections have to be made after classifying glaciers using either NDSI or ratio methods. High resolution panchromatic images used in manual corrections can help to delineate the precise boundaries of glaciers. Similarly, figure 3 also indicates spectral responses of different types of snow and water.

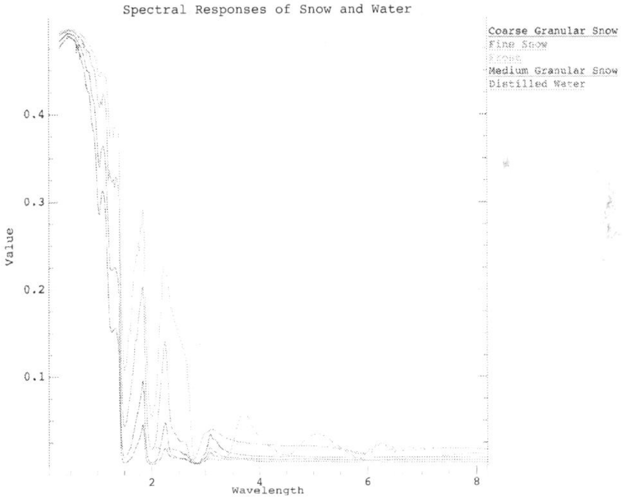

(Data Source: ENVI Spectral Library derived from John Hopkins University http://asterweb.jpl.nasa.gov/speclib/)

Fig. 3. Spectral responses of different types of snow and water

3. Digital elevation models in glacier mapping

Debris-covered glaciers can be mapped effectively by using morphometric parameters derived from DEM and thermal bands (Ranzi et al., 2004). The surface temperatures of the debris on glacier ice are cooler than the debris outside the ice, which can be identified by a thermal image. AVHRR and MODIS satellites have coarse resolution thermal bands; the Landsat TM/ETM+ satellite, on the other hand, has a higher resolution thermal band. Using morphometric characteristics like slope can help to separate a debris-covered glacier from its surrounding moraines (Bishop et al., 2001). DEMs help in estimating morphometric characteristics such as slope, glacier profile curvature, and aspect. DEMs are most important for the estimation of volume change for inaccessible glacier regions (Bolch et al., 2008).

DEMs are generated from topographic maps, spaceborne optical stereo data, synthetic aperture radar (SAR) imagery, spaceborne radar and laser altimetry, such as LiDAR data. Terra ASTER optical stereo satellite data with a spatial resolution of 15 m has been used successfully in generating DEM to study glaciers (Kargel et al., 2005). SPOT-5, ALOS PRISM, CartoSat I and II, IKONOS, Quickbird and GeoEye-1 satellites also have stereo capability and can be used in creating DEMs. ASTER and ALOS PRISM produce along-track stereo images and are capable of simultaneous image acquisition. Other sensors provide across-track acquisition with a time lag, which causes problems in certain situations, such as under different atmospheric conditions (Bhambri & Boch, 2009). Measuring surface changes using DEMs, known as indirect geodetic methods, is a useful method to calculate a glacier's volume and mass balance changes (Etzelmüller & Sollid, 1997).

Another technology used to study the cryosphere is radar, which is an active microwave system composed of a transmitter and a receiver. Radar imagery is not confined only to daylight, cloud-free conditions, and thus has advantages in estimating glacial change. The first spaceborne SAR was the SEASAT satellite, launched in 1978. ERS-1/2, Envisat - ASAR, Radarsat-1, ALOS PALSAR, Radarsat-2 and TerraSAR-X are some of the satellites with spaceborne SAR launched after the initial success of SEASAT. Due to its all-day, all-weather imaging capabilities, large spatial coverage, and ability to measure minute changes on the earth's surface, SAR has a major advantage in glacier studies. Digital elevation models (DEM) created using SAR images are useful in measuring glacier mass balance, glacier velocity, and snow density.

Light Detection and Ranging (LiDAR) is also used in remote sensing of glaciers. High density measurement with a high vertical accuracy (10 cm) allows a very detailed representation of terrain in LiDAR imagery. LiDAR imagery also comes with accurate positional information because of the global positioning system receiver in the LiDAR instruments on aircraft.

4. The geodetic method of measuring glacier mass balance

The mass balance of a glacier is referred to as the total loss or gain in glacier mass at the end of a hydrological year (Cuffey and Paterson, 2010). Glacier mass balance is the link between climate and glacier dynamics (Kaser et al., 2002). Mass balance change is a direct reaction to atmospheric conditions. The specific mass balance can be compared directly between different glaciers. This makes it easier than using length changes to establish a link to climate data (UNEP, 2008). Glacier mass losses affect local hydrology and are thus important for regional water supplies and assessing global sea level rise.

In the geodetic method, changes of glacier volume are measured from maps and elevation models, unlike the direct or glaciological methods that derive mass changes from ground-based spot measurements. With the development of remote sensing techniques, digital elevation models (DEM) are commonly used in the geodetic method. The difference in DEM values in two years is multiplied by the glacier accumulation area to obtain volume change (dV). The difference in volume (dV) multiplied by the density of ice, generally considered to be 850-900 Kg/m^3, would provide the mass balance (Cuffey & Paterson, 2010).

The accumulation and ablation area in a typical glacier is separated by an equilibrium line altitude (ELA). This line divides the accumulation zone (the higher reaches of a glacier where there is a net mass gain) and the ablation zone (the lower reaches where mass is lost). ELA is the elevation at which mass is neither gained nor lost through the course of a hydrological year. At the end of the season, the elevation at which there was no net gain or loss is identified as the ELA. The snow line altitude (SLA) divides the ice in the ablation zone from the snow in the accumulation zone. Since ice and firn have lower albedo, SLA can be determined using remote sensing images (Khalsa et al., 2004; Racoviteanu et al., 2007).

The geodetic approach has been used in several studies based on historical topographic maps and DEMs derived from SPOT imagery (Berthier et al., 2007), SRTM (Racoviteanu et al., 2007), ASTER (Rivera & Cassassa, 1999; Kaab, 2007). Studies have also used high resolution DEMs derived from ALOS PRISM and Corona (Narama et al., 2007) to estimate mass balances with the geodetic method.

The two bands in ASTER VNIR, 3N and 3B, generate an along-track stereo pair with a base-to-height (B/H) ratio of about 0.6. Studies have found that the DEM accuracy has a linear relation with terrain slopes (Toutin, 2008; Bolch et al., 2004; Racoviteanu et al., 2007). DEMs created using ASTER images on Mt. Fuji, Japan; in the Andes Mountains, Chile-Bolivia; at San Bernardino, CA and Huntsville, AL resulted an accuracy of ±5 m, ±10m, ± 6m and ± 1.5m respectively (Hirano, 2003). Therefore, accuracy, number and distribution of GCPs are required to create an accurate DEM.

The ALOS data have been available since January 2006. ALOS has three remote-sensing instruments. The Panchromatic Remote-Sensing Instrument for Stereo Mapping (PRISM) sensor of ALOS data is suitable for digital elevation mapping. It consists of three independent telescopes for forward, nadir and backward view, and each telescope provides 2.5m spatial resolution. The Advanced Visible and Near Infrared Radiometer type 2 (AVNIR-2) sensor collects data with 10m resolution and is suitable for glacier mapping. The accuracy of elevation extracted from ALOS PRISM is 5 m (Racoviteanu et al., 2008).

The three images (forward, nadir and backward) that ALOS provides are required for creation of DEM (Ye, 2010). The base-to-height (B/H) of PRISM on ALOS is set to 1.0

(forward view + backward view) and 0.5 (sidelong view + nadir view). ALOS images have been used to produce highly accurate DEMS. The DEMs created for the northern slope of Qomolangma in the Mt. Everest region had a mean elevation difference of 1.7m with a DEM created using topographic maps in non-glaciated areas. The mean difference between Aster and ALOS images was found to be about 45m (Ye, 2010).

Once the outlines of the glaciers are delineated, they can be combined with DEM to derive glacier parameters such as length, termini elevations, and volume. DEMs derived from SPOT5, ASTER, CORONA or ALOS PRISM can be used in mass balance studies.

5. Global Land Ice Measurements from Space (GLIMS)

In an effort to analyze the glacial change throughout the world, a global-level consortium, the Global Land Ice Measurements (GLIMS), has established a database at the National Snow and Ice Data Center (NSIDC) in Boulder, Colorado (Raup et al., 2007). Under GLIMS, 12 regional centers are working to acquire satellite images, analyze them for glacial extent and changes, and assess change data for causes and implications for people and the environment.

GLIMS is an international consortium established to monitor the world's glaciers. Although GLIMS is making use of multiple remote-sensing systems, ASTER (Advanced Spaceborne Thermal Emission and Reflection Radiometer) satellite images are the major data input in GLIMS database. The GLIMS team has put together a network of international collaborators who analyze imagery of glaciers in their regions of expertise. Collaborators provide digital glacier outlines and metadata. Data also include snow lines, center flow lines, hypsometry data, surface velocity fields, and literature references. The National Snow and Ice Data Center archives the data provided by the regional centers.

The GLIMS team also developed tools to aid in glacier mapping, such as GLIMSView, which is an open-source, cross-platform application designed to support and standardize the glacier digitization process. It allows regional centers to transfer data to the National Snow and Ice Data Center for incorporation into the GLIMS glacier database. Users can view various types of satellite imagery, digitize glacier outlines and other material units within the images, attach GLIMS-specific attributes to segments of these outlines, and save the outlines to ESRI shapefiles. GLIMSView is free and available at http://www.glims.org/glimsview/.

6. Conclusions

Satellite remote sensing of the cryosphere has progressed over the last five decades. It has helped us to understand the global distribution of the cryosphere, variation and trends in snow cover, sea ice, and glaciers. We have a pretty decent map of the cryosphere. Remote sensing has helped in rapid assessment of glaciers in hostile ground conditions in areas such as Antarctica, the Artic and alpine glaciers.

There are several challenges in remote sensing of the cryosphere. Acquiring cloud-free satellite imagery is still challenging. Synthetic aperture radar (SAR) imagery has received a great deal of attention in recent years as it can provide cloud-free data. SAR interferometry has been used successfully in areas such as glacier motion and topographical mapping. The use of radar has been gaining more attention recently. Ground penetrating radars are being used to study the internal structure and bedrock configuration of glaciers.

For most part, *in situ* measurement of the cryosphere is often not a viable option, so the focus of cryosphere study remains on the use of remote sensing techniques. The World Glacier Monitoring Service (WGMS) coordinates the global glacier observation strategy with the help of the Global Land Ice Measurement from Space project and the European Space Agency's Global Glacier Project.

Spaceborne remote sensing techniques in the last five decades have shown tremendous advancement. From Landsat to InSAR imagery, the remote sensing technology has helped in understanding and mapping the cryosphere. Many of these data are available free or for low cost; some of them are very expensive, and using them requires specialized skill. With the increase in computer processing power, the potential for the collection, storage, transmission and processing of remotely sensed data on the cryosphere has improved.

Changes in glaciers provide evidence of climate change, and therefore glaciers play a key role in early detection of global climate-related observations (WGMS, 2011). Glacier change will impact global sea level fluctuations and other natural hazards. These environmental changes require international glacier monitoring efforts to make use of remote sensing and geo-informatics along with the more traditional field observations.

7. References

Ambinakudige, S. (2010). A study of the Gangotri glacier retreat in the Himalayas using Landsat satellite images. *International Journal of Geoinformatics* 6 (3), pp. 7-12.

Andreassen, L.M., Elvehoy, H. & Kjollmoen, B. (2002). Using aerial photography to study glacier changes in Norway. *Annals of Glaciology* 34: pp. 343-348.

Bajracharya, S.R., Mool, P.K., & Shrestha, B. (2007). *Impact of climate change on Himalayan glaciers and glacial lakes: case studies on GLOF and associated hazards in Nepal and Bhutan*. Kathmandu, International Centre for Integrated Mountain Development and United Nations Environmental Programme Regional Office Asia and the Pacific. (ICIMOD Publication 169).

Berthier, E., Arnaud, Y., Kumar, R., Ahmad, S., Wangnon, P., & Chevallier, P. (2007). Remote sensing estimates of glacier mass balances in the Himachal Pradesh (Western Himalaya, India). *Remote Sensing of Environment* 108:327-338.

Bhambri, R., & Bolch, T. (2009). Glacier Mapping: A Review with special reference to the Indian Himalayas, *Progress in Physical Geography* 33(5), 672–704.

Bolch, T., Buchroithner, M. F., Pieczonka, T., & Kunert, A. (2008). Planimetric and volumetric glacier changes in Khumbu Himalaya since 1962 using Corona, Landsat TM and ASTER data, *Journal of Glaciology*, 54, pp. 592–600.

Bishop, M.P., Bonk, R., Kamp, U. & Shroder, J.F. (2001): Topographic analysis and modeling for alpine glacier mapping. *Polar Geography*, 25: 182-201.

Cuffey, K.M & Paterson, W.S.B. (2010). *The Physics of Glaciers*. 4th ed. Academic Press.

Dozier, J., (1984). Snow reflectance from Landsat-4 Thematic Mapper. *IEEE Transactions on Geoscience and Remote Sensing*. 22(3), pp. 323-328.

Etzelmüller, B. & Sollid, J.L. (1997). Glacier geomorphometry - an approach for analysing long-term glacier surface changes using grid-based digital elevation models. *Annals of Glaciology*, 24, pp. 135-141.

Fujita, K. Kadota, T. Rana, B., Kayastha, R.B., & Ageta, Y. (2001). Shrinkage of glacier Ax010 in Shorong region, Nepal Himalayas in the 1990s. *Bulletin of Glacier Research* 17, pp. 51-54.

Grabs, W. & Hanisch, J. (1993). Objectives and prevention methods for glacier lake outburst floods (GLOFs). *Snow and Glacier Hydrology*: Proceedings of the Kathmandu symposium, November 1992. IAHS Publication, 218.

Hall, D.K., Williams, R.S., & Bayr, K.J. (1992). Glacier récession in Iceland and Austria. *EOS*, 73(12), 129.

Hirano, A. (2003). Mapping from ASTER stereo image data: DEM validation and *accuracy* assessment. *ISPRS Journal of Photogrammetry and Remote Sensing*, 57(5-6), pp. 356-370.

Huggel, C., Kääb, A., Haeberli, W., Teysseire, P. & Paul, F. (2002). Remote sensing based *assessment* of hazards from glacier lake outbursts: a case study in the Swiss Alps. *Canadian Geotechnical Journal*, 39, pp.316-330.

Huggel, C., Kääb, A., & Salzmann, N. (2004). GIS-based modeling of glacial hazards and their interactions using Landsat-TM and IKONOS imagery. *Norsk Geografisk Tidsskrift - Norwegian Journal of Geography*, 58(2), pp 61-73.

International Space Station. (2000). Colonia glacier: from the glacier photograph *collection*. Boulder, Colorado USA: National Snow and Ice Data Center/World Data Center for Glaciology. Digital media.
http://eol.jsc.nasa.gov/scripts/sseop/photo.pl?mission=ISS001&roll=E&frame=5107

Kargel, J.S, Abrams, M., Bishop, M., Bush, A., Hamilton, G., Jiskoot, H., Kaab, A., Kieffer, H.H., Lee, E., Frank, P., Rau, F., Raup, B., Shroder, J.F., Soltesz, D., Stainforth, D., Leigh, S, & Wessels, R. (2005). Multispectral imaging contributions to global land ice measurements from space. *Remote Sensing of Environment*, 99(1-2), pp.187-219.

Kaab, A. (2007). Glacier volume changes using ASTER optical stereo. A test study in Eastern Svalbard. *IEEE Transactions on Geosciences and Remote Sensing* 10:3994 -3996.

Kaab, A, Paul, M. Maisch, M., Hoelzle, M. & Haeberli, W. (2002). The new remote sensing derived Swiss glacier inventory: II. first results. *Annals of Glaciology*, 34.

Kaser, G., Fountain, A., & Jansson, P. (2002). A manual for monitoring the mass balance of mountain glaciers. *International Hydrological Programme*, 69.

Khalsa, S.J.S., Dyurgerov, M.B., Khromova, T., Raup, B. H., & Barry, R.G. (2004). Space-based mapping of glacier changes using aster and gis tools. *IEEE transactions on geosciences and remote sensing*, 42, pp. 2177-2183.

Kulkarni, A.V., Rathore, B.P., Singh, S. K., & Bahuguna, I. M. (2011). Understanding changes in the Himalayan cryosphere using remote sensing techniques. *International journal of remote sensing*, 32: 3, 601-615.

Narama, C., Kaab, A., Kajiura, T., & Abdrkhmatov, K. (2007). Spatial variability of recent glacier area and volume changes in central Asia using corona, Landsat, Aster and ALOS optical satellite data. *Geophysics Research Abstract*, 9.

NSIDC (2011). Education resources. http://nsidc.org/pubs/education_resources/

Oerlemans, J., Anderson, B., Hubbard, A., Huybrechts, P., Jóhannesson, T., Knap, W. H., Schmeits, M., Stroeven, A. P., van de Wal S.R.W., & Wallinga, J. (1998). Modelling the response of glaciers to climate warming. *Climate Dynamics* 14, pp.267-274.

Paul, F & Kaab, A. (2005). Perspectives on the production of a glacier inventory from multispectral satellite data in the Canadian Arctic : Cumberland Peninsula, Baffin Island. *Annals of Glaciology*. 42, pp. 59-66.

Pellika P. & Rees, W.G., (2010). *Remote Sensing of Glaciers*. CRC Press. London.

Ranzi, R., Grossi, R., Iacovelli, l. & Taschner, T. (2004). Use of multispectral ASTER images for mapping debris-covered glaciers within the GLIMS Project. *IEEE International Geoscience and Remote Sensing Symposium* 2, 1144-1147.

Racoviteanu, A., Manley, W.F., Arnaud, Y. & Williams, M. (2007). Evaluating digital elevation models for glaciologic applications: An example from Nevado Coropuna, Peruvian Andes. *Global and Planetary Change*, 59(1-4), pp.110-125.

Racoviteanu, A.E., Williams, M.W., & Barry, R.G. (2008). Optical Remote Sensing of Glacier Characteristics: A Review with Focus on the Himalaya. *Sensors*, 8, pp 3355-3383.

Raup, B., Racoviteanu, A., Khalsa, S. J. S., Helm, C., Armstrong, R. & Arnaud, Y. (2007). The GLIMS geospatial glacier database: A new tool for studying glacier change. *Global and Planetary Change*, 56, pp. 101-110.

Richardson, S.D. & Reynolds. J.M. (2000). An overview of glacial hazards in the Himalayas. *Quaternary International*, 65-66, pp.31-47.

Rivera, A. & Casassa, G. (1999). Volume changes on Pio Xi Glacier, Patagonia. *Global Planet Change* 22:233-244.

Schmidt, S. (2009). Fluctuations of Raikot Glacier during the past 70 years: a case study from the Nanga Parbat massif, northern Pakistan. *Journal of Glaciology*, 55 (194), pp. 949-959.

Shiyin, L., Wenxin, S., Yongping, S., & Gang, L. (2003). Glacier changes since the little ice age maximum in the western Qilian Shan, Northwest China, and consequences of glacier runoff for water supply. *Journal of Glaciology*, 49, pp.117-124.

Toutin, T. (2008). ASTER DEMs for geomatic and geoscientific applications: a review. *International Journal of Remote Sensing*, 29(7), pp. 1855-1875.

UNEP. (2008). *World Heritage Site: Sagarmatha National Park, Nepal*. United Nations Environment Programme.

Vogel, S.W. (2002). Usage of high-resolution Landsat-7 band 8 for single band snow cover classification. *Annals of Glaciology*, 34, pp.53-57.

WGMS (2011). World glacier monitoring service. http://www.wgms.ch/

Wessels, R., Kargel, J. S. & Kiefffer. H.H. (2002). Aster measurement of supraglacial lakes in the Mount Everest region of the Himalaya. *Annals of Glaciology*, 34, pp. 399-408.

Williams, R. (1987). Satellite remote sensing of Vatnajökull, Iceland, *Annals of Glaciology*, 9, pp. 119-125.

Ye, Q. (2010). The generation of DEM from ALOS/PRISM and ice volume change in Mt. Qomolangma region. *Geophysical Research Abstracts*, 12.

17

Remote Sensing Application in the Maritime Search and Rescue

Jing Peng and Chaojian Shi
Shanghai Maritime University
P.R. China

1. Introduction

Maritime search and rescue (MSR- In the maritime publications, the abbreviation for search and rescue is also SAR. Here we use MSR to distinguish it from the abbreviation for Synthetic Aperture Radar.) became an enormous task with the vast growth of marine transportation and other marine activities. In the year of 2006, the MSR centers and maritime authorities in China organized and coordinated 1620 MSR operations, which involved 5322 vessels and 17498 human lives. The past few years have witnessed tremendous changes in the organizations of maritime rescue. A large part of this evolution stems from the involvement on an international scope and the contribution of the advanced technology. However, current maritime search operation, especially searching people over board, depends mostly on human eyes.

SOLAS (International convention for safety of life at sea) convention prescribes that ships must be equipped with GMDSS (Global maritime distress and safety system) equipments, which have improved the search and rescue. However, for many non SOLAS convention ships, such as fishing boats and small crafts, the detection results are not very much satisfied. With the complex sea environment, the searching of distress vessel becomes a nail-biting task. Because of the physiological characteristics of human eyes, it is difficult for the rescuer to find small target in the adverse background lighting, night or dark condition, wave or clustered seas. Continuous long time observation also causes fatigue of human eyes, resulting poor sensitivity of detection. All those factors decay the results of searching operation.

In order to improve the effect of MSR operations during the dark hours or in adverse lighting or sea conditions, remote sensing technique is a potential approach to overcome the limitation of human eyes in MSR, and thereby may hopefully improve the searching performance in complex environment or in a fatigued state of human being. Regarding ship monitoring, compared with shore-base, shipboard or airborne detecting devices, and other visible visible or infrared monitoring methods, the Synthetic Aperture Radar (SAR) remote sensing system possesses the capability of all-time, all weather, extensive and high resolution for detecting ships on the sea. Especially due to its working characteristics of not being limited by the sea surface, weather or human factors, it can detect the sea areas with geographical remote positions and hostile environment which cannot be entered directly.

In this chapter, some remote sensing techniques and algorithms concerned with the MSR are introduced. A Remote Sensing Monitoring System for Maritime Search and Rescue (RS-MSR) is presented. This work is a part of our project—Vision Enhancement System for Maritime Search and Rescue. The main task for the RS-MSR is to acquire general information in a wider scale. The distress ship is detected and located for guiding the search operation. Surrounding ships are also distinguished to coordinate the MSR operation. Some important data such as current and sea state are retrieved to help decision-making of the operation. Section 2 proposes the outline of the remote sensing methods for maritime search and rescue; Guided by the systematic functions and structures of the RS-MSR described in Section 2, Section 3 introduces the related algorithms used in RS-MSR. Section 4 describes the architecture of the remote sensing aided system for maritime search and rescue. The experiment design and the implementation performance are given in Section 5. Finally, Section 6 concludes the paper.

2. Outline

The primary role of the remote sensing is to provide a secondary source of information for the MSR operations. A remote sensing monitoring system can, to some extent, overcome the shortcomings and inadequacies of human eyes. It can also improve the searching speed and accuracy, and is of significance in promoting rescue success rate and efficiency. It can aid the rescuers to fulfill the task of search and rescue, especially for small targets, such as persons in distress and life boats, and could provide a good detection and identification result.

2.1 Introduction to the system functions

According to the requirement of search and rescue, a Remote Sensing Monitoring System for Maritime Search and Rescue (RS-MSR) is designed. Table 1 illustrates the main functions of a RS-MSR system.

RS-MSR System demand	Function
Satellite transit inquiry	Satellite transit inquiry
Ship detection	1) Ship position detection 2) Ship type identification/classification 3) Ship size estimation 4) Ship heading/direction estimation
Sea state analysis	Wave direction and estimation
Integrated processing	1. Non-remote sensing data fusion - VTS (Vessel Traffic Services) report - AIS (Automatic Identification System) report 2. Performance analysis - Recognition rate - Identification/classification rate - Position error 3. Rescue position prediction - Time prediction - Ship speed and velocity - Predict the rescue area according to the heading/direction of the distress ship and the sea state

Table 1. The main functions of a RS-MSR system required

2.2 Algorithm introduction

According to the requirement of the maritime search and rescue, the Synthetic Aperture Radar (SAR) imageries are used for this purpose. The algorithms concerned with the system demand are introduced as follows.

2.2.1 Distributed target detection method in the Gaussian scale-space

In the SAR images with high resolution, each target occupies several resolution units to form area target. So detecting the ship target in the SAR images with high resolution should regard the target as distributive target, and the assumption of point target under the traditional radar is not suitable any more. The project here proposes a distributed target detection method in the Gaussian scale-space. The distance relationship among the detected objects is adopted to identify the distributed target. In the situation of hardly estimating the background's scattering distribution or of low SNR (signal-to-noise ratio), this method can realize the distributive target detection more effectively than CFAR method.

2.2.2 Ship size category estimation model

In the SAR images with high resolution, the ship target can be divided into three categories according to their dimension, among which small ships(≤50m) are represented as point target, while the middle size ship(≤100m) as distributive target of single corner and big size ship(≥100m) as double corner distributive target. According to the relative positions of the corner and combining the orbit information (resolution ratio and incident angle) of the SAR image, the length, height and the direction of the ship will be worked out.

2.2.3 Ship location correction and ship direction estimation method

The imaging geometry of the SAR imagery is slant-range projection. So, due to the geometric distortions, such as layover, foreshortening, and shadows, there exits measurement error between the observed position and its actual location. In this method, using the directional texture of the wake of ship, the convergent point of the wake pattern can be calculated, which is the actual location of the stern. Then the position correcting parameters can be worked out.

A bow wave is the wave that forms at the bow of a ship when it moves through the water. As the bow wave spreads out, it defines the outer limits of a ship's wake. Theoretically, the convergent point of the bow wave' outline must at the extended line of the ship's heading. The vanishing point can be calculated by the Hough transform, and the heading direction of the ship can be calculated according to the coordinates of the bow and the stern.

2.2.4 Wave direction estimation based on the partial energy direction

Wave direction estimation can be used to analyze the sea state of the target area and supply basis for search and rescue area decision. This method is set up on the basis of steerable filter, which is a filter set composed of an even-symmetric filter and an odd-symmetric filter. When the orthogonal filter set is rotated to the same orientation of the local texture, the oriented energy reaches its maximum. The orientation corresponding to the maximum oriented energy is defined as the dominant orientation of the local energy at that point. And the main direction of the wave can be estimated.

2.2.5 The registration algorithm of SAR image and nautical chart based on Gaussian principle curve

To ensure precise detection and location of the distressed ship in the MSR, the navigational chart and the remote sensing image should be matched beforehand. Because the SAR image and the electronic chart are data from different sensors, the content and intensities of these images are much different from each other. Coastline is a stable and reliable feature for navigation in coast area. However, the deformation between edges extracted from different signals may produce position errors, and the noise in radar signals may greatly influence the edge extraction result. And how to obtain reliable control-points and how to obtain the correct correspondence are the key issues in the registration algorithm. In this chapter, a multi-scaled registration algorithm for SAR image and electronic chart is presented. Based on the scale-space theory, coastlines from the two images are matched in both frequency domain and image domain with continuous scale level.

3. Algorithms and methods in the maritime search and rescue

3.1 Distributed target detection method in the Gaussian scale-space

In the high resolution SAR images, the large-scale ship and super large-scale ship (>100 meters) are presented as distributed targets. In this algorithm, these targets are detected by the distance relationship between the echo intensity of the masthead light and the ship hull. The detection of distributed target based on location-dependent information can be completed by two-step detection. In the first step, the ship target is characterized in the Gaussian scale-space. This transforms the signal range value into binary. Then, the detection of the singular objects is implemented using constant false alarm rate(CFAR), and record the location of the pixel point whose value is 1. The second step aims at finding the distributed target from the result of the first step detection by applying location-dependent information.

3.1.1 Ship characteristic description

Under the ideal condition, in the $z(x)$ of the image which has its background gray value as 0, there is a maxima value h and strong scattered point $f_b(x)$ with width as w. Considering the edge effect of the image, we build the mathematical model as follows:

$$f_b(x) = \begin{cases} h(1-(x/w)^2), & |x| \le w \\ 0, & |x| > w \end{cases} \tag{1}$$

The responses of these spot-like targets in Gaussian scale space are presented as $r_b(x,\sigma,w,h)$:

$$\begin{aligned} & r_b(x,\sigma,w,h) = g_\sigma(x) * f_b(x) \\ & = \frac{h}{w^2}[(w^2 - x^2 - \sigma^2)(\phi_\sigma(x+w) - \phi_\sigma(x-w)) \\ & -2\sigma^2 x(g_\sigma(x+w) - g_\sigma(x-w)) - \sigma^4(g'_\sigma(x+w) - g'_\sigma(x-w))] \end{aligned} \tag{2}$$

Among them, $\phi_{\sigma}(x) = \int_{-\infty}^{x} e^{-\frac{x^2}{2\sigma^2}} dt$.

In order to departure the ship target from the grey scale space, we define a Gaussian comparison function $e_b(x)$. If the responses of $f(x)$ in Gaussian scale-space is $r(x,\sigma, w, h)$, then we have:

$$e_b(x) = \begin{cases} 1, & r(x,\sigma,w,h) \le f(x), \text{ and } |x| \le w \\ 0, & \text{other} \end{cases} \tag{3}$$

Then, the detection of the singular target is implemented by way of constant false alarm rate(CFAR).

3.1.2 Distributed target detection based on location-dependent

Set the SAR image as a $N_R \times N_A$ matrix, where N_R and N_A represent the dimension of range and azimuth respectively, and use $\{i_{R1}, i_{R2}, \cdots, i_{RK}\}$ as range coordinate and $\{i_{A1}, i_{A2}, \cdots, i_{AK}\}$ as azimuth coordinate separately for convenience.

The detected relative distance among different scattered units can be defined.

$$d(j,k) = \|i_j - j_k\|, k > j, j = 1,2,\cdots,K-1 \tag{4}$$

where $\| \ \|$ means norm. Equation(4) represents the relative distance between two scattered units j and k, and it expresses the location relations among each target pixel point after the Gaussian scale-space detection.

When the range and the azimuth are considered separately, the distance can be defined as

$$d(j,k) = \left(\left|i_{Rk} - i_{Rj}\right|, \left|i_{Ak} - i_{Aj}\right|\right) \tag{5}$$

Due to the geometric distortions of the SAR image, the ship hull and the mast will occupy several resolution units in the image. Set a candidate ship target M_0 (M_0 includes the ship hull and the mast), where M_{0R} and M_{0A} represent the size of M_0 in the range direction and the azimuth direction respectively. Set a bounding box with size of $M_{0R} \times M_{0A}$, which is the smallest rectangle containing the detecting ship. Assume that the size of the detecting target is $S_R \times S_A$ and the radar resolution is $\Delta R \times \Delta A$, then $M_{0R} \times M_{0A} = \frac{S_R}{\Delta R} \times \frac{S_A}{\Delta A}$.

Regarding to the distributed targets, at least M scattering points will be detected after the second detection step, and then M can be defined as the distance threshold d_{th} within the distributed target.

$$d_{th} = M = \mu M_{0R} \times M_{0A} \tag{6}$$

Here $\mu \in (0,1)$ is the confident coefficient and it is determined by the empirical data of the radar echo.

Define the size of the distributed target T_s as:

$$T_s = d(M_0,(0,0)) \tag{7}$$

Equation (7) defines the distance from the position of the bounding box M_0 to the origin point (0,0). The distance describes the size of the detecting target, which is related to the form of the distance definition.

According to the size of the bounding box M_0 and Equation (6), the number of scattered points (marked as u) of each target reference window and their locations (i.e. target location) can be calculated as follow.

$$\left.\begin{array}{ll} u \geq d_{th}, & \text{target} \\ u < d_{th}, & \text{non-target} \end{array}\right\} \tag{8}$$

Under the definition in Equation (7), $d(T_{position}, i_k)$ and T_s are all two-dimension data, we need define its output. Set $T_s = (T_{sR}, T_{sA})$, $d = (T_{position}, i_k) = (d_R, d_A)$, and then the export definition of $d(T_{position}, i_k) > T_s$ is:

$$\mathrm{d} = (\mathrm{T}_{\mathrm{position}}, \mathrm{i}_\mathrm{k}) > \mathrm{T}_\mathrm{s} = \begin{cases} 1, & \text{other} \\ 0, & \mathrm{d}_\mathrm{R} \leq \mathrm{T}_{\mathrm{sR}} \text{ and } \mathrm{d}_\mathrm{A} \leq \mathrm{T}_{\mathrm{sA}} \end{cases} \tag{9}$$

3.1.3 Algorithm realization

According to the above description of the algorithm, to design the integrated detection algorithm based on two-dimension location-dependent information, the procedure is shown as follows:

1. Initially set $k = 1$, $u = 1$, and the position of the target centre is $T_{position} = i_k$;
2. If $d = (T_{position}, i_k) \leq T_s$, the number of pixels contained in the current target is $u = u + 1$, and renew the central position of the current target $T_{position} = \dfrac{(u-1)T_{position} + i_k}{u}$;
3. If $d = (T_{position}, i_k) > T_s$, then create a new target with its central pixel location as $T_{position} = i_k$, $u = 1$;
4. Traversal searches the image.

3.2 Ship size category estimation model

In the SAR images with high resolution, the ship target can be divided into two categories according to their dimension. The small ships are represented as single corner target while big ships as double corner distributed targets. According to the relative plane location of the double corner and combining the orbit information (resolution ratio and incident angle) of SAR images, the parameters of the ship's status, such as the length, height and the heading direction can be calculated. Table 2 shows the corresponding incident angle of ENVISAT ASAR.

Image Position	Breadth (km)	Distance to the Substellar Point (km)	Incident Angle (°)
IS1	104.8	187.2—292.0	15.0—22.9
IS2	104.8	242.0—346.9	19.2—26.7
IS3	81.5	337.2—418.7	26.0—31.4
IS4	88.1	412.0—500.1	31.0—36.3
IS5	64.2	490.4—554.6	35.8—39.4
IS6	70.1	549.7—619.8	39.1—42.8
IS7	56.5	614.7—671.1	42.5—45.2

Table 2. Corresponding incident angle range of ENVISAT ASAR

Build a space coordinates as Fig.1 with the true north (Y axis) as the reference direction. The ship length is l, and the width is w, height is h, azimuthal angle of the bow is α (the angle between the bow and the North), and then a ship model can be descripted by a set of geometric parameters as $P=[\alpha,l,w,h]$. For example, if the satellite is descending, for ENVISAT ASAR it is a right view image. Set radar incident angle as θ, the azimuthal angle as β (the angle between the satellite and the true north) and the height as H, which is shown as in Fig.1. According to SAR imaging mechanism, the shadow in the figure shows the ship's scattering image, which is determined by the ship's ground position, shape and the radar's scattering orientation. Assume that the ship hull scattering length is s and the width is d. The range is

$$w = b\cos(\beta - \alpha)$$

$$b = d\tan^2\theta \qquad (10)$$

$$w = d\tan^2\theta\cos(\beta - \alpha)$$

Here w is the ship width, b is the width of the ship at the radar range, d is the scattering width of hull target, θ is the radar incident angle, β is the radar azimuthal angle and α is the bow azimuthal angle. The azimuth is

$$s = l\cos(\beta - \alpha)$$

$$l = \frac{s}{\cos(\beta - \alpha)} \qquad (11)$$

$$w = d\tan^2\theta\cos(\beta - \alpha)$$

Here l is the ship length and s is the scattering length of the ship target images at the radar range, β is the radar azimuthal angle and α is the bow azimuthal angle.

According to the length of the ship ships can be divided into small-sized (<50 meters), mid-sized (50-100 meters), big-sized (100-200 meters) and extra big-sized ship (>200 meters).

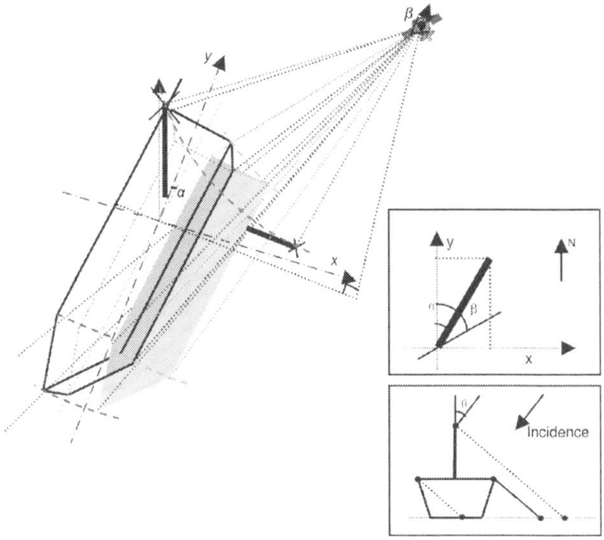

Fig. 1. The ship geometric projection model in the SAR image

3.3 Ship positioning correction and direction estimation

3.3.1 Ship positioning correction

Due to the geometric distortions of the SAR imaging, there exits measurement error between the observed position and its actual location. With this method here we can get the real position of the stern by way of detecting the ship's wake region with directional texture and calculating the convergent point of the ship's wake. Thereby, the position correcting parameter will be worked out. This method enables us to get the information of wave-making of the distressed ship when the wake profile is invisible or in violent sea status, which improves the flexibility and automaticity for understanding the marine remote sensor images.

3.3.1.1 Wake region determination

A wake is the directional texture formed on the water surface immediately behind the ship. Therefore, the mean direction of the wake is consistent with the ship heading. In this algorithm, regarding the ship as the center, we divide the sea surface of the ship's neighborhood into several partially overlapped sectors. Define the angle between the edge of the sector and the positive direction of X axis as the edge direction angle of the sector, and the two edge direction angles formed by two edges of sectors are defined the direction range of the sector areas, and the median of the two edge angles is the main direction of the sector area. Calculate the textural energy of the main direction in each sector area of the ship, and set the sector area with the most textural energy as the wake areas. The direction range of the sector is regarded as the wake direction range, and the target area close to the wake range is the stern in the image.

3.3.1.2 Course calculation

In the wake area, calculate the textural direction pixel by pixel, and set the pixels whose textural directions belong to the wake direction interval as the wake points. Use the least-square method to calculate the mean direction of the wake, and the result is defined as the direction of the ship's wake, i.e. the course.

3.3.1.3 Actual lactation of the stern

Theoretically, a wake is the region of disturbed flow immediately behind the ship, and the texture of the wake pattern should converge to a point which is the real position of the stern. Collect the wake points in the wake region. Use the Hough transform to calculate the vanishing point of the wake texture. And this point is regarded as the actual location of the stern.

3.3.1.4 Projection offset calculation

According to the actual position of the stern and its image position, the projecting offset can be calculated. Use the stern offset, the projecting model is built. After calibration, the original detected result can be modified and the actual location of the ship target can be obtained.

3.3.2 Heading direction estimation method

A bow wave is the wave that forms at the bow of a ship when it moves through the water. As the bow wave spreads out, it defines the outer limits of a ship's wake. Theoretically, the convergent point of the bow wave' outline must be on the extending line of the ship's heading. The convergent point can be obtained by the Hough transform, and the heading direction of the ship can be calculated by the positions of the bow and the stern. Using the orbit information (resolution ratio and incident angle) of SAR images, the length, height and the direction of the ship can be worked out. The heading direction vector is defined in accordance with the bow wave's outline, which points from the actual location of the stern to the convergent point of the bow wave. Then, the heading direction is defined as the direction between the heading direction vector and the true north in anti-clockwise.

3.4 The registration algorithm of SAR image and nautical chart based on Gaussian principle curve

To implement registration to remote sensing images with navigational radar image and the chart, the detection results will be directly showed both on the remote sensing image and the chart, and then do contrast verification among the remote sensing detection results and the data of radar and AIS. A multi-scale matching algorithm of radar image and chart is proposed in this project, transforming the coastline into a set of smooth curves in the Gaussian scale space, and making coarse to fine image registration to radar image and the coastline in the chart separately in the frequency field and the spatial field.

3.4.1 Curve feature representation

In the extracting coastlines, many near-shore objects such as ships and navigation marks that also have strong echoes may be merged by mistakes, leading to Ω-shaped spurs of the

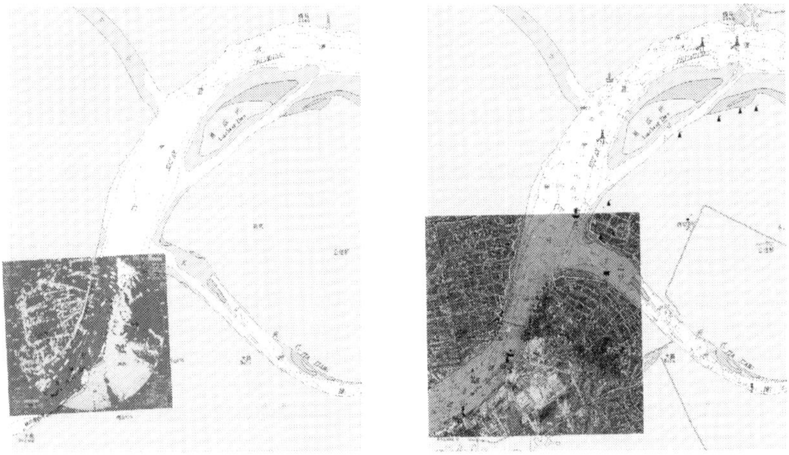

Fig. 6. Registered image pairs. (left) radar image and nautical chart, (right) remote sensing image and nautical chart

$$E(x,y)=\sqrt{(M_e * f(x,y))^2+(M_o * f(x,y))^2} \tag{23}$$

The steerable filter is the linear combination of a set of base filters, which are partially overlapped in the frequency domain, and can be rotated. An orthogonal filter pair is a combination of a steerable filter and its Hilbert transformation, which is designed to detect precisely the features of the edge, texture and singular point of the target. To obtain the 2-D local energy in continuous frequency space, the Wavelet Transform is used to decompose the signal into a series of sub-band signals with particular frequencies. Here we use the Mexico-hat wavelet G_2 to build the steerable filter G_2^{θ}:

$$G_2^{\theta}=k_1(\theta)G_2^{0}+k_2(\theta)G_2^{\pi/3}+k_3(\theta)G_2^{2\pi/3} \tag{24}$$

The Mexican-hat wavelet $G_2(x,y)=\partial^2\exp[-(x^2+y^2)]/\partial x^2$ is a symmetric filter with sharp narrow bandwidth, thus it can effectively restrain noise and enhance the signal in particular frequency, and it is common used in multi-scale edge detection. The fundamental filters G_2^0, $G_2^{\pi/3}$, $G_2^{2\pi/3}$ represent the forms of G_2 rotating to 0, $\pi/3$, $2\pi/3$, respectively. $k_i(\theta)$ is the interpolation function corresponding to the fundamental filters. Then, the form of G_2 in any orientation is represented by the linear combination of G_2^0, $G_2^{\pi/3}$, $G_2^{2\pi/3}$. We can get the direction energy of arbitrary pixel (x,y) of the image in an arbitrary direction θ by using the orthogonal filter bank formed by steerable filter G_2^{θ} and its Hilbert transformation H_2^{θ}.

$$E^{\theta}(x,y)=\sqrt{(G_2^{\theta} * f(x,y))^2+(H_2^{\theta} * f(x,y))^2} \tag{25}$$

As for the singular characteristics, e.g. the edge, when the orthogonal filter moves to the same direction with this characteristic, the direction energy reaches maximum value. The corresponding direction of the local orientation energy is called the principal direction of the pixel's local energy.

The wave image is filtered in this algorithm to eliminate speckles by way of Lee filter, and on this basis the principal energy direction of the wave can be estimated.

The experiment uses the satellite ENVISAT-1 ASAR data of 30th Sep. to 19th Oct, and the experimental area covers 30°48'N ~ 31°20'N, 122°10'E ~ 122°47'E. We use the wave direction estimation based on local energy direction to calculate the wave direction for AP polarization data. The calculation results are compared with the JMH wave analysis chart from Japan Meteorological Agency. Table 3 shows the experiment result of this wave

Acquisition Time (UTC)	Image	Polarization Mode	Wave Direction	Direction Energy	Incident Angle	Estimated Direction	JMH Wave Analysis
2008-09-30 13:53		VV	0°	124.3043	41.1016	75.1857	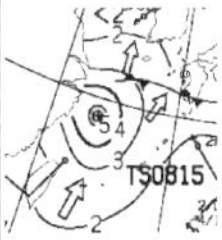
			30	117.9377			
			60	132.5371			
			90	135.8590			
			120	131.9720			
			150	118.6072			
		VH	0	111.2229			
			30	116.8233			
			60	133.8221			
			90	136.7364			
			120	132.8555			
			150	116.2534			
20081008 01:50		HH	0	78.0928	33.9364	104.2056	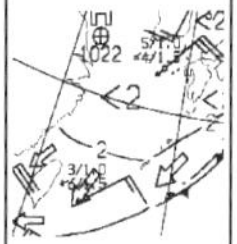
			30	97.9058			
			60	130.5053			
			90	145.1742			
			120	130.5707			
			150	99.4210			
		VV	0	77.5663			
			30	99.3430			
			60	133.5955			
			90	146.8708			
			120	133.2834			
			150	99.9747			
20081010 13:39		HH	0	14.0134	19.2636	99.9323	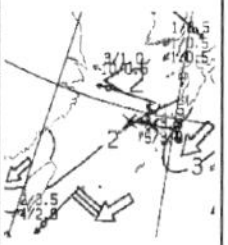
			30	33.8942			
			60	133.1682			
			90	268.4664			
			120	132.8741			
			150	34.1923			

Acquisition Time (UTC)	Image	Polarization Mode	Wave Direction	Direction Energy	Incident Angle	Estimated Direction	JMH Wave Analysis
		HV	0	7.7873			
			30	19.1000			
			60	74.9717			
			90	153.2269			
			120	75.1100			
			150	18.9022			
20081019 13:56		HH	0	120.3329	44.0092	104.8161	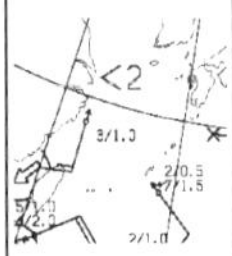
			30	115.1276			
			60	120.3069			
			90	116.8428			
			120	120.2216			
			150	118.1973			
		HV	0	66.6512			
			30	69.6033			
			60	73.6283			
			90	71.8446			
			120	73.2198			
			150	69.0945			

Table 3. The experiment result of the wave direction estimation algorithm

direction estimation algorithm. The experimental result analysis shows that VV polarization mode is the best way for wave analysis, and the following is HH, while cross polarization VH and HV mode are not ideal.

4. The architecture of the remote sensing aided maritime search and rescue system

The Remote Sensing Monitoring System for Maritime Search and Rescue (RS-MSR) consists of four modules including satellite transit inquiry module, vessel detection module, sea state analysis module and integrated processing module. Ship detection module has three functions and they are ship location, ship type identification/classification and ship movement direction estimation. Sea state analysis mainly estimates the wave direction. The integrated processing module receives the detection results from ship detection module and sea state analysis module. According to the distressd ownship' position, heading and the wave direction, combining the time used for data receiving, it estimates the position of the distressed ship, and combining the satellite parameter, it can revise the result obtained through ship detection. The analytic result by way of integrated processing module can be transmitted to the Maritime Safety Administration (MSA) and the rescue vessel on the working field, providing assisting decisions of areas for the rescue work. Fig. 7 describes the architecture of RS-MSR.

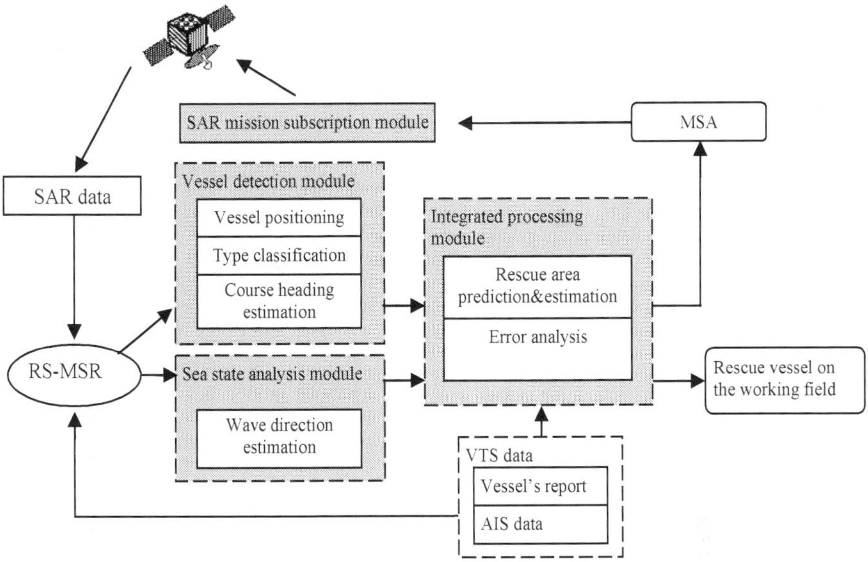

Fig. 7. The architecture of the Remote Sensing Monitoring System for Maritime Search and Rescue (RS-MSR)

4.1 Satellite transit inquiry module

With the development of astronavigation, the number of satellites installed with SAR sensors is increasing. Facing with so many satellites with different purposes, it has become a tough problem for clients to judge and select what they want quickly. RS-MSR sets up a real-time satellite coverage inquiry system including the commonly used satellites around the world, such as RadarSat, Envisat, ERS, CosmoSAR, TerraSAR, helping clients to retrieve quickly the crossing time and the orbit data of these satellites at specific area.

4.2 Ship detection module

Ship detection module is the core unit of the whole RS-MSR system. Using the micro-area images of distressed areas supplied by way of satellite, it can detect and monitor the ships and the accidental areas, supplying clue of the distressed ship for search and rescue and helping to determine the areas quickly. This module consists three parts including: (1) ship detection; (2) ship classification/identification; and (3) ship direction and course estimation.

4.3 Sea state analysis module

Sea state analysis module can perform initial analysis on the situation of the distressed area by estimating the wave direction and supply foundation for search and rescue decision, which is useful for estimating the floating direction and location of the distressed ships.

4.4 Integrated processing module

The integrated processing module is mainly used for follow up processes of the detection result and search and rescue assisting forecast. It has two main functions. Firstly, search and rescue range estimation. According to the current location, navigation direction and the wave direction, estimate the potential searching areas of the distressed ship under the settled speed of the ship and velocity of flow. Secondly, ship location correction. Combine the satellite parameter to revise the error caused by slant-range projection imaging of the SAR images. The analysis data obtained by way of integrated processing module can be saved as *.dat or *.mat form and transmitted to marine department and the rescue spot, supplying assistant for the rescue areas determination.

5. System experimental performance

5.1 Experiment design

In our experiment, the *Yangtze River* estuary (Changjiangkou) precautionary (30°48'N ~ 31°20'N, 122°10'E ~ 122°47'E) is selected as the experimental working zone (Fig. 8). This region includes the inward and outward fairway of the *Yangtze River* estuary and the 1# and 2# anchorages. The *Yangtze River* estuary is a tide-coordinated region, and the tide rise and fall twice per day. The tidal range is up to 4 meters. And the flow is rapid. Both anchored ships and underway ships are aggregated at this place. Casualties happen often, and it is a key region for the monitoring of maritime search and rescue. Therefore, this place is an excellent experimental region.

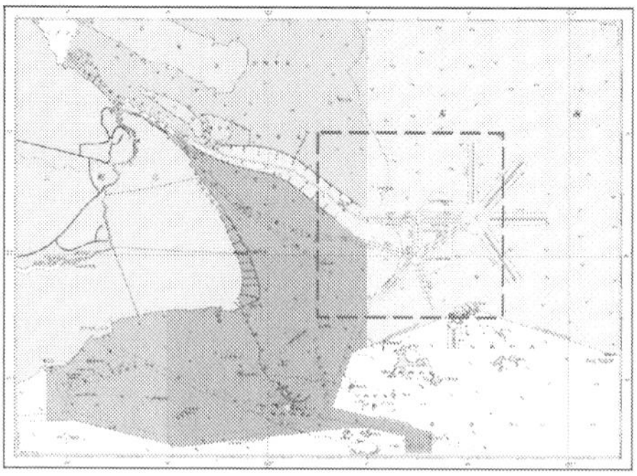

Fig. 8. The experimental working zone in the nautical chart

5.2 Experiment data

From 2008.09.30 to 2008.10.19, we took four times of experiments. EnviSat-1 ASAR data is used in this experiment. Data acquisition is implemented considering the three tidal conditions: rise, fall and still. And the data includes: VTS maritime Radar, AIS, weather information, tide and flow, etc. The details are listed in Table 4. Here we present the experiment on 2008.10.19 as an example.

- Experiment region: 30°48'N~31°20'N, 122°10'E~122°47'E
- Experiment time: from 2008.09.30 to 2008.10.19

- Tide reference: Jigujiao *tide*-gauge station (31°10'24″ N, 122°22'54″ E)
- Flow reference: Xinkaihe *tide*-gauge station (31°14'36″ N, 121°29'12″ E)

No.	Experiment Time (UTC)	Tide reference (Jigujiao)	Flow reference (Xinkaihe)
1	2008-09-30, 13:53:24	14:00 L, 431cm	10:00 L, 105cm/s
2	2008-10-08, 01:50:45	02:00 L, 244cm	10:00 L, 105cm/s
3	2008-10-10, 13:39:00	14:00 L, 303cm	22:00 L, -076cm/s
4	2008-10-19, 13:56:17	14:00 L, 246cm	22:00 L, 077cm/s

Table 4. The hydrological information in the experiment

5.2.1 The remote sensing data

Data: ASA_APP_1PNBEI20081019_135613_000000202073_00082_34705_4038.N1
Polarization Mode: HH/HV
The detail information is listed in Table 5.

Mode	Track	Frame	Lower Left longitude	Upper Right longitude	Upper Left latitude	Lower Right latitude	Swath	Passing direction	Start Date /Time (UTC)
APP_HH/HV	82	596	122.167	122.921	31.6478	30.7285	S7	Descending	2008-10-19 13:56:18.802

Table 5. The detail information of the remote sensing data

5.2.2 The VTS reference data

The radar data and AIS data are received from the VTS center of Shanghai port. Fig. 9 is the VTS shore-based radar detection picture simultaneously at the acquisition time of the remote sensing data. The VTS report shows the name and position of all the vessels equipped with AIS in the Yangtze River estuary. The Radar data and AIS data provide a reference for the performance analysis of the system experiment.

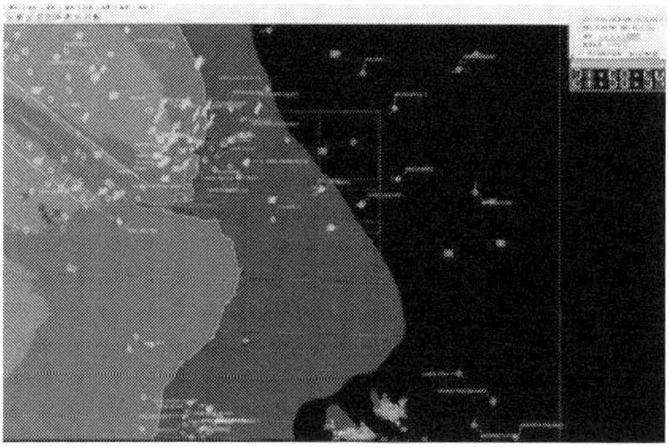

Fig. 9. The VTS report at the passing time of the satellite (2008-10-19, 13:56:17 UTC)

5.2.3 The weather information

The weather information includes the JMH weather chart, wave height, flow rate, and the tidal data. Fig. 10 shows the track of the No.15 typhoon (30 September, 2008).

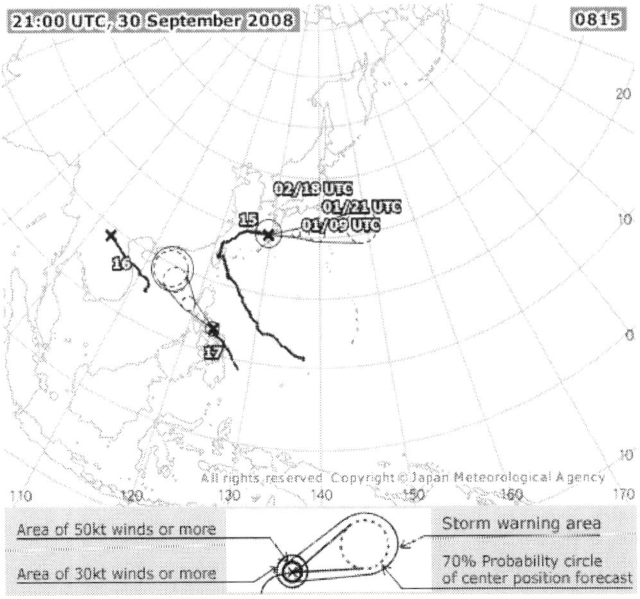

Fig. 10. The track of the No.15 typhoon (30 September,2008)

5.3 System experiment

Here we set detecting zone as follow:

- UpperLeft longitude: 122.338E
- UpperLeft latitude: 30.9261N
- Range in longitude: 9.8276 nmile
- Range in latitude: 8.96046 nmile

Fig. 11 presents the working interface of the Remote Sensing Monitoring System for Maritime Search & Rescue (RS-MSR). And Fig. 12 shows the original SAR image.

5.3.1 Vessel detection

Vessel detection includes: ship position detection, type identification, length estimation and heading estimation. The detection results are shown in Fig. 13(middle) and Table 6. The ship type classified into four categories: small, middle, large, and extra-large. The ship heading direction is the angle between the ship heading and the real north in clockwise.

In Fig. 13(right), the red circle represents small ship (point target) and the green arrow indicates the estimated heading direction of big ship (distributed corners)

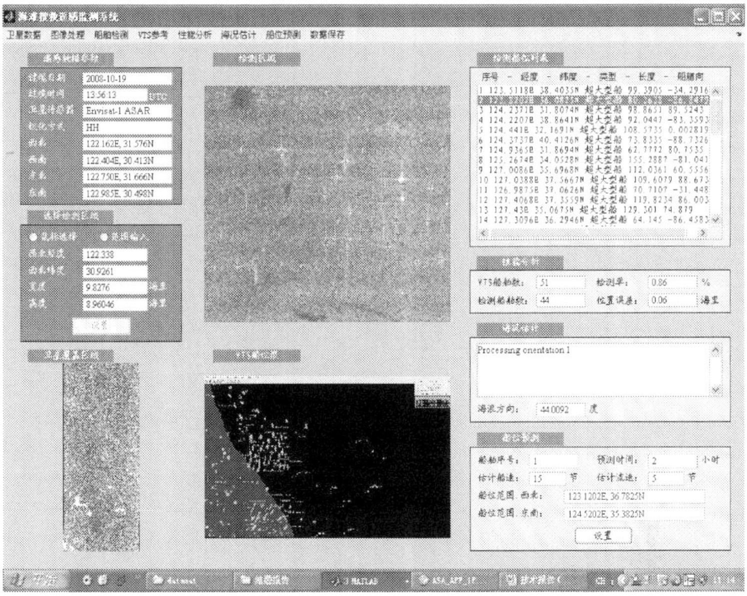

Fig. 11. The working interface of the Remote Sensing Monitoring System for Maritime Search & Rescue (RS-MSR)

Fig. 12. The original SAR image (2008-10-19, Changjiangkou precautionary, Envisat-1 ASAR HH)

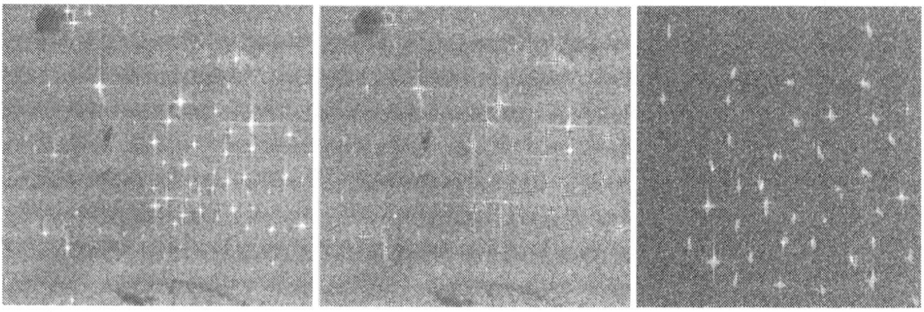

Fig. 13. (left) The selected detecting zone (original image), (middle) The vessel detection results (the red cross), (right) Ship size identification and the heading direction estimation result

No.	Position	Type	Length(m)	Heading angle (degree)
1	123.5118E 38.4035N	M	99.3905	124.2916
2	123.8202E 36.0825N	M	80.3638	176.8479
3	124.2371E 31.8074N	M	98.8651	0.4757
4	124.2207E 38.8641N	M	92.0447	173.3593
5	124.441E 32.1691N	M	108.5735	89.9972
6	129.429E 35.3644N	M	95.7427	233.4018
7	129.5839E 33.1727N	L	140.0645	10.0805
8	129.8325E 36.7182N	L	116.6667	179.5634
9	130.0705E 35.3129N	XL	219.6904	0.0432
10	129.9275E 38.2278N	M	73.48	7.3131

Note: There are totally 44 vessels are detected, and only 10 are listed here.

Table 6. The vessel detection results

5.3.2 Performance analysis

- VTS vessel report: 51
- Detected vessel: 44
- Detecting rate: 86%
- Position error: 0.06 nmile

Among the detection, six small vessels and one middle sized vessel were missed. The performance is better in the detection of big and extra big vessels.

5.3.3 Sea state estimation

The wave direction estimation result is shown in Table 7.

Acquisition Time (UTC)	Polarization Mode	Incident Angle	Estimated Direction	JMH Wave Analysis
20081019 13:56	HH/HV	44.0092	104.8161	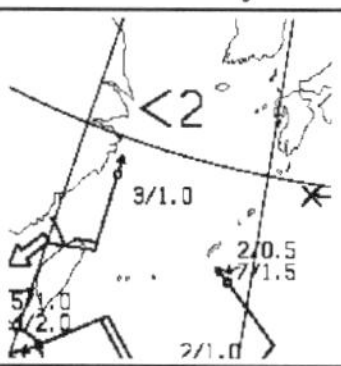

Table 7. The wave direction estimation result

5.3.4 Ship position prediction

- Ship number: 1 #
- Predicting time: 2 hours After
- Ship speed: 15 knot
- Flow speed: 3 knot
- MSR Searching region prediction:
  - Northwest: 123.1202E, 36.7825N;
  - Southeast: 124.5202E, 35.3825N.

5.4 The system performance

The complete maritime search and rescue system supplies the function of ship detection, identification and location algorithm. It can be also used to inquire the satellite transit time and its orbit data. The precision of ship size estimation reached over 80% and the position estimate the position detecting error is within 0.5 nautical miles.

6. Conclusion

In this chapter, a remote sensing monitoring system for maritime search and rescue (RS-MSR) is presented. Some related algorithms are introduced. The satellite remote sensing imageries of large scale water area are acquired to detect and locate distress ships for guiding the search operation. Some important data such as current and sea state are retrieved to assist decision-making of the operation. System experiment design and test are presented, and the performance shows that this system can effectively improve the searching speed and accuracy, and is of significance in promoting rescue success rate and efficiency.

7. Acknowledgment

The research work in this paper is partially sponsored by the Shanghai Leading Academic Discipline Project (grant number: S30602), and the Natural Science Foundation of China (grant number: 40801174), and the Program of Shanghai Subject Chief Scientist (grant number: 10QA1403100).

8. References

Dare P., Dowman I. (2001). An Improved Model for Automatic Feature-based Registration of SAR and SPOT Images, *ISPRS Journal of Photogrammetry & Remote Sensing*, 2001, Vol.56, pp.13-28

Gerkacg J. (1999). Spatially Distributed Target Detection in Non-Gaussian Clutter. *IEEE Trans. on AES*, 1999, Vol.35, No.3, pp.926-934

Huang D.S & Han Y.Q. (1997). A Detection Method of High Resolution Radar Targets Based on Position Correction. *Journal of Electronics*, 1997, Vol.19, No.5, pp.584-590.

Kapoor R.; Banerjee A.; Tsihrintzis G.A. & et al. (1999). UWB Radar Detection of Target in Foliage using Alpha-stable Clutter Models. *IEEE Trans. on AES*, 1999, Vol.35, No.3, pp.819-833

Kuttikkad S. & Chllappa R. (1994). Non-Gaussian CFAR Techniques for Target Detection in High Resolution SAR Images. *IEEE, International Radar Conference 1994, pp.910-914*, Austin, TX, USA, 1994

Morrone, M.C. & Owens, R.A. (1987). Feature Detection from Local Energy, *Pattern Recognition Letters*, Vol. 6, pp. 303-313

Novak L.M. & Hesse S.R. (1993). Optimal Polarization for Radar Detection and Recognition of Targets in Clutter, *IEEE, National Radar Conference*, pp.79-83, Lynnfield, MA, USA, 1993

Steger, C. (1996). An Unbiased Detector of Curvilinear Structures, *Technical Report FGBV-96-03*, Forschungsgruppe Bildverstehen (FG BV), Informatik IX, Technische Universität München

Uratsuka S. & et al. (2002). High-resolution Dual-bands Interferometric and Polarimetric Airborne SAR (Pi-SAR) and Its Applications, *Proceedings of IGARSS'02*, Vol.3, pp.24-28

Section 4

Human Activity Assessment

18

Object-Based Image Analysis of VHR Satellite Imagery for Population Estimation in Informal Settlement Kibera-Nairobi, Kenya

Tatjana Veljanovski[1,2], Urša Kanjir[1], Peter Pehani[1,2], Krištof Oštir[1,2] and Primož Kovačič[3]
[1]*Scientific Research Centre of the Slovenian Academy of Sciences and Arts*
[2]*Space-SI - Centre of Excellence for Space Science and Technologies*
[3]*Map Kibera Trust*
[1,2]*Slovenia*
[3]*Kenya*

1. Introduction

Cities in Africa and developing countries in general are having a difficult time coping with the influx of people arriving every day. Informal settlements are growing, and governments are struggling to provide even the most fundamental services to their urban populations.

Kibera (edge region within the Nairobi) is the biggest informal settlement in Kenya, and one of the biggest in Africa. The population estimates vary between 170,000 and 1 million and are highly debatable. What is certain is that the area is large (roughly 2.5 km^2), host at least hundreds of thousands people, is informal and self-organized, stricken by poverty, disease, population increase, environmental degradation, corruption, lack of security and - often overlooked but extremely important - lack of information which all contribute to lack of basic services such as access to safe water, sanitation, health care and formal education.

In Africa, but also in other continents, urban growth has reached alarming figures. Informal settlements formation has been associated with the rapid growth of urban population caused by rural immigration, triggered by difficult livelihood, civil wars and internal disturbances. The result of this very rapid and unplanned urban growth is that 30% to 60% of residents of most large cities in developing countries live in informal settlements (UNHSP, 2005). Nowadays, informal residential environments (slums) are an important component reflecting fast urban expansion in poor living conditions.

Densely populated urban areas in developing countries often lack any kind of data that would enable the monitoring systems. Monitoring systems joining spatial (location) and social data can be used for the monitoring, planning and management purposes. New methods of monitoring are required to generate adequate data to help link the location and socioeconomic data in urban systems to local policies and controlling actions. In the past, rapid urban growth was quite difficult to manage and regulate when processes were in progress. Available census data barely accounts for the reality, as in most cases, they are based on figures extrapolated from old census, carried out in the 1970s or, if recent, they are obtained with poor accuracy, as informal settlements are difficult to survey (Sartori et al., 2002). More can now be done at least to monitor the extent and consequences of rapid urban growth. Where accurate maps of informal settlements and relevant census data completely lack, answers can be found using independent survey, derived from satellite or aerial technologies. Usage of satellite imagery nowadays enables rather quick answers to questions such as: where informal settlements are, what was the dynamics of their growth, how many people potentially live there, what basic services inhabitants need. Among the main issues to be addressed in informal settlements are the needs for potable water, waste evacuation, energy, education and health care facilities, and crime control. It is believed these actions can be planned based on quality mapping of the phenomena.

The spatial resolution of space-borne remote sensing has improved to such extent that their products are comparable with the ones provided by aerial photography. Satellite images taken with very high resolution (VHR) sensors, i.e. resolution around and below 1 m, enable skilled user to identify and extract buildings, trees, narrow paths and other objects of comparable size. A side effect of higher resolution is larger quantity of data which require more storage capacities and processing costs. Detection of informal residential settlements from satellite imagery is especially challenging task due to the microstructure, merged/overlapping rooftops and irregular shapes of buildings in slum-like areas. High spatial resolution is essential to facilitate extraction of individual buildings that are characterized by small, densely packed shantics and other structures. Informal settlement Kibera is composed of varying sizes of houses, where roofs can be a combination of many different materials, and mainly unpaved road and path network. Typically this can produce a spectral response on satellite imagery that is difficult to interpret and makes it difficult for traditional classification strategies to differentiate across object class type.

Various approaches enable to extract data from imagery in urban environments. Simultaneously with expansion of VHR satellite systems an object-based image analysis (OBIA) was developed to answer new technological opportunities. OBIA approach works in similar way as human brain perceives nature/environment, namely (high detailed) image is segmented into homogeneous regions called segments or "image objects" (Benz et al., 2004), which are then classified into meaningful classes, following the specific context of the study.

1.1 Objectives of the research

Objective of the work perform was to help Map Kibera Trust initiative with satellite data processing. Studies on Kibera informal settlement had two aims: first, to derive detailed land use/cover map that can further supply population estimation, and second, to analyse the potential of VHR imagery for detecting changes and settlement growth in recent past.

Since object-based classification of VHR satellite data has been argued as the most appropriate method to obtain information from urban remote sensing applications, this approach was used to derive accurate land cover map. The study involved GeoEye and QuickBird satellite images acquired between 2006 and 2009. Object-based approach was used to determine detailed urban structure in informal settlements area. Urban expansion was analyzed through comparison of images taken on different dates, using contextual multi-level pixel based approach. The results of object-based analysis based on morphology attributes were further explored to estimate the potential population. There is a big discrepancy among estimations on Kibera population, thus different density parameters were tested to approach the potential population scenario.

The first, introductory chapter sets the informal residential settlement issue in the wider context of the remote sensing possibilities framework, highlighting the methodology of the study. Chapter 2 gives an overview of research and applications of informal residential environments monitoring. Chapter 3 reviews existing conditions in Kibera, Nairobi's informal residential settlement, bringing into perspective the historical development of the slum, and its current characteristics. Chapter 4 consists of a set of specific procedures performed at two spatial extents, to attain both aims of the study. Entire Kibera settlement was being reviewed, to map the general state and dynamics of housing (change detection) between years 2006 and 2009. Raila village was studied in detail using object-based analysis to derive precise map of the village land cover/use to derive population estimation models in a given situation. Chapter 5 collects the results of mapping and population estimations. Chapter 6 discusses the data and analyses involved in managing monitoring aspects of the slums. The last chapter concludes the study with some suggestions for future work.

2. Informal residential environments monitoring

Although there is a strong need to obtain spatial information about informal settlements in order to increase living conditions for its residents and regarding the fact that remote sensing images offer a well suited data source, studies on informal settlements with VHR data are not frequent. Nevertheless, in Hoffman (2001), first results of detecting informal settlements from IKONOS data in Cape Town showed the principle feasibilities using object-oriented approach. The results were promising but seemed to be very dependent on the data. Later on Hoffman et al. (2006) showed that several adaptations were necessary to OBIA algorithm improvement when applying their extraction methods to the QuickBird scene. Automatic image analysis procedures for a rapid and reliable identification of refugee tents from IKONOS imagery over the Lukole refugee camp in Tanzania was made by Giada et al. (2002). Sliuzas and Kuffer (2008) analyzed the spatial heterogeneity of informal settlements using selected high resolution remote sensing based spatial indicators such as roof coverage densities and a lack of proper road network characterized by the irregular layout of settlements. Cooperation between KeyObs, UNOSAT, OCHA and Metria resulted in digitalization of VHR GeoEye satellite image of Afgooye corridor (Somalia) from 2009, where all temporary shelters were identified (UNHCR, 2010). Different methods to detect and monitor spatial behaviour of informal settlements were presented also by Lemma et al. (2005), Radnaabazar et al. (2004), Kuffer (2003), Sartori et al. (2002), Dare & Fraser (2001) and Mason et al. (1998).

3. Study area description

Kibera is a division of Nairobi area, Kenya, within Langata constituency. Located southwest of the city centre of Nairobi, Kibera encompasses an area of 2.5 km^2, accounting for less than percent of Nairobi's total area while containing more than 25% of its population. It is the largest informal settlement in Nairobi, and the second largest urban slum in Africa, with population number varying with the season. The settlement is divided into a number of villages, including Kianda, Soweto West, Raila, Gatwekera, Kisumu Ndogo, Lindi, Laini Saba, Siranga, Kamdi Muru, Makina, Mashimoni and Soweto East (Fig. 1).

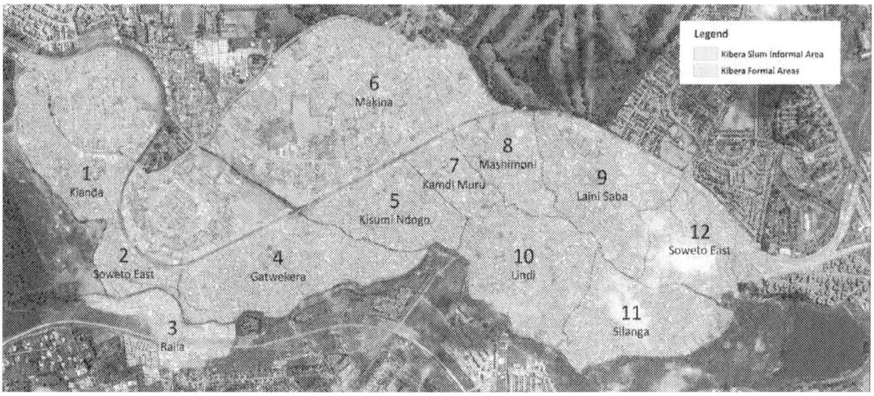

Fig. 1. Kibera settlement is divided into three formal and 12 informal villages.

3.1 General background of Kibera, Nairobi

Kibera emerged in 1912 when the British East African army, known as the King's African Rifles, granted temporary rights to a group of 300 former soldiers from the Nubian community, who had served in the army, to settle on a small piece of land near Nairobi's city centre. Temporary structures were put in place but as the Nubian soldiers grew older and became unable to continue their military service, they began to set up more permanent residence on the land (A history of Kibera, 2011). Turbulent years after the independence combined with socioeconomic factors brought a dramatic increase in the population of Kibera's residents.

Today Kibera consists of 15 villages out of which just 3 are formal and thus connected to the city's utility grids (water, sewage, electricity, waste collection etc.), however the rest (12) are informal and "disconnected" from the rest of the city. Apart from lacking basic services and adequate infrastructure it is also affected by population growth, the illegal construction of infrastructure, and the increasing degradation of the environment. Unclear land-tenure arrangements in informal settlements discourage investments in proper infrastructure and repair; structures are often owned and rented by people, who mostly do not have any rights to the land on which the structures stand. This leads to the lack of legal security of tenure for most of the residents.

Because of this lack of the legal security of tenure and neglectfulness from the city and the government there's little initiative from the residents to improve their living conditions. That is why most of the structures in Kibera are temporary, wooden, mud houses covered with corrugated iron sheets (Fig. 2) and most of the service providers are self-organized groups or cartels which drive up the prices of service delivery - in some cases residents pay 10 times as much as those in the rest of the city.

Fig. 2. View over Kibera informal settlement (Photo: Primož Kovačič).

All these reasons lead to, as one resident of Kibera put it, "survival tactics". These "survival tactics" engulf communities, the provincial administration and the government, leading them into a vicious cycle of under the table dealings, vandalism, lack of engagement, threats, and price controls with no clear perspective or solutions.

3.2 Map Kibera Project and Map Kibera Trust

Kibera is likely one of the most photographed, researched, and well-known slums in the world but the complete and mapped information was not shared and not easily (if at all) accessible. Before October 2009, Kibera did not even appear on any of the online maps. Map Kibera Project (MKP) was first initiated in response to the lack of available data. The initiative wanted to produce reliable data and maps showing the actual physical and socio-demographic features of the Kibera informal settlement, making them publicly available through a digital geo-referenced data base (MKP, 2011).

Map Kibera Project trained 13 youth from the slum in GPS system and basic GIS techniques to map points of interest in their community: clinics, schools, water sources, toilets, street lights, hot spots, businesses and other landmarks. The youth uploaded the data themselves to OpenStreetMap (www.openstreetmap.org), a volunteer-built map of the world.

The Map Kibera Trust (MKT) offers organizational support for the mapping work, as well as other youth driven programmes such as video production and other new media tools (blog, twitter, SMS platforms). The mission of the MKT is to contribute to a culture where digital story-telling, open data and geographic information lead to greater influence and representation for marginalized communities in Kenya. MKT has since grown into a platform specializing in community-driven data for informal settlements and on community based development.

American Association for the Advancement of Science supports the operation of MKT and other NGO activities and has donated several satellite images of the area. MKT activities procedures of co-registration could not be applied. The non-linear rubber-sheeting method was the only possibility to obtain mutually aligned images. This procedure is effective, but very time consuming, since it demands manual selection of hundreds of control points for each image.

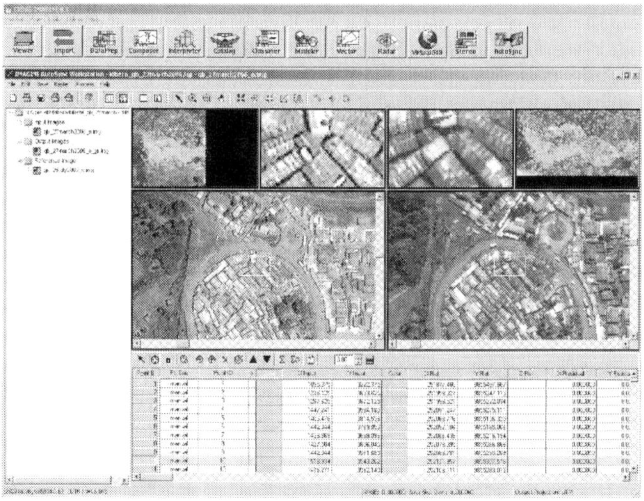

Fig. 4. Selecting the control points for rubber-sheeting method for image geo-referencing (AutoSync module of ERDAS Imagine).

GeoEye image taken on 25th of July 2009 was selected as the reference image considering its highest spatial resolution, good matching with GPS path-network tracks and the fact it is most recent. Then QuickBird images selected for analysis were manually registered to the reference, based on cca. 1,400 manually selected control points per image and using a piecewise transformation based on triangles formed from the tie points (Fig. 4). Resampling was nearest neighbour. An average RMSE is not reported as this is local approximation technique. Geo-corrected images were evaluated through detailed visual control.

Geometrically matched GeoEye and QuickBird images were then used for OBIA. Finally three images were selected for change detection due to their best results in geo-correction phase: GE2009-07-25, QB2008-08-10 and QB2006-03-27. For change detection analysis images were resampled to uniform 1 m resolution, to be prepared for radiometric standardisation.

After geometric adjustments there are still differences amongst the spectral properties of satellite images (spectral bands from the same or different images are not adjusted to each other). Hence, before pixel-based image comparisons (image differencing) the radiometric standardisation is needed. Most standardisation procedures derive from adjustments of invariant objects or from the least squares method (linear regression). The problem with the first method is that invariant areas should be verified with field measurements. Furthermore, the generally recognised invariant objects, such as deserts or light sandy beaches, do not come into play, since they cannot be found in Kibera area. The principle of the second approach is that it tries to globally adjust the given (to-be-adjusted) image or a chosen area with the reference image or a subset through a statistical approach. We applied linear regression for the relative adjustment of spectral bands between the images. Relative radiometric normalisation was done through local adjustments of QuickBird images onto GeoEye reference image, for Kibera settlement with 30 m buffer subset only.

4.2 Land cover classification

Since GeoEye image of the whole Kibera informal settlement contains lots of information and the analysis of the total area would be too demanding in terms of computer processing, we divided image of Kibera into 12 smaller parts (according to 12 informal villages). The complete process of segmentation and classification was applied systematically for each informal village separately, with same parameter settings at each phase. This was possible due to relatively homogenous landscape over Kibera settlement. Thus we obtained 12 regional classification outputs, which were then merged in the final stage. To avoid erroneous classification on the edges of the splitted images, we applied 30 meter buffer when masking the village fragments out from the whole GeoEye image (Fig. 5).

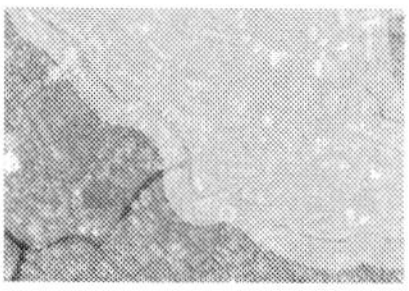

Fig. 5. A buffer of 30 meters around the village border.

Object based classification consists of two stages. Image is first segmented into a set of segments (regions) that are considered to be homogeneous in terms of one or more spectral or spatial properties. Then follows classification where each segment is classified into belonging object class.

Supervised segmentation within software used (ENVI EX, Feature Extraction module) is defined by two segmentation parameters that influence an average size of segments: segmentation and merging. Setting different values for these two parameters causes change of size of segments, allowing for an image to be segmented at many different scales, so both parameter values influence classification results. Since structure of the Earth's surface is similar throughout the whole settlement, same general segmentation parameters were used for each of the 12 villages. Visual example of segment structures are shown on Fig.6.

While classifying Raila village we adapted segmentation parameters to best extract shapes of individual buildings inside informal settlement. Since segmentation parameters were adequate for one particular land use only (i.e. buildings), others were expected to be under- or over-segmented. There is no single "optimal" scale for analysis of remote sensing images, rather there are many optimal scales that are specific to the image-objects that exist within a scale (Hay et al., 2003) and this is why using a multi-scale approach may often be preferable (Johnson and Xie, 2011). All spectral bands were used and given equal weight for image segmentation and all available attributes were calculated for all segments.

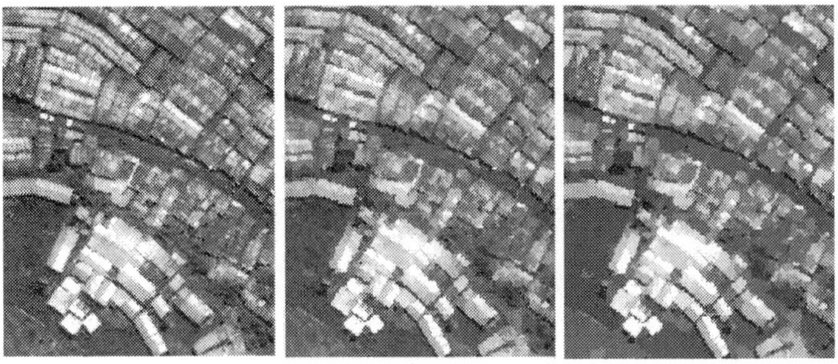

Fig. 6. Original GeoEye image (a), objects segmented (b), objects merged (c) (Feature Extraction module of ENVI EX).

The objects extracted during the segmentation were then classified using Support Vector Machine (SVM) classification algorithm in an object-oriented framework along with training sets, selected by experienced user. Nine land cover classes were used all together (Table 2).

Land use / cover classes	Description
Buildings_blue	Residential houses with blue spectral reflectance on image.
Buildings_light	Residential houses with white or bright spectral reflectance.
Buildings_brown	Residential houses with brown or dark spectral reflectance.
Buildings_red	Residential houses with red spectral reflectance.
Roads	Traffic connection between villages, usually unpaved.
Shadows	Shadowed areas around high objects (high vegetation and buildings).
Soil	Areas of unvegetated soil, mudded ground.
Vegetation 1	Green vegetated areas, low vegetation (grass).
Vegetation 2	Green vegetated areas, high vegetation (trees).

Table 2. Land cover classes anticipated with object-based classification.

These nine urban land use/cover classes included four types of residential housing. Subclasses were choosed because of their different spectral signature inside the same land cover class (e.g. instead of selecting only class "buildings" we selected four subclasses "buildings_blue", "buildings_light", „buildings_brown" and "buildings_red", Fig. 7). This way we obtained better results than we would have using one general class only. More detailed classification of residential housing was made only in Raila village.

Classification results were obtained as a raster image and a vector file. Vectors were exported to a single layer and later processed for the need of post-classification in ESRI ArcMap software (all polygons smaller than 2 m^2 were merged with neighbouring larger polygons).

Fig. 7. Selecting training samples for supervised classification (Feature Extraction module of ENVI EX).

4.3 Change detection

Satellite data offers unique utility for monitoring and quantifying land cover change over time. Consequently, change detection has become a significant part of the remote sensing research over the last decades. The goal of remote sensing change detection is to detect the geographic location of changes, identify their type (if possible) and quantify their amount. If long term imagery time series are handled, trends can be recognised. A large number of change detection methods have been evolved and they differ in their refinement, robustness and complexity (Hall and Hay, 2003). Nowadays a three level systematisation system is proposed that differentiates change detection methods by introducing the notion of pixel, feature and object level image processing (Deer, 1998). In general change detection techniques can be grouped into two major types (Jianya et al., 2008; Coppin et al., 2004; Lu et al. 2004; Singh 1989): image differencing techniques and post-classification comparison techniques. The main difference between the two types is that image differencing methods can identify the location and the magnitude of change but can not identify the type of land use or surface changes taken place in the area. Post-classification techniques can identify the location and provide the change character. Recent advances in change detection mostly involve high resolution data and consequently object-oriented and/or multi-scale approaches, with a range of techniques to approach contextual modelling (Lang et al., 2006; Blaschke et al., 2008; Addink & Van Coillie, 2010).

4.3.1 Change detection of Kibera informal settlement

The use of VHR satellite image time series may provide a reliable approach to detect dense urban growth in detail (Hofmann et al., 2006). A generically applicable and rapid operational land cover mapping of these settlements has generally proven difficult (Netzband & Rahman, 2010). Object-based classification and land use mapping of Kibera settlement from GeoEye image (section 4.2 and 5.2) highlighted some typical problems for object delineation in slum-like areas that can be corrected only with a lot of manual work. Main difficulties are associated with informal area outer-homogenity but inner-heterogenity due to the microstructure of urban agglomeration. Object-based classification is thus very demanding in terms of methodology adaptation to informal residential areas specifics, especially when accounting for their direct relation to representation on different satellite data sources. Thus within the limited framework of this case study pixel-based approach to identify outline of urban growth was preferred. The procedure was implemented on radiometrically adjusted time series (section 4.1). GeoEye 2009-07-25 and QuickBird 2006-03-27 images were compared for the changes over whole Kibera, and GeoEye 2009-07-25, QucikBird 2008-08-10 and 2006-03-27 images were analysed for the observation of sequential urban growth of Raila village.

The simple thresholding of difference images is a well-known method that leads to the delimitation of changes and no-changes. The advantage of this method is that it can be fully automatic. However main disadvantage is that it heavily depends on consistency of datasets. Regardless of the carefully performed data preparations certain unwanted effects remains, which may drastically burden the imagery comparisons. This data variability behaves as a detected change and may well be enclosed within identified pattern of changes, although not all of the identified differences belong to real changes (false, non-intrinsic changes). Such false effects result in over-estimation of change pattern and can cause the quantitative evaluation fail. Since this data noise originates from the pre-processing algorithms as well as the natural and technological conditions during data acquisition, it can not be completely removed with data radiometric corrections (Veljanovski & Oštir, 2011).

To overcome this drawback a contextual multi-level change detection approach was applied that can efficiently treat most of the unwanted differences and suppress sensor related noise (Veljanovski, 2008). Taking into account the neighbourhood and change information by joining two spatial scales (Fig. 8), approach reduces amount of small size false differences.

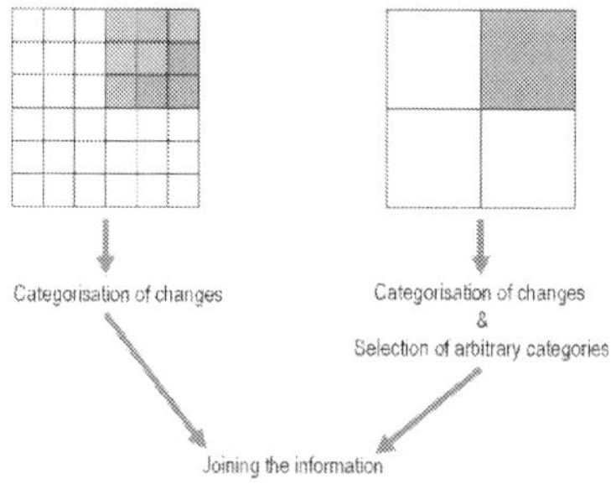

Fig. 8. Change detection approach takes into account the neighbourhood and change information by joining different spatial scales.

The model is based on focal information logic that gives averaged change information in a slightly reduced spatial scale - within the specified neighbourhood. It is based on the fact that in a larger geographic area (e.g. a 3 x 3 pixel window or 3 m spatial resolution for resampled VHR data) the information of the changes will tend to level abrupt change information if a small spatial scale change is present, and will show the averaged difference if the majority of pixels in the observed window are subjected to change. Computing the piecewise change information between two time-successive data sets provides valuable information regarding the location and numeric change value derived from contextual information within the specified neighbourhood.

The procedure was implemented as follows. First, spectral information for a slightly coarser scale (i.e. 3 m spatial resolution) was computed for images or areas of interest. This may be accomplished with a specified neighbourhood mean value annotation. Second, change differentiation between images is performed on a coarser resolution scale and change magnitude categorisation is applied (see below for categorisation classes and their definition). Third, upper positive and negative changes are reclassified so that the mask of important changes based on the neighbourhood context characteristics is prepared. Fourth, change differentiation is calculated on the original data scale (i.e. GeoEye and QuickBird data in 1 m spatial resolution), then categorisation is applied, and finally a mask (or a mask with a buffer) of an arbitrary specified magnitude of changes (obtained in the previous step) is overlaid in order to restrict merely the contextually supported changes.

Normally, if there is no substantial unwanted effects (noise) due to meteorological or sensor influences in images, changes obtained from the difference image (image differentiation) are distributed normally and symmetrically, with the average at 0. Abrupt changes (objects or land cover transformations) can then be defined by thresholding the distribution tails, giving the pattern of positive and negative changes in reflectance (spectral) space. We have calculated change magnitude (transition class category) intervals for every 0.5 standard deviation. Then the criterion of 2.5 standard deviations for negative changes and 2.0 for positive changes was applied to enclose the majority of detected transformations in both real and spectral world situations. Result of such categorisation of image difference is a pattern of abrupt changes (locations of appearance, disappearance of objects), with no association to change characterisation (type of change, from-to).

Each output from the above automatic threshold procedure is finally refined to an arbitrary degree, depending on case study objective. For entire Kibera informal settlement changes were obtained from comparison of GeoEye 2009-07-25 and QuickBird 2006-03-27 images. Change patches smaller than 5 m^2 were eliminated for the purpose of this example, mainly to reduce the impact of change artefacts belonging to small patches of rooftop renovations (Fig. 3). False changes identified due to differences in viewing angle (location of buildings and trees shadows, buildings boundaries due to different original resolution of imagery) were also removed using the results obtained from object based classification (state of land use in 2009, section 4.2). Where shadow or vegetation object class were present, change pattern was corrected in the given context. Described and implemented step-wise change pattern refinement is shown in Fig. 9, for a subset of Raila village and northern neighbourhood. In other words, with simple generalisation we could control many aspects of the change pattern for study's specific aims.

Results were visually examined and evaluation of change pattern characteristics (over- and under-estimations) was done throughout the area using complementary comparison of satellite images involved.

4.3.2 Raila village change detection

Raila village is known to have undergone extensive development during recent years (Kibera Wikipedia, 2011; MKP, 2011). Thus temporally detailed examination was performed observing its urbanisation. Change detection procedure described in section 4.3.1 was in addition implemented for Raila village only, but for two time sequences: 2006-2008 and 2008-2009. QB 2006-03-27, 2008-08-10 and GE 2009-07-25 images were used for this example.

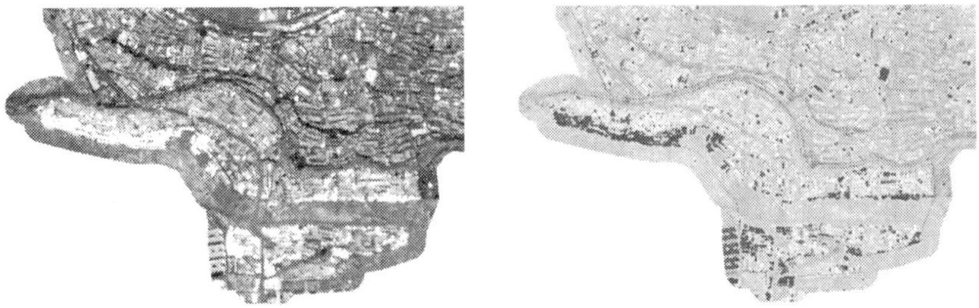

2006-2009 3-band difference image for Kibera subset (Raila village). Change pattern over 2006-2009 3-band difference image.

Change pattern after automatic multi-level threshold procedure. Change pattern after eliminating false changes due to shadows differences. Change pattern after eliminating small patches due to rooftop renovations.

Fig. 9. Overview of change pattern intermediate results through implemented processing steps.

4.4 Population estimation

Population statistics gives very important information for understanding of modern society. Demographic research is one of the main research directions of social science for a better understanding of the interactions between population growth and social, economic and environmental conditions. The collection of population data depends mainly on the census, which is labour-intensive, time-consuming and demands high financial resources. The 2009 Kenya Population and Housing Census reported Kibera's population to be 170,070 (Karanja, 2010). This report was far from the belief of that time that Kibera slum was of the biggest informal urban settlements in the world. Several actors had provided and published over the years increasing estimations of the size of its population, most of them stating that it was the largest slum in Africa with the population exceeding 1 million. According to Davis (2006), a well known expert on urban slums, Kibera had a population of about 800,000 people. International Housing Coalition (IHC, 2007) talked about more than half a million people. UN-Habitat (2004) had released several estimations ranging between 350,000 and 1 million people. These statistics mainly come from analyses of aerial images of the area. IRIN (2006) estimated a population density of 2000 residents per hectare. In 2008 an independent team of researchers began a door-by-door survey named Map Kibera Project (MKP, 2011). A trained team of locals, after having developed an ad-hoc surveying methodology, has so far gathered census data of over 15,000 people and completed the mapping of 5000 structures, services (public toilets, schools), and infrastructures (drainage system, water and electricity supply) in the Kianda village. Considering data collected for Kianda village, the population of the whole Kibera slum can be estimated between 235,000 and 270,000 people.

Nonetheless, no estimation so far guessed by the MKP, or the UN, or the Government of Kenya or by other actors can be taken for granted and does not represent the real dimension of the population of Kibera. In general, no estimation can be proved nor refuted until an exhaustive census will be taken throughout the whole slum (Kibera Wikipedia, 2011).

Because population is not directly related to land cover surface reflectance, population estimation is still a challenging task based purely on remote sensing spectral signatures. Although population is not directly measurable on the remote sensing images this technology may provide good approximation of population estimation by measurement of visible variables, e.g. the number of residential buildings and/or the area of build-up zone (Zhang, 2003). There exist many studies using different approaches on remote sensing data for population estimation. Studies date from the early 1970s onward, where air photos were utilized for manual counts of dwelling units. There are three most used methods of population estimation by remote sensing: residence count method, area (density) method and regression model method (Zhang, 2003). Residence count method was mostly done in first period of studies on this topic on the western urban environment (Horton, 1974, Barrett & Curtis, 1986). Area density method was used by H.H. Wang (1990), F.Z. Wang (1990), P. Sautton (1998), Langford et al. (1994), Z.J. Lin (2001) and others. Regression model method is currently also often used (Galeon, 2010, Dengsheng et al., 2006, Zhang, 2003). With each type of method some ancillary field survey data are needed.

Considering the above situation and the fact that for Kibera we lack other potential socio-geographic data (elsewhere applied to predict population with regression technique), we decided to assess the population on residential land cover class information obtained from object-based classification with density per area method solely. For each village a total area of buildings was calculated and different occupation scenarios (i.e. persons/living area) were tested to observe the range of possible population fluctuation.

5. Results

With object-based (contextual) classification performed on GeoEye image with Feature Extraction module of ENVI EX, it was possible to obtain accurate land cover map and following this, total residential area of Kibera slum and its divisions (villages) with very high accuracy. From this data, those related to build-up areas, were used for population estimation. With multi-level contextual change detection implemented in Erdas Imagine, it was possible to obtain representative change pattern reflecting where in informal settlement intensive urbanisation processes have taken place. Results are presented in the following order: land cover mapping, change pattern identification and population estimation, for Kibera informal settlement (section 5.1) and Raila village (section 5.2), respectively.

5.1 Kibera

5.1.1 Kibera land cover map 2009

Object based classification and post-classification on GeoEye 2009-07-25 image was performed for each village in Kibera informal settlement separately. Finally individual results were joined in a land cover map of Kibera informal settlement (Fig. 10).

Fig. 10. Merged final classification results of GeoEye image where all 6 selected land cover classes are shown for all 12 Kibera villages (ESRI ArcGIS).

Vectorized classes of entire Kibera were merged together in order to be able to calculate area of total land use/land cover. As it is seen from Table 3, residential areas cover 2/3 (66%) of the whole Kibera area and are prevailing when compared to other land uses. This can be well confirmed from the visual examination of the satellite images of the discussed area.

Accuracy assessment was done by comparing results of supervised classification with manually digitalized objects. Comparison was done on the area 200 x 300 m in the village

Land use [m²] / village	Residential	Trees	Green areas	Soils (roads, bare ground)	Shadows	Total
Kianda	117,710	8,144	9,511	16,775	18,005	170,145
Soweto West	49,620	4,089	6,455	15,001	6,177	81,343
Raila	45,657	8,835	17,066	30,886	6,139	108,583
Gatwekera	234,609	17,634	11,718	19,682	23,878	307,520
Kisumi Ndogo	111,472	6,979	6,944	10,891	28,432	164,717
Makina	303,599	53,844	11,992	40,474	34,801	444,710
Kamdi Muru	51,248	1,829	7,541	12,123	8,670	81,412
Mashimoni	96,287	5,520	2,018	7,628	16,900	128,355
Laini Saba	181,211	17,400	13,141	45,545	20,490	277,786
Lindi	159,112	2,5826	11,885	48,215	26,630	271,668
Silanga	150,058	21,569	19,975	34,833	17,298	243,733
Soweto East	174,200	6,238	15,618	29,302	22,823	248,181
SUM [m²]	**1.674,784**	**177,906**	**133,863**	**311,354**	**230,245**	**2.528,152**
SUM (%)	**66.25**	**7.04**	**5.29**	**12.32**	**9.11**	**100,00**

Table 3. Area of different land use types for 12 informal villages and the total sum in Kibera.

Lindi. Only residential segments were estimated. Since with ENVI EX classification the outline of individual residential objects could not be extracted, we compared only the total sums of areas classified as *residential*. Results are shown in Table 4. The best result (error of 3%) was obtained when choosing parameter values for segmentation/merge: 85/85.

All (semi)automatic classification methods display some errors, but as an approximate solution object based classification on VHR data yielded very good results upon selection of proper segmentation parameter values. Different shapes and colours in informal settlements determine a complex urban formation, which is difficult to differentiate from other land cover types, especially from bare soils and unpaved streets. With the object-based classification dwelling zones were found with high accuracy all over the image, in spite of their spectral similarities with streets and other urban features, especially bare soils.

	Area [m²]	Coverage [%]
Testing rectangle (200m x 300m)	60,000	
Manual digitalization	37,913	
Supervised classification ENVI EX (segmentation 85, merge 85)	**36,663**	**97**
Supervised classification ENVI EX (segmentation 85, merge 65)	42,172	111
Supervised classification ENVI EX (segmentation 85, merge 45)	41,174	109

Table 4. Comparison of results of ENVI EX supervised classification using different segmentation parameters with manual digitalization. The total areas were compared for residential classes only.

5.1.2 Kibera change detection

The multi-level contextual change detection method we used in this case study was first developed for monitoring various greater and intensive processes on the Earth's surface with middle resolution imagery (i.e. Landsat, SPOT). The result on Kibera proved same contextual logic is efficient also when implemented on very high spatial resolution imagery.

Change detection applied to Kibera informal settlements aimed to obtain an outline of the distribution and extent of major urbanisation processes. Comparison was made for QB 2006-03-27 and GE 2009-07-25 images. Method has proven to be suitable for monitoring changes related to various processes (buildings construction, buildings collapse or disappearance, rooftop renovation, increase/decrease in vegetation) and/or of the coincident description of their trends (see section 5.2.2 for identified informal settlement growth in Raila village area). Although quantitative assessment of change occurrences was not performed due to lack of independent reference data, a detailed visual control was made through comparison of before and after images and, nevertheless, several conclusions can be derived.

Identified pattern of changes (Fig. 11) clearly draw attention to spots where urbanisation between years 2006 and 2009 was most intensive: Kianda north-east, Raila southern border, Mashimoni and Laini Saba northern border and Soweto East eastern tail. According to the density of change pattern elements, in addition to the edges of Kibera informal settlement, several changes occurred in the eastern part of Kibera. These are mainly due to larger new buildings constructions or complete rooftops renovations. Additional socio-economic data or bigger events information (like flooding) would be welcomed to associate the rate of more intensive rooftops renovations at some of Kibera villages and compared to the other parts of the settlements.

With external data, for example land cover for a reference point in time, a fairly reliable "from-to" change statistics could be extracted. However in rather homogenous land use areas like Kibera informal settlement is, where most of the land use is residential (section 5.1.1) and where urbanisation direction limits are well known due to formal residential settlement boundaries, such information would not additionally characterize the change pattern. On the example of Raila village change detection in section 5.2.2 we comment some difficulties observed related to detection of change in slum-like areas in a more detail.

Fig. 11. Identified changes between years 2006 and 2009 (Erdas Imagine). Change pattern enclose changes due to new buildings construction, buildings disappearance and larger rooftop renovations.

5.1.3 Kibera population estimation in 2009

Total surface of Kibera was calculated as a sum of housing areas of all 12 villages throughout whole Kibera from vector results obtained with object-based classification. All the vectorized structures have been computed using ArcGIS software and we assumed that all the structures are used for habitat purposes.

Sum of the residential areas show that total residential area of Kibera is around 1.616,883 m^2, which is 358,706 m^2 more than total surface from year 1993 given by Sartori et al. (2002), where residential area was 1.333,834 m^2. We can see that informal settlement population density has increased from year 1993. Since Kibera is surrounded by well-defined residential boundaries and its drastical expansion therefore was not possible, we can assume that the increase of residential area can attribute to denser housing inside the informal settlements through years.

As we can see from Table 5, population estimation of Kibera can vary from 150 thousand up to 650 thousand people living in the informal settlements according to different population density sources. Nevertheless, both limits are high taking into account Kibera is spread on 2.5 km^2 only.

Source	Density [people/m^2]	People living in Kibera
MapKibera Project (2008)	0.0951	156,652
IRIN (2006)	0.2000	329,377
AHI US (2005)	0.3000	494,065
Sartori et al. 1 (2002)	0.3300	543,471
Sartori et al. 2 (2002)	0.3900	642,284

Table 5. The estimation of population according to the density acquired from different sources. Housing area of Kibera was calculated from results obtained with object classification of GeoEye image from 2009.

5.2 Raila village

5.2.1 Detailed land cover map of Raila village 2009

We made a detailed classification focused on residential housing only on image subset covering the area of Raila village. The main idea was to automatically obtain polygon shapes of each individual residential object or settlement. Although on VHR image individual objects in general can be distinguished, due to the complexity and variety of dense roof surface there was impossible to extract the shape of every individual residential object or each group of objects (settlements). Roofs and their elements themselves present heterogeneity, resulting in distinct spectral variation within areas of homogeneous land cover. Class „buildings_bright" were possible to detect since they had a very distinct morphological pattern that was contrasted by surrounding (built-up) land cover. Segments of class „buildings_brown" were sometimes integrated with their surroundings (soil) so they could not be readily distinguished. Some urban elements (like rooftops) are a combination of many different surface materials; this produced a spectral response that was difficult to interpret with routine procedures.

Automatic shape detection of each individual residential object would enable good total population estimation. Two approaches to get classified vector shapes closer to individual buildings were considered:

- Manual correction of the shapes of obtained polygons: Small test showed that this is extremely time consuming. Correction took more time than it would take if one would manually digitize the whole image.
- Use of additional attributes that are automatically created in the phase of ENVI EX segmentation: Investigation showed that there is no useful relation within additional attributes of all sub-classes of buildings. Therefore additional attributes could not assist to semi-automatic reclassification of some segments in order to obtain better shapes of individual object in build-up zones.

Finally, focused to residential build-up objects in Raila area four major object classes were modelled and mapped (Fig. 12).

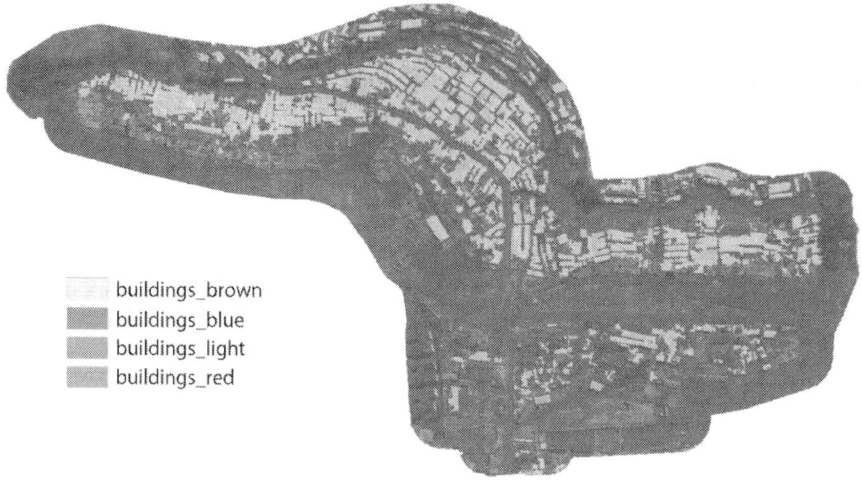

Fig. 12. Detailed object based classification results for residential areas in Raila village.

5.2.2 Raila village urban growth

Multi-level contextual change detection approach was applied to Raila village to highlight the extent of major urbanisation processes that can take action even in a short time span. As expected the development on southern border of the village is well recognised (Fig. 13) and different time sequences analysed outlined where and when the growth has taken place.

Method incorporates spatial neighbourhood dependence to control the false change information (i.e. inherent changes due to locally based variability in data). As a result it gives a change pattern of positive and negative changes in imagery spectral space. Positive changes correspond to an increase in digital number values at the same location and negative changes to a decrease in spectral values. Method was developed for middle resolution imagery, but its application to Kibera informal settlement data proved this concept to be efficient on VHR imagery as well. In general, majority of objects transformations regardless of their size (small patches to entire buildings) was identified. The results were evaluated with visual control or before-after imagery comparison. Although quantitative assessment was not possible as more detailed independent data were not available, we estimate that more than 90% of changes associated to buildings appearance and disappearance are captured.

Hence, detailed inspection of identified changes outlined several difficulties related to detect change in slum-like areas. First of them is preservation of shape. Because of the use of multi-level information the objects edges may be shrinked to some degree, causing that change is not recognised at entire object extent but only at part of it. Second, the level of change recognition (detection rate) is different in positive and negative spectral space. New materials (i.e. metal plates) used to cover houses have very strong reflection, so difference from low reflecting brown soil to new buildings with bright cover is obvious and unproblematic. In contrast, for example from rusty rooftops to bare soil this difference is not

QB2006-03-27 QB2008-08-10 GE2009-07-25

Change pattern from years 2006-2008.

Change pattern from years 2008-2009.

Urban growth pattern through years 2006, 2008 and 2009 for Raila village.
Dark red corresponds to 2006-2008 and light red to 2008-2009 period.

Fig. 13. Results of change detecion for urban growth pattern through years 2006, 2008 and 2009 for Raila village, Kibera.

great and consequently buildings disappearance is more difficult to detect. Same applies if new building has dark colour rooftop covers (for example, often used blue colour plates). Main change missing occurrences therefore apply to the just described situations, making difficulties for automatic approaches. To some extent this problem was solved so that the threshold for negative changes was reduced during transformation categorisation step.

5.2.3 Raila village population estimation in 2009

According to the total residential area in Raila village obtained from object based classification (section 5.1.1, Table 3), population estimation was calculated based on different density per area parameters (Table 6). Additionaly we illustrate this information also in terms of typical size of houses in the Raila village (Fig. 14).

Source	Density [people/m²]	Density [people/32m²]	Estimated population of Raila
MapKibera Project (2008)	0.0951	3.04	4,343
IRIN (2006)	0.2000	6.40	9,131
AHI US (2005)	0.3000	9.60	13,697
Sartori et al. 1 (2002)	0.3300	10.56	15,067
Sartori et al. 2 (2002)	0.3900	12.48	17,806

Table 6. Density of people living in Raila village according to different sources and total residential area obtained from object-based classification of GeoEye 2009-07-25 image.

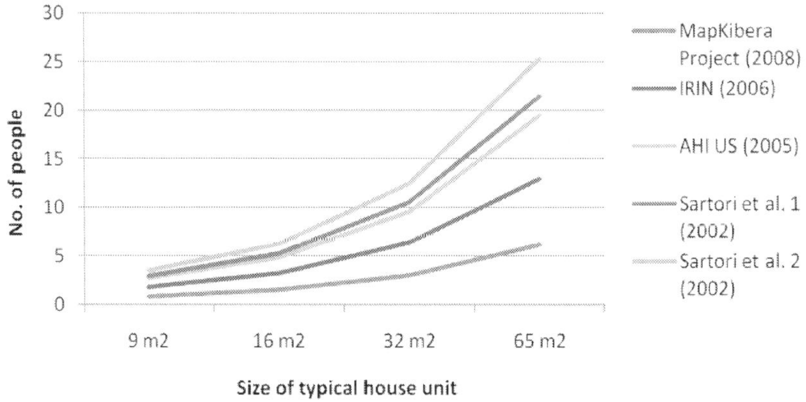

Fig. 14. Density of people according to different sources per typical size of houses in Raila village.

In Table 6 we presented also the density on 32 m^2 unit since this size of house was measured to be a most typical and frequent (medium range size) housing unit in Raila village. It was also observed that size of small buildings is approximately 16 m^2, while large buildings reach on average 65 m^2. Fig. 14 summarise in addition the above relationship.

When perfoming such population estimation it is assumed that all of the build-up area is used for dwelling, which is not allways true as houses may be also used for other activities. For the possible calibration of the above population densities field survey sampling would be required. Nevertheless, Table 6 clearly outlines how population estimation based on critera of simple density per area based modelling can propose up to 4-times higher (different) population estimation.

6. Discussion

Informal settlements are a very dynamic phenomenon in space and time and the number of people living in these areas is growing worldwide. The reasons for this are many-sided and were not under detailed examination of this study. Informal settlements represent a particular housing and living conditions which is from a humanitarian point of view in most cases below acceptable standards (UN-HABITAT, 2004; MKP, 2011; Sartory et al. 2002). Due to informal character and low governmental management services in the past, reliable and accurate data about informal settlements and their population is rarely available. On the other side there is a strong need to transform informal into formal settlements and to gain more control about the actual urbanisation progress. Thus, obtaining spatially and temporally accurate information is a first step to establish proper actions in terms of local or regional planning. For these tasks, conventional data sources, such as maps, statistics or even GIS data are usually obsolete, not available, not as accurate as needed or do not hold the information needed (MKP, 2011; Sartory et al. 2002). The case study presented demonstrates how informal settlements can be approached from VHR satellite image data. Using an object based approach of image analysis detailed land cover/use within informal settlements can be obtained to facilitate GIS-based management tasks and population modelling. The application of automatic, even if simple pixel-based, change detection proved to support real-time observation of informal settlement areas whenever appropriate VHR satellite data are available with relatively low processing costs.

Merits of object-based image analysis in dense informal settlements analysis with VHR remote sensing data have been confirmed in several studies (section 2). However some drawbacks still resist. In case of residential land cover/use map derivation main unsolved difficulty is automatic detection and separation of individual houses. Although small differences in heights of rooftops create visually well distinguished boundaries of objects, the heterogeneity of rooftop material and its small scale changeability often overrule the value of neighbourhood houses boundaries relationship. Current limit of object based analysis is also that still requires substantial post-classification routines and check-over that can be done mainly manually through visual control. What characterise this procedure as time demanding whenever geometrically and semantically correct information is aimed.

Change detection applied revealed great potential for long-term monitoring and informal settlements urbanisation growth analysis. Hence more research is needed to provide sufficient detection rate of spectrally lower magnitude changes that are typical for informal settlements specifics and its reflectance intensity representation on satellite imagery. Due to rooftop material used bright (metal) materials are unproblematic to distinguish from soils, however dark rooftop materials (blue, brown colour) are spectrally closer to bare, unpaved (brown) or vegetated (green) soils. Here object based approaches would prove better option as sub-object attributes could be explored and used.

Valuable population estimation can be made with a relatively low cost if residential area if accurately estimated from high-resolution images, although some considerations exist. Area based population estimation model can be used for the informal settlements in other images of similar resolution knowing the number of people living per surface unit. Zhang (2002) exposed some problems of selection of the scales of remote sensing imagery, reduction of influence of plant cover on remote sensing data, stability of the correlation between population and remote sensing indicator variables and correction of building count. In this research we met all mentioned problems in order to accurately estimate population out of VHR satellite data. Nevertheless, if a clear understanding of mentioned issues is considered, reliable population model of population estimation by remote sensing data can be created. Thus, the application of satellite data information (such as accurate information on land use extent and other measures of surface or environmental characteristics) along with socio-economic data may well facilitate complex modelling to estimate population trends.

In terms of remote sensing technology contribution it is necessary to continue to develop new techniques for complex densely packed urban environments such as informal settlements. Emphasis on spectral properties should be considered but also emphasis on the characteristics of the shape, texture, context, and relationship with neighbouring pixels (and/or objects) information needs to be enhanced; as well as integration of the knowledge on corresponding socio-economic drivers should not be neglected.

7. Conclusion

Effective methods of monitoring informal settlements are required to generate appropriate data fast enough to assist to local policies and their controlling actions. Remote sensing data are especially powerful in that respect since, apart they are up-to date, they assist to link the geographic location with the accurate socio-economic data.

The results of change detection confirmed that VHR imagery is very promising for immediate monitoring of dense informal residences in the areas where much information is lacking. The results of object-based (contextual) classification of the land use in informal settlements of Kibera were highly accurate, especially if taking into consideration that informal settlements are difficult to be interpreted with automatic or semi-automatic routines. On the other side, the results indicate the problem of the ratio between spectral and spatial heterogeneity of objects in slum-like areas when viewed only from the above (satellite) perspective. Overall, the use of the object-based image analysis holds great promise for dense urban environments and was proved useful for studies of urban change structure and corresponding population estimation.

Satellite derived information can greatly complement the information that is traditionally collected by field observations (UNHCR, 2000). Quantitative information that can be derived from it should not be underestimated. The production of maps with geometrical shapes of settlements can contribute to recover the management of informal settlements, especially when interfaced with database that has information collected on the field. Although several challenges have not been yet solved adequately, e.g. delimitation of individual objects in slum-like areas, we can notice that applications are being developed. Thus (automatic) analysis of objects enables tremendous opportunities for population estimation in informal settlements.

8. Acknowledgment

The Slovenian Research Agency has sponsored the post-doctorate research project (Z2-2127). The Centre of Excellence for Space Sciences and Technologies Space-SI is an operation partly financed by the European Union, European Regional Development Fund, and Republic of Slovenia, Ministry of Higher Education, Science and Technology. American Association for the Advancement of Science supported the operation of Map Kibera Trust with donated satellite images of the area.

9. References

A History of Kibera (2011). Available from www.brianekdale.com/?p=230.

Addink, E.A., Van Coillie, F.M.B. (Eds.) (2010). *GEOBIA 2010: Geographic Object-Based Image Analysis*, Ghent, Belgium, 29. June - 2. July 2010. ISPRS Vol.No. XXXVIII-4/C7, Archives ISSN No 1682-1777. Available from http://geobia.ugent.be/proceedings/html/papers.html.

AHI US (2005). Affordable housing institute. Available from http://affordablehousinginstitute.org/blogs/us/2005/07/kibera_africas.html.

Barrett, E., Curtis, L. (1986). *Introduction to Environmental Remote Sensing*, pp. 238-241. London: Chapman and Hall.

Benz, U.C., Hofmann, P., Willhauch, G., Lingenfelder, I., Heynen, M. (2004). Multi-resolution, object-oriented fuzzy analysis of remote sensing data for GIS-ready information. *ISPRS Journal of Photogrammetry and Remote Sensing*, Vol. 58, pp. 239-258.

Blaschke, T., Burnett, C., Pekkarinen, A. (2004). New contextual approaches using image segmentation for object-based classification. In: *Remote Sensing Image Analysis: Including the spatial domain*, De Meer, F. and de Jong, S. (Eds.), Kluver Academic Publishers, Dordrecht, pp. 211-236.

Blaschke, T., Lang, S., Hay G.J. (eds.) (2008). *Object-Based Image Analysis: Spatial Concepts for Knowledge-Driven Remote Sensing Applications* (Lecture Notes in Geoinformation and Cartography). Springer.

Dare, P.M., Fraser, C.S. (2001). Mapping informal settlements using high resolution satellite imagery. In: *Int. J. of Remote Sensomg*, Vol. 22, No. 8 pp. 1399-1301.

Davis, M. (2006). The Planet of Slums.

Deer, P. (1998). *Digital change detection in remotely sensed imagery using fuzzy set theory*. Ph.D. thesis, Departments of Geography and Department of Computer Scoence, University of Adelaide, Australia.

Dengsheng, L., Qihao, W., Guiying, L. (2006). Residential population estimation using a remote sensing derived impervious surface approach. *International Journal of Remote Sensing*, Vol. 27, No. 16, pp. 3553-3570.

Galeon F. (2008). Estimation of population in informal settlement communities using high resolution satellite image. *The International Archives of the Photogrammetry: remote sensing and spatial information sciences*, Vol. 37, Beijing.

Giada, S., De Groeve, T., Ehrlich, D. (2003). Information extraction from very high resolution satellite imagery over Lukole refugee camp, Tanzania. *International Journal of Remote Sensing*, Vol. 24, pp. 4251-4266.

Hay, G. J., Blaschke, T., Marceau, D.J., Bouchard, A. (2003). A comparison of three image-object methods for the multiscale analysis of landscape structure. *ISPRS Journal of Photogrammetry & Remote Sensing 57*, pp. 327-345.

Wang H.H. (1990). City population estimation method with satellite imagery. *Remote Sensing technology*, pp. 48-54.

Zhang, B. (2003). Application of remote sensing technology to population estimation. *Chinese geographical science*, Vol. 13, No.3, pp.267-271.

19

Remote Sensing Applications in Archaeological Research

Dimitrios D. Alexakis[1], Athos Agapiou[1],
Diofantos G. Hadjimitsis[1] and Apostolos Sarris[2]
[1]*Cyprus University of Technology, Department of Civil Engineering and Geomatics*
[2]*Foundation for Research and Technology, Institute for Mediterranean Studies, Laboratory of Geophysical, Satellite Remote Sensing and Archaeoenvironment*
[1]*Cyprus*
[2]*Greece*

1. Introduction

The spectral capability of early satellite sensors opened new perspectives in the field of archaeological research. The recent availability of hyperspectral and multispectral satellite imageries has established a valid and low cost alternative to aerial imagery in the field of archaeological remote sensing. The high spatial resolution and spectral capability can make the VHR satellite images a valuable data source for archaeological investigation, ranging from synoptic views to small details. Since the beginning of the 20th century, aerial photography has been used in archaeology primarily to view features on the earth's surface, which are difficult if not impossible to visualize from the ground level (Rowland and Sarris, 2006 ; Vermeulen, F. and Verhoeven, G., 2004). Archaeology is a recent application area of satellite remote sensing and features such as ancient settlements can be detected with remote sensing procedures, provided that the spatial resolution of the sensor is adequate enough to detect the features (Menze et al., 2006). A number of different satellite sensors have been employed in a variety of archaeological applications to the mapping of subsurface remains and the management and protection of archaeological sites (Liu et al., 2003). The advantage of satellite imagery over aerial photography is the greater spectral range, due to the capabilities of the various on-board sensors.

Most satellite multi-spectral sensors have the ability to capture data within the visible and non-visible spectrum, encompassing a portion of the ultraviolet region, the visible, and the IR region, enabling a more comprehensive analysis (Paulidis, L., 2005). Multispectral imagery such as Landsat or ASTER is considered to be a standard means for the classification of ground cover and soil types (Fowler M.J.F., 2002). Concerning the detection of settlement mounds the above sensors have been proved to be helpful for the identification of un-vegetated and eroded sites. In recent years the high spatial resolution imageries of IKONOS and Quickbird have been used for the detection of settlements and shallow depth monuments (De Laet et al., 2007; 36 Massini et al., 2007; Sarris, A., 2005). Hyperspectral imagery (both airborne and satellite) has been also applied in archaeological investigations on an experimental basis and need further investigation (Cavalli et al., 2008; Merola et al., 2006).

The record of electromagnetic radiation can be achieved using special sensors. Such kind of sensors are used to record the electromagnetic radiation from satellites while handheld sensors can be used for field measurements. The ground radiometry and spectroscopy involves the study of the spectral characteristics of objects according to their physical properties (Milton, 1987). Indeed, data from portable radiometer are often refer in the literature as "ground truth data", due to the fact that measurements are collected in a relatively short distance from the object so that any noise is minimized (Jonhson, 2006). However, as Curran and Williamson (1986) emphasizes even these ground "true" data are subject to errors, which researchers should take into account.

The spectral signature diagram, from different materials or objects, is an easy way to plot radiation against wavelength, in a graphical form. Curves of spectral signature (reflectance curves) and the so-called critical spectral bands (critical spectral regions) are used in many applications of Remote Sensing (e.g. vegetation indices). The way of how measurements are collected by radiometers can be explained through physical laws. Already, by the 1970s Nicodemous et al. (1977) have proposed the basis for the model of "bidirectional reflectance distribution function" (BRDF), which describes the relationship of the incident radiation from a given address in the reflected radiation in another direction. Nevertheless, the use of the results of the Nicodemous et al. (1977) study was not appreciated and understood by the scientific community (Schaepman-Strub et al., 2006; Milton et al., 2009). Their study has been used several years later by Martonchik et al. (2000) and Schaepman-Strub et al. (2006). The original classification proposed by Nicodemous et al. (1977), depending on the geometry of the radiation which included nine categories was reduced to only four which are actually encountered (Martonchik et al., 2000; Milton et al., 2009).

Milton et al. (2009) stated that all spectroscopy measurements in a strict physical sense can be categorized within the "hemispherical-conical reflectance function" (HCRF case). It should also be noted that natural materials do not follow the rules of a diffuse Lambertian surface, since the intensity of the reflected radiation varies regarding the angle of refraction.

Ground spectroradiometer may be used to provide calibrated measurements, since these instruments are often accompanied by special Lambertian targets. Milton et al. (2009) emphasizes that a critical factor for good results is the calibration of a specific target. The only disadvantage, apart from the price of handheld spectroradiometer is that it is difficult to cover a large area (such as an archaeological site) (Atkinson et al. 1992; Milton et al., 2009).

Apart from the purchase of the ground radiometry, there is still an important limitation that should be taken into account. Most spectroradiometers which are found in the market are "single-beam": the same instrument used to measure the radiation to a specific target (reference panel) used to measure the targets of interest (target). In the interval of these measurements the atmospheric conditions are assumed to be the same.

Spectroradiometer may be used for archaeological research in order to retrieve characteristics of vegetation and to calculate vegetation indices. Such indices are quantitative measures, based on vegetation spectral properties that attempt to measure biomass or vegetative vigor. Theoretical analyses and field studies have shown that VIs are near-linearly related to photosynthetically active radiation absorbed by plant canopy, and therefore to light-dependent physiological processes, such as photosynthesis, occurring in the upper canopy (Glenn et al. 2008).

Concerning geophysical techniques they offer a non-invasive way of providing valuable information regarding the subsurface context of the archaeological sites and contribute significantly in leading archaeological excavations, reconstructing the past landscapes, suggesting directives for the cultural heritage management and preservation of sites and historical buildings and providing a prior strategy for large construction works. Employing a suite of methods measuring the physical properties of the soils with high efficiency, reliability and resolution, geophysical prospection has been designated as a valuable tool in the domain of archaeological research, especially in the study of the ancient landscapes. A very detailed review of the physical properties of each method and the fundamentals of the operation of the corresponding instrumentation is provided by Linford (2006) and Scollar et al. (1990).

Methods that involved measurements of the soil's electrical resistance and of the local intensity of the magnetic field of the earth have been the earliest that were applied in the field of archaeological prospection. Making use of Ohm's law, soil resistance meters, acting as active prospection methods, introduce a current within the ground through the use of metal electrodes and measure the electrical resistance of the soil, as this is influenced by the features located within the ground and below the electrode array (Clark, 1990). Soil resistance measurements can be carried out in two main ways, either by moving a fixed electrode spacing (corresponding to a specific penetration depth) array along profiles or through vertical electrical soundings (VES) (Sarris, 2008). In the latter, the current electrode spacing is increased with respect to a fixed centre of the electrode array, providing plots of the apparent resistivity versus electrode spacing that are ultimately compared to theoretical curves to provide information about the layering of the subsoil's strata. Soil resistance values are modified depending of the resistivity contrast of the targets (eg. a high resistance wall structure or a high conductive ditch) with the surrounding soil matrix. The measured apparent resistivity is also affected by the type of the configuration of electrodes (e.g. Twin probe, Square, Wenner, pole-pole, dipole-pole) and the distance between the electrodes, which is relatively proportional to the penetration depth.

This chapter seeks to address applications of remote sensing and GIS in archaeological research in a three-fold way. Initially, potential of satellite remote sensing is highlighted through a multi - sensor case study in Thessaly, Greece, where different satellite image processing techniques contributed to the detection of Neolithic tells (the so called 'magoules') that are found in the Thessaly plain. Four satellite remote sensing images with different spatial resolutions (ASTER, Landsat, HYPERION, IKONOS) were examined in order to search their potential for automatic extraction of Neolithic settlements, by means of pixel - based (RGB composites, spatial and radiometric enhancement, vegetation indices, data fusion, classification methods, data fusion etc). The satellite data were statistically analyzed, together with other environmental parameters, to examine any kind of correlation between environmental, archaeological and satellite data. Moreover, different methods were compared and integrated methodologies for the detection of Neolithic settlements were extracted.

Concerning ground spectroradiometer contribution to archaeological research, new innovative tools and methodologies are also presented in this chapter. Specifically, ground "truth" data, presented as spectral signatures libraries, were provided from different spectro-radiometric campaigns at archaeological environments (e.g. Tombs of the Kings and Sikyon archaeological site, C. Greece). Moreover field spectroscopy was used to detect buried archaeological sites in similar ways as applied in remote sensing applications in Neolithic tells in Thessaly. In addition the comparison of the phenological cycle 5 profile of similar crops - under same meteorological and soil conditions - is also searched over archaeological and non archaeological sites concerning for different case studies in Cyprus.

At the end, in order to highlight the potential of geophysical remote sensing in archaeological research certain case studies from surveys held in Greece and Europe are presented such as magnetic surveys in the Neolithic settlement of Veszto -Bikeri in Hungary and the Byzantine walls of ancient Nicopolis in Greece, the ground penetrating radar methods in the ancient Agora of Feres (Velestino) in Thessaly, in Agora of Sikyon at NE Peloponesse - Greece and in the area of the hypothesized amphitheatre of Ierapetra (SE Crete and finally the electrical resistivity tomographies from the Nemea, Peloponesse.

2. Application of remote sensing, GIS and geomorphology to the reconstruction of habitation in Neolithic Thessaly

2.1 Introduction

The aim of this study is to highlight the contribution of different approaches such as Remote Sensing, GIS and geomorphology analysis for the detection of Neolithic settlements and the modeling of habitation in the area of Thessaly - Greece.

The Neolithic settlement mud mounds in the area of Thessaly, Greece are called Magoules. They are low hills of 1-5 meters height and mean diameter 300 meters. The vast majority of Magoules are laid on Larisa basin and a smaller number is distributed in west Thessaly (Karditsa basin) (Fig.1). Both of these plains consist of Quaternary alluvial deposits (Alexakis et al. 2008). In order to achieve the goals of the research it was necessary to proceed with the topographic mapping of the settlements through the use of GPS, digitize 1:50,000 scale topographic and geological maps and construct a detailed archaeological and environmental database in SQL environment.

Fig. 1. Color composite RGB→3,2,1 of the mosaic of the 4 ASTER images used to cover the whole area of interest. The dots represent the location of the Neolithic magoules (left). Magoula Aerino (right).

2.2 Landscape reconstruction

Prior to the landscape reconstruction of Thessaly during different Neolithic periods, the reliability of the existing Digital Elevation Model (DEM) was evaluated compared to other digital elevation models, such as the 90m pixel size DEM (from the Shuttle Radar Topography Mission (SRTM) and the 30m pixel size DEMs provided by ASTER images or constructed by the L1-A stereoscopic products (3N and 3B) of ASTER. The results indicated that the RMSE for the DEM created through the digitization of the contour lines of the topographic maps was the lowest. The two major plains of Thessaly contain 181 out of the 342 known registered "magoules", stressing the important role of reconstruction of the relief of each basin during each Neolithic sub-period. Both geological (stratigraphic data from boreholes and past geomorphologic studies) and archaeological data were placed under consideration for achieving this task and a reconstructed DEM for each basin and each Neolithic period was created.

2.3 Satellite image processing

2.3.1 Data and preprocessing of satellite images

Concerning this study different multispectral images were used. Specifically, 4 IKONOS images of 1m pixel size, 1 Landsat ETM+ 30m pixel size image, one 30m pixel size Hyperion image and 4 ASTER images (15, 30 and 90m pixel size). Masking of the sea, clouds and snow areas in all images preceded the processing of the images in order to focus to the mainland and the areas that provided useful information. Image mosaics were created accordingly depending on the types of sensors and both image mosaics and isolated images were rectified to a common projection system (EGSA87/HGSR87). Digital numbers of images were also converted to reflectance values according to specific conversion equations. The last step was necessary in order to have a uniformity in the values of images originating from different sensors.

2.3.2 Spectral enhancement techniques

Several RGB composites were constructed in an effort to examine their efficiency in the detection of the Neolithic settlements. For the ASTER image with acquisition date 19-03-2003, where most of the magoules were registered, the RGB→1,2,3, RGB→3,2,5 and RGB→2,3,7 composites were the most successful for the visual detection of the Neolithic settlements (39 out of 239 settlements were highly visible, 49 average visible and 151 poorly visible). All these composites appeared to have the highest Optimum Index Factor (Alexakis et al, 2009). Similarly, RGB composites of IKONOS images were able to detect 27 out of 48 settlements within their area of coverage. It is worth mentioning that 19 of the detectable magoules, namely the highest of all corresponding to an average altitude of 4.6m, were highly visible in all RGB composites. On the other hand, RGB composites of Landsat and HYPERION images were not very promising (for HYPERION composites only 5 out of 21 settlements were detected). Due to their high spatial resolution, all the 5 settlements that felt within the spatial limits of the aerial photo mosaic were easily detectable. As a general conclusion however, the acquisition date of the images proved to be the most crucial factor for the detection of magoules mainly due to the intensive cultivation (mainly soft and shallow cultivation) of the landscape both on the top and the surroundings of magoules. Principal Component Analysis (PCA) was applied to ASTER, Landsat and Hyperion images, being especially effective for ASTER images where 39 and 47 out of 247 settlements were highly or medium discriminated correspondingly. Image fusion techniques through the combination of high spatial resolution images such as IKONOS (1m) and high spectral resolution images such as Hyperion (30m) concluded to very promising results (Fig 2). Finally, a spectral mixer utility (Erdas Imagine 9.1 software) contributed to exploit the dynamic range of all the multispectral information of the Hyperion image by combining more than three bands to an RGB composite. Using the specific utility and assigning a weighting coefficient for each band, a RGB composite of 23 bands (38, 42, 48, 49, 50, 51, 52), (85, 86, 87, 88, 89, 90, 91, 92,) & (93, 94, 108, 109, 110, 111, 113, 114) was constructed that enhanced the visual appearance of the magoules.

2.3.3 Spectral enhancement techniques

Radiometric enhancement was vital for the appearance of the images. After applying radiometric enhancement to ASTER images (acquisition date of 19-03-2003) we managed to detect 57 settlements (Fig. 2). A non-linear radiometric enhancement of the HYPERION PCA image, followed by an inversion of brightness was able to highlight 8 settlements from a total of 9. (Melia 1, Melia 2, Anagennisi 2, Moshohori 3, Kipseli 2, Prodromos 1 of Larisa, Nikaia 17 and Kuparissia 2). Similar type of non-linear radiometric enhancement of the high resolution IKONOS images through the modification of the histogram outlined the round shape of known magoules, as well as outlined 10 more targets of similar geometry that need to be verified by the ground truthing activities that will follow (Fig. 2).

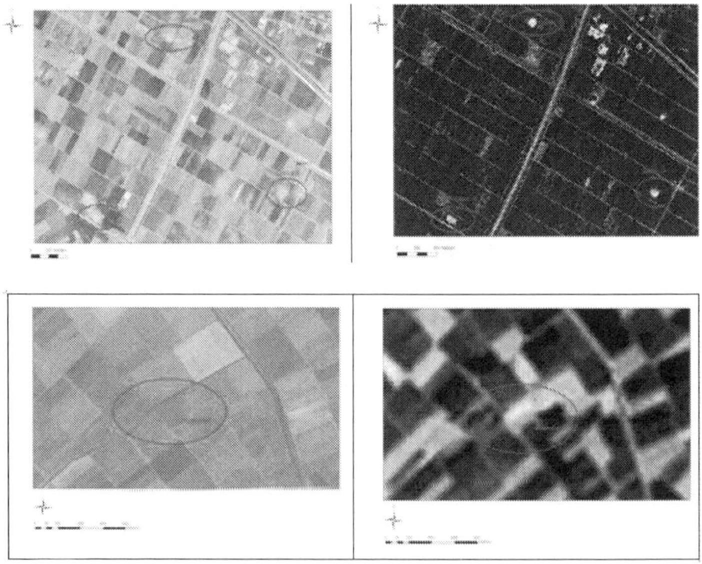

Fig. 2. Appearance of 3 settlements in the original IKONOS image (left) and the radiometrically enhanced image where three Neolithic settlements are highlighted (right). To the north of Galini-3 settlement, shown at the lower right of the image), another smaller potential magoula is suggested.

2.3.4 Land classification, vegetation indices, spatial enhancement

A spectral signature database was constructed to provide the basic spectral information about tells, especially at the plain areas of Thessaly. Several classification methods were applied to Landsat and ASTER images in order to investigate the land use regime around the magoules. Examination of the overall accuracy of the various algorithms tested (based on the error matrix), proved that the Mahalanobis algorithm was the most efficient for the exact classification of the images (Fig. 3a). Additionally, object based segmentation techniques were applied to ASTER images and 15 settlements in total of 234 were detected easily (Alexakis, 2009). The computation of the Normalized Difference Vegetation Index (NDVI) was used to highlight the vegetation differences during different periods of time, in an effort to pinpoint any vague indications for the detection of magoules. As expected, the NDVI of the "spring" ASTER image was higher than the "summer" Landsat image, but still the vegetation differences of the spring time favored the detection of magoules mainly due to the differentiations in the soil's humidity (Fig. 3b).

Application of certain spatial high pass filters contributed further to the spatial enhancement of smaller features such as the magoules. The most reliable of them proved to be Sobel Right Diagonal 3x3 and Laplace 3x3, both of which outlined clearly the limits of the most prominent of them (Fig. 3c).

Fig. 3. (a) Land classification of ASTER image through the use of Mahalanobis distance (fuzzy) algorithm. (b) Detail from the application of NDVI to ASTER image. (c) ASTER image around Halki area after the application of Sobel Right Diagonal filter. Neolithic magoules are indicated within the ellipses.

2.4 Analysis in GIS environment

An extensive spatial analysis of the magoules distribution was carried out in GIS environment using the reconstructed DEMs. Besides the extraction of statistics regarding the relation of settlements to the aspect, slope and relief height, the distance of settlements from natural resources was calculated by applying buffer zones around the quarries and the water springs (mainly springs existing on the mountainous areas). Watersheds were constructed and the distance of each settlement from its neighbor watershed was calculated. Density maps of the settlements were created for each Neolithic period. The calculation of the density of the settlements was accomplished through the use of a non-parametric Kernel technique. The spatial territorial limits of the settlements were explored using the Thiessen polygons analysis. The site catchment of Neolithic settlements was studied through least cost surface analysis. Cost surfaces contributed also to the exploration of communication routes between the different settlements (Alexakis, 2011).

Finally, GIS tools were employed to construct predictive habitation models for each phase of the Neolithic period in an effort to locate areas that could possibly host similar type of settlements. The specific predictive models were based on the use of a multi-parametric spatial analysis method of geographic elements and other information (statistical, archaeological, a.o.). All the environmental factors (height, aspect, slope, distance from watersheds, distance from water springs, distance from quarries, geology, viewshed, distance from chert sources, least cost paths, a.o.) that could affected the choice of habitation in Neolithic Thessaly were statistically examined and certain weight factors were applied to each one of them. At a final stage, a fuzzy logic algorithm and a normalization equation were also applied a more efficient tuning of the results and for rating the final probability from 0 to 1.

2.5 Application of sophisticated fields to the Digital Elevation Models

The final approach of the particular project involved the detection of Neolithic settlements through the analysis of DEMs with the use of three different semi-automated methodologies. Three different DEMs (90m pixel size SRTM DEM, 30m ASTER DEM and a 20m DEM from the digitization of contours of topographic maps) were tested in all procedures to attest for their potential in the detection of the magoules.

The first methodology involved the estimation of the index of convexity (CI) to the three different DEMs according to Fry *et al.* (2004):

$$CI = (x - x_{med}) / (x_{max} \ x_{med}), \qquad (1)$$

where x is the initial DEM, x_{med} is the DEM after the application of median 7x7 filter and x_{max} is the DEM after the application of maximum 7x7 filter.

Although the index of convexity seemed to be ideal for the detection of low hills such as magoules, only 35 (28%) of them at Larisa plain and 28 (47%) at Karditsa plain were detected by this method.

The second methodology is related with the design and application of customized filters similar to those used by Menze and Sherratt (2006). The optimal filter for the detection of a signal with a well known shape is the matched filter (Fig. 4). For the construction of the matched filter, an area of 5x5 pixel was cut around the DEM of each settlement for a total of 50 settlements. Then, the value of the central pixel was subtracted by each pixel, followed by stacking of all the 50 local DEMs (through the layer stack utility of Erdas Imagine software) to form a final multilayer image. The particular image was imposed to Principal Component Analysis and the first 5 principal components were summed. The negative sum replaced the value of the central pixel of each of the above 5 filters, which they were then applied individually to Larisa and Karditsa basins as a detection filter (Fig. 14). The statistics for these filters proved that the specific methodology is really promising especially for the SRTM DEM in the area of Larisa. More specifically, about 60% of the settlements were detected through the application of the first and second filters.

The third methodology followed the approach of Iwahashi and Kamiya (1995) for the estimation of the geometric signatures of DEM through a combined study of slope gradient, surface convexity and texture (Fig. 4). Specifically, binary files were formed through estimation of the mean value of slope gradient: all pixels with value above the mean value took a value 1 and all the rest took a value of 0. The same binary archives were created after the subtraction of the initial DEM from the one that has been processed through the application of a median filter. The last binary image was created after the application of a Laplace 3x3 filter to the initial DEM. In the end, the three binary archives were summed and the final map highlighted the areas of high local convexity. Fuzzy logic algorithms were applied to the final results of the filtered DEM in order to produce a better classification scheme (Pixel values equal to 0 formed the first group, values from 0 to 3 formed the second group and pixels with value equal to 3 formed the third group) (Fig. 15). The application of this methodology to SRTM resulted to the detection of 35% of the magoules in Larisa plain and only 15% of the magoules in Karditsa plain.

The results obtained through the Menze & Sherratt (2006) approach to the SRTM DEM were also implemented to the predictive modelling, together with other subproducts of the satellite image analysis, such as the NDVI map, land use classification and spectral signatures library of magoules. A similar methodology of significant weights and factors was considered and results were subjected to fuzzy logic and normalization techniques. Still, the results of predictive modelling did not alter significantly from the previous approach, signifying a state of saturation for the parameters considered.

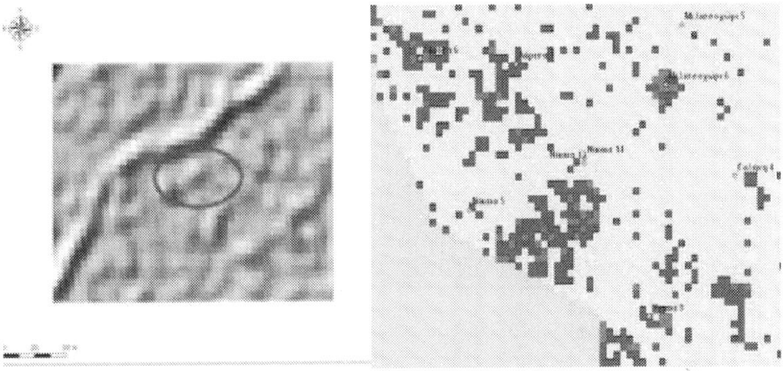

Fig. 4. Magoula Turnavos 6 after the application of matched filter (left). Application of geometric signatures methodology to SRTM DEM in the area of Larisa. With the red color are indicated the areas of higher height where magoules could be established (right).

3. Application of field spectroscopy and satellite remote sensing to archaeology

3.1 Introduction

This chapter aims to introduce the capabilities and the potentials of field spectroscopy to archaeological research (Fig. 5). Field spectroscopy involves not only the acquisition of accurate measurements (e.g. spectral signatures profiles) but also the study of the interrelationships between the spectral characteristics of objects and their biophysical attributes in their field environment. Therefore, field spectroscopy can provide valuable information for an area if we consider the fact that human eye senses only a small part of the electromagnetic spectrum, from approximately 0.4 to 0.7 nm, whereas field spectroscopy in support of remote sensing operates in a wider spectrum range including near infrared as well. In this section, ground "truth" data, presented as spectral signatures libraries, are provided from different spectroradiometric campaigns at archaeological environments (e.g. Tombs of the Kings and Nea Paphos at SW Cyprus, Sikyon archaeological site, C. Greece). Furthermore, spectral libraries include vegetation profiles, mainly over barley crops (from the Palaepaphos - Cyprus archaeological site and from Neolithic tells at Thessaly - Greece). Such libraries are used in order to examine either the seasonal changes of vegetation, or the anomalies of vegetation profiles due to buried archaeological remains. Moreover, spectral libraries are used for the atmospheric correction of satellite imagery. Finally, the theoretical background of scaling up ground narrow bands taken from handheld spectroradiometers to bandwidths satellite imagery, using the Relative Response Filters, is presented.

Ground spectroscopy may be used as a fast detection method in order to evaluate positive or negative crop marks. In this case sections over archaeological areas are taken and evaluated in terms of vegetation indices. Different Neolithic sites at Thessaly (central Greece) are examined with the use of this approach. Finally, an alternative method for the detection of archaeological remains is presented in this chapter. This method is based on the comparison of the phenological cycle profile of similar crops -under same meteorological and soil conditions -over archaeological and non archaeological sites. The case of Palaepaphos site in Cyprus is presented with the use of medium resolution images (Landsat TM/ETM+) and the support of ground spectroradiometric measurements.

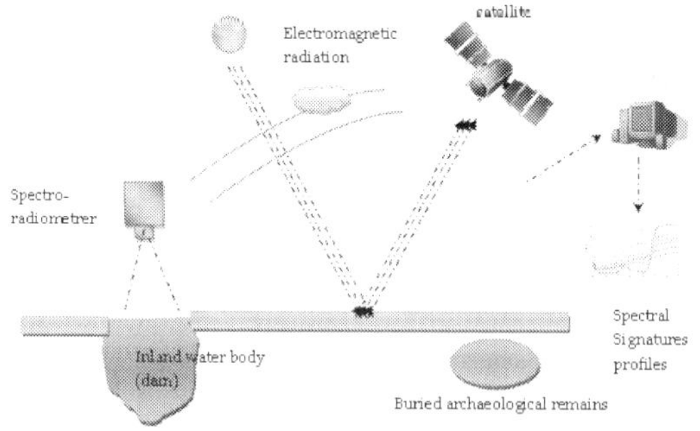

Fig. 5. Potentials of ground field spectroscopy for archaeological purposes

3.2 Collection of field measurements

Field measurements were carried out in different archaeological areas in Cyprus and Greece. The areas investigated and presented in this chapter were fully vegetated with crops. Moreover ground spectroradiometric measurements were taken: a) over visible monuments in order to develop an archaeological spectral signature database and b) at dams (inland clear water) for applying atmospheric correction to satellite images.

The spectroradiometric instrument that was used to register the spectral signature was GER 1500 (Fig. 6). This instrument may record electromagnetic radiation from a range of 350 nm up to 1050 nm. It includes more than 500 different channels and each channel cover a range of about 1.5 nm. The field of view (FOV) of the instrument was set to 4° ($\approx$ 0,02 m^2).

Fig. 6. GER 1500 used in this study with its calibration target (Agapiou et al. 2010)

A reference spectralon panel was used to measure the incoming solar radiation. The Labertian spectralon panel (≈100% reflectance) measurement was used as references while the measurement over vegetated areas or archaeological sites as a target. Therefore reflectance for each measurement can be calculated using the following equation (2):

$$\text{Reflectance} = (\text{Target Radiance} / \text{Panel Radiance}) \times \text{Calibration of the panel} \quad (2)$$

In order to examine the use of broadband vegetation indices such as NDVI, narrow band reflectance (from the spectroradiometer) needed to be recalculated according to the spectral characteristics of a specific satellite sensor. The authors selected to simulate these data to Landsat TM /ETM+ satellite imagery based on Relative Spectral Response (RSR) filters. RSR filters describe the instrument relative sensitivity to radiance at various part of the electromagnetic spectrum (Wu et al. 2010). These spectral responses have a value of 0 to 1 and have no units since they are relative to the peak response (Fig. 7, left). Bandpass filters are used in the same way in spectroradiometers in order to transmit a certain wavelength band, and block others. The reflectance from the spectroradiometer was calculated based on the wavelength of each sensor and the RSR filter as follows:

$$\text{Rband} = \Sigma\,(\text{Ri} * \text{RSRi}) / \, \Sigma\text{RSRi} \quad (3)$$

Where: Rband = reflectance at a range of wavelength (e.g. Band 1)
Ri = reflectance at a specific wavelength (e.g R 450 nm)
RSRi = Relative Response value at the specific wavelength

To avoid any errors due to significant changes in the prevailing atmospheric conditions, the measurements over the panel and the target are taken in a short time. In this case it is assumed that irradiance had not significant change which is true for non hazy days (Milton et al. 2009). Finally the measurements were carried out between 10:00 and 14:00 (local time) in order to minimize the impact of illumination changes on the spectral responses (Milton, 1987) at a height of 1.2 m (Fig. 7, right)

3.3 Spectral libraries

Spectral signature diagram is an easy way to plot target reflectance against wavelength, in a graphical form. Therefore ground field measurements from archaeological sites may be used in order to create an "archaeological" digital spectral signature library. Even though several remote sensing applications investigate the correlation of the spectral signature of an object,

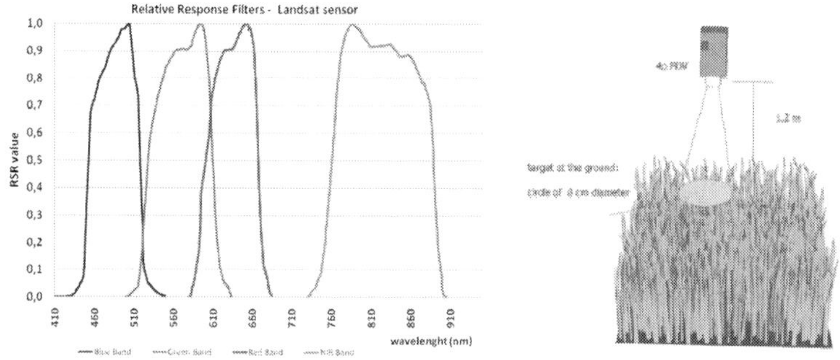

Fig. 7. Relative Response filters for Bands 1-4 of Landat TM sensor (left) and typical diagram of the in-situ spectroradiometric measurements.

in the majority these applications the aim is exactly the opposite: the study and identification of "unknown" targets through the spectral signature. Therefore "archaeological" spectral libraries may be used for identification or correlation of different archaeological sites remotely.

Different spectral signatures from the archaeological site "Tombs of the Kings" (Agapiou et al. 2011a) and "Sikyon" archaeological sites were taken. Spectral profiles indicate that there is great potential for detecting archaeological remains in the spectral range between 550 - to 850 nm (from the green visible part of spectrum to near infrared) because of the extremely different spectral response of the archaeological material compared to sand and local marl/carbonate sandstone in the archaeological site "Tombs of the Kings"(Fig. 8). Similar results were found for "Sikyona" site also (Fig. 9).

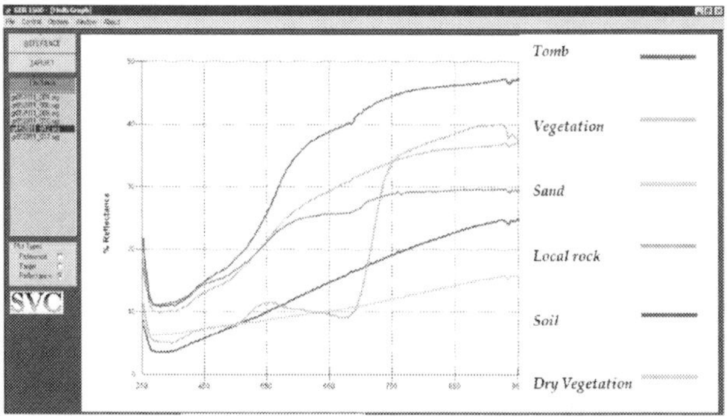

Fig. 8. Spectral signatures profiles from different targets at the archaeological site "Tombs of the Kings".

Spectral signatures libraries proved to be really efficient for any potential researcher that may use satellite imagery in order to detect archaeological relics in the area because it highlights the high correlation of spectral response of archaeological material, sand and local geological formations in the area of red visible band.

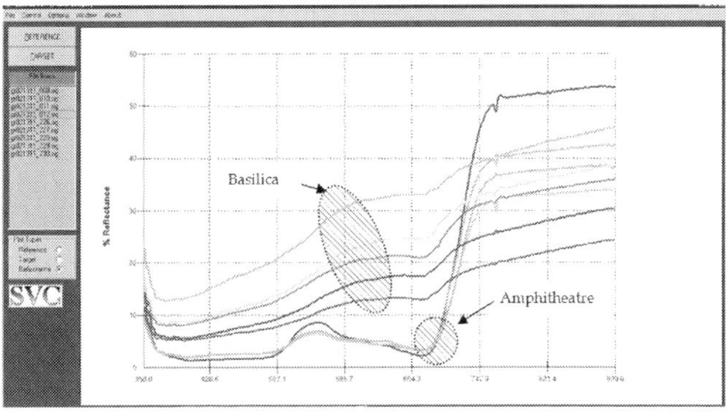

Fig. 9. Spectral signatures from the archaeological site of Sikyon.

3.4 Atmospheric correction of satellite images based on ground data

Atmospheric correction of satellite images is a necessary pre-processing step before any image analysis. Earth's surface radiation, undergoes significant interaction with the atmosphere before it reaches the satellite sensor. This interaction is stronger when the target surfaces consist of non-bright objects, such as vegetated areas examined in many archaeological studies. This problem is especially significant when using multi-temporal satellite data for monitoring purposes (Hadjimitsis et al. 2010). As Lillesand et al. (2004) argue satellite images need to be atmospherically corrected before being subjected to any post-processing techniques.

Atmospheric effects are a result of molecular scattering and absorption of the incoming radiation and influence the quality of the information extracted from remote sensing images. Such errors occurred by atmospheric effects can increase the uncertainty up to 10%, depending on the spectral channel (Che and Price, 1992). Hadjimitsis et al. (2010) have also highlighted the importance of considering atmospheric effects when several vegetation indices, such as NDVI were applied to Landsat TM/ETM+ images for agricultural applications. In their study a mean difference of 18% for the NDVI was recorded before and after the application of darkest pixel method. Therefore atmospheric correction is an important pre-processing step required in many remote sensing applications since is needed to convert the at-satellite spectral radiances of satellite imagery to their at-surface counterparts.

The modified Darkest Pixel (DP) atmospheric correction method (Hadjimitsis et al., 2004) was applied to multi-series Landsat images (Agapiou et al., 2011b). The surface radiance of the dark targets is assumed to have approximated zero surface radiance or reflectance. Instead of assuming $L_{darkest\ target}$ to be zero value, the modified DP considers the 'true' ground radiance or reflectance value over dark targets as the $L_{darkest\ target}$.

For this reason ground spectroradiometric measurements were taken in inland clear water target (Asprokremmos Dam in Paphos). The GER-1500 field spectroradiometer was equipped with a fibre optic probe was used in order to retrieve the spectral signatures from the dam. After using the RSR filters for Landsat TM/ETM+ images the spectral reflectance after the atmospheric correction was calcualated. The results have shown that satellite images were slightly improved after the removal of atmospheric effects. Indeed crop and soil marks from archaeological areas were enhanced. Photo interpretation quality was enchanced at images with low water vapour optical thickness and in general for images with water vapour optical thickness less than 0.05, the quality of the images after the atmospheric correction was improved. In the case of higher values the quality was not improved sufficiently (Agapiou, 2011a). Fig. 10 shows some typical histograms before and after the atmospheric correction. As it is shown the initial histogram of the image is stretched and therefore interpretation is improved.

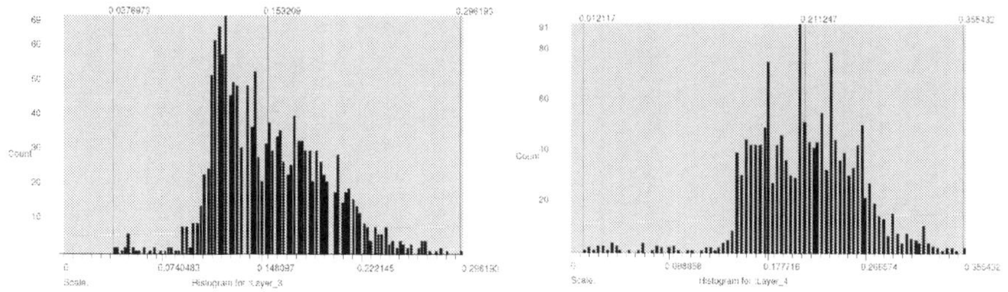

Fig. 10. Histogram for Band 3 before (left) and after (right) atmospheric correction (image: Landsat TM, 25-09-2009).

Generally the interpretation showed that in cloud-free image with low water vapour optical thickness (≈ < 0.05) atmospheric correction can increase the quality of the satellite image and therefore improve the interpretation.

3.5 Verification of buried archaeological sites

Field spectroscopy can be also used for detection of buried archaeological remains. The advantage of using ground spectroscopy against satellite remote sensing, is based on the fact that the researcher may repeatable use such methodology in contrast to the temporal resolution of satellite images (e.g. 16 days for Landsat images). Although spatial resolution is increased (few cm) the extent and scale of spectroscopy is limited compared to the area coverage of a single satellite image.

For the verification of known archaeological sites using ground spectroscopy, GER 1500 spectroradiometer was used in several vegetated archaeological sites. In this chapter results from field campaigns over Neolithic tells in Thessaly (central Greece) and buried remains in the Palaepaphos area (SW Cyprus) are presented (Agapiou et al., 2010; Agapiou and Hadjimitsis,2011 ; Agapiou et al., 2012a). In each archaeological site several sections were carried out. For the first site, along each section, more than 50 ground spectroradiometric measurements were taken while in each consecutive 5th measurement the calibration spectralon panel was used in order to minimize sun changes illuminations. At the second case study measurements were taken over known geophysical anomalies (potential subsurface monuments). To avoid differences due to variations in cultivation techniques, all measurements were carried out within the same parcel. As it shown in Fig. 11, vegetation indices such NDVI and Simple Ratio tend to give higher values at the highest peak of the tell, similar to other flat - healthy crops of the area, while the slope of the tell gives lowest NDVI and SR values. This is due to the fact that top of the tell seems to have similar hydrological behaviour as the flat healthy region (e.g. same level of water surface run off and similar inclination ≈ 0%) in contrast to the slope of the tell. The sloping part of the tell seems to behave differently due to rainfall erosion processes. All these results denote the correlation between the morphology and the spectral response of canopy on the magoules (Agapiou et al., 2012a). Moreover ground spectral signatures at Palaepaphos area (Fig. 11) indicated a stress condition for crops over the geophysical anomaly in contrast to the rest of the measurements. This stress condition was detected from ground spectroradiometric measurements as shown in Fig. 11.

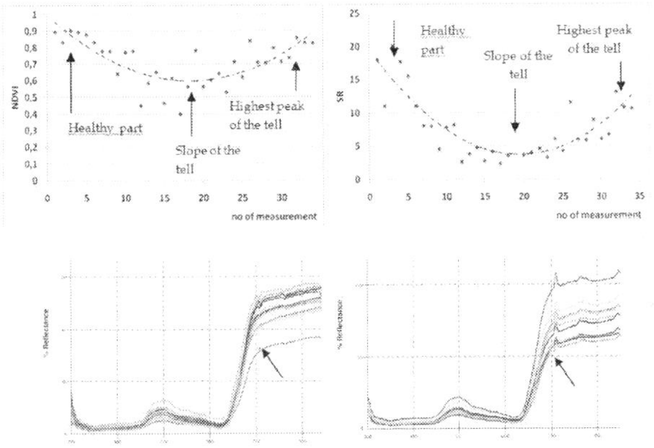

Fig. 11. NDVI (Top left) and Simple Ratio (Top right) profile over archaeological site at Thessaly. Spectral signature profiles over geophysical anomaly (indicated with arrow) at Palaepaphos (Bottom).

The results of Thessaly were able to be confirmed using Landsat TM/ETM+ images. Indeed as it was found the similar characteristics were observed and in satellite images (Fig. 12). Therefore using this experience of the spectroradiometer, where ground hyperspectral data were collected, a researcher focusing in satellite imagery can seek and search for similarly spectral characteristics as those in the spectroradiometric campaign.

3.6 Monitoring phenological cycle of crops

Monitoring the phenological cycle of crops for archaeological sites has been very limited discussed in the literature. Nevertheless as Agapiou and Hadjimitsis (2011) argue, that this approach may be used -under some assumptions- for the detection of buried archaeological remains. This methodology may be used in cases were spatial resolution of satellite imagery is very low or the cost of high multispectral satellite imagery is forbidden for an archaeological research. The basic theory of the applied method is based on the different spectral signature characteristics of 'stressed' (negative crop mark due to buried walls) and 'non-stressed' (i.e. healthy) vegetation based on the following two criteria: (a) similar soil characteristics & (b) similar climatic characteristics. The determination of spectral signatures of barley can be also verified using field spectro-radiometric measurements.

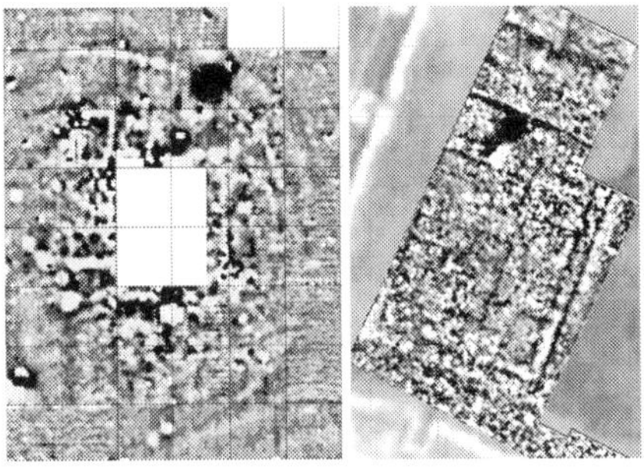

Fig. 14. Results of the usage of Geoscan FM256 and Bartington G601 fluxgate gradiometers for the mapping of the Early Copper Age (ca. 4,500-3,900 BC) settlement of Veszto-Bikeri, in the Great Hungarian Plain (left image) and the urban center of ancient Nikopolis (Epiros, NW Greece) (right image). The magnetic survey on the Hungarian Bronze Age tell revealed three circular ditches encircling the settlement, which consists of a dense cluster of structural remains (rectangular houses, pits, kilns, etc) (Sarris et al. 2004). In the case of ancient Nikopolis, a large complex building (45x70m) with a number of internal divisions appeared right at the edge of the SW corner of the Byzantine Paleochristian walls (Sarris et al., 2010).

Electromagnetic methods (EM), including the ground penetrating radar (GPR) and the soil conductivity techniques (SC), have been also employed for the prospection of archaeological sites and the reconstruction of the ancient terrains. The Slingram type of soil conductivity devices (such as Geonics EM31 or EM38) operate at low frequencies (usually at the range of 50-300kHz), make use of the electromagnetic induction and are capable of providing measurements of both soil's apparent electrical conductivity (quadrature component) and apparent magnetic susceptibility (in-phase component), with various penetration depths depending on the orientation (vertical coplanar (VCP) and horizontal coplanar (HCP) orientations), the frequency of operation and separation of the transmitter and receiver coils (Dalan, 2006; Gaffney & Gater, 2003; Cheethman, 2010). The strength of EM signals that are registered by the receiver system depends on the conductivity of the soils, the magnetic permeability and the dielectric permittivity (especially for the GPR). Operating within the range of radio frequencies, GPR systems consist of a transmitter antenna that sends a signal (~30-1000MHz) which propagates through the different strata or features (reflectors) of the subsurface and a receiver antenna that registers all the secondary reflections (with a modified amplitude) that arrive to it after a time delay which is converted to the depth. Signal attenuation and penetration depth of the GPR decreases with the increase of the frequency of the antennas and the conductivity of the soils. GPR signals are collected with high sampling rate along transects and the resulting reflection sections (radargrams) represent the variation of the amplitude of the reflected signal with depth and thus they depict an image of the stratigraphy of the subsurface. GPR parallel transects are usually combined to created 3D volumetric maps of the subsurface and through the isolation of specific time or depth slices it is possible to allocate the horizontal extent of archaeological features at different depths, which is of importance especially in cases that one wants to have information regarding the vertical extent of the features or to construct 3D models of them. In this sense, the GPR survey can be valuable in mapping different occupation strata and resolving features that are located at various depths (Fig. 15) (Conyers, 2004; Conyers & Goodman, 1997).

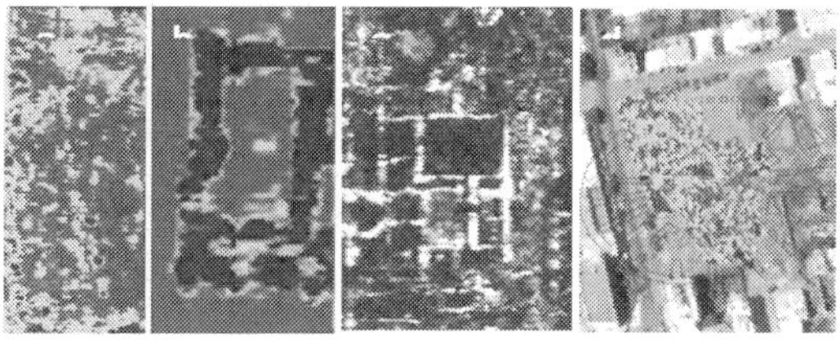

(a) Section of a Stoa from the ancient Agora of Feres (Velestino) in Thessaly. The (image dimensions: ExN=16x40m).
(b) Monumental structural remains SE of the Zeus temple in the archaeological site of Nemea, Peloponesse. (image dimensions: ExN=7x15m) (Papadopoulos *et al.* 2011).
(c) Architectural complex at a depth of 90-100cm below the surface located to the east of the Agora of Sikyon at NE Peloponesse (image dimensions: ExN=30x50m).
(d) The 2.5-3m GPR depth slice from the area of the hypothesized amphitheatre of Ierapetra (SE Crete). The survey was carried out in the area that was suggested through the rectification of the map of the British Vice-Admiral Thomas Spratt (1811-1888) (which was depicting the approximate location of the amphitheatre) on the satellite (Quickbird) image of the region. The deeper GPR slices in combination to ERT measurements provided evidence for the underlying relics of the amphitheatre (Sarris et al., 2011).

Fig. 15. Examples of GPR time slices obtained by Nogin Plus (Sensors & Software) GPR system using a 250MHz antenna.

Even microgravity measurements have been carried out for the detection of features that have a substantial mass density contrast with respect to their surrounding geological domain, creating a difference at the local gravitational acceleration. Measurements of earth's gravitational acceleration are carried out though the use of gravimeters that measure the acceleration of gravity within a hundredth of a mGal (1Gal=1cm/sec^2) or less. As such, the resolution of the method is dependent on the size and volume of the targets and requires tedious corrections and processing (such as drift correction, latitude correction, free air correction, Bouguer correction, etc) as measurements are influenced by the regional or even local trends. A recent review of archaeological, environmental and geological microgravity applications has been provided recently by Eppelbaum (2011).

Mainly used for landscape reconstruction, large monumental structure detection and deep prospection surveys, seismic techniques exploit acoustical waves generated either by a sledge hammer or an explosion discharge. In seismic refraction, the acoustic wave sensors (geophones) are laid along specific distances and record the refracted signals with respect to their arrival times and in this way their velocity of propagation (increasing with depth) is measured. Seismic reflection techniques require a smaller distance between the source and the geophones and through the examination of the arrival times, amplitude and shape of the reflected waves, we can conclude on the types of the subsurface interfaces (Metwaly et al, 2005 ; Scott & Markiewich, 1990).

If the above techniques are capable of providing a mean of detection and localization of architectural features within an archaeological site, magnetic susceptibility (MS) measurements and chemical analyses can contribute in providing a further tool for investigating the land use patterns at a specific area. Magnetic susceptibility provides not only a measure of the effectiveness of the potential application of magnetic surveys (through the estimation of the normalized Le Borgne Contrast, namely the variation of the magnetic susceptibility with depth), but also an index of the past workshop activities in an area. Measurements of the magnetic susceptibility and the frequency dependent susceptibility (namely the variation of MS with the frequency of an induced magnetic field) are capable in distinguishing soils enriched in single domain magnetic particles (from the geological origin multidomain particles) which are indicative of the intensity of the occupation of a site (Clark, 1990; Mullins & Tite, 1973; Thompson & Oldfield,1986). Coupled with results of chemical properties of soils (especially those dealing with phosphate analysis or heavy metals tracing) it is possible to characterize the type of workshop activities (eg. increase of manganese content can be associated to glass workshop activities) or differentiate areas used for animal husbandry, midden deposits, foundation trenches, cultivation, cooking, etc. Even chemical stability of certain organic chemical compounds (e.g. coprastanol) may act as a biomarker of the human presence at a particular locatio (Sarris, 2008).

The choice of the technique depends mainly on a number of factors: the type of the targets, their lateral and vertical dimensions, their deposition depth and type/properties of the surrounding soils (to be able to create a significant signal, contrast or "anomaly"). Architectural features such as stone/brick structures, roads, walls, built/chamber or rock-cut tombs, can be relatively easily resolved through soil resistance or GPR surveys. Brick structures or architectural features that are either burnt or contain residues of heating/burning, kilns, workshop facilities, slag deposits, metal concentrations, and sometimes roads, walls and fortifications can be detected through magnetic and electromagnetic techniques. The use of ERT, GPR and microgravity is especially useful for the identification of vaults, caves, chamber tombs and fissures. Shallow depth surveys usually employ magnetic, soil resistance techniques and the use of GPR. In cases where deeper penetration is required GPR, ERT and seismic approaches are more appropriate (Fig. 3) (Sarris 2008, Linford 2006).

Ground based prospection techniques are not only limited to the survey of archaeological sites in an open/rural context (Fig. 16). They can also be applied within an urbanized environment, but in such a case only specific techniques can be used (such as ERT and GPR) that are influenced as less as possible by the modern interventions and structures that exist in the urban matrix. A lot of these applications do not only involve the mapping of the subsurface (eg. below asphalt roads, pavements, concrete blocks, etc), but sometimes they are oriented towards the stability or structural damage assessment of monuments or historical structures aiming towards their architectural restoration (Bertroli et al., 2011; Pettinelli et al., 2011; Utsi, 2010; Masini et al., 2010).

Currently there are two different tendencies in archaeological prospection: the integration of different geophysical techniques for maximizing the information content and the employment of multi-sensor methods for the rapid coverage of sites. In most cases, the integrated use of various techniques is employed to extract more information about an archaeological site, allowing the interpretation of various measurements that are dealing with different properties of the soil. The fusion of this information permits a more holistic approach as the data can complement each other and provide a more integral plan of the subsurface features. One the other hand, the recent development of new multi-sensor (for magnetics), multi-antennas (for GPR) or multi-electrode (for soil resistivity) motorized systems carrying DGPS allow the fast and detailed assessment of large regions, although

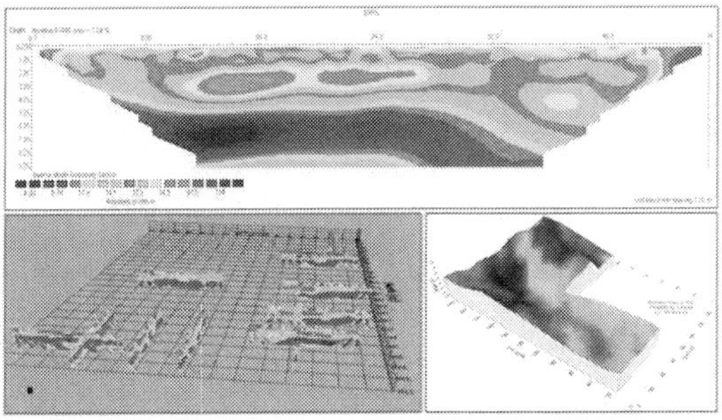

Fig. 16. Example of the application of ERT in the area of the assumed hippodrome in the archaeological site of Nemea, Peloponesse. A number of ERT transects was materialized reaching the depth of about 9m below the surface. Upon the synthesis of all the 2D resistivity inverted sections (a), it was possible to create a scatter plot of all the sections (b) and through interpolation techniques the 3D volumetric resistivity map. The isolation of the various strata of the subsurface was based on the range of their resistivity values (c). In the particular case, the iso-resistivity surfaces that resulted from the ERT transects did not identify any specific leveling of the subsurface at the west side of the archaeological site where the hippodrome was expected, suggesting that the original hypothesis of the archaeologists has to be rejected (Papadopoulos et al., 2011).

they are restricted by the surface coverage and the terrain morphology (Linford et al., 2011; Doneus et al., 2011). Although the particular systems offer increased sampling density and rates of coverage, they often suffer from positioning errors due to high measuring velocities and the introduction of noise due to the non uniform balancing of the sensors or multichannel GPR systems (Zollner et al., 2011; Verdonck & Vermeulen , 2011).

Finally, image processing techniques play a significant role in the visualization of the results of the geophysical surveys as the ultimate goal is to provide images that they depict the underlying features at their exact location and horizontal/vertical extent in a way that can approach the results of an after-the-excavation plan. This objective can be achieved through the use of a number of filtering/convolution processes, the employment of synthetic models or inversion algorithms, or other image processing functions (Papadopoulos & Sarris, 2011; Sarris, 2008; Loke & Barker, 1996; Scott & Markiewich, 1990). In cases that multiple datasets are available for the same region, composites can be made using visualization techniques similar to those used in satellite remote sensing (Böniger & Tronicke, 2010). Even more impressive visualization can be created through the fusion of geophysical data with satellite remote sensing or aero-photogrammetric data and lidar or terrestrial 3D laser scanning (Bem et al., 2011). Indeed, the continuous improvement of high resolution satellite remote sensing sensors has made possible their simultaneous utilization with conventional geophysical data affecting their resolution and potential in the detection and mapping of underground features (Crespi et al., 2011).

5. Conclusions

The various approaches applied on different satellite images for the detection of Neolithic settlements in Thessaly illustrated the benefits that satellite remote sensing can provide in archaeological investigation. It was proven that an integration of images from different satellite sensors can contribute to a faster and more accurate and qualitative detection of archaeological sites. In addition, the GIS spatial analysis and DEM processing contributed substantially to the detection and monitoring of settlements and modeling of Neolithic habitation partners in Thessaly.

Moreover, it was proved that spectroradiometric measurements can be used as an alternative approach in order to identify buried archaeological remains, since they can provide accurate spectral signatures for a wide spectral region. Anomalies of the crop spectral signatures due to buried archaeological remains can be recorded in detail and contribute to the construction of a predictive archaeological model in the future. However, the real benefit of this instrument is when it is used in conjunction with satellite images. Moreover the spectroradiometric measurements highlight the high correlation of spectral response of archaeological material, sand and local geological formations in the area of red visible band. Finally it was proved that the monitoring of the phenological cycle of crops can be used for the detection of buried archaeological sites.

At the end geophysical survey, ground and space remote sensing methods have been gradually adopted in archaeological research in an effort to capture the residues of the past anthropogenic activity underlying below the current surface of the ground and to provide a more synthetic and holistic image of the archaeological landscapes. Magnetic and resistivity techniques, electromagnetic, gravity and seismic methods, measurements of the chemical and magnetic properties of the ground, have been all mobilized to produce an accurate picture of the underlying monuments, contributing to a variety of applications in both urban and rural environmental settings. In this way all these different methodologies Have been applied to various areas of potential archaeological interest throughout Europe. The tuning of the methods and the corresponding instrumentation, together with the development of specific processing algorithms, are necessary in order to enhance the shallow depth signals that are registered within the increased noise levels of upper horizons of the soil. In this way, shallow depth prospection techniques have been used to map architectural relics, to guide excavations, and to identify craft, workshop, agricultural or animal husbandry activities.

6. Acknowledgments

Part of this chapter is part of Dr. Dimitrios Alexakis Phd thesis. This research project was co-financed by INSTAP (Institute for Aegean Prehistory) and PENED (E.U.-European Social Fund (75%) and the Greek Ministry of Development-GSRT (25%))."Field Spectroradiometer and Archaeology" chapter results are part of the PhD thesis of Mr. Athos Agapiou. The authors would like to express their appreciation to the Alexander Onassis Foundation for funding the PhD study. Thanks are given to the archaeologists Dr. K. Vouzaxakis and Dr. J. Lolos, for their valuable assistance during field campaigns at the Neolithic tells (Thessaly) and the archaeological site of Sikyona, Corinthos and to the Department of Antiquities of Cyprus for their permission to carry out filed measurements at different archaeological sites of Cyprus. The whole project has also been co-financed by the internal programs "Integration" and "Monitoring"of Cyprus University of Technology. Finally thanks are given to the Remote Sensing Laboratory of the Department of Civil Engineering & Geomatics at the Cyprus University of Technology for the support (http://www.cut.ac.cy).

7. References

Agapiou, A., Hadjimitsis, D.G., Alexakis, D., Sarris, A. (2012a). Observatory validation of Neolithic tells ("Magoules") in the Thessalian plain, central Greece, using hyperspectral spectroradiometric data. *Journal of Archaeological Science*, 39 (5): 1499-1512, 10.1016/j.jas.2012.01.001.

Agapiou, A., Hadjimitsis, D.G., Alexakis, D., Papadavid G. (2012b). Examining the Phenological Cycle of Barley (Hordeum vulgare) for the detection of buried archaeological remains: the case studies of the Thessalian plain (Greece) and the Alampra (Cyprus) archaeological test fields. GIScience and Remote Sensing (in press).

Agapiou A.and Hadjimitsis D. G.,(2011). Vegetation indices and field spectro-radiometric measurements for validation of buried architectural remains: verification under area surveyed with geophysical campaigns, Journal of Applied Remote Sensing, 5: 053554-1.

Agapiou A.; Alexakis D.; Hadjimitsis D.G. & Themistocleous K. (2011a). Earth Observations and Ground Measurements for Cultural Heritage Protection: the Case Study of Cyprus, *1st International Conference on Safety and Crisis Management in the Construction, SME and Tourism Sectors*, June 24th - 28th 2011, Nicosia, Cyprus (in press).

Agapiou A., Hadjimitsis D. G., Papoutsa C., Alexakis D. D., Papadavid, G. (2011b). The Importance of Accounting for Atmospheric Effects in the Application of NDVI and Interpretation of Satellite Imagery Supporting Archaeological Research: The Case Studies of Palaepaphos and Nea Paphos Sites in Cyprus. Remote Sensing, 3: 2605-2629, doi: 10.3390/rs3122605.

Agapiou A.; Hadjimitsis G. D.; Sarris A.; Themistocleous K. & Papadavid G. (2010). Hyperspectral ground truth data for the detection of buried architectural remains, *Lecture Notes in Computer Science* 6436, pp. 318–331, 2010.

Aitken, M. (1974). *Physics and Archaeology*, 2nd ed., Oxford: Clarendon Press.

Alexakis, D., Astaras, Th., Sarris, A., Vouzaxakis, K., Karimali, L., (2008). Reconstructing the Neolithic Landscape of Thessaly through a GIS and Geological Approach. In: Polsluchn, K. Lambers and I. Herzog (eds.), *Layers of Perception. Proceedings of the 35th International Conference on Computer Applications and QuantitativeMethods in Archaeology/(CAA),* Berlin, Germany.

Alexakis, D., Sarris, A., Astaras, T., Albanakis, K., (2011). Integrated GIS, remote sensing and geomorphologic approaches for the reconstruction of the landscape habitation of Thessaly during the neolithic period. *Journal of Archaeological Science* 38, 89-100.

Alexakis, D.; Sarris, A.; Astaras, T.;Albanakis, K. (2009). Detection of Neolithic settlements in Thessaly (Greece) through multispectral and Hyperspectral satellite Imagery. Sensors 9 (2), 1167- 1187. doi:10.3390/s90201167.

Atkinson P. M.; Webster R. & Curran P. J., (1992). Cokriging with Ground-Based Radiometry, *Remote Sensing of Environment*, 41, pp. 45–60.

Bartroli, S. R., Garsia, G. E. & Tamba, R. (2011). GPR Imaging of Structural Elements. Case Study of the Restoration Project of the Modernist Historic Site of Sant Pau, in Archaeological Prospection - Extended Abstracts, ed. by M. G. Drahor & M. A. Berge, 9th *International Conference of Archaeological Prospection*, Izmir, Turkey, pp. 149-151.

Bem, C.; Bem, C; Asandulesei, A.; Venedict, B. & Cotiuga, V., (2011). Identity in Diversity, in Archaeological Prospection - Extended Abstracts, ed. by M. G. Drahor & M. A. Berge, 9th *International Conference of Archaeological Prospection*, Izmir, Turkey, pp. 25-28.

Böniger, U. & Tronicke, J. (2010). Integrated data analysis at an archaeological site: A case study using 3D GPR, magnetic, and high-resolution topographic data, *Geophysics*, Vol. 75, No. 4, pp. B169-B176.

Cavalli, R.M.; Colosi, F.; Palombo, A; Pignatti, S. &Poscolieri, M. (2007).Remote Hyperspectral Imageryas a support to Archaeological Prospection, *J. Cult. Herit., 8*, 272-283.

Che, N. & Price, J. C. (1992) Survey of Radiometric calibration results and methods for visible and near infrared channels of NOAA-7, -9, and -11 AVHRRs. *Remote Sensing of Environment*, 41, 19 - 27.

Cheetham, P. (2010). An Empirical Reassessment Of The Utility Of The Geonics Em38B, Together With Suggested Methodologies For Its Application In Archaeological Investigations, *Meeting of the New Surface Geophysics Group - Recent Work in Archaeological Geophysics*, London.

Clark, A. J. (1990). *Seeing Beneath the Soil. Prospecting Methods in Archaeology*. London: B.T. Batsford Ltd.

Conghe, S. & Woodcock, E. C. (2003) Monitoring Forest Succession With Multitemporal Landsat Images: Factors of Uncertainty, *IEEE Transactions on Geoscience and Remote Sensing*, 41 (11), 2557 -2567.

Conyers, B. L. & Goodman, D. (1997). *Ground Penetrating Radar: An Introduction for Archaeologists*, Walnut Creek, CA., Altamira Press.

Conyers, L B. (2004). *Ground Penetrating Radar for Archaeology*, Walnut Creek, CA: AltaMira Press.

Crespi, M.; Dore, N.; Partuno, J.; Piro, S. & Zamuner, D. (2011). Comparison of SAR data, Optical Satellite Images and GPR Investigations for Archaeological Site Detection, in Archaeological Prospection - Extended Abstracts, ed. by M. G. Drahor & M. A. Berge, 9th *International Conference of Archaeological Prospection*, Izmir, Turkey, pp. 169-173.

Curran P. J. & Williamson H. D., (1986). Sample Size for Ground and Remotely Sensed Data, *Remote Sensing of Environment*, 20, pp. 31−41.

Dalan, R.A. (2006). Magnetic Susceptibility in Johnson, J.K. (ed.) *Remote Sensing in Archaeology: An Explicitly North American Perspective*, Alabama Press: Tuscaloosa, 162-203.

De Laet, V.; Paulissen, E. & Waelkens, M. (2007). Methods for the extraction of archaeological featuresfrom very high-resolution IKONOS-2 remote sensing imagery, Hisar (southwest Turkey) *J Archeol. Sci., 34*, 830 841.

Demitrack, A. (1986). The Late Quaternary Geologic History of the Larisa Plain,Thessaly, Greece: Tectonic, Climatic and Human Impact on the Landscape, Ph.D.thesis, Stanford University.

Doneus, N.; Flory, S.; Hinterleitner, A.; Kastowsky. K.; Kucera, M.; Nau, E.; Neubauer, W.; Scherzer, D.; Schreg, R.; Trinks, I.; Wallner, M. & Sitz, T. (2011). Integrative Archaeological Prospection - Case Study Stubersheimer Alb. Bridging the Gap Between Geophysical Prospection and Archaeological Interpretation, in Archaeological Prospection - Extended Abstracts, ed. by M. G. Drahor & M. A. Berge, 9th *International Conference of Archaeological Prospection*, Izmir, Turkey, pp. 166-168.

Eppelbaum, V. L. 2011. Review of Environmental and Geological Microgravity Applications and Feasibility of Its Employment at Archaeological Sites in Israel, *International Journal of Geophysics*, vol. 2011, issue 1, pp. 1-9.

Fowler, M.J.F. (2002). Satellite Remote Sensing and Archaeology: a Comparative Study of Satellite Imagery of the Environs of Figsbury Ring, Wiltshire. *Archaeol.Prospec., 9*, 55-69.

Fry, G.L.A.; Skar, B.; Jerpansen, G.; Bakkestuen, V. & Erikstad, L., (2004). Locating archaeological sites in the landscape: a hierarchical approach based on landscape indicators. Landscape and Urban Planning 67, 97-107.

Gaffney, C. & Gater, J. (2003). *Revealing the Buried Past*, Tempus: Stroud.

Glenn E. P.; Huete A. R. ; Nagler P. L. & Nelson S. G., (2008). Relationship Between Remotely-sensed Vegetation Indices, Canopy Attributes and Plant Physiological Processes: What Vegetation Indices Can and Cannot Tell Us About the Landscape. *Sensors*, 8, 2136-2160.

Hadjimitsis, D.G.; Papadavid G.; Agapiou A..; Themistocleous K.; Hadjimitsis M. G.; Retalis A.; Michaelides S.; Chrysoulakis N.; Toulios L. & Clayton C. R. I. (2010). Atmospheric correction for satellite remotely sensed data intended for agricultural applications: impact on vegetation indices, *Nat. Hazards Earth Syst. Sci.*, 10, 89-95 .

Hadjimitsis, D.G.; Clayton, C.R.I. & Hope, V.S. (2004). An assessment of the effectiveness of atmospheric correction algorithms through the remote sensing of some reservoirs. *International Journal of Remote Sensing.*, 25, 3651-3674.

Iwahashi, J., & Kamiya, I., (1995). Landform classification using digital elevation model by the skills of image procesingdmainly using the Digital National Land Information. *Geoinformatics* 6 (2), 97 - 108 (in Japanese with English abstract).

Jonhson J. K., (2006). *Remote Sensing in Archaeology*, The University of Alabama Press, Tuscaloosa.

Kauth R. J. & Thomas G. S. (1976). The tasseled Cap - A Graphic Description of the Spectral-Temporal Development of Agricultural Crops as Seen by LANDSAT". In Proceedings of the Symposium on Machine Processing of Remotely Sensed Data, Purdue University of West Lafayette, Indiana, 4B,44-51.

Lillesand, T. M.; Kiefer, R. W. & Chipman, J. W. (2004). *Remote Sensing and image interpretation*, Wiley International Edition.

Linford, N. (2006). The Application Of Geophysical Methods To Archaeological Prospection. *Reports on Progress in Physics*, v.69, pp 2205–2257.

Linford, N.; Linford, P.; Payne. A.; David, A.; Martin, L. & Sala, J. (2011). Stonehenge: Recent Results from a Ground Penetrating Radar Survey of the Monument, in Archaeological Prospection - Extended Abstracts, ed. by M. G. Drahor & M. A. Berge, 9th *International Conference of Archaeological Prospection*, Izmir, Turkey, pp. 86-89.

Loke, M. H. & Barker, R. D. (1996). Rapid Least-Squares Inversion of Apparent Resistivity Pseudo-Sections using Quasi-Newton method. *Geophysical Prospecting*, v.48, pp. 181-152.

Martonchik, J. V.; Bruegge, C. J. & Strahler, A. H. (2000). A review of reflectance nomenclature used in remote sensing. *Remote Sensing Reviews*, 19, pp. 9–20.

Masini, N. & Lasaponara, R. (2007). Investigating the spectral capability of Quickbird data to detectarchaeological remains buried under vegetated and not vegetated areas, *J. Cult. Herit., 8*, 53-60.

Masini, N.; Persico, R. & Rizzo, E. (2010). Some Examples Of GPR Prospecting For Monitoring Of The Monumental Heritage, *Journal Of Geophysics And Engineering*, vol. 7, pp. 190–199.

Menze, B.H. & Sherratt, A.G. (2006). Detection of Ancient Settlement Mounds: Archaeological SurveyBased on the SRTM Terrain Model. *Photogramm. Eng. Remote Sens., 72*, 321-327.

Merola, P.; Allegrini, A.; Guglierra, D. & Sampieri, S. (2006). Buried Archaeological Structures Detection Using MIVIS Hyperspectral Airborne Data, In *Proceedings of SPIE, the International Society for Optical Engineering*, 2006; pp. 62970Z.1-62970Z.12.

Metwaly, M.; Green, A. G.; Horstmeyer, H.; Maurer, H.; Abbas, A. M. & Hassaneen, A. G. (2005). Combined Seismic Tomographic And Ultrashallow Seismic Reflection Study Of An Early Dynastic Mastaba, Saqqara, Egypt. *Archaeological Prospection*, 12, pp. 245–56.

Milton E. J. & Rollin E. M. (2006). Estimating the irradiance spectrum from measurments in a limited number of spectral bands, Remote Sensing of Environment 100, pp. 348-355.

Milton, E. J. (1987). Principles of Field Spectroscopy. *Remote Sensing of Environment* 8 (12), 1807--1827.

Milton, E. J.; Schaepman, M. E.; Anderson, K.; Kneubühler, M & Fox, N. (2009). Progress in Field Spectroscopy. *Remote Sensing of Environment* 113, 92--109.

Mullins, C. E. & Tite, M. S. (1973). Magnetic Viscocity, Quadrature Susceptibility and Frequency dependence of Susceptibility in Single-Domain Assemblies of Magnetite and Maghaemite, *Journal of Geophysical Research*, no. 78, pp. 804-809.

Nicodemus, F. F.; Richmond, J. C.; Hsia, J. J.; Ginsberg, I.W. & Limperis, T. L. (1977). Geometrical considerations and nomenclature for reflectance. *National Bureau of Standards Monograph*, 160, Washington, D.C U.S. Govt. Printing Office.

Nishimura, Y. (2001). Geophysical prospection in Archaeology in D. R. Brothwell and A. M. Pollard (eds.), *Handbook of Archaeological Sciences*, pp.543-553.

Papadopoulos N.G.; Sarris, A.; Michalopoulou, S. & Salvi, M.C. (2011). Integrated Geophysical Investigations in Nemea and Tsoungiza, *9th International Conference of Archaeological Prospection*, September 19 - 24, Izmir, Turkey.

Papadopoulos, N. & Sarris, A. (2011). An Algorithm for the fast 3-D Invesrion of Direct Current Resistivity Data Using LSMR, in Archaeological Prospection - Extended Abstracts, ed. by M. G. Drahor & M. A. Berge, 9th *International Conference of Archaeological Prospection*, Izmir, Turkey, pp. 197-200.

Pavlidis, L. (2005). High resolution satellite imagery for archaeological application. www.fungis.org/images/newsletter/205-1.pdf

Pettinelli, E.; Barone, P.M.; Mattei, E. & Lauro, S.E. (2011). Radio Waves Technique for Nondestructive Archaeological Investigations. *Contemporary Physics*, Taylor & Francis Group, London, Volume 52, Issue 2, pp. 121-130.

Rowlands, A. & Sarris, A. (2006). Detection of exposed and subsurface archaeological remains using multi- sensor remote sensing. *J. Archaeol. Sci.*, *34*, 795-803.

Sarris, A.; Papadopoulos N.G.; Salvi, M.C. & Dederix, S. (2011). Preservation Assessment of Ancient Theatres through Integrated Geophysical Technologies, *XVIth Congress of the International Union of Prehistoric and Protohistoric Sciences (UISPP)*, Florianopolis, Brazil.

Sarris, A.; Teichmann, M.; Seferou, P. & Kokkinou, E. (2010). Investigation of the Urban-Suburban Center of Ancient Nikopolis (Greece) through the Employment of Geophysical Prospection Techniques, Computer Applications and Quantitative Methods in Archeology "*Fusion of Cultures*" CAA'2010, Fco. Javier Melero & Pedro Cano (Editors), Granada, Spain.

Sarris, A. (2008). Remote Sensing Approaches / Geophysical, in *Encyclopedia of Archaeology*, ed. By Deborah M. Rearsall, Academic Press, New York, vol. 3, pp. 1912-1921.

Sarris, A. (2005). Use of remote sensing for archaeology: state of the art. Presented at the *International Conference on the Use of Space Technologies for the Conservation of Natural and Cultural Heritage, Campeche, Mexico.*

Sarris, A., Galaty, M. L.; Yerkes,R. W.; Parkinson, W. A.; Gyucha, A.; Billingsley, D. M. & Tate, R. (2004). Geophysical prospection and soil chemistry at the Early Copper Age settlement of Vésztő-Bikeri, Southeastern Hungary , *Journal of Archaeological Science*, volume 31, Issue 7, pp. 927-939.

Sarris, A. & Jones, R. E. (2000). Geophysical and Related Techniques Applied to Archaeological Survey in the Mediterranean: A Review, *Journal of Mediterranean Archaeology* (JMA), v.13, no.1, pp.3-75.

Schaepman-Strub, G.; Schaepman, M. E.; Painter, T. H.; Dangel, S., & Martonchik, J. V. (2006). Reflectance quantities in optical remote sensing-Definitions and case studies. *Remote Sensing of Environment*, 103, pp. 27−42.

Scott, J. H. & Markiewich, R. D. (1990). Dips and chips-PC programs for analyzing seismic refraction data: *Proceedings, SAGEEP 1990*, Golden, Colorado, pp. 175-200.

Thompson, R. & Oldfield, F. (1986). Environmental Magnetism. London: Allen and Unwin.

Utsi, E. (2010). Reflections From Westminster Abbey, *Meeting of the New Surface Geophysics Group - Recent Work in Archaeological Geophysics*, London.

Verdonck, L & Vermeulen, F. (2011). 3-D GPR Survey with a Modular System: Reducing Positioning Inaccuracies and Linear Noise, in Archaeological Prospection - Extended Abstracts, ed. by M. G. Drahor & M. A. Berge, 9th *International Conference of Archaeological Prospection*, Izmir, Turkey, pp. 204-212.

Vermeulen, F. & Verhoeven, G. (2004). The contribution of aerial photography and field survey to thestudy of urbanization in the Potenza valley (Picenum). *J. Roman Archaeol., 17*, 57-82.

Wu X., Sullivan T. J. & Heidinger K. A. (2010). Operational calibration of the Advanced Very High Resolution Radiometer (AVHRR) visible and near-infrared channels. Canadian *Journal of Remote Sensing*, 36 (5), 602–616.

Zollner, H.; Kniess, R. & Meyer, C. (2011). Efficient Large-scale Magnetic Prospection Using Multichannel Fluxgate Arrays and the New Digitizer LEA D2, in Archaeological Prospection - Extended Abstracts, ed. by M. G. Drahor & M. A. Berge, 9th *International Conference of Archaeological Prospection*, Izmir, Turkey, pp. 201-203.

20

The Mapping of the Urban Growth of Kinshasa (DRC) Through High Resolution Remote Sensing Between 1995 and 2005

Kayembe wa Kayembe Matthieu[1],
Mathieu De Maeyer[2] and Eléonore Wolff[2]
[1]*University of Lubumbashi*
[2]*Université Libre de Bruxelles*
[1]*D.R. Congo*
[2]*Belgium*

1. Introduction

The study of urban growth in Kinshasa is not a new topic, as shown by the work on the dynamics of housing in the 1970s (Flouriot et al., 1975; Pain, 1978). These authors followed the spatial extension of Kinshasa by collecting old cartographic documents and comparing them. Flouriot (1975) combined a cartographic approach with household surveys to follow the long-term housing growth.

The advent and development of remote sensing and Geographic Information Systems (GIS) have changed the methods, making it now possible to map and quantify urban growth quickly and easily. More recently in Kinshasa, Tshibangu et al. (1997) have integrated into a GIS a vegetation map drawn by Compere in 1960. This was possible thanks to the interpretation of aerial photographs and Landsat and SPOT images conducted respectively in 1982 by Wilmet and 1987 by Nsekera to quantify the urban sprawl. Delbart and Wolff (2002) evaluated the extension of the city of Kinshasa from an old map (1969) and the delineation of the city in 1995 observed on a SPOT image (from 1995). The current extension of the city (between 1995 and 2005) is not precisely known, but the figures are around 600km^2 (Lelo Nzuzi, 2008). The purpose of this chapter is to map and quantify urban growth between 1995 and 2005 using a time series of high resolution satellite images.

2. Study area

The city of Kinshasa province, located between 4 ° and 5 ° south and between 15 ° and 17 ° east, is the largest city in the Democratic Republic of Congo. It covers an area of 9965 km^2 (De Saint Moulin, 2005), about 600 km^2 being only urbanized. The city had 400,000 inhabitants in 1960 and reached more than six million in 2008, the average annual growth rate between 1960 and 2003 would therefore be about 6.80% (Lelo Nzuzi, 2008).

Kinshasa has grown in the plains bordering the Congo River. The plain 300 metres above sea level covers about 200 km^2. This is the most industrialized area and formerly the most densely inhabited, commonly called the "ville basse" (low city). After independence in 1960, the city has spread into the complex hills surrounding the city and low peaks around 600 m above sea level. This area is mainly occupied by slums, called the "high".

3. Data

Two SPOT images dating respectively from March 31, 1995 and July 1, 2005 were used. They are recorded in panchromatic and multispectral modes. Their radiometric quality is variable. The 1995 images have a cloud cover of 7% in multispectral and 5% in panchromatic, while the 2005 images have 6% of cloud cover in multispectral mode and 10% in panchromatic mode. The presence of these clouds is evidence of the difficulty of obtaining cloud-free images for areas located in the sub-equatorial climate. To make the different images comparable, a radiometric correction was performed. Unfortunately, due to the low correlation between the red and the green bands, it did not yield good results and was abandoned.

Other data were collected, digitized and georeferenced, if necessary, to analyse urban growth of the city. This entailed using the old cards to map the growth of Kinshasa over the long-term, population data and the relief and major roads.

In addition, to map the dynamics of the habitat of the Atlas of Kinshasa (Flouriot, 1975), the map "District Urban Leopoldville 1/60 000" presents the urban area in 1920. Plan Leopoldville (map 1/15 500 published by the bookseller Congo Leopoldville) gives the limit of the city in 1954. The map "Plan of Commons of Kinshasa and its Environs" to 1/20 000 published in 1959 by the Geographic Institute of Zaire is the drawing of municipal boundaries of the urbanized area in 1959. The map "City of Kinshasa-health zones" (Card 1/20 000 published in 1969 and revised in 1997 from the bottom of the base map of Kinshasa), provisional edition, published by the Geographical Institute of Zaire has the delineation of municipal boundaries of the urbanized area in 1969. All these documents are completely overwhelmed by the current situation (Delbart et al., 2002; Fox et al., 1997) and require updating.

The population data used suffers from both a paucity of quality and reliability in a country where the offices of the civil state are characterized by operating failure and where the general census of the population is not regularly organized. With the exception of the 1984 population numbers from the 1984 census, the others are mere projections of the National Institute of Statistics.

Coverage maps scale 1 / 10 000 by the Geographical Institute of the Belgian Congo (IGCB) dating from 1958 covering the city of Kinshasa have been scanned. The contours at a contour interval of 5 metres were digitized by students from MA1 geography at the university, corrected and interpolated by Mathieu De Maeyer (IGEAT / ULB) by the spline technique to produce a digital terrain model and derive the slope.

Some roads (in the north of the city and the far east, after the airport) were digitized from the SPOT panchromatic band (of 10 April 2000) and a plan of the city of Kinshasa (1 / 10 000) of March 1970 created by the Geographical Institute of the Congo. The roads in the west and south were measured and corrected by DGPS Pathfinder software. The railway was also digitized from the map of the city of Kinshasa. The roads of the southern part were digitized using only the SPOT panchromatic band of the 10 April 2000.

4. Methodology

Two approaches for change detection exist. "Image-image" comparison methods imply a radiometric normalization; this standardization is difficult to implement on data from different seasons and radiometric quality is also variable (Singh, 1986; Alphan, 2003; Coppin et al., 2004; Yuan et al., 2005). In addition, they do not identify the nature of change. Comparison methods compare the post-classification classifications of land produced independently at different dates (Gupta et al., 1985). The other group of methods is less sensitive to differences in season and they identify the nature of change but are susceptible to misclassification. To detect changes, classifications are compared in pairs. From this comparison, a map where the changes can be located and a change matrix that summarizes the amount and the nature of these changes are derived.

4.1 Geometric correction and cutting recovery images

To detect changes, it is essential that the SPOT images are properly stowed from the geometrical point of view.

This is why the latest panchromatic image has been corrected from an image of higher resolution. This is a panchromatic IKONOS image from 2002 of a resolution of 1 m corrected itself with control points measured in absolute mode with a Garmin GP60 GPS. Root mean square errors of 9.46 m on the hilly part and 4.14 m on the plain were obtained.

Then all the other images SPOT (panchromatic and multispectral mode) were corrected on the panchromatic SPOT image, corrected with a polynomial function of first order and the nearest neighbour method. All are projected onto the ellipsoid WGS 84 UTM coordinates, zone 33 south.

Geometric corrections lead to RMS errors smaller than the size of a pixel with 29 to 35 control points (Table 1), which is acceptable according to Moller-Jensen (1990) and is suitable for a detect changes study.

Image	Cell (m)	Control points number	XRMS (cell)	XRMS (m)	YRMS (cell	YRMS (m)
Spot panchromatic 1995	10	30	0.44	4.4	0.57	5.7
Spot multispectral 1995	20	34	0.49	9.8	0.54	10.8
Spot multispectral 2005	20	29	0.53	10.6	0.44	8.8

Table 1. RMS errors after geometric correction

Not all SPOT images have the same spatial extension. In addition, their size being 60 km on each side, is wider than the extension of the city of Kinshasa. The images of 1995 and 2005 were cut to the same extension.

4.2 Land use classification

Given the uneven quality of SPOT images and the strong texture of the buildings, they were classified by a supervised method and object-oriented software using eCognition.

4.2.1 Legend

The legend distinguishes four categories: the built-up, the non built-up (vegetation and bare soil), water and clouds.

Some classes are difficult to discriminate using only spectral characteristics, especially so in countries in sub-Saharan Africa. The spectral confusions are numerous, for example, the fields are easily confused with the built-up. Production facilities and services, and the buildings for residential use in some places have the same spectral signature as the sand and burned areas. To overcome these problems, we have enriched the description of spectral regions of texture parameters (see 4.2.3).

4.2.2 Selection of training and validation areas

Training and validation areas were selected based on a visual interpretation of SPOT images supported by a consultation of Google Earth and the plan of the city of Kinshasa, and edited by Aquaterra Kin Art in 1997, ensuring changes due to differences between dates of these documents. 68 areas were selected in common areas of the SPOT images. To ensure an equivalent content of classes on each date, only areas unchanged between 1995 and 2005 were selected. The sample was divided into two, 34 areas for training and 34 for validation.

4.2.3 Choice of attributes

The attributes used in the classifications were chosen on the basis of visual interpretation. The regions are described in terms of spectral averages in each spectral band and the NDVI and the textural point of view, by the standard deviations on the green and red bands, and two textural parameters of Haralick (1973), such as homogeneity and entropy of the panchromatic band.

4.2.4 Segmentation and classification

eCognition was used to perform segmentation and classification. This software can simultaneously use a variety of data, panchromatic and multispectral images or vector data bases, and can create multiple levels of segmentation using a hierarchical approach.

The segmentation algorithm is the "multiresolution segmentation." According to the "Definiens Developer 7 User Guide" (2007), this algorithm merges the pixels into segments of image by minimizing the average heterogeneity and maximizing their respective homogeneities. It can do the same with image segments from a previous segment. The procedure iteratively merges the pixels or segments, as long as the maximum threshold of heterogeneity is not exceeded. Homogeneity is defined as a combination of spectral properties and form. The spectral homogeneity is based on the standard deviation of the distribution of the colour and consistency of form is based on the deviation from a compact or smooth (Cantou et al., 2006). The procedure can be influenced by the scale factor that limits the size of the resulting segments. The segmentation was performed on the image of 2005 spectral bands of green, red and near infrared respectively, giving them a weight of 2, 1 and 1. The scale parameter was chosen by trial and ,error and set at 20 with the shape parameter 0.1 (0.5 for compactness and 0.5 for smoothing).

The algorithm for supervised classification of the nearest neighbour was used. It ranks the regions according to their proximity to areas of statistical training.

4.2.5 Validation

The classifications are evaluated by comparing 34 areas of validation within the matrix of confusion. Indices are calculated to assess the quality of results (Richards, 1993):

- The overall accuracy,
- The overall Kappa,
- The Kappa class.

The overall accuracy is good (> 80%) obtained for the different classifications (Table 2). The Kappa coefficient is only acceptable for the classification of 1995 (85%) and 2005 (92%). The classification of 2000 has a poor Kappa (64%) caused by the fog that covers the southwest of the city. This result will therefore not be used subsequently.

Years	Overall accuracy (%)	Kappa Coefficient (%)
1995	93	85
2005	96	92

Table 2. Classification accuracy

Extensive field visits conducted in late January 2009 to the end March 2009 in the extension zones of Kinshasa to understand the factors of urbanization has revealed the existence of different confusions and omissions in the class "building". For example, here are some for the image of 1995 and 2005. They are located in Figure 1 and identified in Table 3.

	Confusions errors in 1995
Zone 1	Canoes to the east of Industrial Limete
Zone 2	Island Mimosa
Zone 3	Fields and sand pit in the southwest in the commune of Mont Ngafula
	Confusions errors in 2005
Zone 1	Sand pit area behind the camp CETA and fields of vegetable crops
Zone 2	Field burned to the east of the city
Zone 4	Island Mimosa with large rocks and a mining company in building materials
Zone 5	Sand bank to the west
	Errors of omission in 2005
Zone 3	Residences in the area of the general staff of the Congolese armed forces (abundance of vegetation)

Table 3. Confusion and omission errors for the class built-up in 1995 and 2005

Fig. 1. (a and b): Location of misclassification (Source: Delbart and Wolff, 2002 for the map of municipal boundaries)

These errors will not be included in the analysis.

The results obtained at different dates are generalized by removing polygons classified as "built" with an area less than 1km² and the inclusion of less than 1 km² within the urban sprawl.

The superposition of classes "built" on two successive dates can map the evolution of the building when the matrix changes can be quantified.

5. Results

5.1 Urban growth

The location change was analysed using:

- Field visits conducted in-depth from the end of January 2009 through to March 2009 in the extension zones of Kinshasa to understand the factors of urbanization,
- The layout of the lines of roads and railways digitized,
- Maps and plans of the city of Kinshasa,
- Population data,
- The digital elevation model and slope map.

The map resulting from the comparison of land use classifications in 1995 and 2005 shows that the extensions of the city is concentrated in the southwest and northeast of Kinshasa (Figure 2).

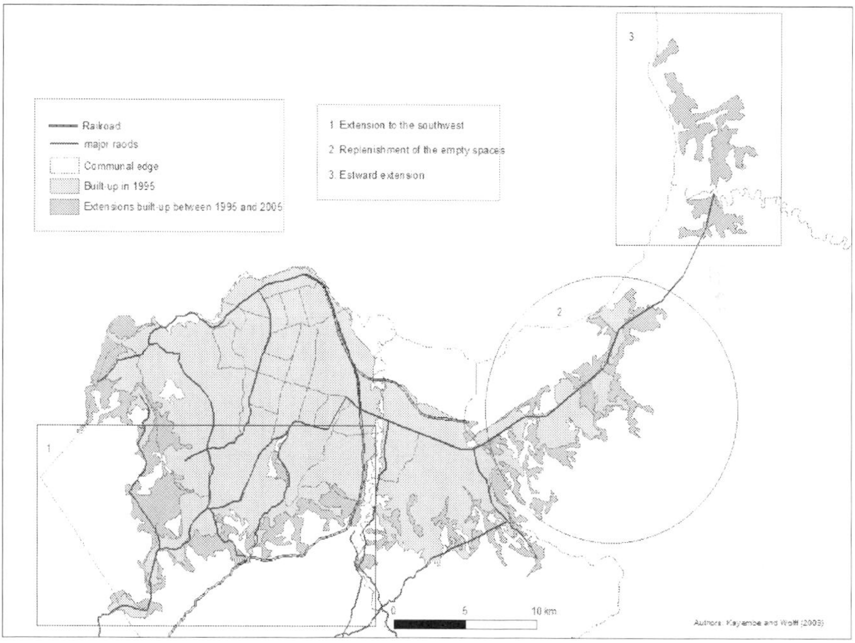

Fig. 2. Evolution of urban sprawl between 1995 and 2005 (Sources: Images Spot KJ 3 096-358 March 31, 1995 and KJ 4 096-358 July 01, 2005)

The spatial extension in the southwest took place mainly along the roads of Matadi and Lutendele (Zone 1). This process concerns the cities of Benseke, Kimbondo, Sans Fil and Matadi Mayo on the road to Matadi, and the cities of Lutendele, Kimbala, Zamba and Mazanza on the road to Lutendele. Cities such as Benseke and Kimbondo or Kimbala and Zamba have even joined in 2005.

To the east, there is a filling of interstitial spaces (Zone 2) and extension (Zone 3). Indeed, in neighbourhoods Mpasa I, II and III, Mikonga and the Badara camp, to the east of the River Ndjili, the blanks were filled. While in the far east, across the river Ndjili (Kinkole), the built-up was extended.

Urban growth can be explained by a population growth (5.1.1). Its spatial location can be explained by two main geographical factors beyond the simple distance to downtown, also an employment centre, the relief (5.1.2) and lines of communication (5.1.3).

5.1.1 Urban growth and population growth

Table 4 shows the evolution of the population of the extent of the city of Kinshasa and its density. In 45 years, the population rose from 400,000 to 7.5 million inhabitants in 2005, while the building area covered 6800 ha in 1960 against 43,400 in 2005. The population density tripled between 1960 and 2005 from about 60 inhabitants / ha in 1960 to 170 inhabitants / ha in 2005, on the whole, the city has expanded and become denser.

Years	Population	Surface (ha)	Density (hab/ha)
1960	400000	6800	59
1967	901520	9470	95
1969	1051000	12903	81
1973	1323039	14600	91
1975	1679091	17992	93
1981	2567166	20160	127
1984	2653558	26000	102
1995	4719862	31007	152
2000	6000000	39518	151
2005	7500000	43414	173

Table 4. Evolution of the population, the extent of Kinshasa and its density (Sources: Lelo Nzuzi, 2008; Yebe Musieme, 2004; Delbart et al., 2002; Mbuila Matot, 2001)

When reporting on data on population and built-up areas in 1960, one can compare the growth in urban population. Figure 3 shows that the extension of buildings characterized by an index of 600 in 2005, while the population has an index of nearly 1900.

To compare growth rates, there is data on a semi-logarithmic graph (Figure 4).

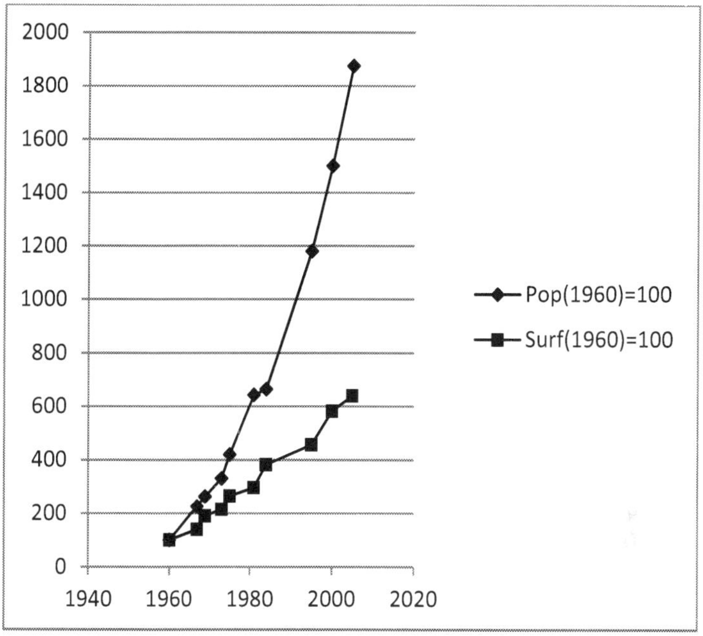

Fig. 3. Urban growth and population growth (Sources: Lelo Nzuzi, 2008; Yebe Musieme Beni, 2004; Delbart et al., 2002; Mbuila Matot, 2001)

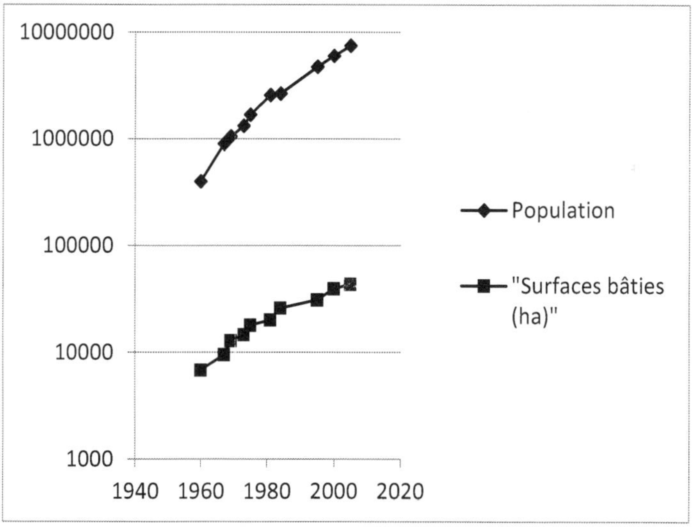

Fig. 4. Semi-logarithmic graph of urban growth and population growth (Sources: Lelo Nzuzi, 2008; Yebe Musieme Beni, 2004; Delbart et al., 2002; Mbuila Matot, 2001)

Figure 4 shows that the average growth rate of the population is less than the extension of the city. The average growth rate of the population over the period 1960-2005 is 6.73%, while that of the built area is 4.21%. Applying this growth rate to the built area of 2005 to calculate the extension of the city in 2009, we do not get the 600 km^2 regularly cited, but only about 510 km^2.

By analysing the slopes, one can compare the growth rates, they both appear to decline in 1981 (Table 5). This result confirms the observations of Bruneau (1994).

Growth rate	Population	Urban
1960-1981	9.26%	5.31%
1981-2005	5.07%	3.25%

Table 5. Population growth rate

The first period covers the 20 years after independence (until 1981). It is characterized by a very high population growth and rapid expansion space. During the second period, the population growth rate slows sharply, although it remains high, from 9.26% between 1960 and 1981 to 7.5% between 1981 and 2005. The spatial extension grew at a slower pace and passes from 5.31% between 1960 and 1981 to 3.25% between 1981 and 2005.

5.1.2 Urban growth and relief

According to the observation of existing maps and plans (Figure 5), the extension to Kinkole phases that have characterized the spatial development of Kinshasa since its inception are confirmed.

Born in the west to the Bay of Ngaliema, the city had its first developments to the east with the birth of the ancient cities (Kinshasa, Barumbu, Lingwala, Kintambo) and Gombe (formerly Kalina) in the late 1920s. From then the city grew to the south with the birth of new cities (Kasavubu and Ngiri-Ngiri) between 1930 and 1940. It was during the 1950s that the city took over the management of the east with the merger of Kalina (Leo West) and Leo. Compared to the town of 1959, we find that the city is much more extensive in the south, southwest and east (beyond the communes of Kimbanseke and Masina). Comparing the growth of map altitudes and landforms, we observe that Kinshasa was first extended in the plain corresponding to the extension of the Malebo pool and until independence in 1960 (Figure 6), the colonial authorities strictly prohibited constructions on the hills in the absence of a particular development.

After independence, the city expanded to the southwest on the plateau to the east and the plains. For the period 1995-2005, the growth has continued in the same directions.

To the east, it extends the plain due to the narrow width of flat land in the east (Biyeye, 1997). Indeed, the extension behind the neighbourhoods Mpasa I, II, III and Mikonga did not take place because of steep slopes; this is how urban sprawl has moved beyond the city Kinkole.

To the west of the river Ndjili, areas of flat plains to the south are being built upon, urbanization covers steep slopes (Figure 7), but they are unfit for human settlement in the absence of appropriate management. Indeed, these areas of steep slopes are subject to

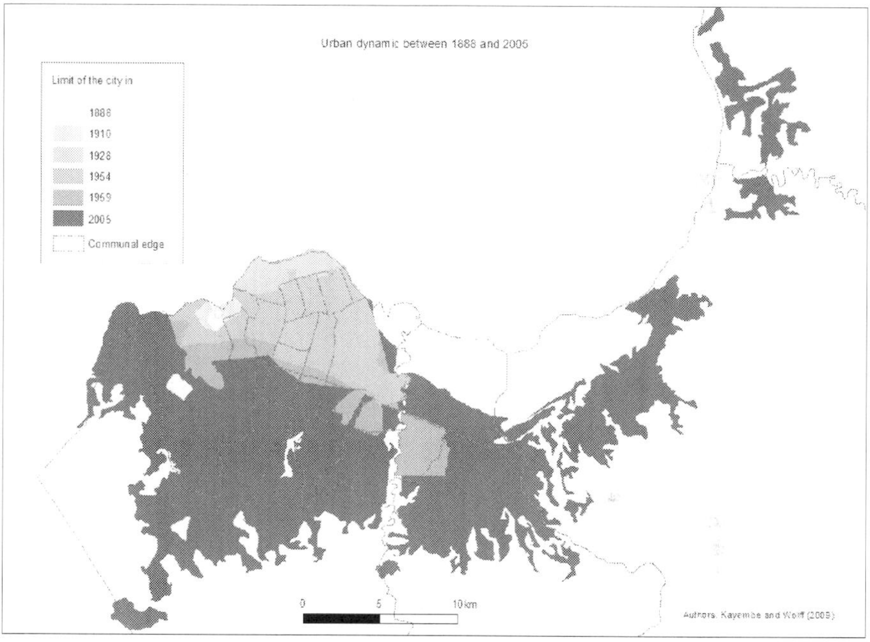

Fig. 5. Urban growth in Kinshasa from 1889 to 2005 (Source: historical maps collected by Johan Lagae, Department of Architecture and Urban Planning, Ghent University)

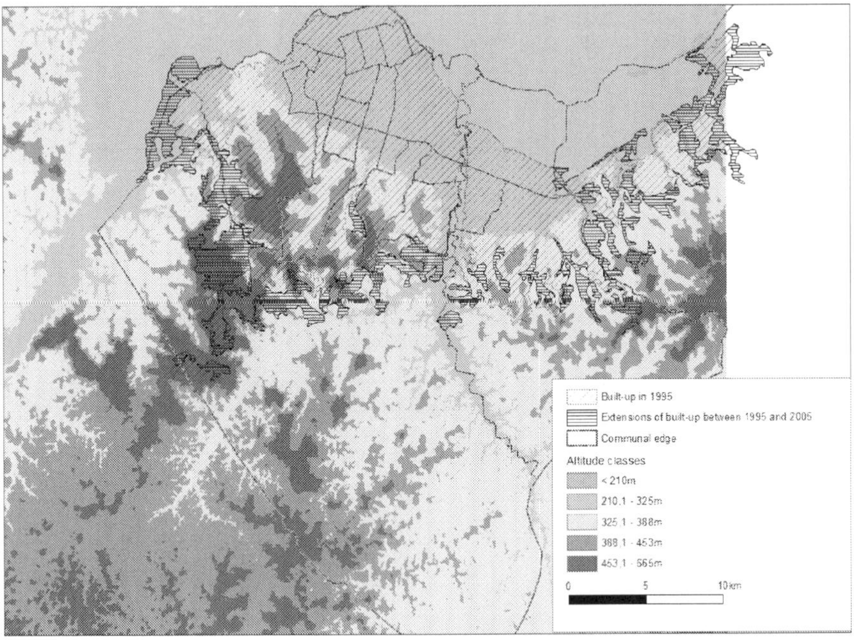

Fig. 6. Urbanization and altitude (Source: Mathieu De MaeyerIGEAT / ULB to the DTM, unpublished)

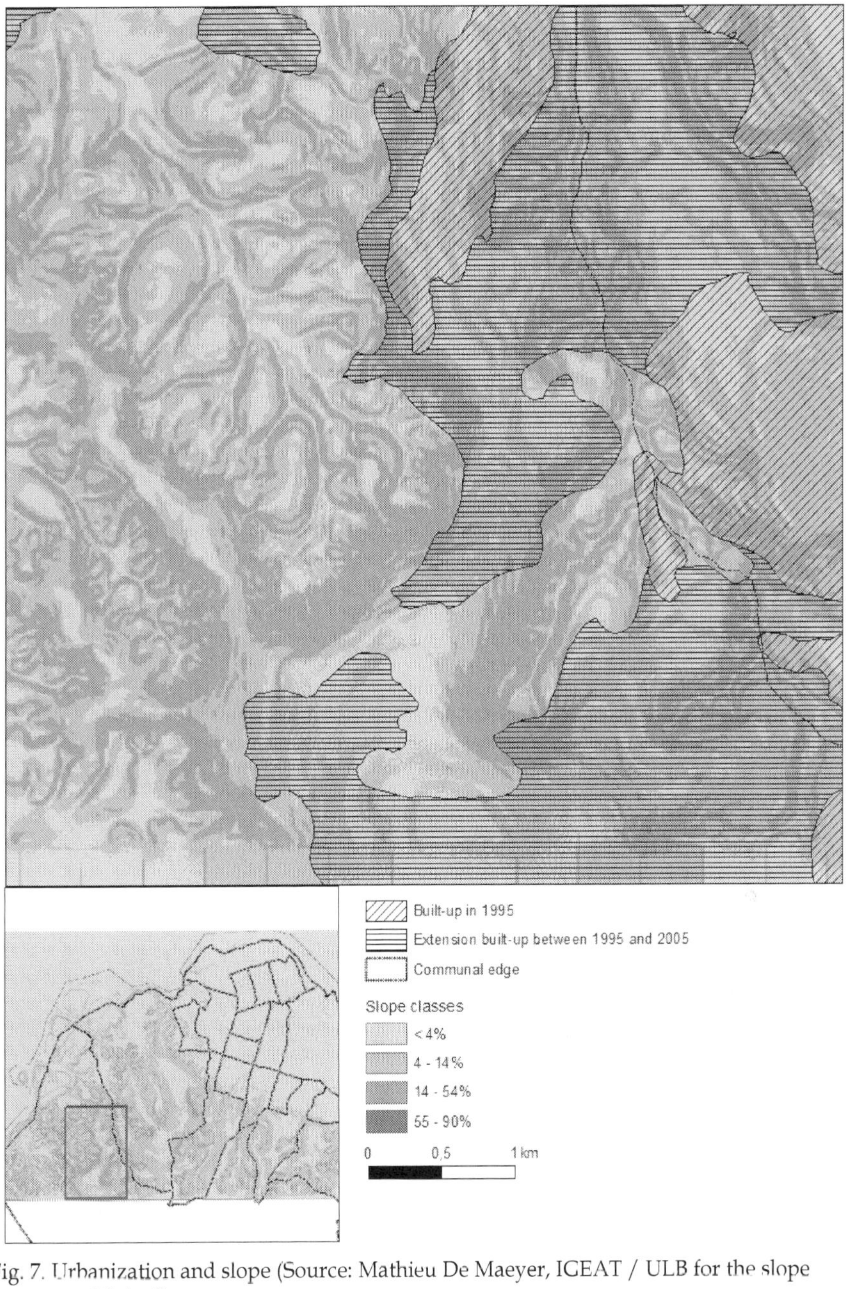

Fig. 7. Urbanization and slope (Source: Mathieu De Maeyer, IGEAT / ULB for the slope map, unpublished)

significant risk of erosion as they are laid bare (Van Caillie, 1990, 1997). In addition, the plains downstream of these steep slopes are affected by floods because of silting. These areas contain steep slopes occupied by the poor. In the future, urbanization will continue to locate in areas of high slope, as is already happening in many places on the hillsides (Camping areas, Kindele, etc.).

5.1.3 Urban growth and major roads

Figure 2 shows that urban growth is more influenced by the roads along the railway. If before independence the railway played a role in the location of industrial areas, the urban railway had not developed, now it has not facilitated access to downtown as a centre of employment and therefore has not polarized urban growth. The roads in Matadi and Bandundu, as well as routes to the drop in Lukaya and to Lake Ma-Vallée, correspond to areas of urban growth today.

Despite the role of highways in urban growth, there are areas that develop latest far from downtown and away from these axes. Indeed, an urban extension area has been observed since the 1990s, south and east of the camp which houses the faculty of the University of Kinshasa. This is the area south Cogelos and neighbourhoods Tchad, Mandela and Department Plateau to the east. These areas develop in the absence of urban amenities. Indeed, they are connected to almost none of the service water supply of REGIDESO. The population is supplied fitted to the sources (Mayi ya Libanga, Mayi ya Niwa, Mayi ya Zamba) or the fountain. This is the case in the district of Mbiti. Where the water is high, people dig a well. The lack of urban amenities does not limit urban growth in Kinshasa. The bottom line for people is to have a home.

The quantitative analysis of urban growth compared to major communication axes, made in the GRASS software, shows that 47.5% of the growth took place at more than 1 km of main roads. Beyond this 1 kilometre threshold, the advantage of proximity to the main roads can be neglected and the neighbourhood effect becomes more important in the sense that people settled near existing neighbourhoods, but at a greater distance from the road.

This helps to highlight the fact that the major communication axes are not always, or are no longer, a major factor in urban growth.

6. Conclusions

Two high-resolution satellite images (SPOT) of 1995 and 2005 were used to map and quantify the urban growth in Kinshasa relatively quickly and with an acceptable reliability. The city spreads very quickly on its margins primarily to the east and southwest along the road to Matadi and Bandundu allowing access by public transport to the city centre which polarizes the bulk of urban employment. However, since the early 1990s, neighbourhoods are growing away from the city centre and transport routes (e.g. district Cogelo, Tchad, Mandela, Department, Plateau), yet they do not benefit from any urban convenience. The extension of the city after 1960 did not spare areas of steep slopes unfit for human settlement in the absence of a particular development. These areas are home to the poor.

By measuring the average growth rate of the population and the extension of the city over 45 years (from 1960 to 2005), it was found that it is 6.73% versus 4.21% for that of the built-up area. The average growth rate of built surface applied to the surface, built in 2005 to calculate the area built in 2009 revealed some significant errors with the figures regularly quoted.

In the future, a study could be carried out to understand the logic which pushes people to occupy the steeply sloping zones where the problem of gully erosion is acute.

7. References

Alphan H. (2003). Land use change and urbanization in Adana, Turkey, *Land Degradation and Development*, Vol.14, No.6, pp. 575-586, Online ISSN 1099-145X.

Biyeye Unzola E. (1997). Urbanisation de Kinshasa: la politique de la partie extrême orientale, Mémoire en Urbanisme et Aménagement, Université Libre de Bruxelles, 85 p.

Bruneau J-C. (1995). Crise et déclin de la croissance des villes au Zaïre - Une image Actualisée, *Revue Belge de Géographie, Belgeo,* 119e année, N° spécial offert au professeur H. Nicolaï, pp. 103-114, ISBN 1377-2368.

Cantou J. P., Maillet G., Flamanc D. & Buissart H. (2006). Preparing the use of pleiades images for mapping purposes: preliminary assessments at ign-france», Topographic Mapping from Space, ISPRS Workshop Commission I, WGI/5 and WG I/6, page 6.

Coppin P., Jonckheere I., Nackaerts K., Muys B. & Lambin E. (2004). Digital change detection methods in ecosystem monitoring: A review, *International Journal of Remote Sensing,* Vol.25, No.9, pp.1565-1596, ISSN 0143-1161.

Delbart, V. & Wolff, E. (2002). Extension urbaine et densité de la population à Kinshasa: contribution de la télédétection satellitaire, *Revue Belge de Géographie, BELGEO,* Vol.2, No.1, pp. 45-59, ISSN 1377-2368.

Flouriot, J., De Maximy, R. & Pain, M. (1975). *Atlas de Kinshasa,* Institut Géographique National du Zaïre.

Guptad, N. & Munshi, M.K. (1985).Urban change detection and land-use mapping of Delhi, *International Journal of Remote Sensing,* Vol.6, No.3-4, pp. 529-534, ISSN 0143-1161.

Haralick R.M., (1979). Statistical and structural approaches to texture, *Proceedings of the IEEE,* Vol. 67, No. 5, pp. 786-804, ISSN 0018-9219.

Lelo Nzuzi, F. (2008). *Kinshasa, ville et environnement,* L'Harmattan, 275p, ISBN 978-2-296-06080-7, Paris.

Moller-Jensen, L. (1990). Knowledge based classification of an urban area using texture an context information in LandSat-TM imagery. *Photogrammetric Engineering and Remote Sensing,* Vol.56, No.6, pp. 899-904, ISSN 0099-1112.

Pain, M. (1978). Kinshasa : Ecologie et organisation urbaines. Thèse de doctorat, Université de Toulouse le Mirail, 470 p+ Annexes.

Singh, A. (1986). Digital change detection techniques using remotely sensed data. *International Journal of Remote Sensing,* Vol.10, pp. 989-1003, ISSN 0143-1161.

Tshibangu, K.W.T., Engels, P. & Malaisse, F. (1997). Evolution du couvert végétal de la région de Kinshasa (1960-1987), *Geo-Eco-Trop,* Vol.21, No.1-4, pp.95-103, Liège.

Van Caillie X.D. (1990). Erodabilité des terrains sableux du Zaïre et contrôle de l'érosion, *Cahier ORSTOM,* série Pédologie, vol. 25, No.1-2, pp. 197-208.

Van Caillie X.D. (1997). La carte des pentes (1/20 000) de la région des collines à Kinshasa. *Cahier ORSTOM*, réseau érosion, bulletin 17: Erosion en montagnes semi-arides et méditerranéennes, pp. 198-204.

Yebe Musieme B. (2004). L'impact des érosions sur l'habitat à Kisenso et les travaux de lutte anti-érosive par la population locale. Travail de fin de cycle, Unikin, Faculté des Sciences, Département des Sciences de la terre, Géographie, 63p, unpublished.

Yuan F., Sawaya K.E., Loeffelholz B.C. & Bauer M.E. (2005). Land cover classification and change analysis of the Twin Cities (Minnesota) Metropolitan Area by multitemporal Landsat remote sensing, *Remote Sensing of Environnement,* Vol.98, No.2 & 3, pp. 317-328, ISSN 0034-4257.

21

Remote Sensing for Medical and Health Care Applications

Satoshi Suzuki and Takemi Matsui
Kansai University, Tokyo Metropolitan University
Japan

1. Introduction

Radar-based remote sensing techniques are typically employed to determine the velocities and positions of targets such as aircraft, ships, and land vehicles. In particular, X- and K-band microwave devices, including oscillators and antennas, have been used to measure the velocity of automobiles and other moving objects in recent years. Microwave devices that are compact, accurate, reliable, and inexpensive are currently commercially available. Over the past few years, there have been increasing attempts to apply such techniques to biomedical measurements. Although some studies have applied these devices to medicine and health care, such research is still in its infancy. This chapter focuses on the mechanisms of and the recent research trends in microwave remote sensing techniques that are used to detect minute vibrations on the body surface induced by heartbeat and respiration.

1.1 Background

The increasing proportion of elderly in the population represents an appreciable problem in developed countries due to social concerns such as increased medical and social welfare costs and a shortage of manpower. Such concerns are expected to worsen in the future. It is thus necessary to focus on preventing illnesses and to promote healthy lifestyles. Consequently, simple equipment that can be used to self-monitor medical conditions and to acquire related data is required for homes as well as medical facilities.

Vital signs are parameters of physiological functions that are used to express the physical condition. They are used by medical professionals for making initial diagnoses. There are four primary vital signs: heart rate, respiratory rate, body temperature, and blood pressure. Thermometers for home use are commercially available and are generally approved by medical bodies. In addition, heart rate and respiratory rate can be easily confirmed by visual and palpation methods. However, there is currently still not spread to home device capable of accurately measuring and recording vital sign data that can be used to make detailed diagnoses. Monitoring cardiac function can be used for diagnosing arrhythmia and mental stress (Akselrod et al., 1981, Singh et al., 1996, Carney et al., 2001). Recently, monitoring mental condition has attracted more attention than monitoring physiological parameters. And also obesity and aging are thought to contribute to the risk of developing sleep apnea syndrome (SAS). Airway obstruction due to fat deposition in the neck is one cause of SAS and it is related to reduced alertness during the daytime (Morriset al., 2008). A simple device that can monitor respiratory activity throughout the night is thus required. These examples show the necessity for monitoring of the vital signs in daily life. Moreover, these sensing technique are presently being studied in the research area on human-machine interfaces that can be applied anywhere (for example, in a car or at the workplace) (Sirevaag et al., 1993, Gould et al., 2009).

In addition, patients who have been exposed to toxic chemicals or infectious diseases are often treated in isolation chambers to prevent secondary exposure to health-care workers. In such cases, a doctor must often make a diagnosis without touching the patient, which is difficult as the vital signs are of primary importance for emergency medical treatment. With the exception of body temperature (which can be measured by infrared radiation), it is difficult to measure vital sign parameters without contact. Consequently, remote sensing of vital signs has attracted much attention.

In this way, several fields require remote sensing of vital signs and various remote sensing methods have been proposed. However, such methods should perform biomedical measurements described as non-invasive, non-restrictive, or non-contact means that can be used without the user being conscious of them. The use of physically attaching sensors should be minimized to reduce the burden on users.

1.2 Purpose and requirements of remote sensing in medicine and health care

Monitoring cardiac and respiratory parameters is useful for health-care management as users go about their everyday lives. However, such daily monitoring needs to overcome many problems. For example, users must have sufficient technical and medical competence to set electrodes to themselves and they must not feel physically restricted by the electrodes and leads. To overcome such problems, research is increasingly being conducted on developing non-invasive and non-restrictive sensing techniques for acquiring vital signs (Jacobs et al., 2004, Wang et al., 2006, Ciaccio et al., 2007). This kind of sensing technique aims to detect and measure vibrations on the body surface induced by cardiac and respiratory activity. In the case of respiratory activity, a person's abdomen expands and contracts during the breathing cycle and this movement can be detected by sensing techniques. Similarly, for cardiac activity, the body surface moves in response to the heartbeat in minute scales. Although the vibration is slight and its amplitude depends on the individual and the part of the body, it has been observed from all parts of the body with an average amplitude of about 0.1–0.2 mm by a high-resolution laser distance meter (Suzuki et al., 2011).

Some studies have measured heart rate by placing a pressure sensor (Jacobs et al., 2004) or polyvinylidene fluoride piezoelectric sensors (Wang et al., 2006) between the user and the mattress on which they sleep. This kind of measurement method measures responses to pressure changes. Other trials have used strain gauges to measure the heart rate (Ciaccio et al., 2007). The size of minute changes due to pressure changes on the body surface induced by the heartbeat and information relating to heartbeat and respiration were obtained. A similar procedure was employed in studies using air mattresses (Watanabe et al., 2005).

Such sensing techniques have the advantages of being inexpensive because of their simple structure and of enabling stable relatively stable data acquisition because they employ direct contact with the body. Some of these sensors are already commercially available. However, they suffer from one drawback: measurement is not possible when the sensor is separated from the body by moving their bodies. This raises the question: "Is a remote sensing method available?"

1.3 Biomedical measurement using microwaves

Radio-frequency sensing techniques were originally developed for military applications and they were used to determine the location and velocity of aircraft and ships. The same technology was then applied to search and rescue; for example, they have been used to locate survivors buried under earthquake rubble (Chen et al., 1986, 2000, Lin et al., 1992). Radar can remotely acquire information on the motion of targets. Additionally, depending on the frequency of the electromagnetic wave used, radar can penetrate barriers. These characteristics of radar have been employed to detect body motion of survivors under earthquake rubble. Such devices initially had very limited effectiveness because of their poor resolution by using low-frequency waves to penetrate rubble; they could only detect relatively large body motion (at best, the abdominal motion due to breathing). However, the permeability is not a problem for everyday applications since microwaves can readily penetrate materials such as clothing, bedding, and mattresses.

A cheap, small unit that is stable and can oscillate at high frequencies has recently been developed and ongoing development is being conducted to produce safer, more flexible devices. As a consequence, higher frequency electromagnetic waves were contributed to enhance the resolution of measurement. At the same time, the output power was reduced to reduce its effect on humans, allowing microwaves to be used in everyday applications. Gradual progress, therfore, has made it possible to detect even human heartbeat.

2. Theory and methods

Here, we describe a system that employs microwaves to remotely measure vital signs by detecting vibrations on the body surface induced by cardiac and respiratory activity. Vibrations induced by heartbeat are particularly small with amplitudes of about 0.1–0.2 mm on average. This section discusses approaches using continuous-wave (CW) Doppler radar and ultra-wideband (UWB) pulse radar, which are generally used for measuring vital signs, and their mechanisms.

2.1 Mechanisms of measurement

While frequency-modulated continuous wave (FMCW) radar is used to identify the exact location of a subject in some reports, UWB or CW Doppler radar are generally used for monitoring vital signs. (Saunders, 1990, Immoreev & Tao, 2008, Li & Lin, 2010)

In a UWB pulse radar, the transmitter sends very short electromagnetic pulses toward the target. A pulse duration of about 200–300 ps and a pulse repetition frequency in the range of 1–10 MHz are typically used for vital sign detection. When the transmitted pulse reaches the chest wall, some of the energy is reflected and captured by the receiver. The nominal round-trip travel time of the pulse is defined as $t = 2d/C$, where d is the nominal detection distance and C is the speed of the electromagnetic wave. If a local replica of the transmitted pulse with a delay close to the nominal round-trip travel time correlates with the received echo, the output correlation function will have the same frequency as the physiological movement.

On the other hand, the CW Doppler radar mechanism is based on following (1);

$$T(t) = \cos\left[2\pi ft + \varphi(t)\right] \tag{1}$$

where an unmodulated signal $T(t)$ with a carrier frequency f and a residual phase $\varphi(t)$, is transmitted toward a human body where it is phase-modulated by the physiological movement $x(t)$. The reflected signal $R(t)$ detected by the radar receiver is given by following (2);

$$R(t) \approx \cos\left[2\pi ft - \frac{4\pi d_0}{\lambda} - \frac{4\pi x(t)}{\lambda} + \varphi\left(t - \frac{2d_0}{c}\right)\right] \tag{2}$$

where $4\pi d_0/\lambda$ is a constant phase shift due to the nominal detection distance d_0 and the $\varphi(t - 2d_0/c)$ is phase noise. Using the same transmitted signal $T(t)$ as the local oscillator signal, the radar receiver down-converts the received signal $R(t)$ to the baseband signal $B(t)$ as following (3);

$$B(t) \approx \cos\left[\frac{4\pi d_0}{\lambda} + \frac{4\pi x(t)}{\lambda} + \theta_0 + \Delta\varphi\right] \tag{3}$$

where $\Delta\varphi$ is determined by the nominal detection distance and the oscillator phase noise.

Since the delay corresponds to the signal round-trip travel time, the detection range of a UWB radar can be varied by controlling the delay between the two inputs of the correlation function block. This makes it possible to eliminate interference caused by reflection from other objects (clutter) and multipath reflection. However, one disadvantage of UWB radar is that the delay needs to be recalibrated when the detection distance is changed; this increases the system complexity and cost. Furthermore, since the correlation function is nonlinear, it is not simple to recover the original movement pattern, even though frequency information can be easily obtained. On the other hand, CW Doppler radar has a low power consumption and a simple radio architecture. These characteristics make it suitable for home-based systems. Moreover, proper adjustment of the radio front-end architecture of a CW radar can cancel clutter (Li & Lin, 2008a, 2008b). In addition, single-input multi-output and multi-input multi-output techniques can be easily implemented with CW radar, enabling the movements of multiple targets to be detected (Boric-Lubecke et al., 2005, Zhou et al., 2006).

2.2 Carrier frequency and output power

The carrier frequency and output power employed must be safe for use on people. Carrier frequencies ranging from hundreds of megahertz to millimeter wave frequencies have been tested for remote vital sign detection using a variety of physiological movements. The carrier frequency should be carefully selected to ensure suitable sensitivities and characteristic response for vital sign measurement. Some studies used extremely high-frequency waves (228 GHz (Petkie et al., 2009)), which have shorter wavelengths and are more sensitive to small displacements. Moreover, a 228 GHz frequency prototype has been extended to perform heart rate and respiration measurements at a distance of 50 m. However, such high frequency waves are not realistic for monitoring vital signs in everyday applications. In many cases, a carrier frequency that does not require a license is often chosen. However, carrier frequencies that do not require a license vary from country to country and some frequency bands are allocated to amateur radio stations. For example, the laws regulating radio frequency use in Japan allow band frequencies 10.525 and 24.15 GHz to be used for detecting moving objects. These devices are marketed as sensors for measuring the speeds of vehicles. Although there are limitations on how they are used (e.g., limited to indoor use), these frequency bands can be used by low-power radio stations without a license provided the output is less than 10 mW. They have been increasing studies on frequency bands for vital sign monitoring. Regarding safety, different countries have different guidelines regarding radio-frequency electromagnetic fields.

The World Health Organization (WHO) and the Scientific Committee on Emerging and Newly Identified Health Risks (SCENIHR) define exposure as *the subjection of a person to electric, magnetic, or electromagnetic fields or to contact currents other than those originating from physiological processes in the body and other natural phenomena* (WHO, 2003, SCENIHR, 2006). The intended frequency band of electromagnetic field intensity differ slightly in different guidelines. For example, the International Commission on Non-Ionizing Radiation Protection (ICNIRP) guidelines (ICNIRP, 1998) specify the frequency band from 300 Hz to 300 GHz, SCENIHR specifies 100 Hz to 300 GHz (SCENIHR, 2006), and the IEEE Standard is from 3 kHz to 300 GHz (IEEE Standard Committee, 1998). Each country employs different methods for determining their criteria. Consequently, it is important consider device development and intended usage.

The carrier frequency of medical applications of ultrasound is a low frequency of about 3 to 10 MHz. Such applications acquire information by penetrating the human body. In comparison, monitoring using microwave frequencies in the range 10.525 to 24.15 GHz is considered to be less invasive and safer. It is not easy to make simple comparisons, but wireless local area networks (WLANs) use 2.4 GHz radio waves and microwaves in the range 10.525 and 24.15 GHz are considered to be safer. Moreover, safety can be further increased by using a lower power than a WLAN.

It seems appropriate to use high frequencies for sensing to ensure a high resolution while considering invasiveness. However, high frequencies are not necessarily ideal for actual applications as increased sensitivity results in increased susceptibility to artifacts; the target motion induced by heartbeat on the body surface is much smaller than the artifacts generated by general movement of the body and arm. Furthermore, people being monitored move freely as they conduct everyday activities, which makes artifacts a significant problem.

3. Examples of applications

While remote sensing is not currently used for medical and health care applications in everyday life, several studies have been conducted. This section discusses the following typical examples of remote sensing in medical and health care settings: (1) monitoring daily health and mental stress by estimating changes in the autonomic nervous system (ANS); (2) welfare and health care for the elderly; and (3) medical diagnosis such as screening of patients with infections.

3.1 Measurement of ANS

To determine stress levels when driving or operating equipment, a 24 GHz compact microwave radar was used to perform remote measurements of the heart rate variability (HRV) under autonomic activation induced by a stressful sound and foods (Suzuki et al., 2008, Gotoh et al., 2009). Changes in the ANS (sympathovagal balance) can be monitored by measuring the HRV. This variability is used as an index of mental stress.

For examples, when an animal is attacked, its sympathetic nervous system spontaneously prepares for fight or flight by elevating the heart rate and increasing the blood pressure and body temperature. In contrast, when it is in a relaxed state, the parasympathetic nervous system is activated, and the opposite phenomena occur. Such processes are constantly in equilibrium in daily life (Akselrod et al., 1981, Derrick,1988).

Sympathovagal balance can be monitored by measuring variations in the heartbeat interval, and the degree of mental stress can be determined. This index is used for diagnosing psychological disorders and as well as the condition of circulatory organs. Moreover, many approaches for measuring the HRV are used in psychology (Vincent et al., 1996), occupational health (Miyake, 2001, Princi et al., 2005), and ergonomics research (Sirevaag et al., 1993, Gould et al., 2009). An estimation technique has been medically established. Thus, if electrocardiograms (ECG) can be replaced by a remote sensing method that allows simple and accurate monitoring in everyday life, mental stress can be assessed at any time and in any place.

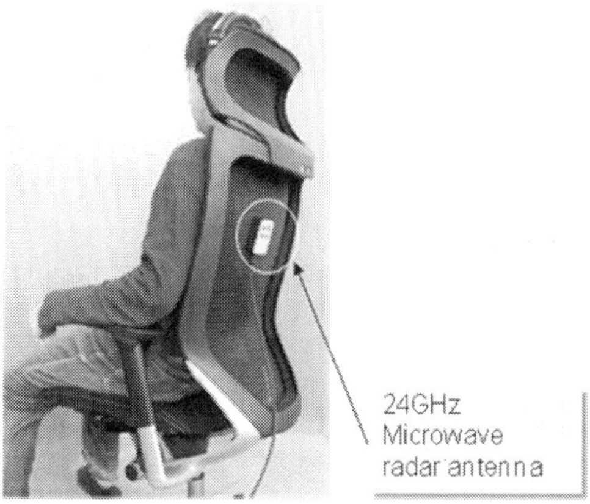

Fig. 1. Prototype chair equipped with a non-contact microwave radar system to monitor mental stress in workers (Suzuki et al., 2008).

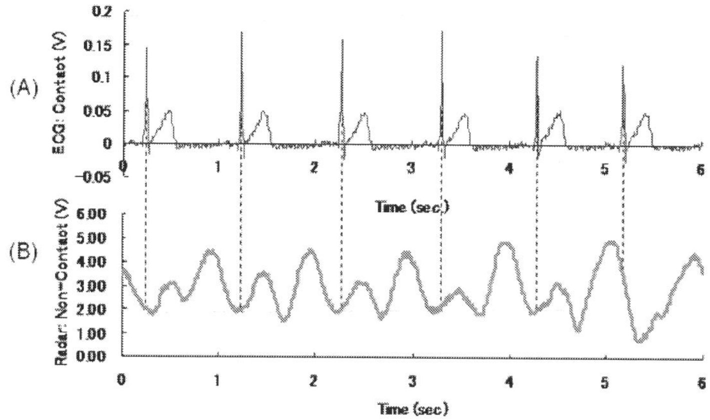

Fig. 2. Sample data of a compact microwave radar output (A) showing a cyclic oscillation that corresponds to cardiac oscillations measured by ECG (B) (Suzuki et al., 2008).

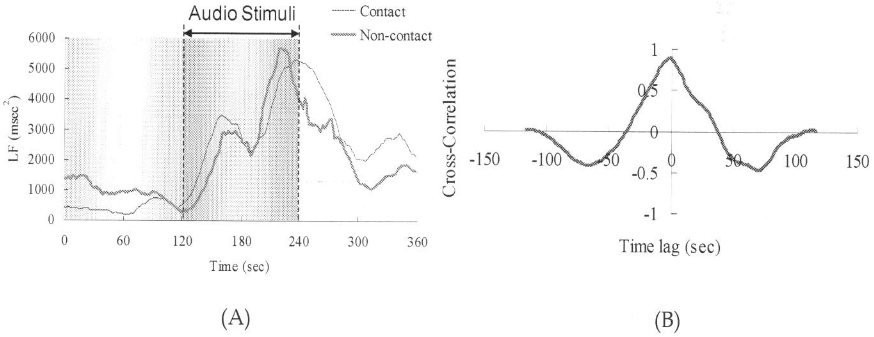

Fig. 3. Example of remote sensing in health care. (A) In both non-contact and contact (ECG) measurements, the LF of a subject (reflecting sympathetic activation) exhibits a peak during audio stimulation. (B) Cross-correlation of the LF between non-contact and contact measurements of the same subject (Suzuki et al., 2008).

The prototype device in this example consists of a compact 24 GHz microwave Doppler radar (8 × 5 × 3 cm) attached to the back of a chair. The low-frequency/high-frequency (LF/HF) components of the HRV (which reflects the sympathovagal balance) of eight volunteers were determined by a prototype system using the maximum entropy method (MEM). The change in the LF and LF/HF components measured by a remote sensor was consistent with that measured by a contact electrocardiography sensor. It is very difficult to extract R-R intervals of heartbeats sufficiently accurately to calculate the HRV; this is thought to be because inaccuracies were introduced due to the microwave radar being susceptible to noise. However, measuring HRV by adopting MEM, which enables HRV to be estimated rapidly, stably, and accurately, was demonstrated to be successful for the first time.

3.2 Elderly care

The next examples were monitoring of the elderly, infants and also animals using a microwave radar, especially focused respiratory monitoring (Li et al., 2009, Suzuki et al.,2009).

There is a real need to reduce the physical and mental work load of care workers and also to immediately detect sudden changes in the condition of a bedridden elderly person, both at home and in hospitals, because nighttime activity of persons with dementia increases the risk of injury and disrupts the sleep patterns of caregivers. Moreover, care of the elderly will increasingly be undertaken by family members at home as the population continues to age. Therefore, devices for long-term monitoring of the elderly that do not interfere with activities of daily life are required.

Various approaches have been proposed in recent years. Rowe (Rowe et al., 2009) proposed a night monitoring system that alerts caregivers when care recipients leave their beds and that tracks them as they move about the house during the night. In addition, another study compared the effects of a self-care and medication compliance device, which was linked to a web-based monitoring system, with the effects of conventional care alone on compliance with recommended self-care behavior (Artinian et al., 2003). Demiris (Demiris et al., 2009) investigated elderly behavior using a video and image processing system while carefully addressing privacy concerns. Another study proposed a non-restrictive, non-invasive vital sign measurement system for measuring heartbeat and respiration to monitor health status at home or in hospitals and nursing facilities (Tanaka et al., 2002). Systems for nursing care should meet the following requirements: (1) monitoring of activity and vital signs must not be a burden for the elderly; (2) accurately monitor changes in physical condition of the patient; and (3) when the physical condition of a patient rapidly changes, the system should promptly notify a doctor or nursing caregiver.

A prototype system using microwaves has been developed and has been applied for monitoring elderly in a nursing home. Although still a trial, the system can effectively perform real-time monitoring and it can acquire SAS respiratory data.

In addition, a baby also monitor using same technology has recently been demonstrated (Li et al., 2009). The baby monitor integrates a low-power Doppler radar that can detect minute movements induced by breathing. If no movement is detected within 20 s, an alarm goes off.

Using same type of sensing devices, respiratory activity of a hibernating black bear has been monitored at a Zoo (Suzuki et al., 2009). Ueno Zoological Gardens in Tokyo made plans to assist a Japanese black bear to enter hibernation, because the bear showed extremely slow movements in winter. Moreover, the staff wanted visitors to understand this instinctive behavior of bears as occurs in the wild. The bear's condition during hibernation must be carefully observed to avoid the risk of long-term fasting. Therefore, to observe the physiological condition of the bear during hibernation, about 3 months, a microwave radar system was set up in the hibernation booth. As a result, the respiratory rate decreased while the bear was entering hibernation, and became extremely low at approximately 2 bpm and showed almost no change. Additionally, a trend similar to a circadian rhythm in the changes of respiratory rate was observed.

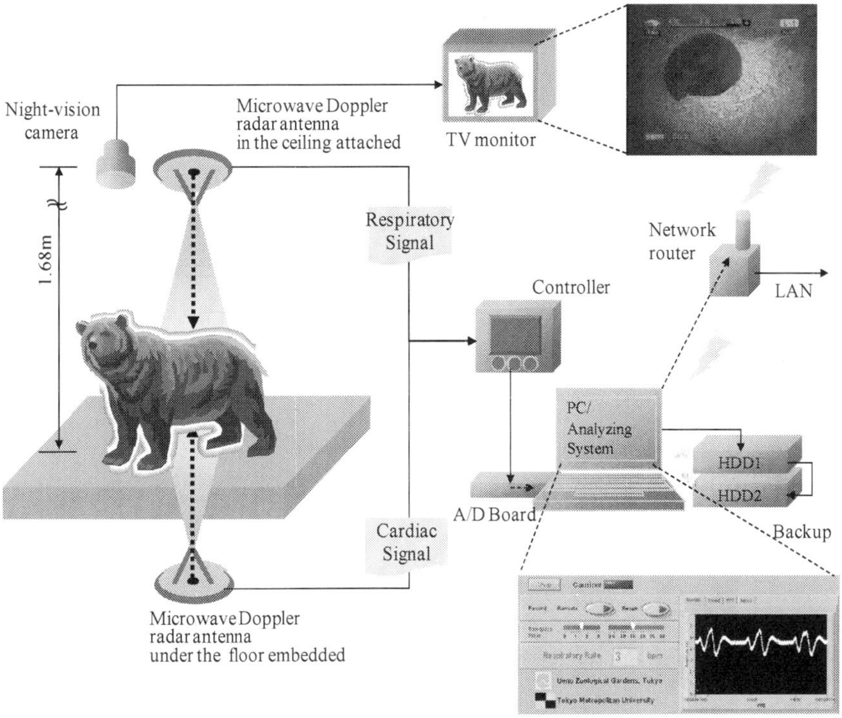

Fig. 4. Example of remote sensing using non-contact vital sign sensors to monitor a hibernating black bear at Ueno Zoo, Tokyo (Suzuki et al., 2009).

3.3 Screening and medical examination of influenza patients

The advantage of monitoring without direct touch and without removing clothing is useful and appealing in the medical field as it is not burdensome to patients and it can reduce the risk of secondary infection. There is also growing interest in the fields of health, life science, and engineering.

In another trial, a new screening system was developed to conduct rapid screening (<5 s) of passengers who may have infectious diseases such as severe acute respiratory syndrome (SARS) or pandemic influenza at quarantine stations. This system enables medical inspection by measuring heart and respiratory rates, as well as body temperature by infrared thermography (Matsui et al., 2009, 2010). A similar system is already being tested at the quarantine station at Naha Airport for domestic flights and at Narita International Airport in Tokyo. The results demonstrate the efficacy of the concept and the system. The system even detected a patient with influenza whose fever was reduced by antifebrile medication. This finding is important as it demonstrates that the parameters for monitoring cardiac and respiratory activity by microwaves are effective. In the future, remote sensing using microwaves is anticipated to attract more interest in the medical field.

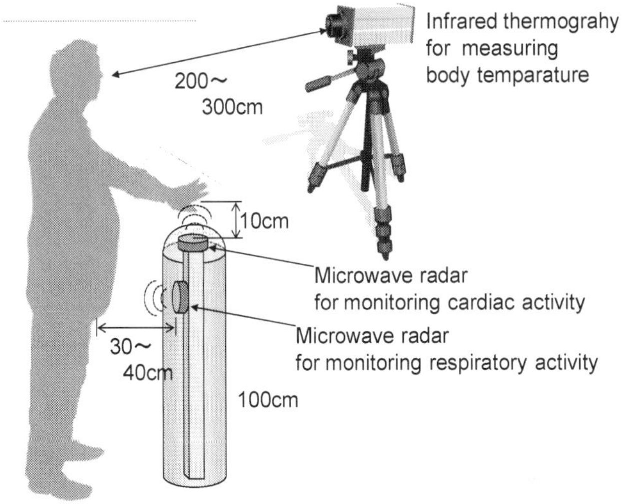

Fig. 5. Example of remote sensing for medical diagnosis; non-contact monitoring for screening systems at airport quarantine stations (Matsui et al., 2009).

4. Topics for future investigation

Remote sensing of biomedical parameters has been described by focusing on monitoring using microwave radar and examples of medical and health care applications have been presented. Microwave-based remote sensing offers the following advantages: movement of an object can be detected from a distance, it means "remotely", and microwaves can pass through many materials (notable exceptions are metals and water). Thus, such systems will enable health care workers to measure motion of a body surface through clothing from a distance.

The most serious obstacle to practical applications of this technology is the need to reduce the effects of noise and artifacts. Vibrations on the body surface caused by heartbeat are quite small (with amplitudes of about 0.1–0.2 mm), whereas motions of arms and the abdomen are much larger. Additionally, movement of the person being monitored while conducting every day activities and data from other body motion will generate noise. The effect of multipass reflection should also be considered. One study performed simultaneous measurements of the vital signs of two people, but it is difficult to apply this technology to real-world applications.

To use this technique in medical fields and daily life, aspects such as non-contact measurement and the non-removal of clothing need to be addressed. Data obtained by remote sensing must be more reliable and accurate than those measured by conventional electrode-based methods. Furthermore, a new index for diagnosing disease is required. Most studies of remote sensing using microwaves only measure the heart rate. However, the heart rate alone has limited diagnostic value; the heart rate only provides information about the tachycardia or bradycardia. It is hoped that such a system will be able to measure parameters for diagnosing cardiovascular disease.

In conclusion, research on remote sensing using microwaves is still in its infancy, but it appears promising not only for medical and health care, but also for other fields such as ergonomics research.

5. References

Akselrod, S. et al. (1981) Power spectrum analysis of heart rate fluctuation: a quantitative probe of beat-to-beat cardiovascular control. Science, 213(4504), pp.220-222.

Artinian, N.T., Harden, J.K., Kronenberg, M.W., Vander Wal, J.S., Daher, E., Stephens, Q., Bazzi, R.I. (2003) Pilot study of a Web-based compliance monitoring device for patients with congestive heart failure. Heart Lung, 32(4), pp.226-233.

Boric-Lubecke, O., Lubecke, V., Host-Madsen, A., Samardzija, D., Cheung, K. (2005) Doppler radar sensing of multiple subjects in single and multiple antenna systems. 7th International Conf. on Telecom. In Modern Satellite, Cable and Broadcasting Services, 1, pp. 7-11.

Carney, R.M., Blumenthal, J.A., Stein, P.K., Watkins, L., Catellier, D., Berkman, L.F., Czajkowski, S.M., O'Connor, C., Stone, P.H., Freedland, K.E. (2001) Depression, heart rate variability, and acute myocardial infarction. Circulation, 104, pp.2024-2028.

Chen, K.M., Misra, D., Wang, H., Chuang, H.R., Postow, E. (1986) An X-band microwave life-detection system. IEEE Trans Biomed Eng, 33, pp.697-702.

Chen, K.M., Huang, Y., Zhang, J. (2000) Microwave Life-Detection Systems for Searching Human Subjects Under Earthquake Rubble or Behind Barrier. IEEE Trans Biomed Eng, 27, pp.105-113.

Ciaccio, E.J., Hiatt, M., Hegyi, T., Drzewiecki, G.M. (2007) Measurement and monitoring of electrocardiogram belt tension in premature infants for assessment of respiratory function. Biomed Eng Online 6, pp.1-11

Demiris, G., Oliver, D.P., Giger, J., Skubic, M., Rantz, M. (2009) Older adults' privacy considerations for vision based recognition methods of eldercare applications. Technol Health Care, 17(1), pp.41-48.

Derrick, W.L. (1988) Dimensions of operator workload. Human Factors, 30, pp.95-110.

Gotoh, S., Suzuki, S., Kagawa, M., Badarch, Z., Matsui, T. (2009) Non-contact determination of parasympathetic activation induced by a full stomach using microwave radar, Medical & Biological Engineering & Computing, 47, pp.1017–1019.

Gould, K.S., Roed, B.K., Saus, E.R., Koefoed, V.F., Bridger, R.S., Moen, B.E. (2009) Effects of navigation method on workload and performance in simulated high-speed ship navigation. Appl Ergon 40, pp.103-114.

ICNIRP Guidelines (1998) For Limiting Exposure to Time-Varying Electric, Magnetic and electromagnetic Fields (UP TO 300 GHZ), Health Physics, 74(4), pp.494-522.

IEEE Standards Coordinating Committee 28 on Non-Ionizing Radiation Hazards (1998) IEEE Standard for Safety Levels With Respect to Human Exposure to Radio Frequency Electromagnetic Fields, 3 kHz to 300 GHz. C95.1-1991 and C95.1a-1998.

Immoreev, I., Tao, T.-H. (2008) UWB radar for patient monitoring. IEEE Aerospace and Electronic Systems Magazine, 23(11), pp.11-18.

International EMF Project of World Health Organization (2003) Framework for Developing EMF Standards.

Jacobs, J., Embree, P., Glei, M., Christensen, S., Sullivan, P. (2004) Characterization of a novel heart and respiratory rate sensor. Conf Proc IEEE Eng Med Biol Soc, 3, pp.2223-2226.

Li, C., Lin, J. (2008) Complex Signal Demodulation and Random Body Movement Cancellation Techniques for Non-contact Vital Sign Detection. IEEE MTT-S International Microwave Symposium Digest, pp.567-570.

Li, C., Lin, J. (2008) Random Body Movement Cancellation in Doppler Radar Vital Sign Detection. IEEE Transactions on Microwave Theory and Techniques, 56(12), pp.3143-3152.

Li, C., Cummings, J., Lam, J., Graves, E., Wu, W. (2009) Radar remote monitoring of vital signs. IEEE Microwave Magazine, 10(1), pp. 47-56.

Li, C., Lin, J. (2010) Recent Advances in Doppler Radar Sensors for Pervasive Healthcare Monitoring. Proceedings of Asia-Pacific Microwave Conference 2010, pp.283-290.

Lin, J. C. (1992) Microwave sensing of physiological movement and volume change: A review. Bioelectromagnetics, 13, pp.557-565.

Matsui, T., Hakozaki, Y., Suzuki, S., Usui, T., Kato, T., Hasegawa, K., Sugiyama, Y., Sugamata, M., Abe, S. (2010) A novel screening method for influenza patients using a newly developed non-contact screening system, Journal of Infection, 60(4), pp.271-277.

Matsui, T., Suzuki, S., Ujikawa, K., Usui, T., Gotoh, S., Sugamata, M., Abe, S. (2009) Development of a non-contact screening system for rapid medical inspection at a quarantine depot using a laser Doppler blood-flow meter, microwave radar, and infrared thermography, Journal of Medical Engineering & Technology, 33(6), pp.481-487.

Miyake, S. (2001) Multivariate workload evaluation combining physiological and subjective measures. Int J Psychophysiol 40, pp.233-238

Morris, L.G., Kleinberger, A., Lee, K.C., Liberatore, L.A., Burschtin, O. (2008) Rapid risk stratification for obstructive sleep apnea, based on snoring severity and body mass index.Otolaryngol Head Neck Surg, 139(5), pp.615-618.

Obeid, D., Issa, G., Sadek, S., Zaharia, G., El Zein, G. (2008) Low power microwave systems for heartbeat rate detection at 2.4, 5.8, 10 and 16 GHz. First International

Symposium on Applied Sciences on Biomedical and Communication Technologies, pp.1-5.

Petkie, D.T., Benton, C., Bryan, E. (2009) Millimeter wave radar for remote measurement of vital signs. IEEE Radar Conference, pp.1-3.

Princi, T., Parco, S., Accardo, A., Radillo, O., DeSeta, F., Guaschino, S. (2005) Parametric evaluation of heart rate variability during the menstrual cycle in young women. Biomed Sci Instrum 41, pp.340-345.

Rowe, M.A., Kelly, A., Horne, C., Lane, S., Campbell, J., Lehman, B., Phipps, C., Keller, M., Benito, A.P. (2009) Reducing dangerous nighttime events in persons with dementia by using a nighttime monitoring system. Alzheimers Dement, 5(5), pp.419-426.

Saunders, W.K. (1990) CW and FM Radar. in Radar Handbook, 2nd ed. M.I.Skolnik) San Francisco, MaGraw-Hill, Inc. pp.14.1-14.45

Scientfic Committee on Emerging and Newly Identified Health Risk (SCENIHR) (2006) Possible effects of Electromagnetic Fields (EMF) on Human Health.

Singh, N., Mironov, D., Armstrong, P.W., Ross, A.M., Langer, A. (1996) Heart rate variability assessment early after acute myocardial infarction. Pathophysiological and prognostic correlates. Circulation, 93, pp. 1388-1395.

Sirevaag, E.J., Kramer, A.F., Wickens, C.D., Reisweber, M., Strayer, D.L., Grenell, J.F. (1993) Assessment of pilot performance and mental workload in rotary wing aircraft. Ergonomics 36, pp.1121-1140.

Suzuki, S., Matsui, T., Imuta, H., Uenoyama, M. (2008) A novel autonomic activation measurement method for stress monitoring: non-contact measurement of heart rate variability using a compact microwave radar. Medical & Biological Engineering & Computing, 46(7), pp. 709-714.

Suzuki, S., Matsui, T., Kawahara, H., Gotoh, S. (2009) Development of a Noncontact and Long-Term Respiration Monitoring System Using Microwave Radar for Hibernating Black Bear. Zoo Biology, 28(3), pp.259–270.

Suzuki, S., Matsui, T., Kawahara, H., Ichiki, H., Shimizu, J., Kondo, Y., Yura, H., Gotoh, S., Takase, B., Ishihara, M. (2009) A non-contact vital sign monitoring system for ambulances using dual-frequency microwave radars, Medical & Biological Engineering & Computing, 47(1), pp.101-105.

Suzuki, S., Matsui, T., Sugawara, K., Asao, T., Kotani, K. (2011) An approach to remote monitoring of heart rate variability (HRV) using microwave radar during a calculation task, Journal of Physiological Anthropology, 30(6), pp.241-249.

Tanaka, S., Matsumoto, Y., Wakimoto, K. (2002) Unconstrained and non-invasive measurement of heart-beat and respiration periods using a phonocardiographic sensor. Med Biol Eng Comput, 40(2), pp.246-252.

Vincent, A., Craik, F.I., Furedy, J.J. (1996) Relations among memory performance, mental workload and cardiovascular responses. Int J Psychophysiol 23, pp.181-98.

Wang, F., Tanaka, M., Chonan, S. (2006) Development of a wearable mental stress evaluation system using PVDF film sensor. J Adv Sci, 18, pp.170-173.

Watanabe, K., Watanabe, T., Watanabe, H., Ando, H., Ishikawam, T., Kobayashi, K. (2005) Noninvasive measurement of heartbeat, respiration, snoring and body movements of a subject in bed via a pneumatic method.IEEE Trans Biomed Eng, 52(12), pp.2100-2107.

Zhou, Q., Liu, J., Host-Madsen, A., Boric-Lubecke, O., Lubecke, V. (2006) Detection of multiple heartbeats using Doppler radar," IEEE ICASSP 2006 Proceedings, 2, pp.1160-1163.

22

Demonstration of Hyperspectral Image Exploitation for Military Applications

Jean-Pierre Ardouin[1], Josée Lévesque[1], Vincent Roy[1], Yves Van Chestein[1] and Anthony Faust[2]
[1]*Defence R&D Canada, Valcartier*
[2]*Defence R&D Canada, Suffield*
Canada

1. Introduction

Airborne hyperspectral imagers have been available from various providers for many years and their performance keeps improving. On the other hand, space-based hyperspectral sensors have only been available from few exploratory missions such as NASA Hyperion on EO-1 (Pearlman et al, 2003) and ESA CHRIS on Proba (Cutter et al, 2003). In recent years, there have been many civilian space missions being planned in different countries (Buckingham & Staenz, 2008), as well as military space demonstrations (Cooley et al, 2006).

Given the increase in potential space-based hyperspectral sensors, Defence R&D Canada (DRDC), which is part of the Canadian department of National defence, began in 2005 a project to demonstrate the military utility of space-based reflective hyperspectral imagery (0.4-2.5 microns) to the Canadian Forces (CF). The project is called HYperspectral iMage EXploitation (HYMEX) and ended its activities in 2010 (Ardouin et al, 2007).

Before the HYMEX project, DRDC had been conducting and sponsoring R&D in the area of hyperspectral image exploitation for a number of years to explore its various possibilities (Davenport & Ressl, 1999; Sentlinger et al, 2003; Webster et al, 2006). The focus of this work was on military target detection applications. In parallel with these activities, the Canadian remote sensing community has also been active in developing hyperspectral applications for various civilian applications related to forestry, agriculture, fisheries, mineral exploration and environmental monitoring (Buckingham et al, 2002). Many hyperspectral techniques developed for civilian applications can be applied to military applications such as terrain characterization.

Building on previous efforts at DRDC and with support from Canadian industry, academic institutions and other government departments, the HYMEX project identified a set of applications and related algorithms to be demonstrated to the Canadian Forces.

This chapter presents an overview of the project, beginning with a description of its main activities (Section 2.0), including field trials, data analysis and algorithms evaluation and the development of an image exploitation software. Then, for each application areas, target detection (Section 3.0), land mapping (Section 4.0) and marine mapping (Section 5.0), we discuss some of the most promising algorithms and show examples of application of these algorithms.

2. HYMEX applications and activities

The HYMEX project addressed applications divided in 3 categories:

1. Target detection and identification: This includes targets such as military vehicles, camouflages and various man-made materials.
2. Land mapping applications: This includes the characterization of soil and vegetation.
3. Marine mapping applications: This includes beach characterization and near-shore bathymetry, as well as water color mapping.

As mentioned in the introduction, DRDC developed expertise and advanced exploitation techniques for target detection over the years but had limited expertise for terrain analysis and water mapping. To identify available techniques for the last 2 categories, HYMEX conducted a survey of about 60 different groups from industry, academia and other government departments forming the Canadian remote sensing community. Most algorithms from the groups that responded have been included in the project.

The project delivered a suite of near-operational hyperspectral image exploitation tools called HOST (Hyperspectral Operational Support Tools). HOST is developed in the IDL language as an add-on to a commercial-off-the-shelf package called ENVI available from ITT Visual Information Solutions. ENVI is already a widely used platform for hyperspectral image analysis. HOST builds onto ENVI to provide tools specifically oriented for the needs of the Canadian Forces. The HYMEX System Integrator responsible for the development of HOST was MacDonald Dettwiler and Associates (Richmond, BC).

HYMEX adopted an algorithms validation strategy that led to the selection of the best algorithms to be integrated into HOST and explore the performance limits of these algorithms. Knowing these limits is an important element of information for transitioning the tools into operational use by the Canadian Forces. The validation strategy included the following components:

Field Trials: The HYMEX project conducted several field trials at different locations across Canada to acquire remotely sensed hyperspectral imagery and ground truth for testing the various algorithms. Each location had specific characteristics likely to affect the performance of the algorithms. Marine trials were conducted in two different locations (East and West coast). Land trials were conducted in an Acadian forest (CFB Gagetown, NB), a boreal forest (CFB Valcartier, QC), an aspen dominant forest (CFB Wainwright, AB) and in prairie grassland (CFB Suffield, AB). One trial was conducted in winter under snow conditions. While some space-based hyperspectral imagery was acquired, the project main data type came from airborne hyperspectral sensors. Airborne data contains similar information as space-based sensors, and can be used in the interim to evaluate algorithms. In fact, even though HYMEX focused on demonstrating the utility of space-based imagery, the results can easily be transposed into the utility of airborne sensors.

Data analysis: All HYMEX algorithm providers analyzed trial data using performance metrics approved by DRDC. In addition to providing quantitative performance evaluation and comparisons of the algorithms, the overall data analysis allowed the demonstration in a limited way of a) how the performance varies in different environments, b) the advantage of hyperspectral (HSI) over multispectral (MSI) imagery and c) the effect of varying ground sampling distance (GSD) via the analysis of airborne data at different altitudes or of different sensors.

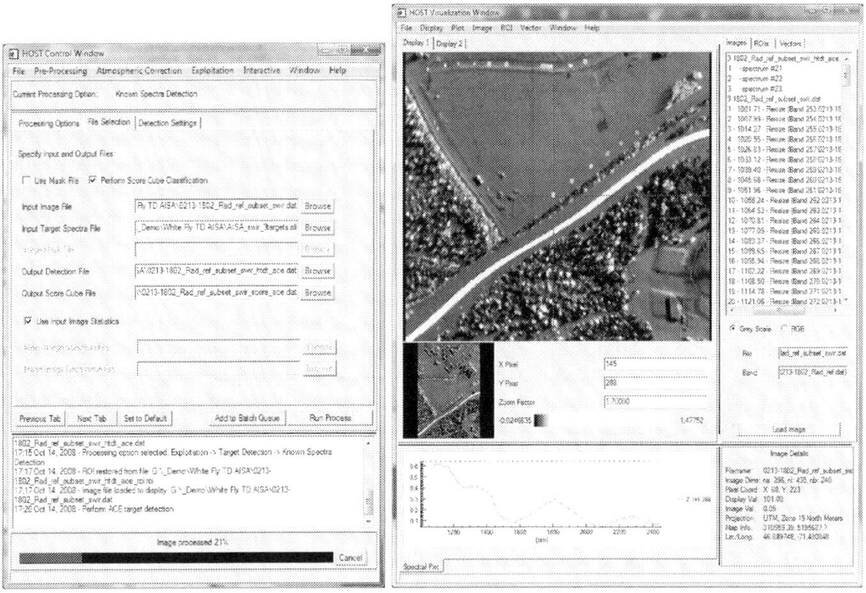

Fig. 1. HOST two main windows: the visualization window on the left and the control window on the right

HOST demonstrations: We developed the HOST software in three successive iterations. Each iteration ended with a live or hands-on demonstration of the tools to CF stakeholders and image analysts. Feedback from the demonstration participants were integrated in the following iterations and the selection of the algorithms to be integrated for the iteration was based on interim data analysis results. As explained above, HOST is an add-on to ENVI. While ENVI is a powerful exploitation package for advanced hyperspectral imagery users, HOST is oriented towards military end-users with introductory knowledge of hyperspectral image exploitation. In order to present a simplified and more uniform user interface than ENVI, HOST is organized in two main windows as illustrated in Figure 1: the visualization window and the control window. The HOST visualization window regroups, in a single window, many of the familiar visualization tools offered by ENVI (image display, plot display, available bands list, region of interest tool and the vector layer manipulation tool). By regrouping these tools in a single window, the user can more easily keep track of these functions as they are applied to specific hyperspectral images. The control window provides, in a single window, an interface to the parameters of the different advanced exploitation algorithms offered by HOST. The user would typically use the control window to setup batch scripts to process many hyperspectral images without user intervention. The HOST control window is organized in different logical categories of algorithms such as: pre-processing, atmospheric correction, exploitation and interactive tools. A Navigation Tool also allows loading customized task descriptions with links to the HOST user interface. This guides the user through the algorithms needed to accomplish a task and the selection of the parameters for those algorithms.

3. Target detection algorithms

Throughout the HYMEX project, DRDC gained experience in applying algorithms for target detection applications. In this section, we describe a typical processing chain (atmospheric correction, detection and target abundance estimation (Roy, 2010)) used in the project and present results from an experiment aimed at evaluating the performance of the target abundance estimation part of the processing chain using data collected in difficult illumination and atmospheric conditions.

In late October and early November 2009, DRDC collected airborne hyperspectral imagery near Suffield, Alberta (50°13'N, 110°10'W) using an Itres SASI-600 SWIR pushbroom imaging system. The sensor was flown at various altitudes ranging from 330m to 1700m above ground level in order to acquire imagery at across-track ground sampling distance (GSD) of 0.4m, 1.0m, and 2.0m, while along-track GSD remained constant at 1.0m. Imagery was collected between 13h00 and 15h00 local time, which resulted in sun elevation between 17 and 25 degrees. Furthermore, thin altostratus clouds and an overcast of altocumulus clouds on the first (29 Oct) and second day (03 Nov) of collect respectively degraded the illumination conditions considerably, as illustrated in Figure 2. Compared to typical reflective hyperspectral field trials usually conducted under clear skies and high solar elevation, this data collection was conducted under significantly adverse environmental conditions not often considered in the hyperspectral literature.

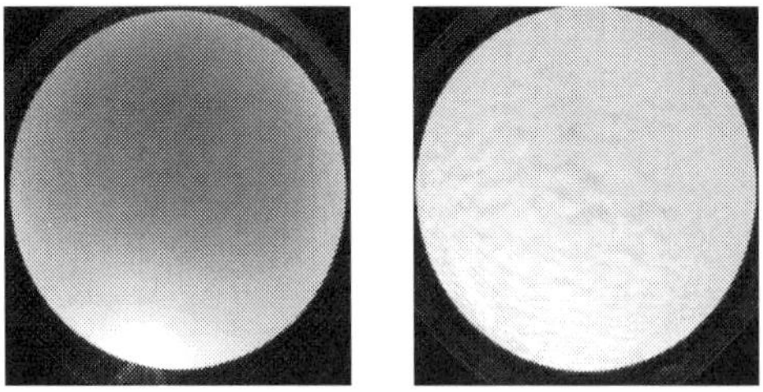

Fig. 2. Typical sky conditions on 29 Oct 2009 (left) and 03 Nov 2009 (right).

One objective of this field trial was the evaluation of the constrained energy minimisation (CEM) algorithm (Settle, 2002) sub-pixel abundance estimation accuracy. For this purpose, we designed targets of known abundances made of thin strips of painted metal, as illustrated in Figure 3 below. The design allowed changes to the abundance level by varying the distance between the strips of metals while their overall size (5m x 5m) ensured that they filled completely at least one pixel in the imagery, as showed in Figure 4. We used two different types of paint to vary the contrast between the target and background, one beige (see Figure 3) and one green (not shown). The base color was mixed with small quantities (2 to 10% per volume) of black feature-less paint to control the spectral features depth and overall signature albedo. A total of 6 targets were used in this experiment.

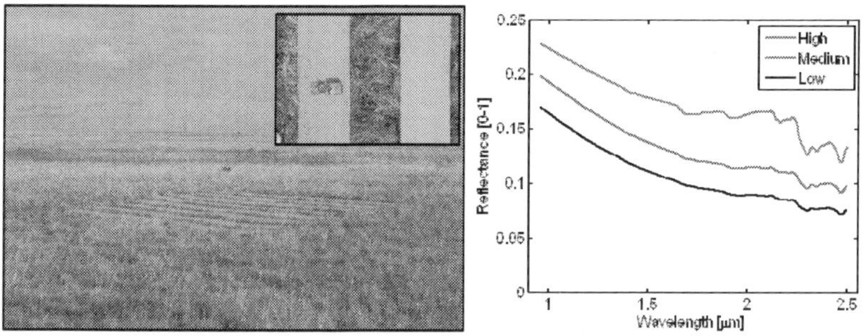

Fig. 3. Left) Example of controlled abundance target. Right) Spectral signatures of the beige paint at different albedo levels, as measured in field conditions using an ASD FieldSpec Pro spectrometer.

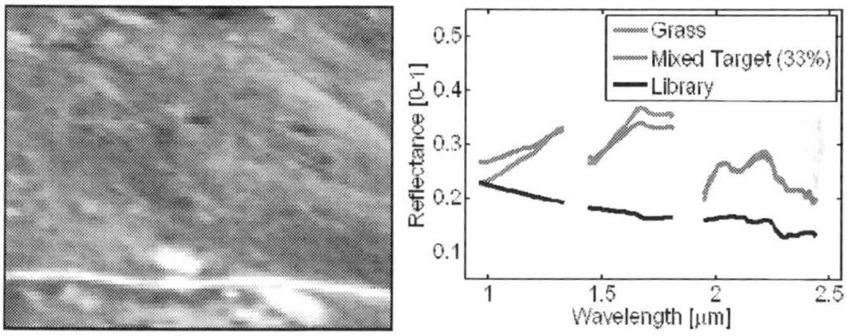

Fig. 4. Left) SWIR 3-colors composite of targets, imaged at a 2.4m GSD (coarsest resolution). Right) Background (red) and target (blue) signatures as measured by the airborne sensor, compared to the target library signature (black).

DRDC favours automated and adaptive approaches to hyperspectral target detection, minimizing user interaction and processing time as much as possible. In this context, we used the following processing chain for this dataset exploitation:

1. We first manually identified and removed 12 out of the original 100 spectral bands due to their low signal to noise level. We also removed unusable part of the imagery due to sensor vignetting. This is typically done only once for a given sensor.
2. We then converted the imagery from at-aperture radiance to apparent reflectance units using the empirical line method (Smith & Milton, 1999), with five very large targets (greater than 10x10 pixels) of known reflectance. This was possible due to the controlled environment of this imagery collection; else we would typically have used a semi-empirical technique such as the QUick Atmospheric Correction (QUAC) method (Bernstein et al, 2005).
3. The CEM algorithm requires a description of the background 2nd order statistics (mean and covariance). For this calculation, we identified a subset of background pixels from the complete image using the RX anomaly detection algorithm (Reed & Yu, 1990) by keeping only the first 90% lowest scoring pixels. Second order statistics were then evaluated using this limited set of pixels.

4. Finally, we calculated the score for all pixels of the image using the CEM algorithm. As described in Settle (2002), when properly normalised, the CEM output is an estimation of the searched target signature abundance in the pixel under test.

The controlled ground targets were collected in 45 different images. On 29 Oct, the abundance was set to 50% while on 03 Nov it was reduced to 33%. All images were manually interpreted to delineate the area of the targets in the images. The CEM scores were averaged over each target area to derive an "average abundance", as shown in Figure 5. This was necessary because as imaged, the targets had inhomogeneous abundance over their physical extent, particularly at the finest GSDs. This suggests that the target design could be improved for future experiments by using thinner strips of material more closely spaced together.

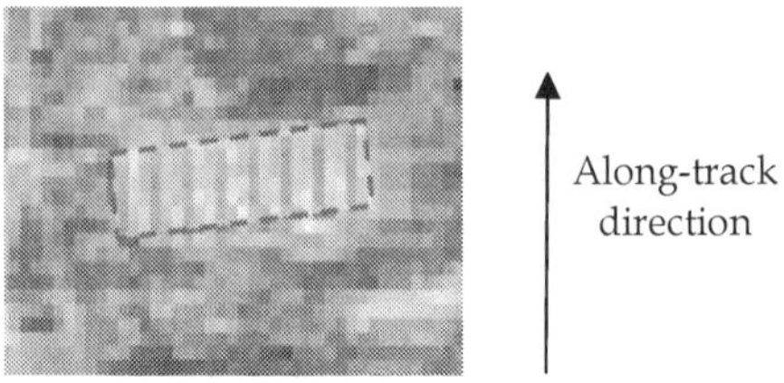

Fig. 5. Example of manual delineation of target area. Target is 5m x 5m, GSD is 0.4m (across-track) by 1.0m (along track).

Estimated abundance error was calculated using the root mean square error (RMSE) and the estimation bias, both normalized by the true abundance in order to get a relative error in percent:

$$\text{Relative RMSE [\%]} = \frac{100\%}{\alpha_{\text{true}}} * \sqrt{\frac{\sum(\alpha_{\text{evaluated}} - \alpha_{\text{true}})^2}{N}} \tag{1}$$

$$\text{Relative bias} = \frac{100\%}{\alpha_{\text{true}}} * \frac{1}{N}\sum(\alpha_{\text{evaluated}} - \alpha_{\text{true}}) \tag{2}$$

RMSE results are presented in Table 1. Overall, the root mean square error for this experiment is between 11.7% and 30%. In absolute terms, the overall RMSE translates to 0.064 and 0.078 for the 0.5 and 0.33 abundance targets respectively. The retrieved abundances were slightly underestimated, with bias of -1.9% and -14.7% again on the 0.5 and 0.33 abundance targets respectively. Since atmospheric conditions degraded between the two collects, it is unclear if the observed increase in error is related to the change in illumination conditions, to the lower abundance level considered, or to a combination of both.

The results achieved are encouraging and show that target abundance can be retrieved at the subpixel level using the CEM algorithm with a high accuracy. The fact that the estimated abundances are generally lower than the true abundances which is consistent with an error that could have been introduced during the manual delineation of targets area, by assigning larger areas to targets than their true area. Also, the imaging system true point spread function has not been characterized and taken into account in this analysis; non-uniform sampling over the GSD could lead to an underestimation of the sub-pixel abundances (Settle, 2004).

Target type		Target configuration	
Contrast type	Albedo type	α = 0.5 (29 Oct 2009)	α = 0.33 (03 Nov 2009)
High (beige)	High	11.7	30.0
High (beige)	Medium	13.4	24.3
High (beige)	Low	12.8	19.6
Low (green)	High	11.9	20.8
Low (green)	Medium	13.2	24.5
Low (green)	Low	13.2	21.7
Average over all targets:		12.7	23.5

Table 1. Relative RMS errors of the retrieved abundances using the CEM algorithm; α denotes the target abundance.

The results demonstrate the robustness of the processing chain; with minimal user interaction and using a simple processing chain suitable for near real-time exploitation, targets can be characterized at the sub-pixel level even under adverse illumination conditions. This demonstrates the processing chain's military utility, and indicates that it could be adapted to the detection and characterization of spectral signatures of interest in a military operational context.

4. Land mapping applications

Land mapping applications were studied in collaboration with the University of New-Brunswick, the University of Alberta, the University of Lethbridge, York University and Laval University. The work was oriented towards soil and vegetation characterization and mapping for trafficability and environmental applications. In Section 4.1, algorithms for classification and the extraction of vegetation canopy attributes (density, structure) were evaluated using airborne hyperspectral data acquired over three Canadian Forces bases (CFB). The resulting validated hyperspectral products were then used to improve a trafficability model developed by the University of New Brunswick for Gagetown military base as well as promote environmentally sustainable training on military bases. Winter airborne images were also acquired over the Montmorency experimental forest (near Quebec City) to investigate the potential of winter imagery to better derive forest information. In Section 4.2 we show that among the classification algorithms that were evaluated, the Mercury algorithm (an evidential-reasoning-based supervised classification algorithm developed by the University of Lethbridge (Peddle & Ferguson, 2002)) achieved the best performance. Finally, Section 4.3 shows results from a laboratory study conducted by the University of Alberta demonstrating how hyperspectral techniques can be used to discriminate between vegetation stresses caused by exposure to different toxic industrial chemicals (Rogge et al, 2008).

4.1 Trafficability and the monitoring of military training areas

This section presents results obtained for the two main land applications of HYMEX, trafficability and the monitoring of training ranges to promote environmentally sustainable training. Table 2 provides a summary of the trials and the objectives sought for each application. The primary objective of each trial was the validation of algorithms used to derive vegetation cover information such as type, density and height, the presence of wetlands and the determination of soil type. These surface features are easily derived from hyperspectral imagery and can contribute to improve knowledge of the terrain for the purposes of trafficability and environmental applications. Each trial was conducted in a different vegetation background ranging from various forest biomes (deciduous, mixte, boreal) to prairie grassland. Details regarding each trial, the available ground truth and the algorithms used to analyse the various datasets can be found in (Ardouin et al, 2007).

Trial location	Date	Sensor	GSD	Background	Objectives	Application
CFB Gagetown, NB	Sep 2005	Probe-1	15m	Acadian forest	Forest parameter algorithms validation (mixed deciduous) & CFB Gagetown trafficability model	trafficability & sustainable training
CFB Wainwright, AB	Sep 2006	AISA	4m	Boreal/grassland	Forest parameter algorithms validation (single deciduous species)	trafficability
Montmorency Experimental Forest, QC	Jun 2004 Feb 2007	AISA	4m 4m	Boreal/summer Boreal/winter	Forest parameter algorithms validation (mixed conifers), wetland mapping & summer/winter dataset investigation	trafficability
CFB Suffield, AB	Sep 2006	AISA	4m	Prairie grassland	Map invasive species, burnt areas, soil disturbance	sustainable training

Table 2. HYMEX land mapping application trials

CFB Gagetown trial. One of the objectives of this trial was to improve the trafficability model used by the Army Meteorological Center (AMC) at CFB Gagetown to plan training exercises, avoid erosion by vehicles and promote environmentally sustainable training. The model currently use as input, the soil moisture content simulated by the University of New Brunswick (UNB) Forest Hydrology Model (ForHyM2) which is based on air and ground temperatures, soil type, the amount of precipitation and the wind speed and direction. Improvement of the trafficability model was achieved by the addition of above ground restrictions such as the forest type (hard/softwood), density and height which can be readily derived from hyperspectral remote sensing. Figure 6b shows a vegetation species classification derived from 15m GSD imagery collected by the Probe-1 sensor (Figure 6a). The overall accuracy (81.8%) and Kappa coefficient (0.78) are based on 533 pixels. These results were obtained with the University of Lethbridge Mercury classification algorithm. Figure 7 shows the shadow fraction of the forest canopy which was derived from spectral mixture analysis (SMA) (Peddle & Smith, 2005) along with the sunlit deciduous fraction, the sunlit conifer fraction and the background fraction. The image shadow fraction was found to correlate the best with LAI, as measured on the ground on 29 plots with hemispherical pictures ($R^2 = 0.55$) with an average difference between the SMA LAI and the ground LAI of less than 0.5 LAI. Figures 8a and 8b show two forest canopy structures, stem density and stand height, as output from the University of Lethbridge Multiple Forward Mode 3-D Canopy Reflectance Model (MFM-3D) applied to the modified geometric optical mutual shadowing model (GOMS) (Peddle et al, 2003). MFM-3D uses a Look-up-tables (LUT) approach based on various ranges and increments of forest structure parameters (density, horizontal & vertical crown radius, crown height and height distribution) as input. The ranges and increments can be determined either from field data or automatically without prior knowledge. Inversion of MFM-3D model produces results when image reflectance values match the modelled reflectance. Field and MFM-3D stand height produced less than 2 m average height difference with the under-estimation of the MFM-3D model attributed to the difficulty in locating neighbouring pixels with similarity to the center pixel.

The addition of above ground restrictions to UNB trafficability model, as determined by the vegetation layers described above, helps produce more refined trafficability classes as illustrated in Figure 9. Figure 10a shows the graphical user interface (GUI) of the route planning tool. Once all the available layers are loaded into the Input Dialog, the user can select from the Interactive Parameters which restrictions to apply for a particular vehicle type. Examples of route planning for four types of military vehicles are shown in Figure 10b with a low environmental concern (not avoiding areas with a high rutting index), and 10c with a high environmental concern (avoiding areas prone to produce ruts).

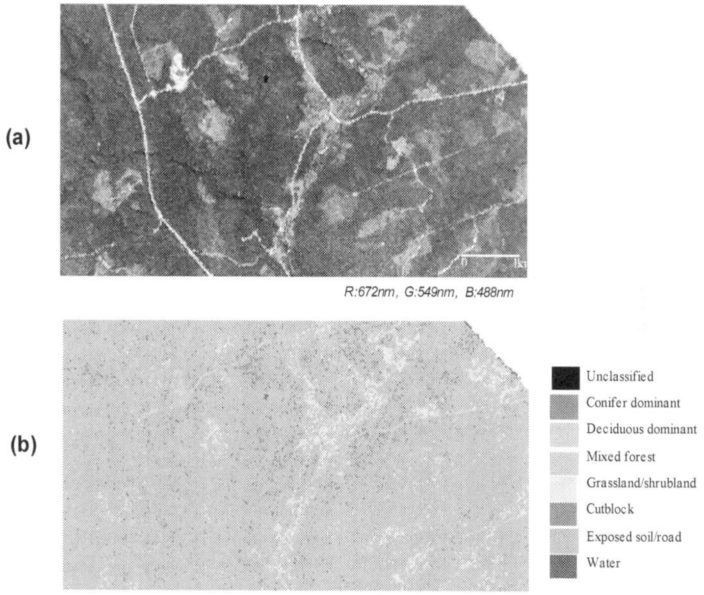

Fig. 6. (a) Sub-image of Probe-1 airborne imagery of CFB Gagetown and (b) landcover classification derived from an evidential reasoning classifier with classes of vegetation and exposed soil and roads. Overall accuracy: 81.8%. Kappa coefficient: 0.78 (D. Peddle, U. of Lethbridge).

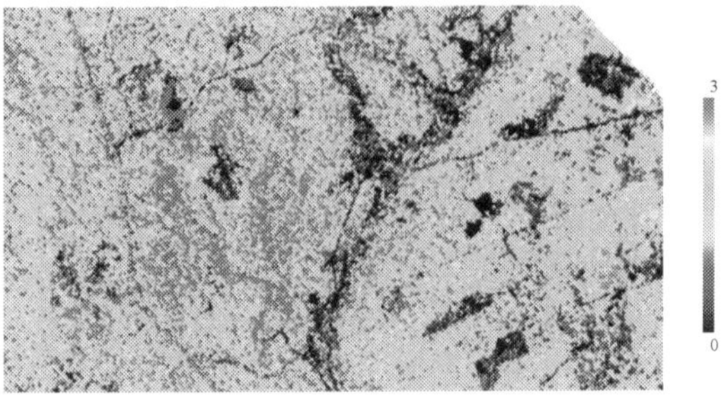

Fig. 7. Sub-image of Probe-1 airborne imagery of CFB Gagetown showing the shadow fraction of the forest canopy derived from spectral mixture analysis (D. Peddle, U. of Lethbridge).

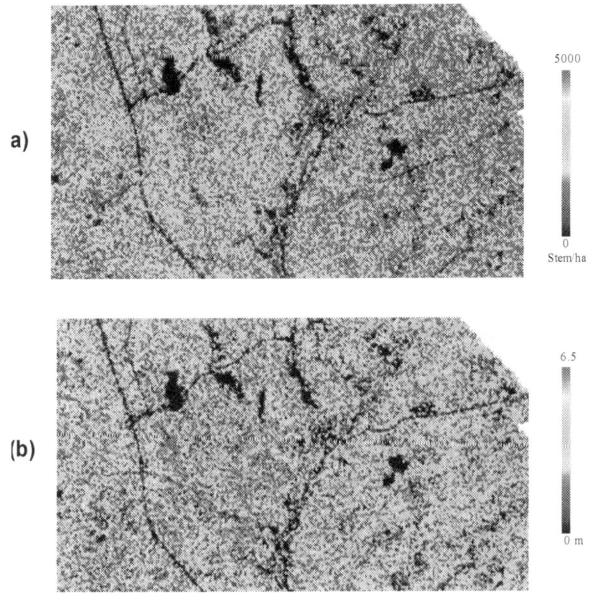

Fig. 8. Sub-image of Probe-1 airborne imagery of CFB Gagetown showing forest structural parameters (a) stand density and (b) stand height as derived from the MFM-3D model (D. Peddle, U. of Lethbridge).

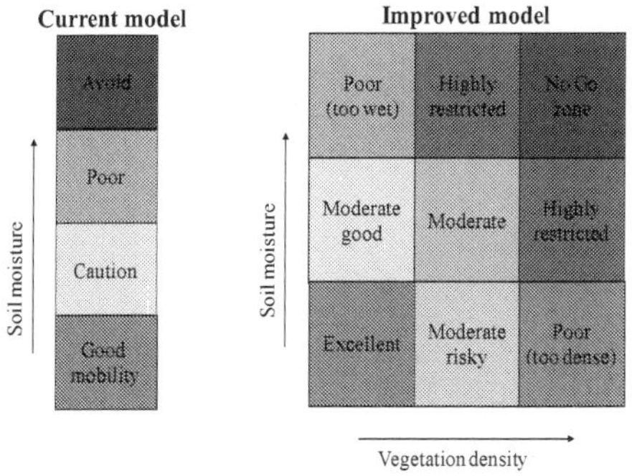

Fig. 9. Off-road trafficability classes for the current (left) and improved models (right).

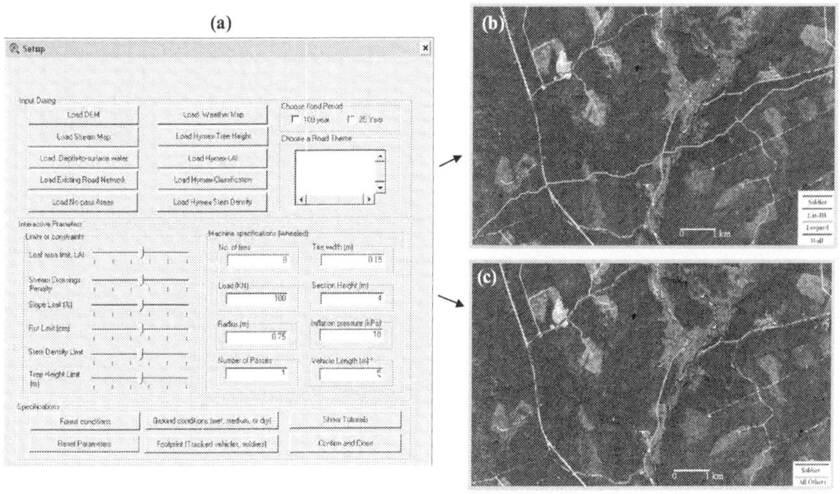

Fig. 10. (a) Route planning for four military vehicles with (b) low and (c) high environmental concern.

CFB Wainwright trial. A similar trafficability model was also constructed by UNB for CFB Wainwright with the objective to demonstrate its portability to different soil and vegetation ecosystems. The above ground trafficability is determined by a semi-arid prairie landscape with well defined dry and wet seasons. The vegetation is composed of grassland with areas of deciduous trees (aspen, balsam poplar and willow) and the topography is gentle. The hyperspectral vegetation products made available to UNB are the land cover classes and the leaf area index (LAI) (Figures 11b and 11c) provided by York University under contract to HYMEX. Figures 11d to 11f show the optimal route planning between point A and point B using a Wolf and a LAV vehicles when constrained by wet areas and vegetation during the wet season (Figure 11d), when constrained by wet areas and vegetation during the dry season (Figure 11e) and when constrained by wet areas and vegetation during the wet season and the requirement to move along tree lines as closely as possible (Figure 11f).

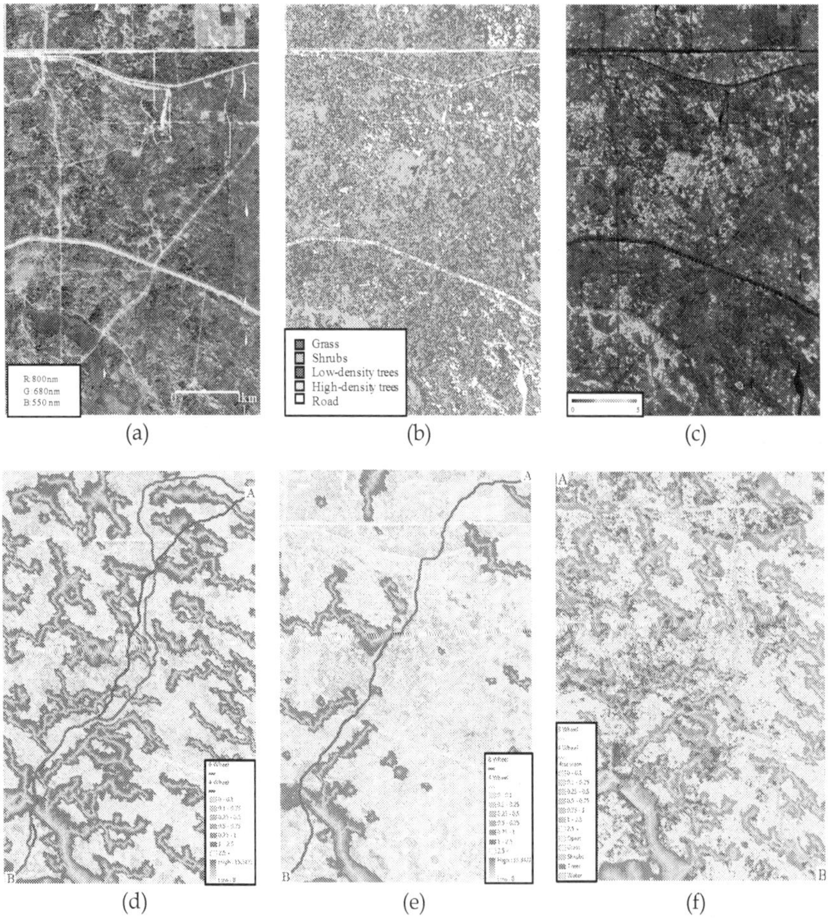

Fig. 11. (a) AISA color composite, (b) land cover classes and (c) leaf area index map of CFB Wainwright (York University). Route between A and B for a Wolf and a LAV vehicles: (d) when constrained by wet season conditions, (e) when constrained by dry season conditions and (f) when constrained by wet season conditions and the requirement to move along tree lines (UNB).

Montmorency Experiment Forest (MEF) trials. Two airborne hyperspectral datasets are available for the MEF site, one from Jun 2004 (summer) and one from Feb 2007 (winter). The objective for imaging this site was to test algorithms for vegetation mapping in a coniferous dominant forest ecosystem. The winter 2007 dataset was acquired for the purpose of investigating the usefulness of summer/winter data to extract relevant terrain information for trafficability in the boreal forest. One first attempt to address this later objective is to map tamarack trees (*Larix laricina*) which can be used as an indicator species for the location of peatlands areas dominated by trees, a wetland type being of interest for trafficability.

Tamarack trees thrive in open areas because of their intolerance to shade and their resentment to compete with other species (Beeftink, 1951). They also adapt very well to poorly drained soil. Their presence is generally associated with peatlands although their absence do not indicate that there is no wetlands. Spectrally, tamaracks are similar to other conifers in summer and to deciduous trees in winter because they loose their leaves (needles) before winter. A combination of summer and winter data allows the exploitation of this unique characteristic of the tamarack trees to locate and map treed wetlands. Figure 12b shows a RGB of AISA data acquired in winter 2007 at the MEF. Figure 12c is a moisture stress vegetation index in which the red color represents exposed bark. When looking closer at the single tree where the red arrow is pointing in Figure 12a, it is easy to recognise from the shape of its shadow on the snow that it is a defoliated coniferous tree. Not having this information on hand it would be difficult to determine whether the trees in the red color class in Figure 12c are deciduous, dead spruce or fir, or dormant tamarack.

Under contract to DRDC, Laval University applied four filters, each made of a band ratio index and a predefined threshold, to classify the tamarack trees in the AISA image with 95% of the tamarack pixels correctly classified and only 1.2% of the remaining pixels misclassified as hardwood trees. Each filter discriminates tamarack trees from other forest features such as other softwood and deciduous. The effects of the application of the first 3 and the first 4 filters are shown in Figure 13a and 13b.

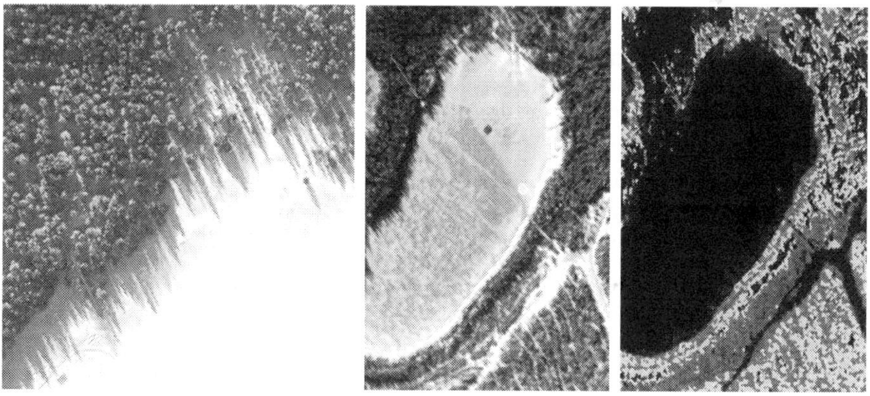

(a) (b) (c)

Fig. 12. (a) high resolution color image, (b) AISA color composite (R:1290 nm , G:1655 nm, B:2189 nm), and (c) a moisture stress vegetation index from the AISA winter image of an area at the Montmorency experimental forest. The red arrow points at a tamarack tree.

CFB Suffield trial. The objectives of this trial were primarily environmentally oriented. Despite a semi-arid climate, prairie grasslands are very sensitive to the introduction of invasive species which are often dispersed during military training and along the maintenance roads of pipelines and gas wells. Leafy spurge and crested wheat are the main invasive species and can easily spread in windward direction into preserved native prairie areas. Moreover, there is a need for monitoring training areas for an environmentally sustainable training. This is to ensure that excessive training does not over stress the soil and

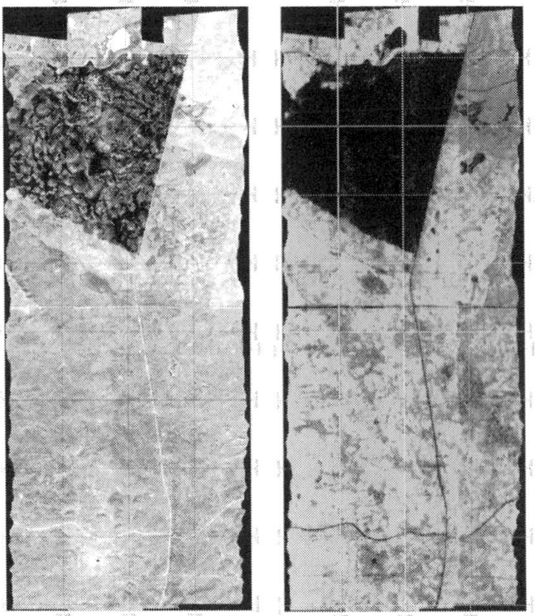

(a) (b)

Fig. 15. (a) RGB "true color" (red: 640nm, green: 550nm, blue: 460nm) AISA image. (b) RGB composite image of the most widely spread green vegetation endmember (Red) and two dominant dry vegetation endmembers (green, blue). (B. Rivard, U. of Alberta).

Class	Train	Test
dirt road	15	15
paved road	52	50
conifer forest	140	62
deciduous forest	263	113
grass	170	80
clearcut area	143	58
water	106	39

(a)

Class	Train	Test
water	49	49
trees	91	98
shrubs	68	65
muskeg	36	50
gravel	40	35
grass	33	37

(b)

Table 3. Number of training and test pixels for each class: (a) classes for CFB Gagetown data (Probe-1, 15m GSD) and (b) classes for CFB Wainwright data, (AISA, 4m GSD).

During the comparison, it was found that the tested algorithms behaved differently as the number of bands used in the classification process increases. Some see classification accuracy increase, others prove unaffected by the number of bands while a third group see the accuracy decrease, albeit slightly.

An immediate advantage in using fewer bands is that processing times are shorter, which is very convenient when analyzing large files. It can also be useful to identify which algorithms are most consistent in accuracy as the number of bands is changed. This way, by using a classification algorithm with known consistency, the optimal band-set can be selected quickly after performing a few tests.

The major finding was that the Mercury algorithm consistently provides very high overall classification accuracy values as illustrated in Table 4. It proves stable and offers the advantage of not requiring that the number of training pixels for each class be at least equal to the number of bands used plus one as is the case with the Maximum Likelihood and Mahalanobis Distance techniques. Mercury's accuracy increased with the number of bands and it offered the highest individual accuracy values in both datasets. Using Mercury on the principal components yielded lower accuracy than with the original dataset. With the Maximum Likelihood algorithm applied to the principal components, results were almost identical to those obtained with the original data. The following table illustrates the findings.

Number of bands →	4	7	20	50	Average
Mercury	82.30%	83.60%	88.20%	91.30%	**86.4%**
Support Vector Machine	84.40%	83.10%	84.40%	85.60%	**84.4%**
Mahalanobis Distance	77.60%	75.20%	86.20%	89.50%	**82.1%**
Neural Network	81.60%	84.20%	88.30%	65.20%	**79.8%**
Maximum Likelihood	87.60%	73.00%	77.40%	75.20%	**78.3%**
Minimum Distance	80.30%	76.70%	76.60%	76.50%	**77.5%**
Parallelepiped	75.60%	74.00%	77.40%	77.00%	**76.0%**
Spectral Angle	67.20%	67.60%	68.80%	69.00%	**68.2%**
Spectral Information Divergence	68.70%	66.60%	68.40%	67.60%	**67.8%**
Binary Encoding	52.00%	51.40%	56.60%	66.70%	**56.7%**

Table 4. Comparison of Mercury and ENVI supervised classification algorithms accuracy. Green is for the algorithm that ranked 1st in classification accuracy and yellow is the algorithm that ranked second.

The study also highlighted the fact that class accuracy varies greatly with the choice of bands in most algorithms. Figures ranging from 0% to 100% accuracy were observed in some algorithms but Mercury came out with very consistent global figures.

In summary, Mercury compares very favourably with ITT's offering for global and class accuracy and for all algorithms, one would be well advised to run a few tests as to the number and choice of bands to ensure optimal feature accuracy.

4.3 Chemical effects on vegetation

4.3.1 Plants as chemical detectors

The Canadian Centre for Mine Action Technologies initiated a study in 2003 to investigate the possibility of exploiting advances in genetic engineering and plant biotechnology to design a process by which plants, local to a region of interest, could be genetically modified

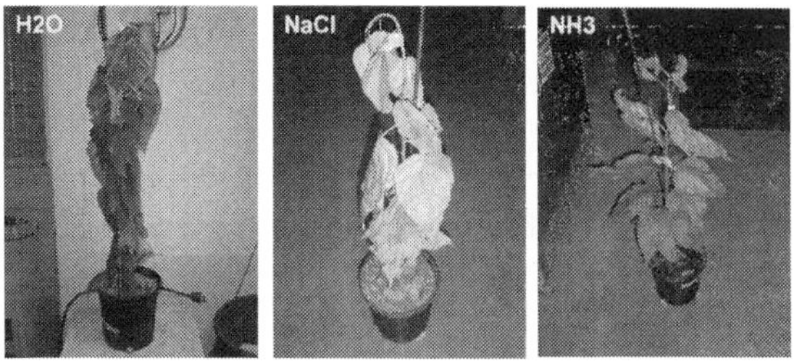

Fig. 17. Examples of environmental (H_2O, NaCl) and chemicals (NH_3) stresses on plants.

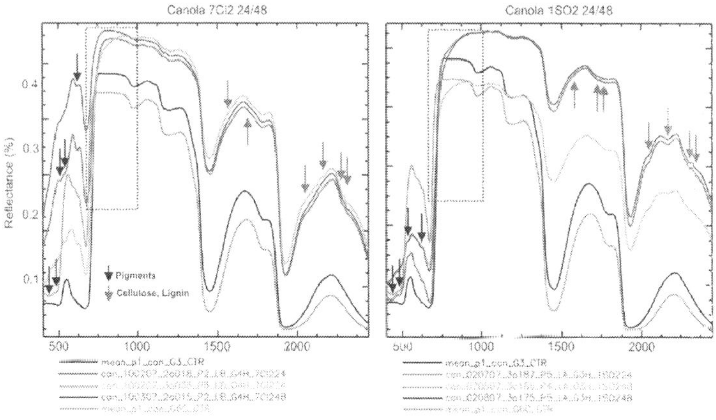

Fig. 18. Representative spectra for canola experiments using Cl_2 and SO_2. Mean spectra of control G1 (black line, mature healthy) and G4C (pink line, high senescence) are included for reference. Arrows denote key absorption features observed in endmember spectra compared with G1. Black dotted box denotes smoothing of red edge. Chemical exposure differences is apparent.

environmental stresses result in diagnostic light reflectance data trends from healthy mature to highly stressed leaves.

Comparison of relevant vegetation indices, such as that depicted in Figure 19, showed that specific combinations could be used to distinguish NH_3, SO_2, Cl_2 consistently across all three species (Rogge et al, 2008). The trends result from the variable leaf response within plants, between plants and between species and it is expected much of the variability observed within species would be preserved or even enhanced in nature. As such it is encouraging for the possible detection of TIC effects on natural vegetation using airborne/spaceborne imagery.

As the detection methodology was developed from leaf-level observations, it is important to note that field trials remain to be conducted in order to test if the findings of this study can be extended to the detection of TICs in the natural environment. The principal unknown is the effect of varying vegetation canopy structural parameters (e.g. canopy gaps, leaf area) and background properties (litter and soil reflectance) on the specific data trends that were identified.

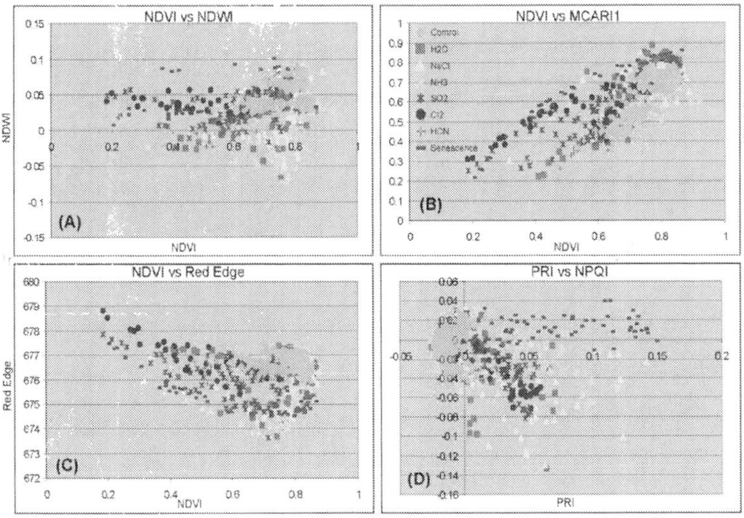

Fig. 19. A selection of vegetation indices across all species for treatments with NaCl, NH_3, SO_2, Cl_2, HCN, and, dehydration (H_2O), senescence and controls plants. The existence of species-specific responses of vegetation to TICs presents both a challenge and an opportunity for regional remote sensing.

While the exact physiological response to each stressor remains to be understood, the existence of species-specific responses of vegetation to TICs presents both a challenge and an opportunity for regional remote sensing.

5. Marine algorithms

The HYMEX project studied several potential marine applications in collaboration with Borstad Associates and the Dept. of Fisheries and Ocean (Institut Maurice Lamontagne and Bedford Institute of Oceanography). We conducted airborne hyperspectral surveys on East and West coasts of Canada to evaluate algorithms for near-shore bathymetry, beach trafficability, near-shore bottom type mapping as well as retrieval of chlorophyll and suspended matter concentrations as indicator of water clarity. More details are provided in (Ardouin, 2007). Through this work it was realized that most of these algorithms can be applied to multispectral imagery and that their experimental validation is difficult. The later is particularly true for products that vary with time (or current) and thus would require many measurement stations (for validation) that would need to operate coincidentally with the airborne survey and be distributed over the area of the survey.

More recently, we tasked OEA Technologies to provide an operational assessment of HYMEX marine algorithms. In this assessment, a distinction was made between dynamic (e.g. water color) and static (e.g. bathymetry) products. It was pointed out that the Canadian Forces needs for off-shore dynamic products (e.g. water colour) is already fulfilled by marine multispectral sensors (MERIS, MODIS) with pixel size > 250m (Williams, 2009). There might however be a niche for hyperspectral sensors (airborne and spaceborne) which typically have better spatial resolution (e.g. from submeter to tens of meter) for near-shore static and dynamic products. The better spatial resolution and increased number of bands of hyperspectral sensors might provide an ability to handle the more complex near-shore environment. Potential static products to consider include target detection and near-shore bottom characterization in support of mine countermeasures and battlespace mapping and possibly submarine operations. To this we can also add near-shore bathymetry in support of route survey, battlespace mapping, anti-submarine warfare and submarine operations. While not requiring hyperspectral sensing, HSI could keep playing a role (e.g. selection of optimal bands) in the development of new dynamic products for both near-shore and off-shore applications. Overall, this assessment point to possible follow-up for marine applications development with hyperspectral sensors.

6. Conclusion

In this chapter, we discussed a wide variety of military applications resulting from the exploitation of reflective hyperspectral imagery. These applications were demonstrated in the DRDC HYMEX project, allowing DRDC and the Canadian Forces stakeholders to get more familiar with the military utility of hyperspectral imagery. While some of these applications such as target detection are relatively mature and are near to operational deployment, others still require further development but are representative of the unique capability of hyperspectral remote sensing. The many datasets that were acquired and the algorithms and exploitation tools that were developed in the project are being used to continue the development of hyperspectral technology at DRDC. One avenue that is being pursued is the development of an airborne hyperspectral real-time target detection demonstration system. We are also looking at opportunities to further develop the land mapping and marine applications areas as well as potential space-based demonstration with international partners.

7. References

Antunes, M.S.; Ha, S.B.; Tewari-Singh, N.; Morey, K.J.; Trofka, A.M.; Kugrens, P.; Deyholos, M. & Medford J.I. (2006) A synthetic de-greening gene circuit provides a reporting system that is remotely detectable and has a re-set capacity, *Plant Biotechnology J.*, Vol. 4, No. 6, (November 2006), pp. 605-622

Ardouin, J.-P.; Lévesque, J. & Rea, T.A. (2007) A Demonstration of Hyperspectral Image Exploitation for Military Applications, *Proc. of the 10th International Conference on Information Fusion (FUSION 2007)*, Quebec, Canada, (9-12 July 2007), pp. 1-8., ISBN 978-0-662-45804-3

Beeftink, H.H. (1951) Some observations on tamarack or eastern larch. *Forestry Chronicle*, Vol. 27, No. 1, (1951), pp. 38-39

Bernstein, L.S.; Adler-Goldstein, S.M.; Sundberg, R.L., Levine, R.Y.; Perkins, T.C.; Berk, A.; Ratkowski, A.J.; Felde, G. & Hoke, M.L. (2005) A new method for atmospheric correction and aerosol optical property retrieval for VIS-SWIR multi- and hyperspectral imaging sensors : QUAC (QUick Atmospheric Correction), *IEEE International Geoscience and Remote Sensing Symposium*, (25-27 July 2005), pp. 3549-3552, ISBN 0-7803-9050-4

Buckingham, R.; Staenz, K. & Hollinger (2002) A., Review of Canadian Airborne and Space Activities in Hyperspcetrai Remote Sensing, *Canadian Aeronautics and Space Journal,* Vol. 48, No. 1, (2002), pp. 115-121

Buckingham, R. & Staenz, K. (2008) Review of current and planned civilian space hyperspectral sensors for EO, *Can. J. Remote Sensing,* Vol. 34, Suppl. 1, (2008), pp. S187-S197

Cooley T.; Davis, M. & Straight, S. (2006) ARTEMIS-Advanced Tactically-Effective Military Imaging Spectrometer: Tactical Satellite 3 for Responsive Space Missions, *International Symposium on Spectral Sensing Research,* Bar Harbor, MN, (May 2006)

Cutter, M.A.; Johns, L.S.; Lobb, D.R.; Williams, T.L. & Settle, J.J. (2003) Flight Experience of the Compact High Resolution Imaging Spectrometer (CHRIS), *Proc. of SPIE Imaging Spectrometry IX.* Vol. 5159, (2003) pp. 392-405

Davenport, M. & Ressl, W. (1999) Shadow Removal in Hyperspectral Imagery, *International Symposium on Spectral Sensing Research (ISSSR 99),* Las Vegas (November 1999)

Deyholos, M.; Faust, A.A.; Miao, M.; Montoya, R. & Donahue, D.A. (2006) Feasibility of landmine detection using transgenic plants, in *Detection and Remediation Technologies for Mines and Minelike Targets XI,* J. Thomas Broach, Russell S. Harmon, John H. Holloway, Jr, eds., *Proc. SPIE,* Vol. 6217, (May 2006), pp. 700-711

Deyholos, M.K.; Rogge, D.; Rivard, B. & Faust, AA. (2007) Plants as sensors for toxic industrial chemicals and munitions: A feasibility analysis, *In Vitro Cellular & Developmental Biology-Animal,* Vol. 43, pp. S7-S7.

Deyholos, M. (2009) *Demonstration of Plant-based Explosives Detection,* Contract Report Defence R&D Canada - Suffield, DRDC Suffield CR-2010-010 (2009).

Pearlman, J.S.; Barry, P.S; Segal, C.C.; Shepanski, J.; Beiso & D. & Carman, S.L. (2003) Hyperion, a Space-Based Imaging Spectrometer, *IEEE Trans. on Geoscience and Remote Sensing,* Vol. 41, No. 6, (June 2003), pp. 1160- 1173, ISSN 0196-2892

Peddle, D.R. & Ferguson, D.T. (2002) Optimization of Multisource Data Analysis using Evidential Reasoning for GIS Data Classification, *Computers & Geosciences,* Vol. 28, No 1, (2002), pp. 45-52

Peddle, D.R.; Franklin, S.E.; Johnson, R.L.; Lavigne, M.B. & Wulder, M.A. (2003) Structural Change Detection in a Disturbed Conifer Forest Using a Geometric Optical Reflectance Model in Multiple-Forward Mode. *IEEE Trans. on Geoscience and Remote Sensing* Vol. 41, No. 1, (Jan 2003), pp. 163-166, ISSN 0196-2892

Peddle, D.R. & Smith, A.M. (2005) Spectral Mixture Analysis of Agricultural Crops: Endmember Validation and Biophysical Estimation in Potato Plots, *International Journal of Remote Sensing,* Vol. 26, No. 22, (2005), pp. 4959–4979

Peddle, D.R.; Boulton, R.B.; Pilger, N.; Bergeron, M. & Hollinger, A. (2008) Hyperspectral detection of chemical vegetation stress: evaluation for the Canadian HERO satellite mission, *Can. J. Remote Sensing,* Vol. 34, Suppl. 1, (2008), pp. S198-S216

Reed, I. S. & Yu, X. (1990) Adaptive multiple-band CFAR detection of an optical pattern with unknown spectral distribution, *IEEE Trans. on Acoustics, Speech, and Signal Processing,* Vol. 38, No. 10, (October 1990), pp. 1760-1770, ISSN 0096-3518

Rivard, B.; Deyholos, M. & Rogge, D. (2008) *Chemical Effects on Vegetation Detectable in optical bands 350-2500 nm,* Contract Report Defence R&D Canada - Suffield , DRDC Suffield CR-2008-234, (2008)

Rogge, D.; Rivard, B.; Deyholos, M.; Lévesque, J. & Faust, A. (2008) Toxic Industrial Chemical Effects on Poplar, Canola, and Wheat detectable over the 450-2500 nm Spectral Range, *Intl. Geoscience & Remote Sensing Symposium (IGARSS 2008)*, Boston, MA, (July 2008)

Roy, V. (2010) Hybrid algorithm for hyperspectral target detection, *Proc of SPIE Algorithms and Technologies for Multispectral, Hyperspectral, and Ultraspectral Imagery XVI*, Vol. 7695, May 2010, pp. 1-10

Sentlinger, G.; Davenport M., & Ardouin, J.-P. (2003) Automated Target Recognition in Hyperspectral Imagery using Subpixel Spatial Information, *Proc. SPIE Aerosense – Automatic Target Recognition XIII*, Vol 5094, (April 2003)

Settle, J. (2002) On constrained energy minimization and the partial unmixing of multispectral images, *IEEE Trans. on Geoscience and Remote Sensing*, Vol. 40, No. 3, (March 2002), pp. 718-721, ISSN 0196-2892

Settle, J. (2004) On the Use of Remotely Sensed Data to Estimate Spatially Averaged Geophysical Variables, *IEEE Trans. on Geoscience and Remote Sensing*, Vol. 42, No. 3, (March 2004), pp. 620-631, ISSN 0196-2892

Smith, G. M. & Milton, E.J. (1999) The use of the empirical line method to calibrate remotely sensed data to reflectance, *International Journal of Remote Sensing*, Vol. 20, (1999), pp. 2653-2662

Van Chestein, Y. (2011) *Comparative evaluation of the Mercury classification algorithm: On the influence of the number of bands on classification accuracy using hyperspectral data*, Defence R&D Canada - Valcartier Technical Memorandun, TM 2010-385, (March 2011)

Webster, A.H.; Davenport, M.R. & Ardouin, J.-P. (2006) 3D Deconvolution of Vibration Corrupted Hyperspectral Images, *International Symposium on Spectral Sensing Research 2006*, Bar Harbor, MA., (May 2006).

Williams, D.; Vachon, P.W; Wolfe, J.; Robson, M.; Renaud, W.; Perrie, W.; Osler, J.; Isenor, A.W.; Larouche, P.; Jones, C. (2009) *Spaceborne Ocean Intelligence Network – SOIN – fiscal year 08/09 year-end summary*, Defence R&D Canada - Ottawa External Client Report, DRDC Ottawa ECR 2009-139, (September 2009)